Organic and Printed Electronics

Organic and Printed Electronics

Fundamentals and Applications

edited by

Giovanni Nisato
Donald Lupo
Simone Ganz

PAN STANFORD PUBLISHING

Published by

Pan Stanford Publishing Pte. Ltd.
Penthouse Level, Suntec Tower 3
8 Temasek Boulevard
Singapore 038988

Email: editorial@panstanford.com
Web: www.panstanford.com

British Library Cataloguing-in-Publication Data
A catalogue record for this book is available from the British Library.

Organic and Printed Electronics: Fundamentals and Applications

Cover image courtesy: Lapp Group, Germany

ISBN 978-981-4669-74-0 (Hardcover)
ISBN 978-981-4669-75-7 (eBook)

Printed in the USA

Contents

Preface

The goal of this book is to provide an introductory knowledge of organic and printed electronics to students and engineers. There are many monographs, dissertations, and research papers available for higher-level studies on the topic. Our motivation for writing this book was to provide a preliminary reference guide and manual for beginners and a support tool for teachers. Such a textbook did not exist two decades ago when some of us first encountered this field, and this situation still seems to continue: we often have to use whatever materials we have at our disposal when introducing newcomers to this field.

We begin the book by presenting the contents and structure of this book, which are the result of a long process as we considered different perspectives including analyses of the educational courses on organic electronics in Europe. Besides more technical chapters, we chose to introduce topics such as entrepreneurship and environmental impact, which are key to making a lasting societal contribution.

This book took over four years to complete, which appears like an eternity in a fast-moving technical field. The process also naturally forced us to focus on topics that have reached a certain level of maturity. This project was a rewarding group effort. The chapters have been contributed by many talented authors, who are all experts in their fields, and we are greatly indebted to them for their trust and patience and for transmitting in their chapters a working knowledge that is connected to practice.

In such a multidisciplinary field, it is not possible to cover all topics within a book of this size. In this book, we focus on printing and solution coating rather than on vacuum deposition. Additionally, we decided to specialize on the handling of liquid deposition and on understanding the physics behind it instead of simply summarizing all deposition processes that are used in organic electronics today. Many concepts for devices are very similar, therefore, vacuum processing is presented indeed briefly in several chapters. Further information on vacuum processing is available in specialized publications. As new organic molecules are being synthesized and reported almost weekly, a thorough introduction to the chemistry of organic and printable semiconductors would require a book in itself. Flexibility and thin-film mechanical properties are very important and play a role in the design, processing, and final functionality of printed electronics. Although these have been mentioned at various instances in the book, they have not been presented in an independent chapter. Further, several devices such as printed memories and optical sensors are not covered in this textbook. Hopefully, there will be a chance to expand on these and other topics in a further edition.

Giovanni Nisato

Donald Lupo

Simone Ganz

January 2016

Acknowledgments

We are grateful to Pan Stanford Publishing for providing the first spark to initiate this project and Shivani Sharma for her patience and support during this endeavor.

Several projects and organizations provided logistic and financial support for this book. We especially thank the COLAE EU project, which delivered the first pan European professional training program of organic electronics, and the Organic Electronic Association's Working Group on Education for providing crucial support in bringing out this book. We also thank the authors of the chapters for their commitment to this project and their excellent work. Co-editors Donald Lupo and Simone Ganz especially thank coordinating editor Giovanni Nisato, who was the primary driving force behind this project.

Chapter 1

Introduction to Organic and Printed Electronics

Giovanni Nisato,[a] Donald Lupo,[b] and Simone Ganz[c]
[a]*Centre Suisse d'Electronique et de Microtechnique, Tramstrasse 99 Muttenz, Switzerland*
[b]*Department of Electronics and Communication Engineering, Tampere University of Technology, PO Box 692, Korkeakoulunkatu 3, FI-33101 Tampere, Finland*
[c]*Institute of Printing Science and Technology, Technische Universität Darmstadt, Magdalenenstraße 2, Darmstadt, 64289, Germany*
giovanni.nisato@csem.ch, donald.lupo@tut.fi, ganz@idd.tu-darmstadt.de

1.1 The Field and the Goals of the Book

Organic and printed electronics is at the same time an established field in terms of academic, scientific, and technological research and still an emerging one in terms of mass industrial application. As we write, hundreds of scientific articles are written on the topic yearly, with dedicated scientific journals. Today, organic light-emitting diodes (OLEDs) are ubiquitous in mobile phones, and it is possible to buy printed organic photovoltaic modules and simple printed

Organic and Printed Electronics: Fundamentals and Applications
Edited by Giovanni Nisato, Donald Lupo, and Simone Ganz

ISBN 978-981-4669-74-0 (Hardcover), 978-981-4669-75-7 (eBook)
www.panstanford.com

memories; and yet, printed electronics still has a lot of potential to make a larger and positive impact on the lives of millions of people.

Organic semiconductor molecules are at the core of these technologies. Classical inorganic semiconductors rely on the specific properties of given elements and their organization in space, for instance, in crystalline form. On the other hand, organic semiconductor molecules can be designed and engineered to create new semiconductor types, for example, with specific color emission or absorption properties. This opens many possibilities. Electrical conductivity and the modulation of charge transport with thin-film transistors make it possible to create electronic circuits and analog and digital electronics. Light absorption and conversion to electricity enable organic photovoltaics for energy collection as well as photodiodes for detection of objects and images. Light emission is used in displays and general lighting sources.

Organic molecules can be formulated in functional inks, which can then be deployed in large-area printing and coating techniques, benefiting from hundreds of years of accumulated know-how about printing presses, paper, and photographic film production for example. These techniques are proven to be remarkably cost effective when deployed in mass quantities. Digital printing (such as inkjet) makes it even possible to produce custom-designed electronics on a massive scale. With the added value of being lightweight and flexible, a myriad of alternative design options are possible for the creation of new products.

In short the combination of electronics and large-area production by printing is exciting and promising but also challenging scientifically, technologically, and economically, as innovation requires far more than science and technology to succeed. The aim of this book is to provide an introduction to organic and printed electronics, its fundamental aspects, and applications.

In the 1980s there were still a number of scientific publications arguing whether organic molecules could actually exhibit semiconducting properties at all. Today these papers can be read on portable tablet computers equipped with OLED displays, showing videos of the production processes and the facilities to manufacture them. One of the key questions today is rather how to effectively introduce the field of organic and printed electronics to an increasing number of students, engineers, and enthusiasts.

There is a large amount of high-level monographs and specialized books covering several essential aspects ranging from organic semiconducting materials chemistry to advanced device processing and physics. However, these books are often very specific and are meant for advanced postgraduate students or experts in the field. This book is an attempt to introduce a vast and varied field of technology, science, and applications to students in their final undergraduate or beginning graduate studies, as well as engineers interested in approaching new fields.

The authors of the chapters are all experts in their respective areas and are involved in organic and printed electronics in their daily work. This book was written by practitioners of the science and technologies presented, many of whom are working in an industrial environment. We feel it is important to transmit a working knowledge that is connected to practice.

The style of the book is meant to support self-learning and reflection, as well as to provide a support to classroom teaching. Each chapter provides questions and simple exercises to check the understanding and apply the concepts presented in the material. The reader willing to deepen his or her knowledge is also provided with a list of further readings at the end of each chapter.

1.2 Contents and Book Organization

The chapter order is meant to introduce at first the core concepts of organic semiconductors and printing technologies (Fig. 1.1). The next section introduces the reader to key device types (transistors, light-emitting diodes, solar cells, power storage). Encapsulation is a separate topic, given its importance in ensuring lifetime performance at the component and at the system level. In the following chapter, two examples of more complex system-level integration (sensors and hybrid printed electronics) are presented. The final chapters are of a broader nature, presenting larger perspectives on the application fields, their evolution, the challenges related to innovation, and environmental impact. Last but not least, real-life practice is for many the best way to get involved. Therefore, the last chapter provides hands-on experiments. As a teaching support, the chapters should cover material for a couple of semesters, depending on class frequency and ambition.

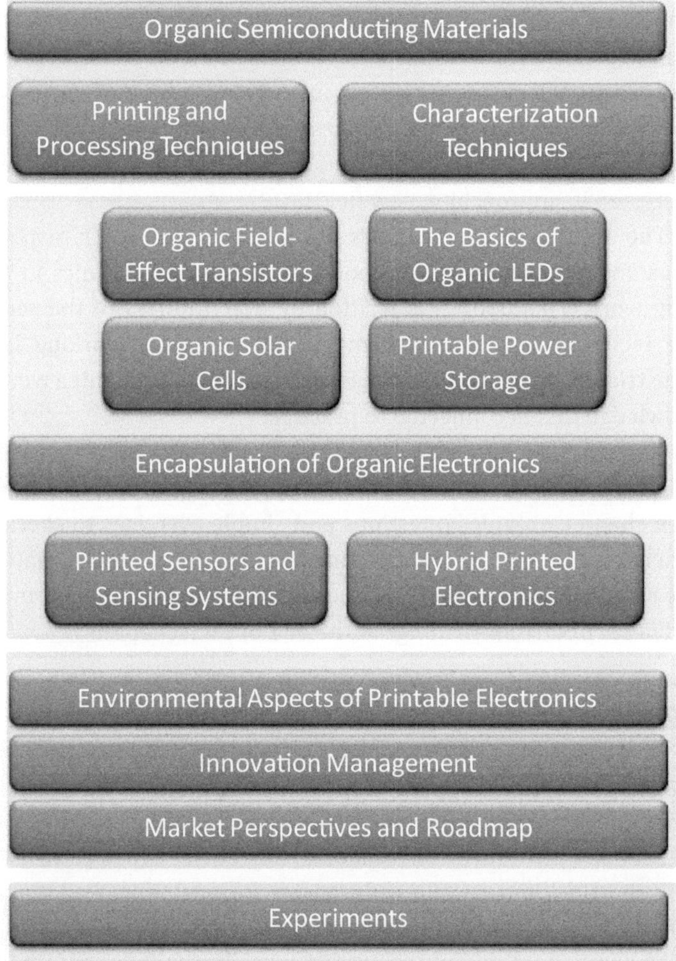

Figure 1.1 Overview of the chapters and structure of the book.

The chapters are often self-contained and are meant to be read independently, so some redundancies in the exposition were necessary in many cases. Depending on the reader's background, or on the presentation needed for a given course, there are naturally different ways to use the text. The book can be, for example, approached from the transversal topics or even backwards for readers already engaged in environmental studies or innovation management and interested to learn more about the modes of operations of specific devices and the fundamentals of organic semiconductors.

1.2.1 Description of the Chapters

The content of the chapters is described briefly next.

Organic Semiconducting Materials

This chapter introduces the basic concepts from the ground up, starting from chemical bonds and showing the link to molecular orbitals (highest occupied molecular orbital–lowest occupied molecular orbital [HOMO-LUMO] levels). Charge injection (ohmic contact, injection barrier) and transport mechanisms (from Peierls instability to organic semiconductors' mobility) in organic semiconductors are introduced, as well as the impact of disorder on conductivity. Important links are made to optical properties for both radiative and nonradiative effects. It is recommended to be read prior to the chapters on organic thin-film transistors (OTFTs), light-emitting diodes, and solar cells.

Printing and Processing Techniques

There is a large step between a solid semiconducting material and a functional ink that can be reliably printed or coated on large areas. The chapter covers the basics of functional ink formulation, of interactions between substrates and fluids (e.g., wetting behavior), and of the complex fluid dynamical processes linking the printed fluid to the finally dried shape. The chapter thoroughly introduces the most used and relevant printing (from gravure to screen printing and inkjet printing), patterning (from lithography to hot embossing), and coating techniques (from spin coating to slot die coating).

Characterization Techniques for Printed Electronics

Traditional printing relies largely on optical inspection to evaluate the quality of the final product. As challenging as this is, printed electronics requires a qualitatively different approach for characterization. The chapter introduces the techniques required to characterize: the properties of inks prior to printing (e.g., viscosity), the structural properties of layers formed (e.g., thickness and roughness), and finally their optical or electronic properties.

Organic Field-Effect Transistors

This chapter introduces the main types of OTFTs, their structure, and their functioning principles; it provides practical knowledge about electrical measurement methods for them and shows how to gain insights into the performance and limitations of the devices being measured. Guidelines on how to correctly interpret the measurements and to provide feedback to the manufacturing process are given. Simple circuits such as logic gates and oscillators are presented, as well as examples of advanced research for devices in the biomedical area.

The Basics of Organic Light-Emitting Diodes

The basic structure and working principle of simplified OLEDs are given. The chapter introduces the main materials and layers contributing to light emission efficiency (light-emitting, transport, and injection layers) as well as the theme of light extraction from the device. Lifetime aspects are presented in relation to materials, processing, chemical stability, and encapsulation. Production processes (vacuum and solution), applications, and perspectives, including flexible devices, are presented.

Organic Solar Cells

The basic principles of organic solar cells are introduced, as well as the basic process and production technologies. Examples of the key materials of organic solar cells are given, including photoactive layers, electrodes, and encapsulation materials. The key parameters and procedures for the electrical characterization of solar cells and modules are given. The reader is also introduced to lifetime aspects and degradation mechanisms of organic solar cells. Current examples of different applications of organic solar modules and perspectives are given in conclusion.

Printable Power Storage: Batteries and Supercapacitors

Many devices including printable electronics are meant to be portable or disconnected from a power grid. The ability to store electrical power is therefore critical. The chapter introduces printable storage cells, batteries, and supercapacitors. The chapter introduces the key performance parameters (from voltage to cycle life considerations), the principles of the devices, and their characterization. The reader is further introduced to the materials (electrodes, liquid/solid electrolytes, high-surface materials) and the different architectures for processing. A comparison of supercapacitors and batteries, as well as perspectives on the challenges of electrochemical systems, is given.

Encapsulation of Organic Electronics

Without protecting encapsulation layers there would be no practical application of electronic devices and especially of printed organic electronic devices. The reader is introduced to the permeation phenomena in solids and thin films as well as to the principles of permeation measurements. The chapter presents the main materials used as encapsulation, including glass, polymer films, sealants, thin-film coatings, and getter materials. Relevant processing techniques to apply encapsulating materials to organic devices are presented as well.

Printed Sensors and Sensing Systems

Sensors are a key to enable interactions between humans, machines, and their environment. This chapter introduces the fundamental aspects of sensors and the working principles of common sensors. Different types of printed sensors and organic sensors (including resistive, electrochemical, capacitive, optical, and ionic) are presented. Sensors can comprise organic devices introduced in previous chapters (e.g., OTFTs). The chapter also gives examples of applications and integration of organic printed sensors in more complex systems.

Hybrid Printed Electronics

On the one hand, fully printed, flexible organic electronic devices are an exciting perspective that remains on the horizon. On the other hand, it is already possible to combine existing electronics with printed and organic components to enable applications in a shorter time frame. The chapter introduces examples of (inorganic) printable inks, substrates, and processes for printing circuitry. The assembly of components in systems, including contacting, pick-and-place, and foil-to-foil lamination, is presented. Finally the reader is introduced to a concrete case study of a hybrid sensor system.

Environmental Aspects of Printable and Organic Electronics

One of the perspectives of organic and printed electronics is the opportunity to produce electronics on a large scale with a reduced environmental footprint. The choices made even at very early research stages about materials and processing can have very far-reaching and long-term consequences leading to public health and regulatory issues or, on the contrary, to a market advantage. Therefore it is important to have an introduction to the methodologies to address key aspects on environmental impact such as life cycle analysis and ecotoxicology.

Innovation Management

The chapter introduces the reader to many nontechnical aspects that nevertheless deeply affect the development, deployment, and, finally, success of new technologies. The chapter also puts in perspective the field from a historical development point of view and the lessons learned, often the hard way, by the pioneers in the field. The reader is further introduced to core innovation processes and innovation management concepts, as well as to practical aspects for commercialization, including business models and intellectual property considerations.

Market Perspectives and Road Map for Organic Electronics

This chapter provides an insight at this point in time into the application perspectives of organic and printed electronics, including OLED lighting, organic photovoltaics, flexible displays, printed electronic components, and integrated smart systems. Making statements about the future, planning, and designing road maps are at best difficult, especially in an emerging technology field, but as Dwight D. Eisenhower already said, "In preparing for battle I have always found that plans are useless, but planning is indispensable." With this in mind, the road-mapping process in itself provides valuable reflection and communication tools.

Experiments

This chapter describes four practical experiments in the field of organic electronics and printed electronics. Two experiments can be performed in an applied science laboratory that is equipped for bachelor education. The other two require more specific equipment. The reader is shown how to make devices including organometallic light-emitting diodes, printed antennas for radio-frequency identification (RFID) tags, and a simple piezoelectric inkjet printing head.

Chapter 2

Organic Semiconducting Materials

Mattias Andersson and Mats Fahlman
Department of Physics, Chemistry and Biology, Linkoping University, 58183 Linkoping, Sweden
mats.fahlman@liu.se

2.1 Introduction

The use of materials in order to enhance our quality of living always has been a key component of the human experience. Early on it consisted of using tools fashioned out of stone, wood, animal hides, and bones, but our choice of materials and the number of applications have grown dramatically over time. Improved materials and/or new materials have had such a dramatic impact on human life that we often describe and name periods of history by the predominate material used at the time, for example, Stone Age, Bronze Age, Iron Age, etc. In the early times of "modern" man, advances in materials and their applications were slow and, consequently, so was the development of overall prosperity in society. The Renaissance, with its focus on science and knowledge, brought about an accelerated

Organic and Printed Electronics: Fundamentals and Applications
Edited by Giovanni Nisato, Donald Lupo, and Simone Ganz

ISBN 978-981-4669-74-0 (Hardcover), 978-981-4669-75-7 (eBook)
www.panstanford.com

development of new materials and in the 1800s the Industrial Revolution was accelerated by improvements in steel. Finally, the groundbreaking work in quantum physics carried out in the early 1900s paved the way for the silicon age, where a host of new materials, besides silicon, enabled technologies that have brought us a whole new way of life. One particular powerful recent trend is nanoelectronics, and nanotechnology in general, that depends upon the creation of functional materials, devices, and systems through control of matter on the nanometer length scale (1–100 nm) and the exploitation of novel phenomena and properties (physical, chemical, biological) at that length scale. Hence, the scientific and technological revolution that nanotechnology represents may very well demand new processing techniques and in fact whole new materials systems. In nature, nanotechnology already was realized millions of years ago and relies on self-organization of organic (carbon-based) molecules. Evolution chose organic materials over inorganic materials, and nature has managed to create highly complex systems such as the human body that comes equipped with a wide variety of sensors, memory functions, and data- and signal-processing devices. So, why shouldn't we mimic nature and try to make electronics out of organic materials and nanoelectronics through self-organizing molecules? Indeed, development of organic electronics is underway, and perhaps the first major breakthrough, electrically conducting and semiconducting polymers (plastics), was awarded the Nobel Prize in chemistry in the year 2000. Already today, electronic components such as light-emitting diodes (LEDs), transistors, and solar cells can be made using semiconducting and conducting polymers and organic molecules. What do we then mean by the term "polymer"?

Popularly known as plastics, polymers are large molecules constructed from smaller structural units (monomers) covalently bonded together in any conceivable pattern (see Fig. 2.1). By modifying the monomer-building blocks and the bonding scheme, the mechanical and thermal properties of the polymers can be controlled. Polymers can be made rubbery or brittle, soft or hard, soluble or nonsoluble, and in practically any color by blending with suitable chromophores. The special properties of polymers such as solubility allow for a variety of convenient processing methods, such as conventional printing and coating techniques (inkjet, flexo, screen, etc.) as well as injection molding, spin coating, and spray painting. This is in striking contrast to the processing of metals and even more so inorganic semiconductors like silicon, as mentioned earlier. Hence, polymers can be found in almost all products in our present-day society: clothes, furniture, home appliances, cars, airplanes, electronics, etc. There also are naturally occurring polymers like proteins and rubber, but most polymers in use today are synthesized and hence are called synthetic polymers. Yet, despite the naturally occurring polymers, the concept of polymers did not in fact achieve general acceptance until the early years of the 20th century.

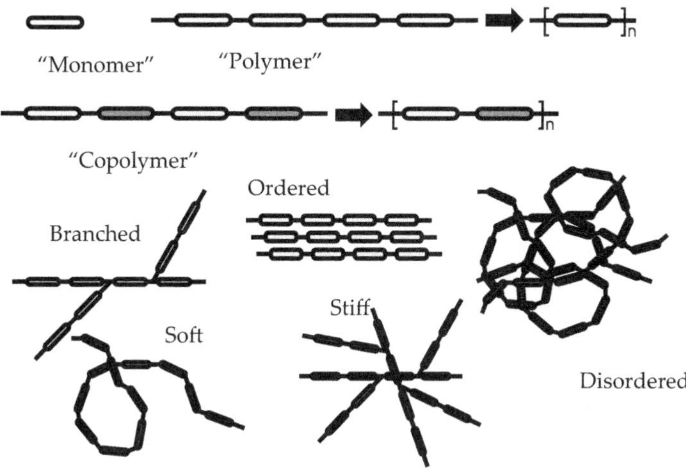

Figure 2.1 Illustration of a monomer and several types/forms of polymers.

Polymers are mainly used for their structural properties, be it the elasticity of plastic food wrap or the hard yet light and easily moldable plastic parts in a car. The idea of utilizing the electrically conducting properties of polymers was proposed as early as in the 1960s, and since then polymers have been used as active components in a variety of electronic applications. For instance, carbon- and metal-filled polymers are used as moldable semiconductors in the electronics industry, and polymers and organic molecules long have served as photoreceptors in electrophotographic copying machines. In this chapter, however, we mainly will concentrate on a particular class of electronic polymers, the so-called π-conjugated polymers.

In 1977 it was discovered that an alternating-bond conjugated polymer, *trans*-polyacetylene, is a semiconductor that can be doped and thereby transformed into a good electrical conductor. This discovery led to the Nobel Prize in chemistry in 2000, as mentioned earlier. Polyacetylene differs from normal household plastics, polyethylene (also known as "plastic wrap"), only in that it has one hydrogen atom per each carbon atom, instead of two, as depicted in Fig. 2.2. Yet this small chemical change gives rise to striking differences in both electronic and physical properties. Indeed, by making subtle changes in the chemistry of a polymer, the organic material can change from insulating to semiconducting to conducting! This is one of the dramatic differences between organic electronics and traditional electronics. In building "normal" electronic components, one chooses metals such as aluminum for the conducting parts, (doped) silicon for the semiconducting parts, and perhaps an oxide for the insulating parts of the component, that is, a combination of a wide variety of different elements that have highly dissimilar processing properties. In organic electronics, we can use basically one type of building block, carbon-based polymers and molecules that have fairly similar processing properties, for the conducting, semiconducting, and insulating parts of the components. Of course, there are situations where adding inorganic materials and/or metals to the devices is desirable, but the ability to tailor the electronic and mechanical properties of the polymers through synthesis is very useful from both device design and fabrication purposes. (If mechanical properties are to be modified, the traditional materials differ from polymers in a similar way as above. For instance, to change the mechanical properties of iron, another element such as manganese, copper, or carbon is added in a melt, creating an alloy. For a polymer, elasticity, shear resistance, etc., can be changed merely by changing the molecular weight (chain length) of the polymer, the alignments of the chains, and/or the amount of cross-linking, that is, no alloying is necessary.) The electronic and optical properties of π-conjugated polymers and molecules used in organic electronics, and how chemical and morphological changes affect those properties, will be discussed in this chapter.

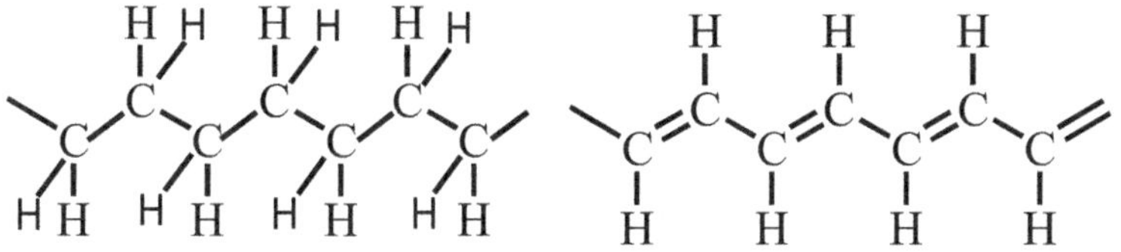

Figure 2.2 Polyethylene (plastic wrap) (left) compared to polyacetylene (right). Note the alternating carbon double-/single-bond structure of the polyacetylene backbone.

2.2 From Chemical Bonds to Bands

We will here describe how atoms can form bonds, becoming molecules; how molecules can bond into polymers; and how the electronic structure then evolves from discrete energy levels corresponding to atomic and molecular orbitals to energy bands in polymers. We will treat this on a purely conceptual level, so readers are referred to standard quantum mechanical textbooks for details.

We start by using hydrogen, the simplest atom available, to introduce one of basic concepts needed: the atomic orbital. The time-dependent Schrödinger equation for the hydrogen atom, which is an electron e^- and in the potential of a proton p^+, can be written as

$$i\hbar\partial\Psi/\partial t = H\Psi \tag{2.1}$$

where

$$H = -(\hbar^2/2m)\nabla^2 + V(r) \tag{2.2}$$

is the Hamiltonian, or the energy operator, and $\hbar$ is Planck's constant divided by 2π, m is the mass of the electron, and $V(r)$ is the potential energy of the electron in the attractive central electrostatic field of the nucleus (proton, in this case). The electronic wave function, Ψ, is related to the probability of finding the electron at a particular point is space around the nucleus at a particular point in time so that the electron density ρ is equal to $\Psi^*\Psi$. Ψ is then referred to as an atomic orbital, since it describes the orbit of the electron around the nuclei. We hereafter will use "electronic wave function" and "orbital" interchangeably in the text. Many solutions to Eq. 2.1 exist with their unique set of integer quantum numbers (n, l, m) that determine the energy and shape of the orbit or "cloud." The so-called principle quantum number n is related to the allowed energies of the electron, the quantum number l corresponds to the angular momentum of the electron in its orbit around the nucleus, and m is called the magnetic quantum number but will not be discussed further here. As per the Pauli principle, each orbital can contain two electrons, one spin-up and one spin-down. There are $2l + 1$ possible solutions with the same energy for a given n, so the maximum number of electrons of this energy is then $2(2l + 1)$. On the basis of the quantum numbers, the orbitals have names that come from the early days of optical spectroscopy. When $l = 0$, the orbital is called an "s" ("s" stands for *sharp*) orbital. For $l = 1$, the orbital is called a "p" (*principal*) orbital; for $l = 2$, the orbital is called a "d" (*diffuse*) orbital; and for $l = 3$, the orbital is called an "f" (*fundamental*) orbital. For the hydrogen ground-state orbital ($n = 1$), then called 1s, the resulting cloud is spherical in shape, and for excited states ($n = 2, 3, 4, \ldots$), the shape of the orbitals becomes increasingly more complex (see Fig. 2.3). Note that all Ψ go toward zero as the distance to the nucleus approaches infinity and that a wave function has $n - 1$ nodes as well. Solving the Schrödinger equation and determining the atomic orbitals for heavier atoms require some additional approximations due to many body effects, but the general concept of atomic orbitals still holds. Hence we can then also describe the electronic structure for heavier atoms such as carbon using this technique, that is, the total wave function of the electrons on an atom can be written as a linear combination of individual atomic orbitals.

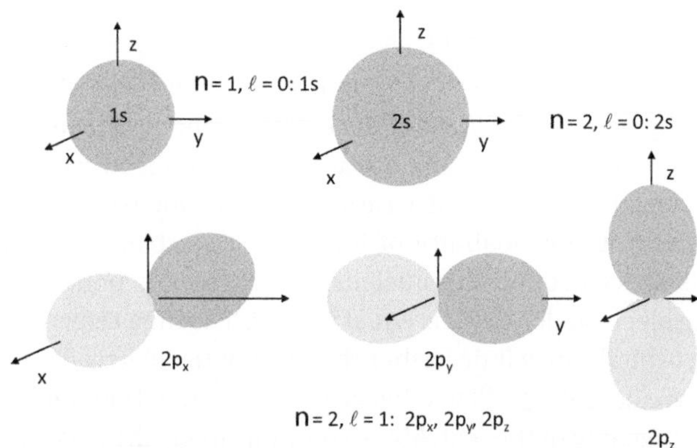

Figure 2.3 Shape of the 1s, 2s, and 2p orbitals and their corresponding quantum numbers. The 2s orbital has a spherical nodal plane (not shown).

Another important concept needed for understanding how carbon atoms bond to form molecules and polymers is the so-called hybrid atomic orbitals. The ground state of a carbon atom with its six electrons can be written $1s^22s^22p^2$, that is, two electrons in the 1s orbital, two electrons in the 2s orbital, and finally two electrons in the three 2p orbitals (remember, $2l + 1$ degeneracy) (see Fig. 2.4). When interacting with nearby atoms, however, the electrons in the frontier occupied $n = 2$ orbitals can reorganize themselves to minimize the overall energy by promoting one of the two 2s electrons into a 2p orbital, subsequently forming linear combinations of the original $n = 2$ orbitals. This can occur in three basic hybridization schemes: sp, sp^2, and sp^3. The sp-hybridized orbital consists of a linear combination of equal weight between the 2s orbital and a 2p orbital, for example, $2p_y$ in Fig. 2.4. The other two 2p orbitals in this case, $2p_x$ and $2p_z$, retain their original shape. For sp^2 hybridization, a linear combination between the 2s and the $2p_x$ and $2p_y$ orbitals occur, with the $2p_z$ orbital unchanged. Finally, for sp^3 hybridization, all orbitals participate in the linear combination. When the atomic orbitals are added together in this fashion, the shape of the resulting orbit or electronic cloud obviously will change (see Fig. 2.4).

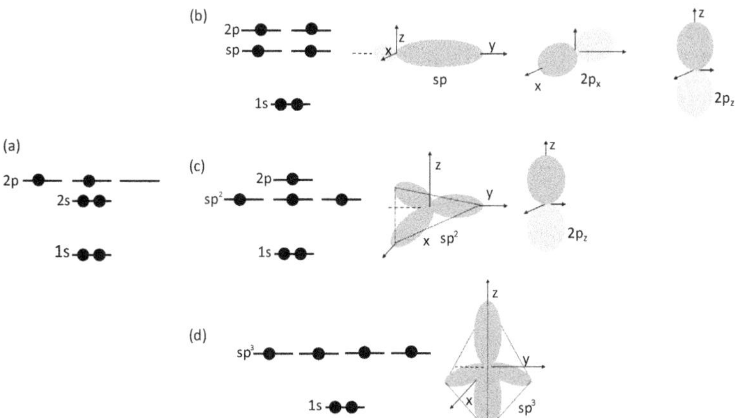

Figure 2.4 Occupied atomic orbitals of (a) the carbon atom, (b) sp-hybridized carbon, (c) sp^2-hybridized carbon, and (d) sp^3-hybridized carbon. Note that the energy difference between the 1s and 2s orbitals is significantly decreased in the figures.

Now, recalling that the atomic orbitals were connected to the electron density around an atom, we can see that the electrons density is not evenly distributed around the carbon atom, but a higher electron density will exist in certain areas, depending on the hybridization (sp, sp^2, or sp^3). The location of the clouds of high electron density in the hybridized orbitals, often called lobes, determines the bonding structure of carbon. As the carbon nuclei carry a positive charge from the six protons, bringing two carbon atoms close to each other will be easier if the approach is along an axis where a lobe of each atom intersects, forming a sort of "handshake." We can conceptually think of it in this way that the high electron density in the intersecting lobes will shield the positive charge on the nuclei, decreasing the coulomb repulsion and allowing them to move closer to each other compared to a case where the approach is along an axis lacking lobes and hence has little contribution to the screening from the electrons in the $n = 2$ shell. The higher the resulting electron density of the intersecting lobes between the nuclei, the stronger the bond. From Fig. 2.4 it should then be apparent that sp^3-hybridized carbon will form chemical bonds with four neighboring atoms in a tetrahedral-like structure, sp^2-hybridized carbon will form chemical bonds with three neighbors in plane at a 120° angle, and sp-hybridized carbon will form chemical bonds with two neighbors in a linear fashion. This type of bond is called covalent, as the two atoms forming the bonds share (valence) electrons. In the discussion before, the bonds formed as a handshake between two lobes of neighboring atoms are called σ bonds as they are circularly symmetrical in cross section when viewed along the axis of the bond (lobe handshake). We have so far left out the 2p orbitals that didn't participate in the sp and sp^2 hybridization, but these too can form bonds, albeit of a weaker kind, so called π bonds. Take for example the case of sp^2 hybridization where the $2p_z$ orbitals on each carbon atom extend perpendicular out of the plane defined by the σ-bonding sp^2 lobes (see Fig. 2.5).

The two weakly overlapping $2p_z$ orbitals do not contribute to the electron density in the plane of the σ bonds and hence are not in the direct line drawn along the bond of the two carbon atoms, but there is still some additional screening of the nuclei added to the σ bond contribution and hence a slightly shorter bond length as compared to the sp^3 case. This is a so-called double bond (a σ and a π bond working together). For sp hybridization, both $2p_x$ and $2p_z$ orbitals can form π bonds, yielding an even shorter bond length, a so-called triple bond (a σ and two π bonds working together).

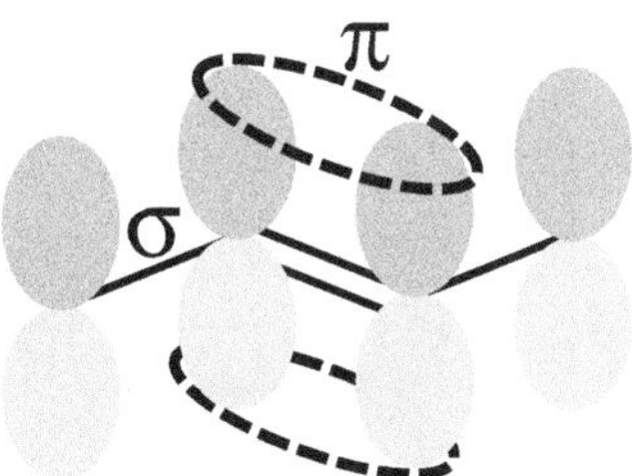

Figure 2.5 Sketch of p_z–p_z overlap forming a p bond. The sp^2 lobes forming the handshake s bonds that define the molecular geometry are not shown.

If we continue with sp^2-hybridized carbon atoms, it is easily realized that the 120° bond angle means one can construct a hexagon-shaped molecule by bonding six carbon atoms together in a plane. The overlapping sp^2 lobes are supplemented by overlapping $2p_z$ orbitals orthogonal to the plane forming π bonds. Each carbon atom in the hexagon has an unused sp^2 lobe (valence electron) left, which then can bond by overlapping with, for example, the 1s orbital of a hydrogen atom (see Fig 2.6). We have thus formed benzene, a π-conjugated molecule, the formal definition of which is a molecule that has a p orbital on an atom adjacent to a double bond. Hence, when we have an unbroken sequence of carbon atoms bonding in the sp^2 hybridized scheme, we have π conjugation, which is in general a prerequisite for obtaining organic semiconducting materials, as the π conjugation allows for delocalized π bonds. Imagine further that instead of adding hydrogen atoms to the 1 and 4 position of the benzene hexagonal ring, we let two other benzene rings missing hydrogen in the 1 and 4 positions attach themselves instead (see Fig. 2.6), and continue the process ad infinitum, we would wind up with an infinitely long chain of repeating benzene molecules: a polymer (*poly*: many; *mer*: parts, "many parts" in Latin). In practice, real polymers are not infinite but do contain a large number of repeating molecular units. If two or more different molecular units are repeated, we have a so-called co-polymer.

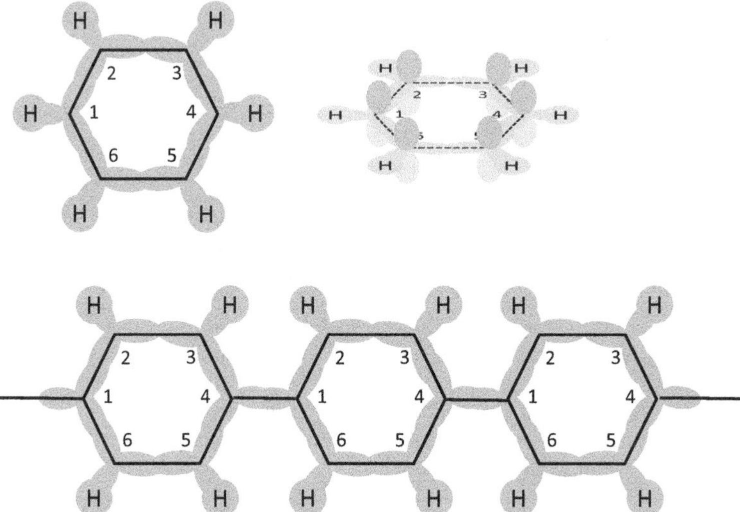

Figure 2.6 Benzene ring formed by σ-bonding sp^2 orbitals (top left) and p_z π-bonding orbitals (top right). Benzene rings repeatedly bonding at the 1 and 4 positions, forming a linear polymer chain: poly(*p*-phenylene) (bottom).

Returning now to the electronic structure, we have discussed how the energy and location in space of the electrons in an atom can be described by a linear combination of atomic orbitals. When σ and π bonds are formed, they form by overlap between (valence) orbitals on adjacent atoms. When these electron clouds coincide in space, it is reasonable to assume that both the energy and the shape of the interacting orbitals are affected, which indeed is the case. We begin by using the atomic orbitals of the hydrogen atom to construct an approximation to the molecular orbitals of the hydrogen molecule, H_2. This derivation appears in a large number of textbooks, so we will not go deeply into the mathematics. The key concept is that the atomic orbitals of the hydrogen atoms are used as a basis for the molecular orbitals of the hydrogen molecule, that is, each molecular orbital of H_2 can be derived from a linear combination of the atomic orbitals. We choose the 1s atomic orbitals of the electron on each of the two hydrogen atoms, located at a constant distance $\mathbf{r}_{12}$ from each other. We then express the molecular orbital as a linear combination of the two atomic orbitals, ϕ (a 1s state, each centered on one of the two protons),

$$\Psi = a\phi_1 + b\phi_2, \tag{2.3}$$

where $\phi_1 = \phi(\mathbf{r} - \mathbf{r}_1)$ is the 1s orbital centered on proton 1 and $\phi_2 = \phi(\mathbf{r} - \mathbf{r}_2)$ is the 1s orbital centered on proton 2. The coefficients a and b must satisfy the condition the $a^2 + b^2 = 1$ in order that Ψ will be normalized to unity, $\int\Psi^*\Psi d\tau = 1$. Also, we take a and b to be real without loss of generality. Then solving the time-independent Schrödinger equation $H\Psi = E\Psi$, where E is energy, gives us two molecular orbitals:

$$\Psi_+ = (2)^{-1/2}(1 + S)^{-1/2}(\phi_1 + \phi_2), \quad \text{and} \quad \Psi_- = (2)^{-1/2}(1 - S)^{-1/2}(\phi_1 - \phi_2) \tag{2.4}$$

with the corresponding energies:

$$E_+ = (\alpha - \beta)/(1 + S), \quad \text{and} \quad E_- = (\alpha + \beta)/(1 - S) \tag{2.5}$$

where α is the so-called Coulomb integral $\int\phi_i{}^*H\phi_i\, d\tau$, β is the so-called resonance integral $-\int\phi_i{}^*H\phi_j\, d\tau$, and S is the so-called overlap integral $\int\phi_i{}^*\phi_i\, d\tau$. The two new molecular orbitals created from the interacting atomic orbitals of the respective hydrogen atom and their energy levels are plotted in Fig. 2.7.

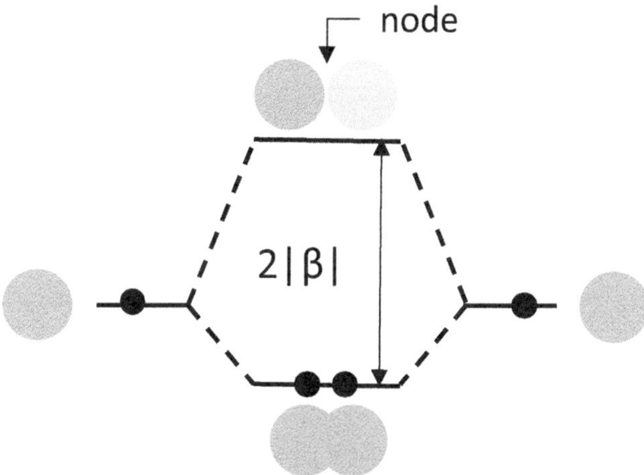

Figure 2.7 Energy level diagram of two hydrogen atoms interacting to form a H_2 molecule. The 1s orbital overlap cause two new molecular orbitals to form: one bonding (Ψ_+), where the electrons from the respective 1s orbitals go, and one antibonding (Ψ_-) that becomes unoccupied. The energy gap between the two molecular orbitals is $2|\beta|$. The simplified shape of the molecular orbitals also is shown.

The electrons from the two singly occupied 1s orbitals of the two hydrogen atoms will then move into the Ψ_+ molecular orbital, and the Ψ_- molecular orbital becomes unoccupied in the H_2 molecule ground state. The gap in energy separating the highest occupied molecular orbital (HOMO) and the lowest unoccupied molecular orbital (LUMO), Ψ_+ and Ψ_-, respectively, for our H_2 molecule, represent the minimum energy needed to create an excited state. The results can be generalized to the interaction between any two atoms where the resulting molecular orbitals can be constructed from linear combinations of the atomic orbitals of the atoms involved in the bond formation.

Building upon the concepts outlined above (and letting $S = 0$ for simplicity), we can consider what happens when we have more atoms in a row, forming a linear molecule and eventually a polymer. If we take the hypothetical case of N hydrogen atoms, each with one electron in one atomic orbital, there then will be N molecular orbitals generated in the linear hydrogen molecule (see Fig. 2.8). The molecular orbitals will be linear combinations of the hydrogen atomic orbitals such that a molecular orbital $\Psi_j = \Sigma c_{ji}\phi_i$. Again solving the time-independent Schrödinger equation using the so-called particle-in-a-box/infinite-potential-well approximation gets us the corresponding energies of the molecular orbitals:

$$E_j = \alpha + 2\beta \cos(j\pi/(N + 1)), j = 1, 2, \ldots, N \tag{2.6}$$

with the **k**-vector (from the orbital electron momentum $\mathbf{p} = \hbar\mathbf{k}$) is then

$$\mathbf{k}_j = j\pi/a(N + 1), j = 1, 2, \ldots, N \tag{2.7}$$

where a is the distance between atoms. If we take Eq. 2.6 and let N become very large (i.e., >>1), such as in a very long polymer chain, we see that a band of energies will appear between $E_1 \approx \alpha + 2\beta$ and $E_N \approx \alpha - 2\beta$, with a negligible energy difference separating the molecular orbitals in sequence. The results can be generalized for any atom from which then follows that a band created from N interacting orbitals of an l quantum number can contain $2(2l + 1)N$ electrons as each orbital can contain $2(2l + 1)$ electrons due to the Pauli principle (see Fig. 2.9). The allowed electron energies are then defined by the bands separated by gaps, similar to the case of atomic and molecular orbitals.

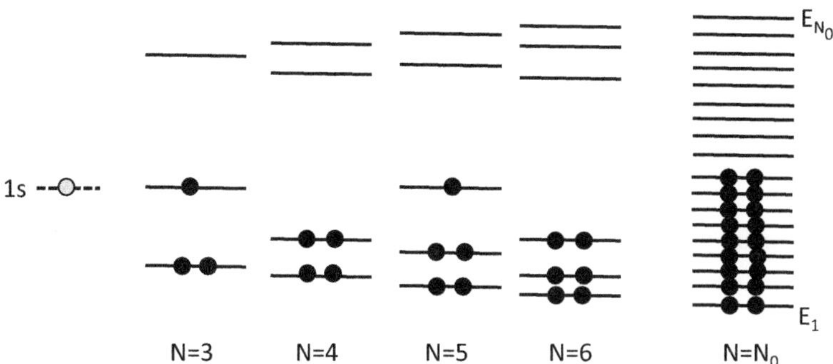

Figure 2.8 Evolution of the molecular orbital energies for a linear chain of *N* interacting singly occupied 1s orbitals (hydrogen atoms).

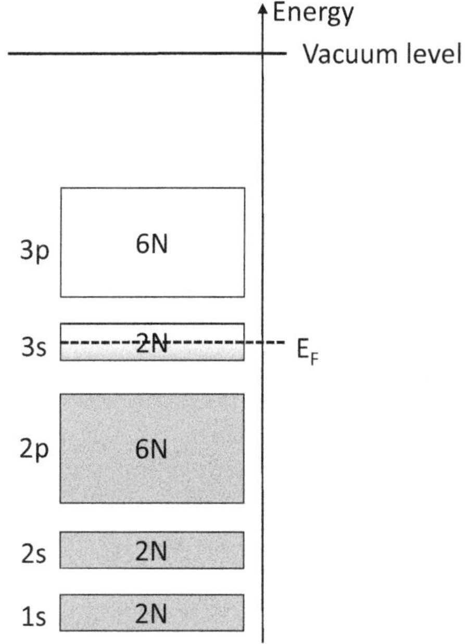

Figure 2.9 Energy bands of a material containing *N* atoms and the number of electrons allowed in each band. The energy scale is referenced to the vacuum level, which signifies an electron that is free but at rest (zero kinetic energy). Any electron with an energy situated below the vacuum level (negative energy on this scale) is then considered bound. The Fermi energy, E_F, defines the highest occupied electronic state at $T = 0$ K and is here positioned in the middle of the 3s band, that is, Na metal.

Whether residing in energy bands or in discrete molecular orbitals, the electrons are bound to the material. To escape the material (atom, molecule, or polymer), energy must be added (absorbed). The energy of the molecular orbitals or energy bands is typically referenced to as the so-called vacuum level (see Fig. 2.9), which corresponds to an electron that has been emitted from the material but has zero kinetic energy, sort of equivalent to a launched rocket just barely escaping the gravitational pull of the earth.

So to wrap up this section, each electron in a material occupies a state determined by its wave function. A material's density of states (DOS) enumerates the number of possible states as a function of energy. Quantum mechanics dictates that only certain energies are allowed, so the DOS can have a lot of features. Electrons are considered indistinguishable and obey the Pauli exclusion principle, so only two electrons can occupy the same state. Filling of the DOS starts at the lowest energy and builds upward. At 0 K, the electrons hence occupy the lowest-possible energy states, and for a metal the energy up to which all the levels are filled is called the Fermi level. (At temperatures above 0 K, thermal excitation occurs resulting in vacancies below, and occupation above, the Fermi level.) The allowed energy levels are usually grouped together in bands, which, in turn, are separated by energy regions without any allowed states at all. If the Fermi level is positioned in the middle of such a band, the material is a conductor; if the Fermi level is positioned between bands and the separation is large, the material is an insulator; and if the Fermi level is positioned between bands but the separation is small, the material is a semiconductor. The requirement for charge conduction is that there be partially filled bands with vacant states available. This is the default state of a conductor but requires, for example, thermal excitation in a semiconductor and is difficult, but not impossible, to achieve in an insulator. Almost all interesting material properties are determined by the electronic structure around the Fermi level. The highest (at 0 K) fully occupied band is commonly referred to as the valence band, the lowest unoccupied band is referred to as the conduction band, and the separation between the valence band and the conduction band is referred to as the band gap. The valence band and conduction band edges thus correspond to HOMO and LUMO, respectively, in a molecule. Crystalline materials have very well-defined bands, and an unambiguous definition of ionization potential (IP) and electron affinity (EA) is possible from the top of the valence band (HOMO) and the bottom of the conduction band (LUMO), respectively. Disordered materials lack well-defined band edges and their positions are difficult to measure. HOMO and LUMO are therefore used rather arbitrarily to define energy levels reasonably close to the true values but where the DOS is significant. This, unfortunately, introduces a measurement aspect where experimentally determined values are somewhat dependent on how they were measured and evaluated, as illustrated in Fig. 2.10.

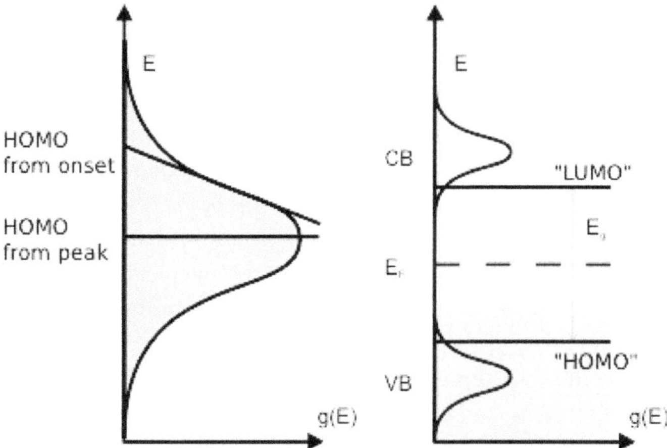

Figure 2.10 A schematic view of the density of states around the Fermi level, together with two of the most common ways to evaluate the frontier orbital energy edges.

2.3 Charge Injection and Transport

The primary function of an organic electronic device generally involves injection and transport of charge, two processes that we will discuss in this section. Often, hybrid devices are fabricated where metals or transparent conducting oxides are used as charge-injecting electrodes for the organic semiconductor films, but organic conductors, typically doped conjugated polymers, are used as well.

Organic molecules are very soft compared to most inorganic materials. They also tend to have a more heterogeneous charge distribution. Even a conjugated polymer chain such as polyacetylene, which at first glance appears to be a one-dimensional crystal, has a nonconstant charge distribution. Peirels showed that such a system needs to dimerize to become stable. This dimerization into alternating carbon single and double bonds (see Fig. 2.11) causes a positional and electronic perturbation along the chain. In general, adding charge (reducing) or subtracting charge (oxidizing) to or from π-conjugated materials causes both positional and electronic distortions as the molecule first reorganizes its electronic density in response to the charge whereupon the nuclei react to the new electron density (more electron density between the nuclei: shorter bonds; less electron density: longer bonds). These distortions generate a potential well that, at the cost of some energy, can travel along with the charge. The combination of a charge with an accompanying potential well is a quasi-particle called a polaron, and they are then inherently localized in space, typically on a single molecule or part of a polymer chain. (The name comes from *polar crystals*, where the phenomenon was first encountered.) Other similar quasi-particles are possible: electronic states occupied by two electrons (or holes), basically two polarons of equal charge interacting that then are called bipolarons (see Fig. 2.11). Note that the polaronic states appear in the gap separating the HOMO and LUMO, so the cost in energy of removing an electron from the HOMO, called the IP, is thus lesser than the HOMO energy versus the vacuum level. Similarly, the energy gained from adding an electron to the LUMO, called EA, is greater than the LUMO versus vacuum-level difference. The energy gained from the reorganization into polaronic species is called the relaxation energy, λ.

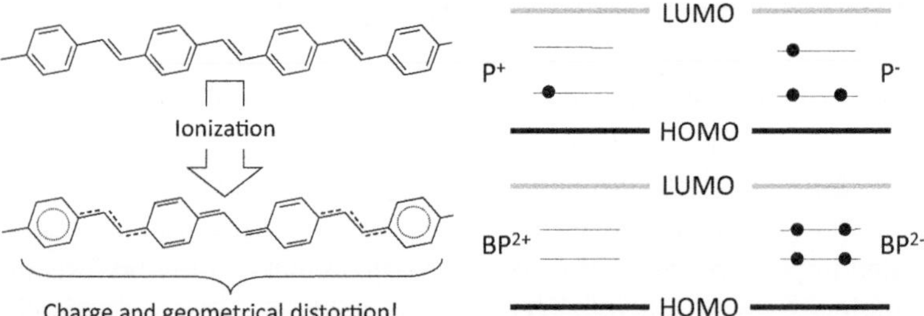

Figure 2.11 (Left) Segment of a poly(*p*-phenylenevinylene) chain undergoing bond alternation upon ionization. (Right) Polaronic species (P: singly charged; BP: doubly charged) appearing in the HOMO–LUMO gap upon ionization.

A further complication arises as we go from a single molecule or polymer chain in vacuum to the solid films typically used in organic electronic devices. In this scenario, if we, for example, remove a charge from the HOMO of a molecule, not only do the molecules reorganize their electron density, as described above, but the electron density on the nearby molecules in the film will also be slightly modified, screening the charge on the ionized molecule. This is called a polarization effect as the local medium is polarized in response to the charge on the molecule. Hence both the gas-phase IP and EA will be modified so that

$$IP_s = IP_g - P_+ \text{ and } EA_s = EA_g + P_- \tag{2.8}$$

where P_+ and P_- are the polarization energies in response to a positive and negative charge, respectively. For a film featuring disorder, the local environment of a molecule varies in the films and hence will also the P_+ and P_- contributions. This means that instead of the well-defined discrete energy levels of the molecular orbitals we will have a distribution of ionization energies for each orbital, resulting then in a broader density of such states as compared to the gas phase (see Fig. 2.12).

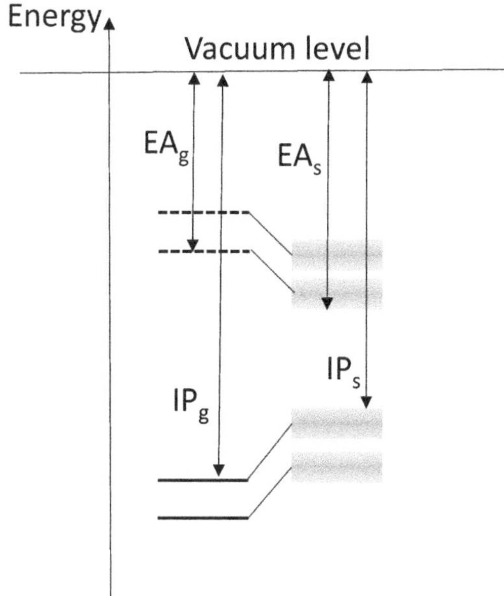

Figure 2.12 Change in ionization energies going from gas phase to a solid.

The process of charge injection in a device (see Fig. 2.13) involves transferring electrons from the HOMO (valence band edge) of the organic semiconductor into the electrode Fermi level, that is, hole injection, or transferring electrons from the electrode Fermi level to the LUMO (conduction band edge), creating the polaronic species in both cases. If the electrode Fermi level position versus the vacuum level, the so-called work function, is such that it is energetically favorable to transfer charges from the electrode to the semiconductor, that is, larger than the positive polaron or smaller than the negative polaron formation energy, no extra energy is needed to transfer a charge and spontaneous transfer will occur until the work function of the electrode and the Fermi level of the semiconductor are equilibrated. We then have a so-called ohmic contact. If, on the other hand, the work function of the electrode is such that no spontaneous charge transfer occurs, the contact will have a barrier for charge injection and the contact is said to be blocking. A device with one ohmic contact of each kind, one for holes and one for electrons, is a typical diode. Under forward bias, that is when the current flow is such that both contacts inject the carrier type for which it is ohmic, the current increases exponentially with the applied voltage. Under reverse bias, where both electrodes present barriers, no (or, in reality, a very small) current flows through the device. If, however, only one of the electrodes is ohmic, current will still flow through the device but there will be a build-up of space charge since only carriers of one polarity is injected into the semiconductor as opposed to the case where both holes and electrons are injected and charge neutrality is maintained. Space charge–limited currents (SCLCs) in single-carrier devices are quite useful; the total amount of charge is that can be maintained in the semiconductor is determined by the geometric capacitance of the device and this information can be used experimentally to characterize the charge transport properties of the material. The current density, J, versus voltage, V, dependence is described by Eq. 2.9:

$$J \approx \varepsilon \mu V^2 / L^3 \tag{2.9}$$

where L is the film thickness, ε is the dielectric constant, and μ is the mobility (more on that later).

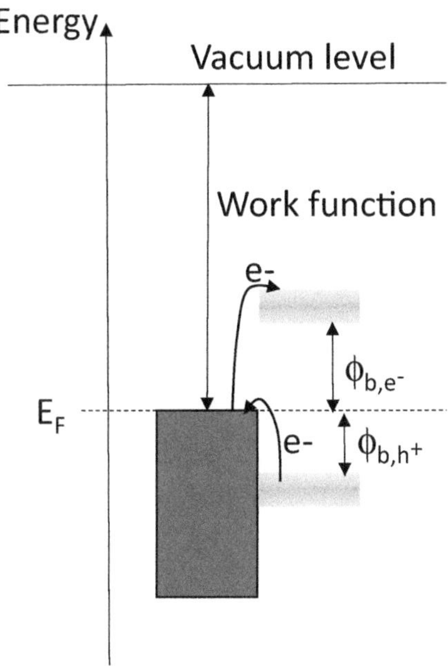

Figure 2.13 Sketch of injection barriers (ϕ_b) involved when injecting an electron into an organic semiconductor, forming a negative polaron, or injecting a hole (extracting an electron), forming a positive polaron. The positive and negative polaron densities of the organic semiconductor film at the interface of the metal electrode are drawn assuming vacuum level alignment.

Even if a contact is not ohmic, it is still possible to inject carriers, but there is an additional cost in energy involved due to the offset between the energy levels involved. The size of that offset, often called injection barrier, ϕ_b, quickly becomes a limiting factor. For such cases, the current density versus voltage relation becomes more complex but typically involves an exponential dependence on ϕ_b such that $J \approx \exp(-\phi_b)$, that is, the larger the barrier, the lower the injected current density, all else being equal.

As we can see, the injection barrier, if any, at an electrode/organic semiconductor interface is of great importance. Tuning the energy level alignment at such interface is then critical for optimizing device performance, and a variety of ways to do just that will be discussed in later chapters in the book. Here we will merely describe what types of energy level alignment can occur and the basic concepts behind them. When an organic semiconductor material is brought into contact with a metallic surface, two basic regimes of energy level alignment will hold, (i) vacuum level alignment and (ii) vacuum level off-set (see Fig. 2.14). For vacuum level alignment to occur there can be no potential step at the interface, meaning that no (significant) amount of charge is transferred across the interface upon contact (and that the organic semiconductor does not form a dipole layer by preferential ordering of individual molecular dipole moments). The consequence of this is that as soon as we have partial charge transfer through covalent bond formation at the interface (metal atoms reacting with the organic molecules), we will be in regime (ii). Even when the metal surface is rendered nonreactive by partial oxidation or adsorbtion of carbohydrogen "dirt" (metals films prepared in low vacuum and/or exposed to air), charge transfer and hence vacuum level offsets can still occur if the work function of the metal surface is high enough to extract electrons from the HOMO or low enough to inject electrons into the LUMO of the molecules adjacent to the metal surface in the absence of an external field (see Fig. 2.14). If additional charge is transferred in the subsequent layers extending away from the interface, the resulting energy gradient is typically referred to as energy level bending or band bending.

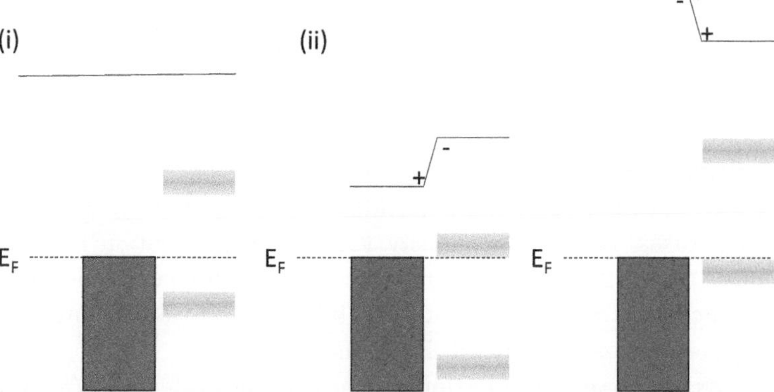

Figure 2.14 Sketch of energy level alignment at a metal/organic interface. (i) Vacuum level alignment (no charge transfer between the organic film and the metal); vacuum level offset due to electron transfer from the low-work function metal to the organic film (ii, left panel) and due to electron transfer from the organic film to the high-work function metal (ii, right panel).

Having injected charge into an organic semiconductor film, the purpose is then to transport it. We will now discuss the properties that affect the charge transport and introduce the main concepts. The number of electrons in the conduction band and holes in the valence band together with their respective mobility determines the electronic conductivity, σ, of a material (Eq. 2.10), with q being the elementary charge, μ_n, μ_p, and n, p, the mobility and concentration of electrons and holes, respectively. Controlling the number of carriers is usually part of the device function, so the important quantity to consider is the mobility, which defines the velocity with which an electron (or hole) moves in response to an electric field. A high mobility allows, for example, transistors a higher modulation range and faster switching speeds. Organic materials generally have a lower mobility than inorganic semiconductors.

$$\sigma = q(n\mu_n + p\mu_p) \tag{2.10}$$

The electronic conductivity of conjugated molecules is a cornerstone of organic electronics. It must be mentioned, though, that most of the organic materials have much higher ion conductivity than common inorganic semiconductor materials. This is due to both a considerably higher prevalence of ions and higher ion mobility. The former is largely due to synthesis and purification issues and the latter largely to the soft and porous nature of organic materials. In some cases, the ions, or ion-dependent processes, are part of the intended function. Electrochemical transistors are one example where the functionality comes from doping and de-doping of the channel. Another one is light-emitting electrochemical cells, where mobile ions enable current injection under operation. For more conventional devices such as field-effect transistors and Schottky-like diodes, the ion conductivity is instead contributing to a nonideal behavior. Purification of organic materials is difficult and can be a significant cost. From a commercial point of view, it might thus make sense to develop technology that is robust to such nonidealities rather than trying to remove them by extensive purification.

One very distinctive feature of charge transport in most organic semiconductors is that it is activated, that is, the charge carrier mobility increases with increasing temperature. This is at variance with most of the common inorganic semiconductors where the mobility instead decreases with increasing temperature, and is due to charge carrier localization. There are fundamentally two factors that contribute to the localization: One is polaronic effects and is due to the very nature of the molecules, and the other is disorder and related to the interaction between molecules (and usually also between molecules and impurities).

Although many conjugated molecules and polymers can crystallize, they do not readily form large, perfect crystals. Some materials are liquid crystalline, and others completely amorphous. In all cases, the processing conditions are important for the end result. A given material can, and usually does, therefore, have a complex phase structure. The concomitant variations in the surroundings of a charge carrier make the system disordered. This disorder can be both positional, that is, when the distance between the areas contributing to the conduction or valence bands in the material is nonuniform, or energetic, for example, due to interactions with intrinsic dipoles or with (nonuniformly distributed) charged impurities. Both the distance and energy difference between localized states influence the rate of charge transfer and can thus cause localization.

Positional disorder is relatively straightforward; the closer the sites, the stronger the electronic coupling gets. On the other hand, if the sites are separated far enough the charge will inevitably localize. Localization can occur even if the sites are very close to each other due to energetic disorder. The important parameter is the site density per unit energy. When the site density per unit energy is high, such as in crystals, the localization radius of an electron or hole is large, and if sufficiently large, the charge is said to be delocalized and is no longer associated with a particular atom or molecule. Such a charge can easily move around and the mobility is therefore high; the charge transport is said to be ballistic. When the site density decreases, so does the localization radius. At some point the localization radius becomes too small to encompass a continuous path of sites within a small-enough energy interval to allow efficient charge transfer and the charges become localized. Since activation is necessary for localized charges to move, the mobility is low. The transition from localized to delocalized states is labeled as "mobility edge," and it marks the transition in charge transport mechanism from hopping between localized sites to ballistic transport of delocalized charges. Figure 2.15 shows the relationship between site density and localization. Very few π-conjugated organic materials are macroscopically dominated by ballistic transport at room temperature mainly due to local variations in intermolecular order, but that does not mean that it never occurs. The complex phase structure of many materials can certainly include very well-ordered regions, but as long as there are disordered regions in the path that a charge carrier needs to traverse the overall behavior will be that of an activated process.

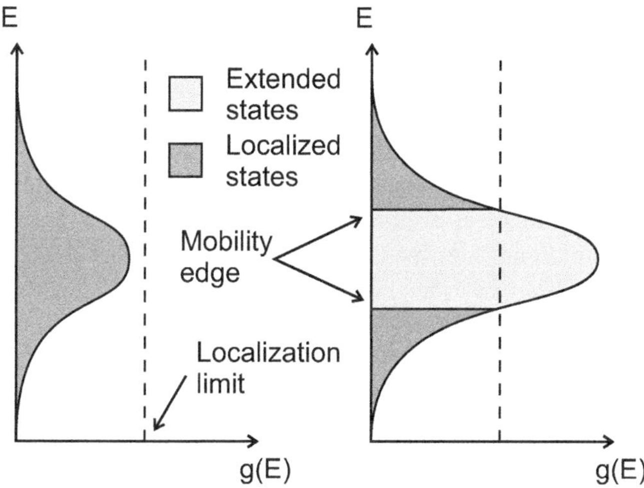

Figure 2.15 Sketch showing the frontier density of states and the relation between the state density and localization (all states to the left of the dashed line are considered localized). The mobility edge refers to the fact that the mobility is considerably higher when the state density is above the localization limit.

As mentioned, even for well-ordered molecular films, the polaronic effects will cause a self-localization of the charge species. Charge transfer of localized charges under the influence of polaronic effects is well described by Marcus theory, which was originally developed for chemical electron transfer reactions between dissolved reactants. Movement of polarons is fundamentally very similar to such a reaction, but instead of solvent molecules it is the molecules themselves (or impurities) whose polarization it is that needs to be realigned. Mathematically, the initial and final states of the charge transfer can be approximated with parabolas whose vertices correspond to the respective reaction coordinate and energy of the relaxed polaron in its initial and final positions. The additional vertical energy difference between the vertex of the initial state and the parabola of the final state is the reorganization energy (see Fig. 2.16). To move from one site to another, the charge must overcome an energy barrier $E_{a,M}$ proportional the energy difference between the sites, ΔE, and the reorganization energy, λ (see Eq. 1.11).

$$E_{a,M} = \frac{(\lambda + \Delta E)^2}{4\lambda} \tag{2.11}$$

If the reorganization energy is small and the average energy difference between neighboring sites is large, the latter will dominate the charge transfer. Such a situation may occur in molecular films featuring significant variation in the intermolecular order. The average energy difference between sites should, based purely on statistical arguments, be proportional to the average energy difference between the occupied and unoccupied states. In the Gaussian disorder model (GDM), the DOS is approximated by a Gaussian function. Apart from empirical success, such an approximation can also be rationalized on, for example, the basis that the molecules in an amorphous molecular solid are randomly oriented with respect to each other. At its inception, the Gaussian shape was also motivated by, for example, the shape of the absorption spectrum of many disordered organic semiconductors, but such arguments can be misleading since different states are usually involved in the optical processes. Fermi–Dirac statistics can be used to calculate the occupancy probability and thus the average energy difference between occupied and unoccupied states. The GDM activation energy is taken to be proportional to this difference. The important parameter in the GDM is the standard deviation of the Gaussian function, σ, sometimes also for convenience expressed as $\hat{\sigma} = \sigma / k_B T$ instead.

The density of occupied states in a Gaussian DOS is, at equilibrium and for a low-enough occupation density, an identical Gaussian shifted in energy by $\sigma^2 / k_B T$ (see Fig. 2.17). To move from one site to another, a charge must, on average, overcome this energy difference, and the energy predicted by the GDM, $E_{a,\,GDM}$, is thus proportional to this number. In practical applications an empirically determined constant of 4/9 is usually included in the GDM activation energy (Eq. 2.12):

$$E_{\mathrm{a,GDM}} = \frac{4}{9}\frac{\sigma^2}{k_{\mathrm{B}}T} \tag{2.12}$$

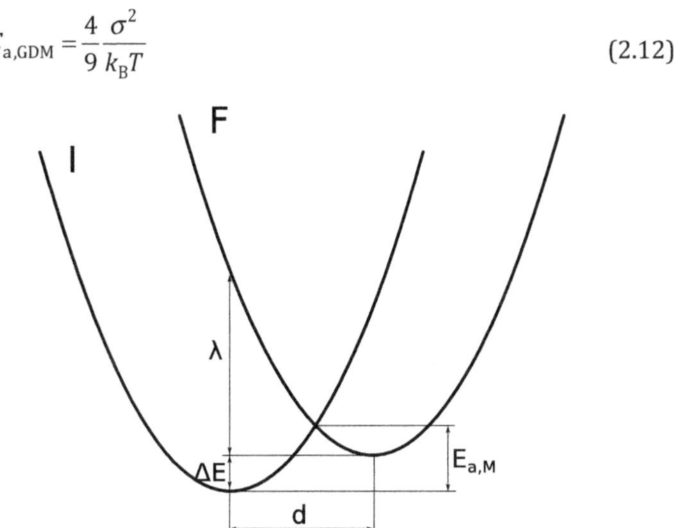

Figure 2.16 The initial (I) and final (F) state energy parabolas of a charge transfer process, where d is the distance between the localized states involved, ΔE is the difference between the initial and final state energy, $E_{\mathrm{a,M}}$ is the required activation energy according to Marcus theory, and λ is the reorganization energy.

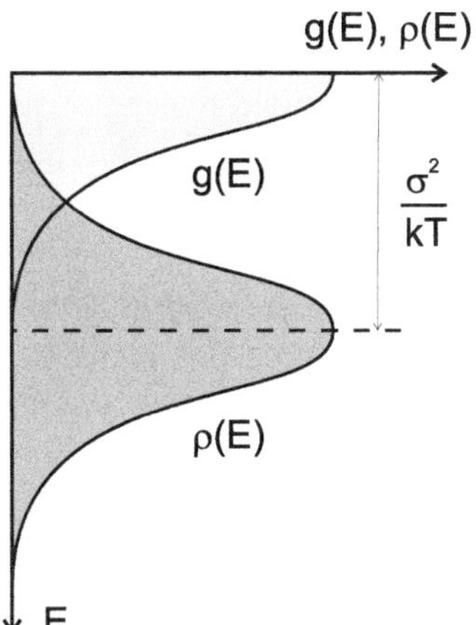

Figure 2.17 Sketch of the occupied and unoccupied state densities at equilibrium for low carrier concentrations assuming a Gaussian density of states. The separation between the peaks of the respective distributions is equal to $\sigma^2/k_{\mathrm{B}}T$, which is the physical motivation for the expression for activation energy used in the GDM.

The slightly lower activation energy suggested by empirical data can easily be rationalized by the fact that it is not necessary for a carrier to, on average, jump to the exact center of the DOS but rather to a slightly lower energy where the state density is high enough for states to be readily available. Although less stringent than Marcus theory, the GDM can offer a more direct and intuitive basis for understanding macroscopic charge carrier transport in disordered materials. Many arguments can be made that a single Gaussian is not a perfect representation of the DOS of these materials, but it is close enough to allow for at least a very good qualitative representation of many properties.

When the carrier concentration increases above a certain level, the offset between the center of the DOS and the average carrier energy will start to decrease due to state filling effects (c.f. a fully occupied band where the offset is zero). The concomitant decrease in the activation energy increases the charge carrier mobility. At very high carrier concentrations the number of available free sites to which a carrier can jump will decrease and so then will the mobility. Most organic systems have a relatively low site density per unit energy, which means that such effects become relevant for many common applications. This carrier concentration dependence is strongly correlated with the amount of disorder in the system or, concomitantly, the width of the DOS. More disorder gives stronger concentration dependence. Field-effect transistors are an example of where the charge carrier concentration dependent mobility can be important due to the high carrier concentrations frequently encountered in such devices.

If carriers are randomly introduced into a system, these will initially not be at equilibrium. The offset between the center of the DOS and the average carrier energy, and thus mobility, then becomes a time-dependent property until equilibrium is reached. Equilibration can be slow, and the mobility immediately after carrier introduction can be orders of magnitude higher than at equilibrium. Since the active layers in many practical devices are very thin, this effect can be very important and have a large influence on the performance of the device.

Externally applied electric fields add a further positional dependence to the energy of a charge carrier. Sites in the appropriate direction (according to polarity) along the net electric field will have their relative site energies reduced. Since the activation energy is proportional to the initial and final site energies of a carrier jump, this means that the mobility will increase with increasing electric field. Experimentally, the mobility has frequently been shown to be proportional to the square root of the electric field, and it is therefore usually referred to as the Poole–Frenkel effect after the theory for electric field–assisted electronic conduction in insulators.

It should now be clear that the degree or intermolecular order plays a huge role in determining charge transport properties in an organic semiconductor film. There is an additional very important aspect to this, mainly that impurities incorporated into the films can further introduce both positional and energetic disorder. Indeed, many organic materials are difficult to purify (a 99% purity is usually considered very good). Impurities can act as dopants, influence the DOS, and/or act as traps for charge carriers. Many (disordered) organic materials are also highly permeable to impurities (e.g., water) so that the impurity levels depend on the environment of the device. Contrary to most inorganic semiconductors where the impurities are deliberately introduced to tune functionality, most impurities in organic materials are involuntary. An appropriate treatment of impurities and their effects is thus very important. Few concepts are so freely used and with as many meanings, as traps within the field of organic electronics. When transport is ballistic the concept is simple; a trapped carrier is just that. It is trapped and does not, at that particular moment, move. All the carriers in a system with localized states are in essence trapped for the majority of the time. While there may certainly be trapping by impurities, it is not necessarily possible to distinguish those from carriers trapped in low energy states of the main DOS. In many cases the impurities simply influence the shape of the main DOS slightly by, for example, broadening it. If there is an increased site density at some narrow energy interval that is possible to link to a specific impurity, and that has a significant impact on the transport properties, it can be useful to talk about traps in the traditional sense, but if there is no obvious and clearly defined trap level, it is, from an engineering perspective, best to consider the impurity-based effects as simply part of the DOS. What is important to know is that the same organic materials can behave differently based on its environment and history.

2.4 Optical Properties

Optical absorption occurs through interactions between an electromagnetic wave (light) and absorbing matter. Organic molecules are frequently asymmetric and have many different internal dipoles. Ultimately, this leads to many different interaction possibilities, which can give rise to anisotropy and several distinct absorption bands. Additionally, for example, positional disorder tends to cause local variations that smear out the macroscopically observable absorption. Amorphous organic materials thus frequently have one or more smooth, Gaussian-like absorption peaks. Highly ordered samples have much sharper features and frequently display vibronic structure (see Fig. 2.18).

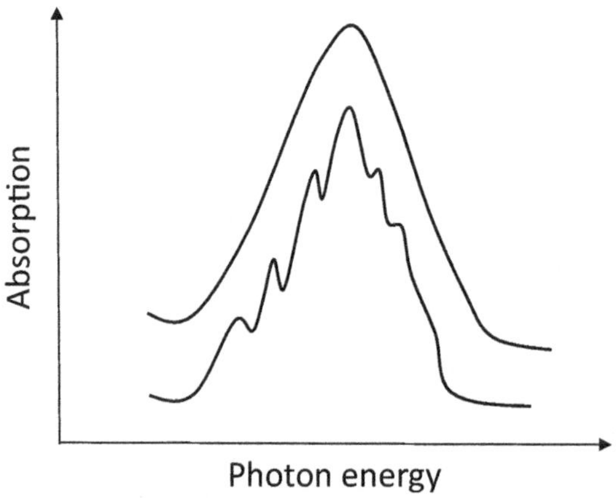

Figure 2.18 Sketch of the frontier parts of the optical absorption spectra of (top) a disordered organic semiconductor film and (bottom) a well-ordered film of the same semiconducting molecule showing distinct vibronic features.

Just as for charge carriers (polarons), the softness and polarizability of organic materials provide additional stabilization of excited electron–hole pairs and such electrostatically bound quasi-particles are called excitons. Excitons are sometimes classified according to their spatial extent and binding energy, with small, tightly bound excitons being referred to as Frenkel excitons, while those that are large and loosely bound are referred to as Wannier–Mott excitons. Large or small is in this context related to the lattice constant. Excitons can move, through either dipole–dipole interactions (Förster resonance energy transfer) or wave function overlap (Dexter energy transfer); recombine radiatively (emitting a photon) or nonradiatively; or separate into free carriers. Almost all organic materials have relatively high (higher than the thermal energy) exciton-binding energies, and this is of fundamental importance for, for example, photovoltaic applications where optical excitations need to be separated into free carriers.

Atomic movement influences the relative positions of the atoms in a molecule and this also influences the electronic states that absorb and emit light. In disordered materials, there are usually too many small variations to be detectable, but in ordered systems there are certain vibrational modes that can even dominate the optical properties. Since both absorption and emission preferentially occurs from the vibronic state with the lowest energy, the absorption and photoluminescence spectra are, in the ideal case, mirrored (Fig. 2.19). In reality, there are usually a lot of defects in the solid state, and some of the vibronic transitions might be forbidden for, for example, symmetry reasons so that the absorption and emission spectra become rather dissimilar. Exciton migration (toward lower energy) also means that ordered domains will have a bigger influence on emission than absorption. Absorption and emission from the lowest vibronic level do usually not occur at the same energy despite the same states being involved. The reason for this is the reorganization energy. Both the energy of the excited electron and that of the resulting hole are shifted into the energy gap by one reorganization energy (see Fig. 2.19).

In total, this means that the emission is shifted by two reorganization energies compared to the absorption. Disorder and the fact that there are many different electronic states that can absorb light in a solid usually make it difficult or impossible to distinguish the appropriate transitions, and thus the reorganization energy, at room temperature. The qualitative correlation remains, however, and materials with a large so-called Stokes shift (i.e., a large red shift of the emission spectrum compared to the absorption spectrum) tend to have large reorganization energies.

As mentioned, excitons can recombine radiatively or non-radiatively. The rate of exciton recombination depends, among other things, on the spatial overlap of the electronic wavefunctions of the constituent hole and electron. Differences in electronegativity due to molecular asymmetry can cause displacements in the electron and hole charge densities and thus increase the exciton lifetime. Excitons are not only important for photoexcitations; they can also be an intermediate in recombination processes. Electrons are spin-1/2 particles, that is, fermions, and the total spin of the exciton can therefore be either 1 or 0. No two fermions can occupy the same quantum state. Excitons with a total spin of 0 are called singlet excitons and excitons with a total spin of 1 triplet excitons. Photoexitations typically yields singlets, which can recombine quickly and radiatively (fluorescence). In a random population, such as in an electroluminescent device, however, there will also be triplets. Mathematically, the ratio between singlets and triplets is 1:3. Triplets are more long lived and do not recombine radiatively in organic semiconductors. Without intersystem crossing, for example, electroluminescent devices are thus limited to a quantum efficiency of at most 25%. Hence, metal–organic molecules capable of emitting light from triplet decay (phosphorescence) are typically used together with fluorescent organic semiconductors in organic light-emitting diodes.

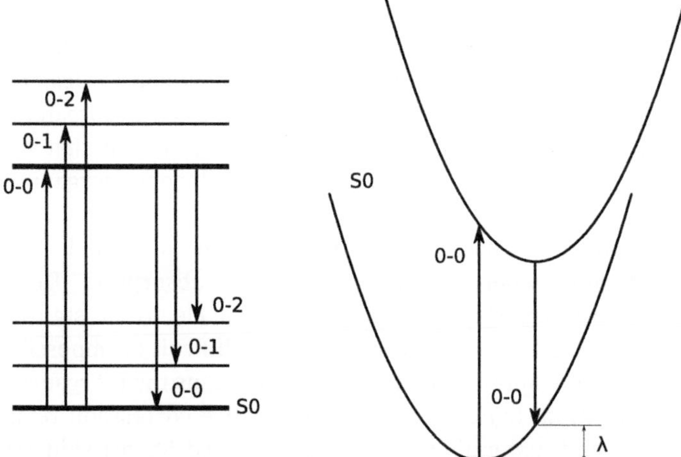

Figure 2.19 (Left) Excitation (light absorption) from the ground state (S0) to a vibronically split excited state (arrows pointing up) and relaxation (light emission) from the excited state's energy minimum to the vibronically split ground state (arrows pointing down). (Right) Energy parabolas of the ground state and the excited state involved, where λ is the reorganization energy. The total shift in energy between absorption and emission is 2λ.

Since the exciton-binding energy typically exceeds the thermal energy in organic semiconductors, separation of optical excitations into free charges is relatively inefficient. The charge separation probability is well described by the Onsager–Braun model, in which it is predicted to be proportional to the relative recombination and dissociation rates. Both of these rates are strongly dependent on the nature of the material and not easily calculated. Qualitatively though, it has merits and the model can be used to predict how various physical properties will influence charge dissociation. Applying an external electric field, for instance, aids the charge dissociation, and the Onsager–Braun model uses a Bessel function of order 1 to describe this dependency (see Fig 2.20). Just as excitons can dissociate into free carriers, free carriers can be reformed into excitons. This is also included in the model and the rate with which free carriers recombine is usually taken to be given by the Langevin factor in low-mobility, low–dielectric constant materials such as those discussed here. The Langevin factor is proportional to the sum of the electron and hole charge carrier mobilities and the permittivity of the material.

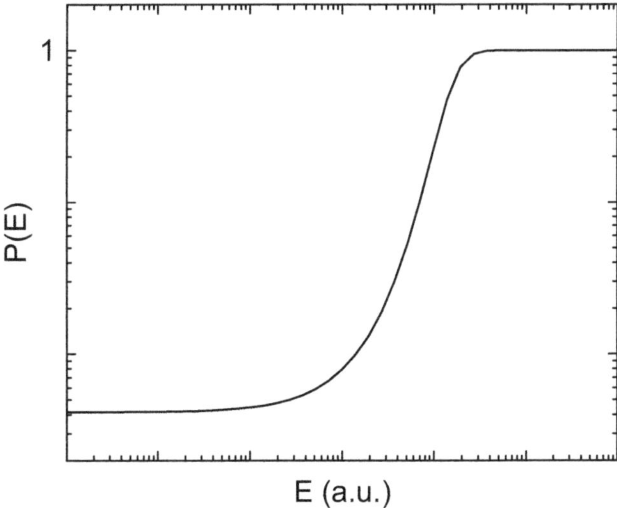

Figure 2.20 The probability (P) of charge separation as a function of an external electric field (E) for excitons according to the Onsager–Braun model. The 0 field probability is determined by the material properties and can be close to 1 in some two-phase systems.

Although relatively slow, the recombination rate is still typically much faster than the dissociation rate at reasonable electric fields. To remedy this, it is possible to use two-phase systems, where there is an (appropriate) energetic offset between the respective frontier orbitals of the two materials (see Fig. 2.21). When one of the materials is excited, either the hole or the electron will transfer to the other phase while still being electrostatically tightly bound. Such a state is called a charge transfer state and the corresponding exciton a charge transfer exciton. The electron and hole of a charge transfer exciton becomes (even more) spatially separated and the wave function overlap reduced, which results in an increased lifetime that is proportional to the frontier orbital offset. Many material combinations can reach close to 100% charge separation efficiency in this way.

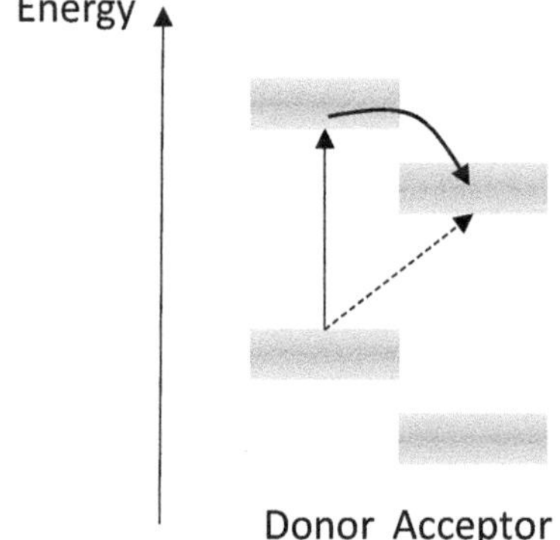

Figure 2.21 Sketch of light absorption-induced charge transfer exciton formation either by formation of an exciton on the donor molecule followed by transfer of the excited electron into the acceptor molecule LUMO (solid arrows) or direct formation where an electron from the donor HOMO is excited into the acceptor LUMO.

2.5 Short Summary

It should now hopefully be clear to the reader that organic semiconductors offer a near limitless opportunity to tune optical and electronic properties through synthesis, both by design of the molecules and by combining one or more molecules into a polymer.

For example, in Fig. 2.22 are depicted a few polymers used in organic photovoltaics, illustrating also the use of mixed monomers (co-polymers) to tailor desired properties. The freedom available through synthesis also presents a challenge. We have learned that the electronic and optical properties are affected by film order, and film order will to a large extent be controlled by the chemical structure of the molecule or polymer as well as the film-forming process conditions. A minute change in the design of a molecule that doesn't change the optical or electronic properties in the gas phase then can still dramatically change the electro-optic functionality of the film if the film-forming properties are affected. Likewise, changing solvent or coating technique can also radically change the performance of a particular material. It thus is hard to predict the functionality of a molecule or polymer in a device solely on the basis of the properties of the isolated system, which makes design and screening of materials for a chosen application a very complex task. The "perfect" material for a given application is thus out there ready to be synthesized, but until computer simulations have progressed to the point where film morphology accurately can be predicted, finding it will require significant trial-and-error work.

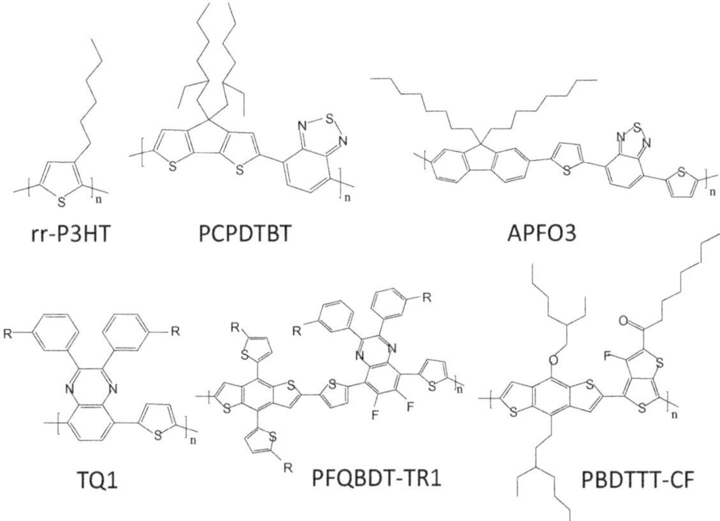

Figure 2.22 Examples of π-conjugated polymers identified by their acronyms. The R-groups signify alkyl or alkoxy chains, which are added to tune solubility and crystallinity.

Exercises

2.1 Will the bandwidths resulting from the intermonomer molecular orbital interactions in a linear polymer chain increase or decrease with the number, N, of monomers in the chain?

2.2 What is the requirement for an ohmic contact to be formed between a conductor and an organic material?

2.3 If you want to decrease the work function of a metal electrode, should you adsorb a monolayer of a donor molecule that donates charge to the metal or an acceptor molecule that withdraws charge from the metal?

2.4 Charge carrier mobilities in organic semiconductors are usually low. Why?

2.5 Will increased film order increase or decrease charge mobility?

2.6 What is the fundamental relationship between optical absorption and photoluminescence in organic material, and why is it so?

Suggested Readings

Bässler, H. (1993). Charge transport in disordered organic photoconductors: a Monte–Carlo simulation study. *Phys. Status Solidi B*, **175**, 15.

Braun, C. L. (1984). Electric field assisted dissociation of charge transfer states as a mechanism of photocarrier production. *J. Chem. Phys.*, **80**, 4157.

Braun, S., Salaneck, W. R., and Fahlman, M. (2009). Energy-level alignment at organic/metal and organic/organic interfaces. *Adv. Mater.*, **21**, 1450.

Bredas, J. L., Beljonne, D., Coropceanu, V., and Cornil, J. (2004). Charge-transfer and energy-transfer processes in π-conjugated oligomers and polymers: a molecular picture. *Chem. Rev.*, **104**, 4971.

Facchetti, A. (2011). π-conjugated polymers for organic electronics and photovoltaic cell applications. *Chem. Mater.*, **23**, 733.

Ishii, H., Sugiyama, K., Ito, E., and Seki, K. (1999). Energy level alignment and interfacial electronic structures at organic metal and organic organic interfaces. *Adv. Mater.*, **11**, 605.

Lampert, M. A., and Mark, P. (1970). *Current Injection in Solids*, Academic Press, New York.

McQuarrie, D. A. (2007). *Quantum Chemistry*, University Science Books, USA.

Chapter 3

Printing and Processing Techniques

Simone Ganz,[a] Hans Martin Sauer,[a] Simon Weißenseel,[a] Julian Zembron,[a] Robert Tone,[a] Edgar Dörsam,[a] Moritz Schaefer,[b] and Malte Schulz-Ruthenberg[b]

[a]*Institute of Printing Science and Technology, Technische Universität Darmstadt, Magdalenenstraße 2, Darmstadt, 64289, Germany*

[b]*Fraunhofer ILT, Steinbachstr. 15, 52074 Aachen, Germany*

doersam@idd.tu-darmstadt.de, moritz.schaefer@ilt.fraunhofer.de

3.1 Introduction

When Shirakawa, Heeger, and MacDiarmid created the new field of polymer electronics in 1977, the connection to the well-established printed media market was an obvious step in direction to mass production. Printing is a quite enticing technique, since it is associated with high speed, precision, and cost efficiency, attributes that this technology has earned in the century-long tradition of the graphical industry.

In spite of this long tradition, it is not easy to establish printing as a production technique for electronics: the requirements on the printing quality are much higher for electronic devices than for

Organic and Printed Electronics: Fundamentals and Applications
Edited by Giovanni Nisato, Donald Lupo, and Simone Ganz

ISBN 978-981-4669-74-0 (Hardcover), 978-981-4669-75-7 (eBook)
www.panstanford.com

graphical applications, while the materials are much more difficult to handle. Printing is not a black box for deposition. Our concept is to subdivide printing into a sequence of processes and to explain each of them on the grounds of physical, thermodynamic, or hydrodynamic models.

We want to convince the reader of the necessity of a systematic access to printing-related challenges that he or she may already have encountered in the lab: Why are some materials harder to process than others? Which technique is optimal for my material? How can I check whether my printed samples are OK or what has gone wrong?

The authors do not focus on printing presses and general descriptions of all printing techniques. In this chapter we explain the printing process by following the printing material through the whole process chain. We start with an introduction to functional fluids. After a listing of requirements and fluid parameters, we introduce the substeps of printing. In the following section we describe the most common printing techniques by showing their working principle, applying the previously defined substeps to the specific system, and sharing practical experiences with the reader.

3.2 Fluid Formulation: Making Functional Materials Printable

Nowadays printing can be defined as the *reproduction of patterns by transfer of matter to a surface by mechanical force or hydrodynamic stress.* For this, functional materials need to be transformed to a physical condition in which they can be mechanically transported, subdivided, deformed, and deposited: the liquid state. This liquid state needs to fulfill specific requirements of the respective printing technique like inkjet, gravure, flexographic, or screen printing. Depending on the physical nature of these materials there are different methods that can be used: dissolving the material in a more or less volatile solvent or creating nano- or microparticulate dispersions. The formulation of these fluids has a vital impact on the printing or coating process, as well as on the drying process.

This section concentrates on the description of the printing fluid during the printing process. Starting with aspects of formulating organic materials, properties of printing liquids as well as their influences on layer formation and drying are explained.

Learning targets:

- Understanding aspects of fluid formulation
- Knowing properties of printing fluids
- Recognizing fluid dynamical phenomena

3.2.1 Creating Printing Fluids

In functional printing, ready-to-print fluids are rarely offered by material manufacturers. Even if they are, quality fluctuations are often an issue the end user has to deal with. Usually solid materials that still need to be brought into the liquid state are provided. Especially in research applications, adjustments of the printing fluids are a daily business. Therefore, important aspects of fluid formulation are explained in the following section.

3.2.2 Properties of Printing Fluids

The transfer of printing fluids to a substrate essentially depends on two relevant fluid parameters, viscosity and surface tension. Additional parameters must be considered for drying. The fluid's viscosity determines its ability to transfer mechanical stress and hydrodynamic pressure to the substrate through the liquid phase at a given printing velocity. On the other hand, capillary forces at the fluid–substrate and fluid–air interfaces, that is, the surface tension, are essential as well, as they are responsible for a constant and reproducible distribution of ink on the substrate after the printing form has been lifted.

In the present section we explain the relevance of viscosity and surface tension for printing, with specific regard to common functional materials.

3.2.2.1 Viscosity of printing fluids

3.2.2.1.1 *Shear rate and shear stress*

Imagine an infinitesimal cube of liquid situated in a reservoir of an *isotropic liquid*, as, for example, water, oil, or any organic solvent. The six faces of the cube have a given size ΔA_i, where the index $i = x, y, z$ denotes the direction of the normal vector on this face with respect to a Cartesian coordinate system. If a small tangential mechanical force ΔF_j is imposed to two of the opposing faces of the cube, a continuous, dissipative shear flow is enforced inside. This shear flow is the gradient of the flow velocity v_k, that is, the *shear rate*

$$\dot{\gamma}_{ij} = \frac{\partial v_i}{\partial x_j} + \frac{\partial v_j}{\partial x_i} \tag{3.1}$$

which is proportional to, or at least monotonically increasing with the *shear stress*, the ratio of tangential force and face size:

$$\tau_{ij} = \frac{\Delta F_j}{\Delta A_i} = \eta \dot{\gamma}_{ij} \tag{3.2}$$

where the constant of proportionality η is called the *dynamic viscosity*, usually given in units of Pa·s. The components of the shear tensor are not fully independent. An often-used simplification is the *incompressibility condition*:

$$\frac{1}{2} \sum_{i=x,y,z} \dot{\gamma}_{ii} = \vec{\nabla} \cdot \vec{v} = 0 \tag{3.3}$$

This equation fits to the majority of liquids with a very good accuracy, but it is usually not applicable for gas flows.

We emphasize that shear force and shear stress are tensor-valued quantities and that $\tau_{ij} = \tau_{ji}$ and $\dot{\gamma}_{ij} = \dot{\gamma}_{ji}$. This symmetry feature is because the forces between two molecules of an isotropic liquid only depend on the length of their mutual distance vector, not on its orientation in space.

3.2.2.1.2 *Newtonian and non-Newtonian fluids*

For *Newtonian fluids,* such as water, the dynamic viscosity η is constant at a given temperature. In contrast, many printing fluids are *non-Newtonian*; their viscosity depends on the shear rate. One distinguishes essentially two classes of non-Newtonian behavior, shear thinning and shear thickening. For *shear-thinning* liquids the viscosity decreases with increasing shear rate. This means that the shear rate increases overproportionally to the increase of the applied shear stress. Common examples are polymer solutions, or slurries and suspensions of solid particles in a carrier liquid. Analogously, the viscosity of *shear-thickening* liquids increases with increasing shear rates.

In many cases, one can describe non-Newtonian behavior by a power-law term:

$$\eta = \eta_{\text{ref}} \left(\frac{\dot{\gamma}}{\dot{\gamma}_{\text{ref}}} \right)^{\alpha} \tag{3.4}$$

where η_{ref} is the nominal viscosity measured at a defined reference shear rate $\dot{\gamma}_{\text{ref}}$. For Newtonian liquids the exponent α equals 0, whereas $\alpha > 0$ indicates shear-thickening and $-1 < \alpha < 0$ a shear-thinning behavior of the liquid.

3.2.2.1.3 *Complex fluids*

There is a wide range of fluid types that are used in printed electronics, such as solutions of linear polymers with a stiff molecular backbone or dispersions of nonspherical particles where the viscous properties cannot be fully characterized by a scalar viscosity.

These are referred to as *anisotropic, complex*, or *mesogenic* liquids. Viscosity does not only depend on the shear rate $\dot{\gamma}$ but is also a tensorial quantity with different independent components.

3.2.2.1.4 *Stokes and Navier–Stokes equations*

The equations of liquid motion can be derived from the laws of momentum conservation combined with a space- and time-dependent velocity field as well as a pressure field, resulting in the famous *Navier–Stokes equation* (NSE)

$$\rho \left(\frac{\partial v_k}{\partial t} + (\vec{v} \cdot \vec{\nabla}) v_k \right) = \vec{\nabla} \cdot (\eta \vec{\nabla} v_k) - \frac{\partial p}{\partial x_k} \tag{3.5}$$

where ρ is the mass density of the liquid and p the hydrostatic pressure. In the case of incompressibility of the liquid, one obtains a pressure equation

$$\nabla^2 p = -\rho \sum_{j,k} \frac{\partial v_j}{\partial x_k} \frac{\partial v_k}{\partial x_j} \tag{3.6}$$

For slow flows, one may linearize the NSE, and obtain the *Stokes equation*. The Stokes equation is much more convenient to solve, as it reduces flow problems to a boundary condition problem

$$\rho \frac{\partial v_k}{\partial t} = \vec{\nabla} \cdot (\eta \vec{\nabla} v_k) - \frac{\partial p}{\partial x_k} \tag{3.7}$$

or when written in terms of the viscous stress tensor

$$\rho\frac{\partial v_j}{\partial t}=\sum_{i=x,y,z}\frac{\partial \tau_{ij}}{\partial x_i} \tag{3.8}$$

Using the incompressibility condition (Eq. 3.6), one also obtains a Laplace equation for the hydrostatic pressure:

$$\Delta p = 0 \tag{3.9}$$

For many aspects of printing the Stokes equation proves to be sufficient, and cases where the nonlinear terms of the NSE are substantial are usually restricted to specific conditions.

3.2.2.2 Surface tension

The surface tension—or surface energy—describes the property that any liquid interface exerts a mechanical stress to the environment, or to a different phase or material, in order to minimize the size of the interface. This refers to liquid–gas, liquid–liquid, and liquid–solid interfaces.

3.2.2.2.1 *Capillary or Laplace pressure*

On a curved interface the surface tension may create a force per surface area that is normal to the surface. This force is directed toward the liquid volume if its curvature is convex and directed into the surrounding gas phase if it is concave.

For this reason the surface of a liquid drop tends to attain a spherical shape, minimizing its surface at a given drop volume. Inside a drop of a liquid with surface tension σ and radius R, a nonzero capillary or Laplace pressure,

$$p_{\mathrm{L}}=\frac{2\sigma}{R}, \tag{3.10}$$

is effective, and equilibrates the surface or capillary forces created by surface tension. More generally, the capillary pressure below an arbitrarily shaped liquid interface is proportional to its Gaussian curvature, the sum of the reciprocal principal curvature radii R_1 and R_2 at the considered point:

$$p_{\mathrm{L}}=\sigma\left(\frac{1}{R_1}+\frac{1}{R_2}\right) \tag{3.11}$$

The principal radii R_1 and R_2 are defined to be positive if the corresponding tangent parabola is convex and negative if it is concave. Under the surface of a liquid volume that is at rest and in thermal equilibrium, and if gravity can be neglected, the hydrodynamic pressure p of the liquid is constant everywhere. As a consequence, the Gaussian curvature of the liquid surface must be constant throughout the interface as well. Spatial differences in capillary forces will cause liquid flows below differently curved parts of the surface, equilibrating the surface toward a more constant curvature.

3.2.2.3 Fluid parameters and printing techniques

For each printing technique—gravure, flexo, inkjet, screen, offset—the preferred range of viscosity and surface tension of the printing fluids can be roughly estimated. This is shown in Fig. 3.1.

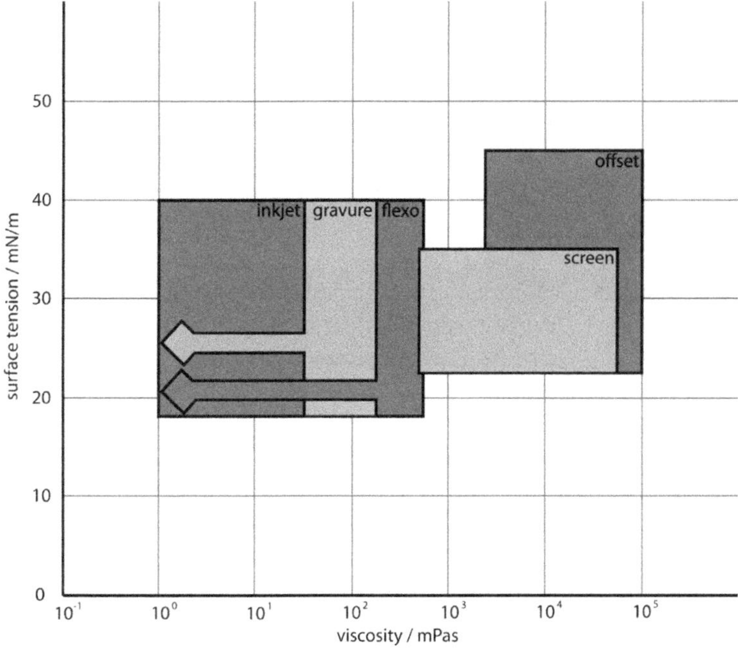

Figure 3.1 Range of surface tensions and viscosities used in the most common printing techniques. Inkjet, gravure, and flexography all cover a range of viscosity that starts at about 1 mPa·s.

We emphasize that even printing fluids that are not in the range indicated for a specific printing technique may be, and in fact have been, successfully processed. In this case, however, the machine and process parameters, such as transfer ratio or printing resolution, may drastically differ from what is typical for a graphical printing process.

3.2.3 Choosing a Solvent

The following aspects should be considered for the solvent:

- The solubility of the material in it
- Its viscosity and the effect of adding solutes
- Surface tension and wetting capabilities on the substrate
- Evaporation from the printed liquid film: boiling point and vapor pressure at drying temperature

In practice, further technical aspects have to be considered, which will only be mentioned but not discussed further in this chapter: the chemical resistivity of the printing machine's components against the solvent; the disposal of its vapor in the drying process, toxicity, flammability, and flash point; and aspects of the subsequent cleaning of parts contaminated with the printing fluid.

3.2.3.1 Polymer materials

Polymers play a dominant role in printable electronics: They are the base for many organic semiconductors, dielectrics, and protective coatings. Compared to common polymer materials, which are mostly branched, functional polymers typically have a linear structure. They can therefore be characterized by the chain length (the number of monomer molecules) forming the chain. Concentration and chain length have a strong impact on the properties of polymer solutions, mainly on their viscosity. One distinguishes different types of polymer solutions:

- *Diluted solutions*: The concentration of the polymer in the solution is low in the sense that the individual polymer chains are independently distributed in the solvent. Driven by Brownian motion, each chain tends to form a sphere, or blob. The viscosity of the solution is essentially that of the solvent.
- *Semidiluted solutions*: With increasing polymer concentration, the interaction between the blobs becomes significant. The

fluid's viscosity is more or less determined by their mutual contact or by microscopic viscous friction between them. If the solution is sheared, the blobs are deformed by shear forces such as to attenuate mutual friction. This is known as shear thinning: The viscosity of a semidilute polymer solution decreases with increasing shear rate, and the fluid clearly exhibits a non-Newtonian behavior.

- *Concentrated solutions*: At high concentrations the polymers cease to form individual blobs. The fluid properties are dominated by the interactions between the polymer chains, leading to a linear dependency between chain length and viscosity. With increasing concentration the solution finally attains a rubber-like, viscoelastic behavior under shear stress.

A few prominent examples of semiconducting polymers are poly-(3-hexyl-thiophene) (P3HT) with a molecular weight of 20,000 to 100,000 g/mole that has a solubility of approximately 1.5 wt% in toluene at room temperature. *SuperYellow*, a yellow emitter for organic light-emitting diodes (OLEDs), is soluble up to 10 wt% in the same solvent, depending on chain length. Polystyrene (PS), a common technical polymer that is also applicable for printed dielectic layers, has no solubility limit in toluene. Increasing the PS content of such a preparation will continuously increase viscosity, leading to a more or less viscoelastic gel, and finally be indistinguishable from the pure solid polymer.

The gel transition is due to percolation of polymer chains. With rising concentration, the polymer chains interconnect to an increasingly extended, elastic network, and viscosity rises unboundedly when approaching the percolation point. The polymer solution becomes thixotropic, that is, its viscosity is dependent on the preceding *shear history*. Solidification of polymer solutions by percolation is a second-order phase transition, and it takes increasingly long times to relax to thermal equilibrium. This is also true for the dissolution of polymers and caused by the profound changes in chain conformation.

Freshly deposited polymer layers are usually amorphous—with a random molecular arrangement—and still contain substantial quantities of solvent. For this reason, a thermal treatment of printed polymer layers should be considered. By this annealing step solvent residuals are removed from the layer, and the polymer has an opportunity to become partially crystalline. There are, however, specific exceptions from this rule: P3HT is known to have a so-called steric transition, even in the solute state. Depending on concentration and temperature, the molecular blobs reversibly transform to a particulate dispersion of ordered molecular clusters. The degree of structural order in these clusters is a further important feature: it has a considerable impact on the charge carrier mobility of the resulting P3HT layer.

Surface tensions of polymer solutions can be roughly estimated as those of the solvent and the solid polymer being weighted with their volume fractions. This is not more than a first qualitative estimate—for example, for PS or polymethylmethacrylate (PMMA) in toluene the solution's surface tension is between that of the pure materials but does not linearly depend on their volume fractions. This is due to the limited chemical affinity between polymer and solvent and to the diffusion of the polymer at the fluid interfaces. This may cause additional effects such as skin formation, impeding the flow and drying of the liquid layer. It is therefore obvious that a deliberate choice of the solvent with respect to solubility, viscous behavior, and layer deposition is of utmost importance for designing a successful printing and drying process for polymer materials.

3.2.3.2 Small molecules

Small-molecule semiconductors play an increasing role for printed electronics. Initially being designed for vacuum deposition, they can be sublimated and evaporated at elevated temperatures without being disintegrated. Most of them are also soluble in specific solvents. Small-molecule materials have molecular weights between 300 and 2000 g/Mol and a strong tendency to condense as crystals or crystalline layers. Finally, this is the key feature of the outstanding performance of small-molecule semiconductors.

The effect of small-molecule semiconductors on the viscosity of liquid solutions is quite distinct from polymers. As for inorganic salts the increase of viscosity caused by the semiconductor is small and is usually negligible even in saturated solutions.

Layer deposition from solution is a first-order phase transition from saturated or even supersaturated solutions, and there is a considerable enthalpy of crystallization associated with this transition. This is critical for the printing process. Once the solution becomes supersaturated within the printing or drying step, the deposition of the solid phase can be initiated by nucleation at any time. Driven by internal crystallization enthalpy, the deposition is not controllable any more and may result in the formation of isolated crystal grains or even in dendritic pattern formation rather than in a smooth and homogeneous layer. In any case, the printer may have to find a possibly difficult compromise toward electronic material performance, and additional measures such as tempering of the semiconductor layers subsequent to the drying process may become necessary.

With respect to surface tension, the same applies as for polymers: Mixing rules approximately apply to some degree, but additional phenomena affecting surface tension, such as solute diffusion at the liquid interfaces, may have a crucial impact.

3.2.3.3 Colloidal dispersions

An important class of materials constituting a backbone of any electronics manufacturing process is inaccessible to the formulation methods described so far: metals and metal oxides. There is no volatile solvent by the aid of which these materials could be converted to a homogeneous liquid phase that is accessible to a printing process. However, it is possible to create colloidal dispersions by finely distributing the material in a carrier liquid. This can be achieved in such a manner as to obtain a composition that forms a macroscopically homogeneous phase, with fluid mechanical properties that can entirely be discerned in terms of viscosity, and with interfacial properties that can be condensed in a surface tension and the phenomena associated with this. Nevertheless, colloidal dispersions such as silver inks or dispersions of semiconducting zinc or titanium oxides have specific features the printer ought to be aware of.

By virtue of van der Waals attractions the particles in a dispersion tend to form clusters, to agglomerate, and to drop out of the solution. To prevent this, chemists have to provide a stabilization mechanism. One distinguishes *electrostatic* and *steric* stabilization.

In electrostatically stabilized dispersions the dispersed particles carry electric charges of identical polarity on their surfaces. The respective countercharges are distributed as ions in the carrier liquid, forming an electric double layer of some width around each particle.

In contrast, steric stabilization is achieved by attaching amphiphilic molecules to the particle surface. The polar groups of such surfactants bind to the particle, whereas the nonpolar parts can enhance the chemical affinity of the particle surface to the liquid and endow the coated particle with a finite amount of dissolution enthalpy. A typical example of a sterically stabilized dispersion is nano–zinc oxide coated with oleic acid. Such particles can be dispersed in nonpolar solvents such as toluene, hexane, or decane. Concerning viscosity, colloidal dispersions resemble polymer solutions in many aspects. Diluted dispersions are usually Newtonian. With increasing concentration a pronounced increase in viscosity is often observed.

Upon drying, a solid layer may form by gelling. This layer may still contain a considerable quantity of carrier liquid. On further extraction of solvent, air will finally be entrained between the deposited particles and agglomerates. This, in turn, gives rise to a large internal interface and intense capillary forces. These may redistribute the remaining colloid on the substrate, leading to a broad class of pattern formation phenomena such as, for example, the coffee stain effect.

In printed electronics microparticulate suspensions of silver and other noble metals or metal oxides with up to 70 wt% of dispersed metal are frequently used. The particle size ranges between a few hundred nanometers to 5 μm. These formulations should be referred to as slurries, as they are usually not in a permanently stable colloidal state but are two-phase or granular mixtures. Applying them to the substrate requires specific deposition techniques, for example, screen printing, and drying must usually be supplemented by a sintering process at elevated temperatures.

3.2.4 Fluid Dynamics of Printed Liquid Films

When a wet printed film is transported into the drying unit, one might think that the printing process is finished and that solidification will not change the structure of the layer. But there are many hydrodynamic processes that occur in a wet film and that can change the printed result enormously. For many applications, the creation of thin, spatially smooth, and homogeneous films is necessary. Liquid film relaxation, driven by capillary forces, may support this but may also lead to a failure of the whole printing process. In this section, we describe the most important aspects of printed wet films.

3.2.4.1 Wetting properties

One of the principal properties of printing fluids is the wetting of the printing form and substrate. Wetting is defined as the ability of a liquid to supplant a competing phase—gas or another liquid—from a solid surface and to bring the interfaces between the three participating phases into a thermodynamically more favorable state. Wetting properties can be characterized by the surface energies or surface tensions σ of the respective interfaces and by the contact angle θ, that is, the angle between the substrate plane and the tangent at the free liquid–gas interface that passes through the three-phase contact point. These are linked by the Young's equation:

$$\sigma_{LG} \cos\theta = \sigma_{SG} - \sigma_{SL} \tag{3.12}$$

Here σ_{LG} is the surface tension of the liquid–gas interface and σ_{SG} and σ_{SL} surface energies of the substrate surface at the interfaces toward the gas and the liquid phase. One characterizes wetting properties by the *wetting parameter*:

$$S = \sigma_{SG} - \sigma_{SL} - \sigma_{LG} \tag{3.13}$$

and by the *energy of adhesion*, the attractive interaction energy (per unit area) between printing fluid and substrate surface:

$$\begin{aligned} W_{LS}^{(adh)} &= \sigma_{SG} + \sigma_{LG} - \sigma_{SL} \\ &= \sigma_{LG}(1 + \cos\theta) \end{aligned} \tag{3.14}$$

$S > 0$ is referred to as perfect wetting. A drop of ink that is deposited on the substrate is distributed over the whole surface by capillary forces. With respect to printing, the case of

$$\sigma_{LG} \gg -S > 0 \tag{3.15}$$

is optimal, as it corresponds to a small contact angle, according to Young's equation, and to a reasonably large energy of adhesion.

3.2.4.2 Dynamic phenomena: The lubrication limit

Printing is a dynamical process, and static models such as wettability can only capture a limited range of phenomena. We therefore consider models based on the NSE and the Stokes equation in order to analyse such fluid dynamical problems. These models have their origin in applications ranging from valve and pipe flows in mechanical engineering to atmospheric dynamics in meteorology to even pyroclastic flows in geology. Nevertheless, fluid dynamical problems that are related to printing are specific due to the fact that they almost always involve a free interface between liquid and air. Besides velocity v and pressure p the interface has an independent set of equations of motion, coupled to the flow velocity and the pressure, which must be solved simultaneously. This makes such problems mathematically much harder to solve and represents the real challenge of the fluid dynamical access to printing.

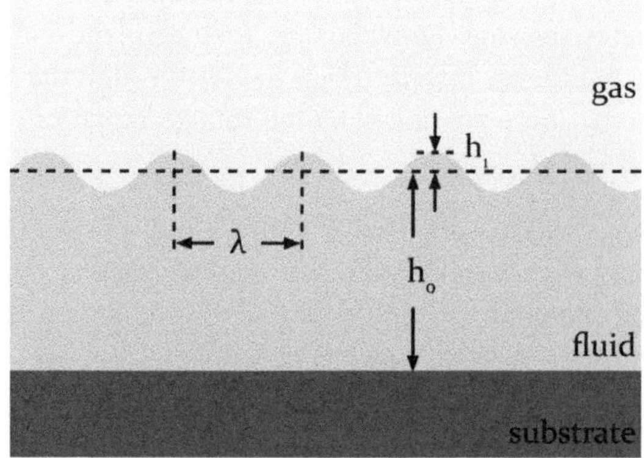

Figure 3.2 Lubrication shear flow.

For thin liquid films deposited on a solid surface a relatively convenient class of solutions exists, known as the *lubrication limit*. We assume that the film thickness h is much smaller than any feature size of the printed pattern. In this case one may apply further simplifications to the Stokes equation and obtain a closed equation for the surface, the liquid flow, and the pressure. This equation is known as the *Landau–Levich equation* (*LLE*). We postpone its derivation to the exercises. The local thickness $h(x, y, t)$ of the liquid film is considered as a function of the position (x, y) on the substrate plane and time t. It satisfies

$$\frac{\partial h}{\partial t} = \vec{\nabla} \cdot \left(-\frac{h^3}{3\eta} \vec{\nabla} [\sigma \nabla^2 h - p_{\text{vol}}] + \frac{h^2}{2\eta} \vec{\nabla} \sigma \right) \tag{3.16}$$

The gradients here are defined as being projected to the 2D substrate plane. The first term in brackets describes the action of surface tension at the liquid–air interface: areas with convex curvature ($\Delta h < 0$) create a positive Laplace pressure in the liquid and areas with concave curvature ($\Delta h > 0$) a negative one, such that the resulting pressure gradient will cause a lateral viscous flow in the film. This flow will transport liquid from the convexly to the concavely shaped areas and thus level out surface inhomogeneities. The second term takes account of volume related pressures p_{vol} such as, for example, gravity. The third term is related to the Marangoni drag.

3.2.4.3 Surface-leveling time

One important consequence of the LLE concerns surface leveling of printed layers. Assume a freshly printed layer of a specific average thickness h_0 that exhibits a surface topography given by a height function $h(x, y, t) = h_0 + h_1(x, y, t)$, with $/h_1/ << h_0$, as indicated in Fig. 3.2. This modulation may result from, for example, the drop-based transfer of the printing fluid to the substrate, or from *viscous fingering*, a hydrodynamical instability that occurs in the ink-splitting process. The surface tension will smooth out the modulation, but the speed of leveling is limited by the viscous friction of the fluid flow under the liquid surface. Using the LLE and expanding to first order in h_1 one finds that leveling proceeds exponentially in time. Consider a sinusoidal height function with a specific wavelength λ and amplitude $A_\lambda(t)$ of a modulation that is oriented in, say, the x direction:

$$h_1\,(x, t) = A_\lambda(t) \sin 2\,\pi x/\lambda.$$

This perturbation will smooth out exponentially in time, with a relaxation timescale of

$$\tau_\lambda^{\mathrm{rel}} = \frac{3\eta\lambda^4}{16\pi^4 \sigma h_0^3} \tag{3.17}$$

As we see, relaxation time increases by fourth order in λ. Capillary forces may therefore be very effective for small wavelengths but rather slow on larger ones. Typical modulations in the micrometer range disappear within milliseconds, but those in the millimeter range need minutes. Moreover, relaxation is much less effective on very thin liquid films than on thicker ones.

This has some consequences: leveling of printed thin films containing volatile solvents should take place well before the film thickness has appreciably changed. We shall therefore briefly discuss the most important types of surface patterns associated to printing. Some of the most relevant printing methods, for example, gravure and inkjet printing, deposit arrays of liquid drops rather than continuous liquid layers. The characteristic size and wavelength of such modulations is defined by the gravure pattern of the printing form or by the nozzle spacing of the inkjet print head. In contrast, patterns created by hydrodynamic *nip* instabilities (viscous fingering) are related to film thickness and printing speed v_p, and the predominant wavelength

$$\lambda_{vf} \propto h_0 \left(\frac{\eta v_p}{\sigma} \right)^{-1/2} \tag{3.18}$$

It defines the length scale for the leveling, which is, according to Eq. 3.17, possible within the timescale

$$\tau_{vf}^{rel} \propto \frac{h_0 \sigma}{\eta} v_p^{-2} \tag{3.19}$$

and is remarkably decreasing as the printing speed is raised. Both types of perturbations may be in conflict with the drying process, and we observe from Eq. 3.19 that the leveling time of the viscous fingering problem can at least be reduced by increasing the printing speed.

3.2.4.4 Marangoni effect

Consider the case that the surface tension σ of the liquid changes with solute concentration ϕ. In the beginning of the liquid film's drying process its thickness h continuously decreases by solvent evaporation. The dissolved material's concentration ϕ in the remaining liquid will rise accordingly: $\Delta\phi/\phi = -\Delta h/h$. As a consequence, surface tension will change if the liquid film thickness is decreasing by solvent evaporation:

$$\phi \frac{\partial \sigma}{\partial \phi} \approx -h \frac{\partial \sigma}{\partial h} \tag{3.20}$$

and the coefficient $\sigma_\phi = \partial\sigma/\partial\phi$ will appear in the LLE of the drying liquid film, giving eventually rise to lateral fluid transport, driven by gradients of capillary forces:

$$\frac{\partial h}{\partial t} = \vec{\nabla} \cdot \left(-\frac{h^3}{3\eta} \vec{\nabla} \left[\sigma \nabla^2 h - \frac{3}{2h} \sigma_\phi \right] \right) \tag{3.21}$$

Analyzing this equation along the same lines as in the leveling problem, that is, by considering the reaction on a small periodic perturbation of a wavelength λ, one finds that a second type of process takes place, with a characteristic timescale of

$$\tau_\lambda^M = \frac{\lambda^2}{2\pi^2 h_0 \sigma_\phi} \tag{3.22}$$

This process can either result in the relaxation of the perturbation, or in spontaneous pattern formation, depending on the sign of σ_ϕ. This, in turn, depends crucially on the materials and the solvents used. In the case of negative σ_ϕ, the leveling process will be inverted, and local thickness differences in the liquid film will grow in time. As a consequence, the Marangoni drag may give rise to spontaneous dewetting, irrespective of the question whether the wettability of the substrate has initially been achieved. We emphasize that practice may comprise more pattern formation phenomena than assumed here and that lateral fluid transport, solute diffusion, and solvent evaporation may cause similar effects as well. Nevertheless, our simplified calculation shows that successful printing is dependent on the deliberate choice of the used materials and material combinations.

3.3 Substeps of Printing

For analyzing the printing process for electronics, we come back to the definition of printing we introduced at the beginning of this chapter: *The reproduction of patterns by transfer of matter (printing ink or fluid, or toner) to a surface by mechanical force (i.e., printing pressure), hydrodynamic stress (e.g., inkjet printing), electric fields, or radiation (e.g., laser structuring).*

This definition comprises much more than only the very act of transfer of a printing liquid to a substrate but implies a specific sequence of substeps to be done. These substeps, although not independent of each other, can be understood by the aid of comparatively closed physical concepts of hydro- and thermodynamics. The printing engineer or machine designer can commonly provide various technical solutions by which each of the substeps can be performed. Which combination of the available methods should be selected depends on the properties of the printing fluid, of the electronic device to be printed, and on the substrate materials.

In this section the substeps of printing are described in detail.

Learning targets:

- Knowing the single steps for understanding different printing techniques
- Getting to know physical phenomena that occur during each step

The process of printing can be subdivided into a sequence of substeps:

Substep 0: Conditioning (0 because it happens before printing)

Substep 1: Fluid Acquisition

Substep 2: Predosing

Substep 3: Dosing

Substep 4: Transfer

Substep 5: Relaxation

Substep 6: Drying

Each of these substeps can be characterized by a set of physical parameters that dominantly determine its success. For this reason care should be taken when defining printing layouts, formulating printing fluids, or defining printing parameters. In the following, all substeps are described in detail.

3.3.1 Substep 0: Conditioning (of the Printing Fluid and Substrate Pretreatment)

Substrates, for example, paper, polymer foils, or glass plates, are usually supplied as sheet stacks or on reels. When entering the printing machine, the substrate surface has to be made available to the components of the printing process: Sheets have to be individually separated or unrolled, and their surface has to be prepared for the deposition of functional materials. This implies that contaminations that may have resulted from the production process or from storage have to be removed. Secondly, as the substrate surface is supposed to form a stable, permanent interface to the printed layers, the wettability of the substrate by the printing liquids must be ensured. This has to be done with regard to the specific printing fluid.

Usually, when a new type of substrate or printing formulation is considered for a printing process, the contact angle of the printing fluid on the substrate should be determined. If the wetting behavior of the fluid is not satisfactory, the situation can be corrected either by using a printing formulation with better wetting capabilities or by applying a plasma or corona pretreatment to the substrate. This serves the purpose of chemical surface activation and increases the surface energy of the substrate.

By *corona pretreatment*, an electrical discharge is generated close to the substrate surface. The intense electrical field as well as the ions and radicals created in the discharge region implant an electrical charge below the substrate surface, break up chemical bonds, or transform crystalline surfaces morphologies to a more amorphous state. This enhances the bonding opportunities to the molecules of a liquid deposited on the surface, thereby improving the wetting properties of the substrate and reducing the contact angle.

Within a *plasma treatment*, an ionized gas of specific composition, usually containing argon, nitrogen, or oxygen, is brought in contact with the surface. Depending on the type of ions contained in the plasma, specific chemical surface reactions can be initiated and hydroxyl- or nitrate-bonding groups are created on the substrate surface. Moreover, the plasma is able to etch material from the exposed surfaces and to increase its microscopical roughness. This not only removes possible residuals effectively from the substrate but also inhibits a possible dewetting of the printing fluid from the substrate: the retraction of the solid–liquid contact line of printed patterns on the substrate is suppressed.

3.3.2 Substep 1: Fluid Acquisition

In preparation of the ink transfer step, a specific amount of fluid is provided by the reservoir for the distribution over the whole width of the printing form.

3.3.3 Substep 2: Predosing

For preparing the fluid transfer, the fluid has to be divided and distributed such that it can be applied to the printing form. In, for example, flexo printing, this is achieved by the anilox roler, which offers a constant volume of printing fluid per area of its surface to the printing form. This step is indispensable in order to guarantee that the printing form constantly acquires the desired amount of printing fluid according to the requirements of the printing layout.

3.3.4 Substep 3: Dosing

Dosing of the printing fluid is the specific task of the printing form. By its specific structure each point of the printing form's surface acquires a specific quantity of printing fluid that corresponds to the locally deposited fluid quantity as defined by the printing layout. This point is also the distinction toward predosing: it is the dosing step that is responsible for the reproduction of the printing information as a pattern of printing fluid.

3.3.5 Substep 4: Transfer

When the desired quantity of printing fluid has been provided on the printing form, the transfer to the intended point onto the substrate takes place. Mechanical pressure, hydrodynamical shear, or intense chemical or thermodynamic forces can be applied to create the contact of the liquid with the solid substrate. A direct interface between fluid and substrate is created, and air and particle contaminations are supplanted from the substrate. For a short moment wetting is enforced irrespective of any equilibrium properties as, for example, the contact angle.

The conditions of enforced wetting are changed when the printing form is removed from the substrate or when the supply of thermal or radiation energy that drives fluid transfer is terminated. If fluid transfer is achieved by the use of solid printing forms, this is associated with the process of ink splitting. During ink splitting, the final interface of the printed liquid layer to the surrounding gas atmosphere is created.

3.3.6 Substep 5: Relaxation

The transfer process implies the creation of a free, highly mobile surface of the liquid film that is exposed to the also extremely mobile gas atmosphere. Naturally, the energy of the liquid film increases by the surface energy of the liquid–gas interface, and liquid film stability issues arise if wettability is insufficient. Moreover, using conventional printing techniques, the phenomenon of viscous fingering is frequently observed: the liquid surface of the deposited film develops finger-like structures that are neither intended nor encoded in the printing form.

The relaxation of the printed layer in the short period of time where the transferred material is still in the liquid phase is mainly due to a hydrodynamical flow driven by capillary forces. We call this the *surface-leveling* process. This refers to the wetting of the substrate as well as to capillarity-driven viscous flows at the liquid–gas interface of the film.

If the conditions for wetting are insufficient, dewetting of the printing fluid from the substrate surface may occur. The borderline of the liquid film at the edges of the printed area may retract. This has several consequences: printed structures as, for example, conductive lines may be disintegrated and decompose to a chain of fluid drops by a Rayleigh-type instability. Moreover, microscopically small holes in continuous films may increase in size. One should be aware that printed liquid patterns, consisting of a liquid film of constant thickness, bounded by a more or less arbitrarily shaped border line, are thermodynamically unstable for any nonzero value of the contact angle. Even at very small contact angles the fluid always tends to minimize its surface toward the gas phase. Depending on the substrate material and on surface roughness, this may be prevented by border line pinning and by contact angle hysteresis.

The second aspect of surface leveling concerns the liquid–gas interface. Surface leveling is extremely fast on small length scales but very slow on large ones. The time available for this process is essentially limited by the evaporation of the solvent. Therefore the time of solidification of the printed liquid film should be balanced with the leveling time such that major surface inhomogeneities have an opportunity to disappear.

As the content of solid material in, for example, printing formulations of organic semiconductors is usually in the range of only 0.2 to 2 wt%, most of the volume of the printing liquid will disappear and the film thickness will shrink from a few micrometers of the wet film to typically few tens of nanometers. The concentration of solute in the liquid phase will rise at the same rate until the point of saturation is reached in the liquid film. From this point on the deposition of a solid phase becomes possible. Solid phase deposition is a first-order phase transition for solutions of materials of small molecular weight such as the small-molecule semiconductors or for a number of organic colorants and dyes. Under adverse conditions, specifically in the case of an even minute particle contamination, the formation and growth of isolated centers of crystal nucleation can be observed, a feature that makes printing of small molecules particularly challenging. For polymer solutions, the solid phase mostly forms gradually by gelation and is less critical from the point of view of liquid film dynamics and relaxation.

The rate of solvent evaporation is defined by the saturation vapor pressure of the respective solvent in the gas atmosphere at the given temperature and ambient pressure and by the diffusion and convection of the solvent vapor in the gas atmosphere. So, solvents with low vapor pressure can be used in printing formulations in order to slow down the solidification process. Active ventilation of the drying atmosphere will naturally accelerate solvent evaporation. Moreover, the gas diffusion constant of the solvent vapor is an important parameter of the evaporation process.

3.3.7 Substep 6: Drying (Curing, Sintering)

The process of drying refers to the essentially complete removal of solvent residuals from the printed film. As the film is already solidified, and capillary transport of printing fluid is impossible now, it is without risk for the film integrity to apply additional heat sources in order to accelerate the process. This is often also referred to as *thermal posttreatment*. Such thermal treatment is quite often mandatory for dispersions, for example, for conductive inks, or for inorganic semiconductors. Here, it serves not only to eliminate solvent residuals (*drying*) from the printed film but also to remove stabilizing agents from the surfaces of the dispersed particles (*curing*). Finally, it is usually intended to initiate some solid-state diffusion in the nanoparticles such that particle growth, the formation of solid and electrically conductive "necks" between adjacent particles, or even the sintering of the porous body to a dense solid layer can take place (*sintering*). The latter frequently requires temperatures on the order of several hundred centigrade, and sintering times between a few minutes and several hours. It is usually reasonable not to assign this task to a continuously working printing machine.

In some cases, the process of sintering can be accelerated by electromagnetic radiation. So-called photonic or flash sintering is available for specific materials. In this type of process a short but intense pulse of light is applied to the printed substrate to only heat up the printed layer, leaving the substrate material almost at its initial temperature.

3.4 Printing Processes

Within the organic and printed electronics community several deposition techniques have been established for processing a large variety of functional materials. In this section we give a general overview of the most common techniques, starting with a short introduction to the main aspects and terms of printing.

Learning targets:

- Getting to know the main printing techniques within printed electronics: flexography, gravure, screen, and inkjet printing
- Getting an introduction to deposition techniques beside printing

Printing techniques are classified by the physical existence of a printing plate. If there is a printing plate included, they are termed *conventional printing techniques*. Techniques that dispense of a printing plate are called *digital printing techniques*, as the printing form is only stored digitally.

Conventional printing techniques are characterized by the arrangement of *image* and *nonimage elements* on the printing plate, that is, surface elements that accomplish or abstain from a deposition of the printed material on the substrate. One distinguishes between four main categories:

- Image elements are recessed compared to the nonimage elements (*gravure printing*, for example, pad printing)
- Image elements are raised above the nonimage elements (*letterpress*, for example, flexographic printing)
- Image elements and nonimage elements are located on the same surface plane but differ in surface polarity (*lithography*, for example, offset printing)
- Image elements and nonimage elements are located on the same surface plane but differ in permeability (*screen printing*)

Digital printing techniques are subclassified by the transformation of the digitally stored printing layout into physically existing information. In other words by whether the printing layout is converted into the printed image directly on the substrate, as in inkjet printing, or if it is generated on some kind of transient printing form and transferred to the substrate in the following, as in electrophotography.

Conventional printing processes and especially the final fluid transfer can be reduced to a simple scheme where only three components participate in the process: One component that carries the fluid, a substrate onto which the fluid is transferred, and an impression component that presses the substrate against the fluid-carrying component. The geometrical character of these components is also used to distinguish printing presses. There is *flat-to-flat printing* (e.g., flatbed screen printing), *cylindrical-to-flat printing* (e.g., CD printing press), and *cylindrical-to-cylindrical printing* (e.g., rotogravure printing). Nowadays, terms like *non-contact-on-flat* are implemented in order to include the digital printing processes into this scheme. Printing presses are additionally classified by the principle of substrate transport. The two most famous ones are *roll-to-roll* (reel-to-reel, R2R, web-fed) and *sheet-to-sheet* (S2S, sheet-fed) presses. Roll-to-roll printing has become the flagship for organic and printed electronics as it stands for high-speed and cost-effective production. Nevertheless roll-to-sheet and sheets-on-shuttle are used as alternative principles for substrate transport in printed electronics, as they both come up with deviating advantages, for example, in drying conditions.

An aspect that should be kept in mind is that the aims of functional and graphical printing are incompatible: within graphical printing an illusion is created to the eye. Several single dots generate a homogeneous picture that is observed by the addressed human being in a certain distance of observation, for example, dots on posters are larger than in prospects as they are observed from larger distances. Within functional printing whole layers need to be created that have to fulfill a certain task, for instance, the conduction of a current. Even if those layers look perfect to the eye, the printer cannot rate their electrical quality without further investigation. Additionally, functional fluids are more challenging in handling than graphical inks and exceed many recommended limits concerning the characteristics (viscosity, surface tension). Therefore, several tips and tricks that are commonly used in the graphical industry are not transferrable into this relatively new area of device manufacturing.

3.4.1 Flexography

3.4.1.1 History and facts

Flexography plays an important role in the packaging industry as it is used for patterning foils, bags, cardboards, etc. The main principle is based on the letterpress technology of Gutenberg, who used single raised metal letters for forming words and whole texts more than 500 years ago. In the last years flexography has become one of the fastest printing techniques with speeds up to 1000 m/min (ca. 16 m/s) at about 2000 mm print width and is the most economical letterpress printing technology today.

3.4.1.2 Principle of flexography

Figure 3.3 shows the principle of flexography. The inking unit consists of a fluid reservoir and an anilox roller, whereas the printing unit consists of a plate cylinder and an impression cylinder.

First, the fluid is prepared and its viscosity is adjusted—*conditioning (0)*. The fluid, stored in some kind of reservoir, wets the rotating anilox roller—*fluid acquisition (1)*. After wiping the excess fluid off, a defined amount of fluid is transported by the cells of the anilox roller—*predosing (2)*. In the following step, the fluid is transferred from the anilox roller onto the printing plate, which is mounted on the plate cylinder and contains the printing layout—*dosing (3)*. As in all letterpress-printing techniques, the image elements are raised above the nonimage elements. The fluid is then *transferred (4)* from the raised patterns onto the substrate, which is pressed against the printing plate by the impression cylinder. Hence, there are two ink splitting steps in total. In the next steps, *relaxation (5)* and *drying (6)* follow.

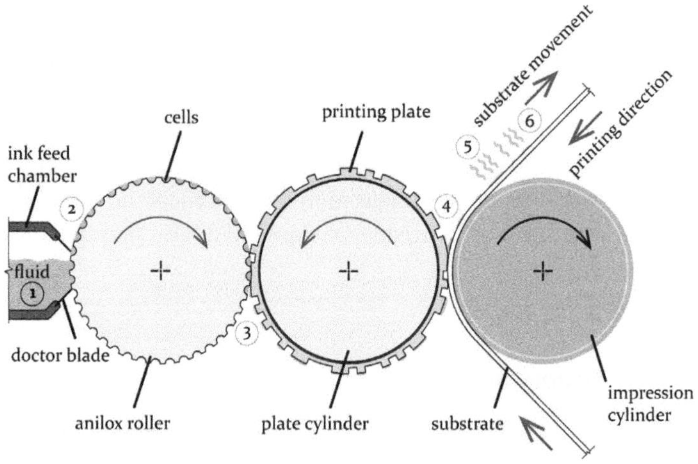

Figure 3.3 Principle of flexography.

3.4.1.3 Characteristics of print equipment

The anilox roller comprises a steel core and a ceramic (see Fig. 3.4) or metal surface with engraved cells. It is an important part of the inking unit as it provides a well-defined amount of ink that will be transferred onto the printing plate. The cells of a ceramic anilox roller, which are responsible for fluid intake and transportation to the printing plate, are created by laser gravure. In comparison to cylinders produced from chrome mostly with mechanical gravure, ceramic cylinders have a longer lifetime and higher robustness.

The gravure angle describes the arrangement of cells on the surface of the anilox roller to the roller axis and varies from 30 to 90 degrees (Fig. 3.5). Important parameters are the screen frequency (lines per cm) and the cell volume (cm^3/m^2) of the anilox roller. The cell volume is the volume of all cells in a defined area. Note that not all of this volume is actually transferred in the printing process.

Figure 3.4 Lab-scale ceramic anilox rollers.

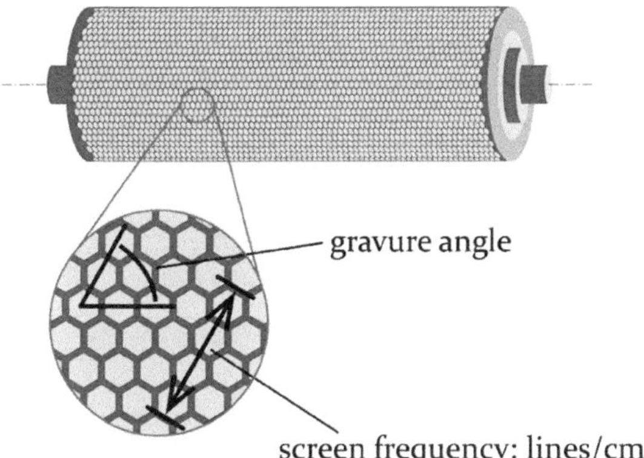

Figure 3.5 Scheme of a hexagonally engraved anilox roller showing screen frequency and gravure angle.

For every new layout, a new printing plate needs to be produced (prepress). Printing plates are available as flat plate (Fig. 3.6), which means that they need to be bent in order to be mounted on the plate cylinder, or as sleeves, that are already produced as hollow shells for the plate cylinder. Printing plates are made of photopolymer or rubber. It should be mentioned that the shelf life of polymeric printing plates is limited to months, even if they are unused.

For printing, the plates are fixed on the plate cylinder by double-faced adhesive tape (see Fig. 3.7). This is a special rubber-like tape providing elastic compression and hence affects the printing quality.

The mounting process of the flat plate onto the round cylinder causes stretching of the partly flexible structures, which can cause slightly distorted printing.

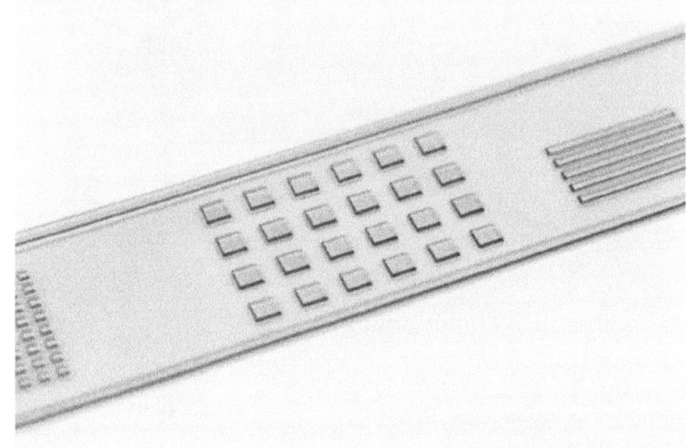

Figure 3.6 Flexographic printing plate in its flat and unbound state.

Figure 3.7 Plate mounted on a plate cylinder with adhesive tape (yellow).

3.4.1.4 Process description

For transfer, the substrate is lightly propped to the printing plate in order to provide the necessary contact. The resulting printing pressure has an important influence on the printing quality and is set as follows: The impression cylinder and the plate cylinder are moved toward each other until they touch. In this case, the advanced printer calls it *kiss printing*, and the flexible printing plate is not distorted. In the next step, the distance between both cylinders is reduced until an acceptable contact area between both cylinders is formed by the deformation of the printing plate, called *engagement* (see Fig. 3.8). The contact area is called *nip*. This low-pressure setting is well suited for multilayer printing. During processing, the printing plate is unrolled onto the substrate in a rotatory movement and forces the fluid to adhere to it. From there on, the printing process is finished by relaxation and drying processes on the substrate.

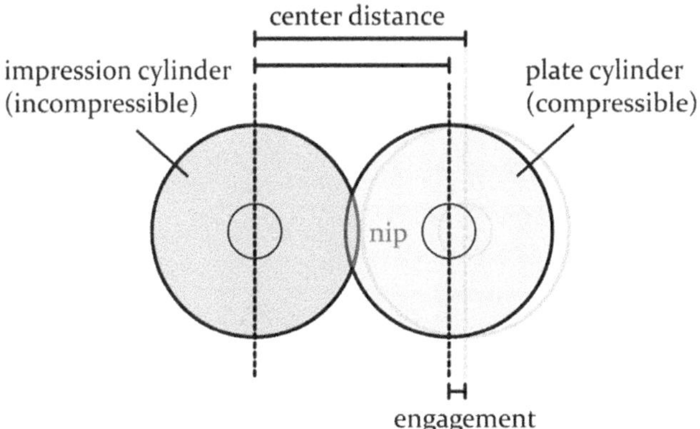

Figure 3.8 Definition of engagement and nip position in flexography.

3.4.1.5 Experiences from a practical point of view

In conventional flexographic printing, layer thicknesses of several tens of nanometers can be reached. But processing those fluids is limited by their viscosity. In graphical industries, there is a rule of thumb: Fluids that have a viscosity below 0.5 mPa·s (1 mPa·s: water @ 20°C) will spin off the printing plate during rotation.

The most typical error in flexography is the so-called *halo effect* (see Fig. 3.9). This effect describes a line that surrounds the contour of a printed structure and that therefore looks like a halo. Due to incorrect settings of the printing pressure, the printing fluid is partly squeezed out at the edges of the compressible printing plate and this surrounding structure arises (see Fig. 3.10).

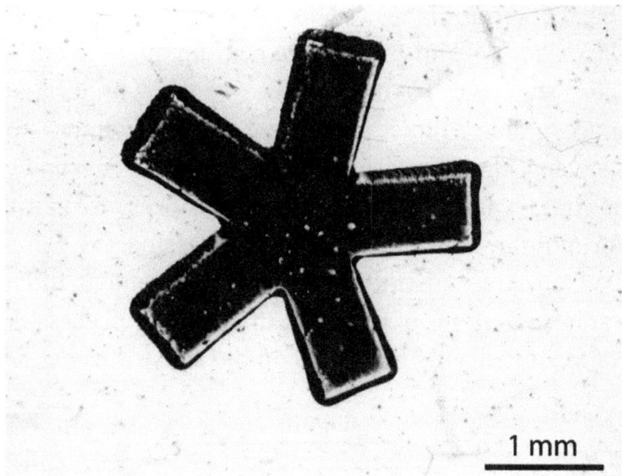

Figure 3.9 Image of a graphical printed star showing the halo effect.

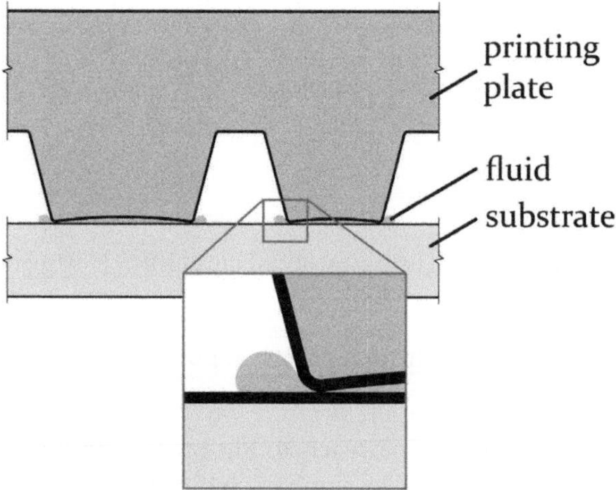

Figure 3.10 Origin of the halo effect.

During printing, the printing plate faces complex stresses. Frequently changing mechanical load can damage it. Especially in combination with the influence of organic solvents, this can lead to a loss of detail after long runtimes. However most significant is the aspect of swelling of the polymer printing form due to aromatic fluids (e.g., toluene) in printed electronics. Besides that, interactions between the printing plate and the solvent-based fluids can cause a contamination of the printing result (printing form residua). Other issues in (industrial) flexography appear, for example, due to incorrect printing contact.

The most important advantage of flexography is the clear distinction between printed and nonprinted areas. Only raised image elements on the printing plate and no other parts are wetted with fluid and thus printed. On the other hand, the maximum amount of fluid that can be transferred is limited due to the twofold ink splitting. However, flexographic printing forms can easily be adapted to smaller lab-scale cylinders and are less expensive in production than lab-scale gravure cylinders.

3.4.1.6 Applications and current research

Flexography is often used in printed electronics for producing electronic circuits or transparent electrodes. For special applications printing plates with a resolution of about 4000 ppi can be produced today. These developments in printing plate manufacturing techniques show the increasing applicability of flexography for producing thin structures in printed electronics.

3.4.2 Gravure Printing

3.4.2.1 History and facts

Gravure printing today covers a wide range of utilizations. It is still used for artworks (e.g., copperplate print, aquatint) as well as in the mass production of high-volume magazines, catalogues, and packages. For economical premium-quality, long print runs, rotogravure printing is the most important technique today as it provides the highest resolution and best edge definition at high process velocities, up to 18 m/s, at 4320 mm print width.

3.4.2.2 Principle of rotogravure

Common rotogravure presses (see Fig. 3.11) consist at least of an ink pan (fluid reservoir), a gravure cylinder with engraved cells (printing form), and an impression cylinder. As opposed to the letterpress printing techniques, gravure printing works with image elements that are not raised above the surface of the printing form, but engraved into it.

The prepared fluid—*conditioning (0)*—is commonly provided in an ink pan into which the gravure cylinder is immersed. As the cylinder rotates, the cells get filled with fluid—*fluid acquisition (1)*. When the filled-up cells run out of the bulk fluid, the whole surface of the gravure cylinder is covered with a roughly even wet film—*predosing (2)*. In the next step, the excess fluid is removed from the nonprinting areas by a doctor blade, resulting in equally filled cells that carry the amount of fluid that is finally available for transfer—*dosing (3)*. Subsequent to this, the dosed fluid is *transferred (4)* onto the substrate by pressing the latter against the cylinder's surface. It is obvious that there is only one ink splitting step. As in all printing processes, the fluid transfer is followed by *relaxation (5)* and *drying (6)*.

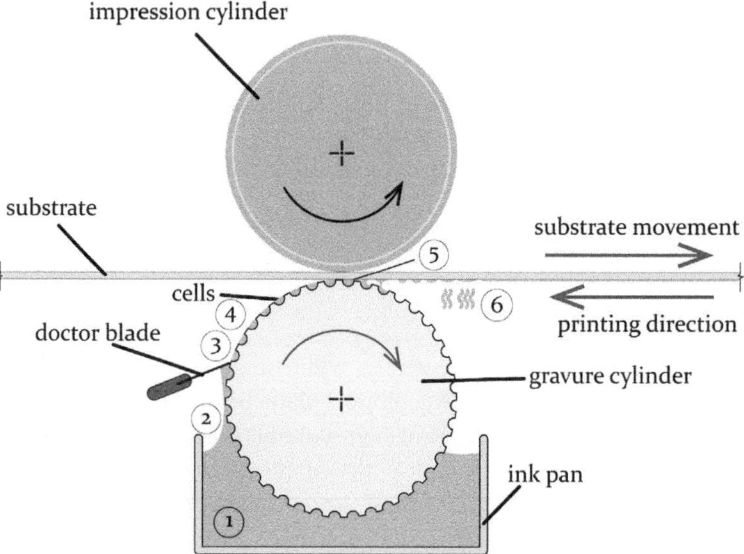

Figure 3.11 Principle of rotogravure.

3.4.2.3 Characteristics of print equipment

Gravure cylinders, see Fig. 3.12, consist of a steel core covered with a skin that carries the image elements. The latter is a stack of a nickel layer, a basic copper layer, and a thin copper engraving layer. After engraving, the surface of the cylinder is chrome plated for more durability and protection against wear. With thicker copper layers it is possible to erase and even the surface by mechanical processes for a new image creation. Copper layers that can be peeled off after use, called *ballard skin*, are also used.

Figure 3.12 Gravure cylinders for lab-scale printing presses.

Engraving can be done by etching, electromechanical gravure, or laser-direct gravure (see Fig. 3.13). All processes lead to well-defined cells that are terminated by the remaining copper material, called cell walls. The cells vary in geometry and are specified by depth and area. Deeper cells can transport more fluid, resulting in thicker layers. Electromechanical gravure works with a diamond stylus (with a given stylus angle), resulting in a pyramidal cell geometry. For the etching procedure, a gelatin-based etch mask is used, leading to cells with a trapezoidal depth profile. Laser-direct gravure induces different cell depths and diameters by modulating beam diameter and energy.

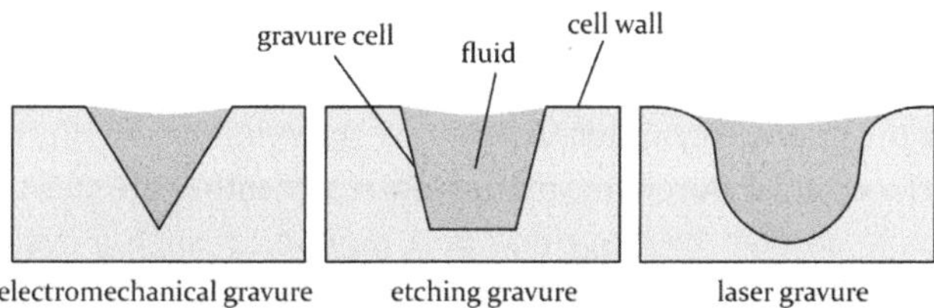

Figure 3.13 Cells of a gravure cylinder manufactured by electromechanical gravure, etching, and laser gravure.

Important parameters for gravure cells are screen frequency (40–140 lines/cm), screen angle (30°–60°), stylus angle (120°–140°), and tone value (10%–100%) (see Fig. 3.14). The volume of a cell is defined by those parameters, but the exact amount of fluid that will finally be transferred onto the substrate can only be roughly predicted, even if those parameters are precisely monitored. Therefore, the fluid's transfer behavior should be determined by the printer before ordering, in the best case by using a testing form that contains a large spectrum of different gravure parameters. An important point that has to be mentioned here is the limited availability of lab-scale printing forms. Production processes are established for standardized cylinder formats (specific diameters). Smaller lab-scale cylinders therefore lead to time- and cost-intensive adjustments.

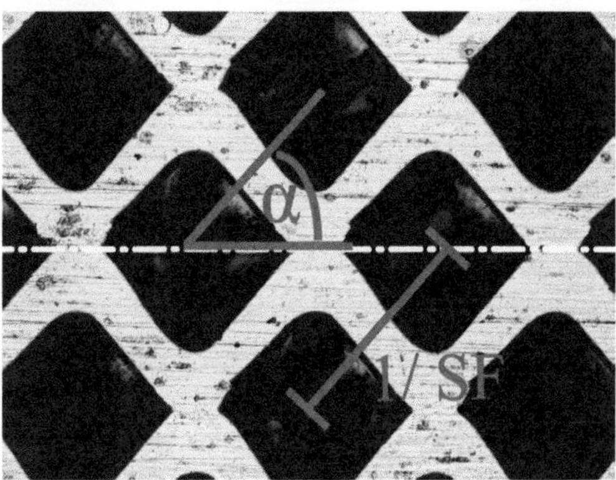

Figure 3.14 Microscopic image of a gravure printing plate showing screen angle α and the distance between two cells, which is defined as the reciprocal of the screen frequency (SF).

3.4.2.4 Process description

Common rotogravure presses work with an ink pan that is mounted below the gravure cylinder. By rotating the cylinder within the fluid, its surface gets covered. In smaller lab-scale testing presses the fluid is usually applied by a syringe or a pipette. In such a system the gravure cylinder is not permanently wetted, so the fluid can dry on it and thus change the result. This has to be prevented by using solvent traps and keeping the latency times short.

After having passed the doctor blade, the now well-defined amount of fluid is transferred from the filled-up cells onto the substrate. The emptying of the gravure cells is influenced by two important characteristics: the cell geometry (engraving process, screen angle), which is fixed for one cylinder, and the process parameters (pressure, velocity), which can be adjusted during processing. The printing pressure influences the emptying of the cells, typically leading to a more complete emptying as it increases. It is therefore very high for graphical printing on rough paper. For printed electronics, the pressure should converge to zero to avoid damaging the substrate.

3.4.2.5 Experiences from a practical point of view

Possible issues in gravure printing will be discussed in this section: Remaining or dried fluid as well as imbedded air bubbles in the cells can reduce the transfer volume, which means that the desired thickness cannot be reached. Furthermore, not correctly filled cells can lead to missing spots in the printing result, called pinholes (in a layer). Another issue is the transfer of fluid from nonprinting areas. As the surface of a gravure cylinder is very rough, the doctor blade is not able to remove all the excess fluid, which leads to a thin film on the nonprinting areas (see Fig. 3.15).

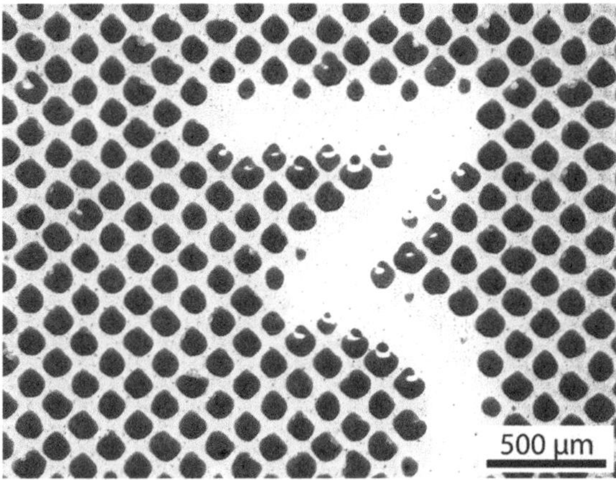

Figure 3.15 Gravure printed image. Due to wrong process parameters the layer hasn't leveled out and single dots are visible. Additionally, the light blue layer surrounding the dots shows the transfer of fluid from nonprinting areas.

The greatest advantage of gravure printing is that the main machine parts (gravure cylinder, doctor blade, ink pan) consist of metal and therefore provide stability even against aromatic solvents, as well as long lifetimes. Another important aspect is the scalability of the printing velocity. Lab-scale as well as industrial presses cover a wide range of printing velocities (~0.1–18 m/s), which simplifies the transfer of processes significantly. In addition to that, gravure printing allows printing of well-defined structures (edge definition) at high resolution in the micrometer range.

3.4.2.6 Applications and current research

Layers for OLEDs, OFETs, and OPV cells have already been successfully printed by rotogravure. Regarding the requirements of those devices, rotogravure is used for printing extended areas (OLED lighting) as well as very small structures (OFET channels). Important aspects in today's research are the optimization of the emptying of gravure cells and the reduction of solvent evaporation from the printing forms.

3.4.3 Screen Printing

3.4.3.1 History and facts

Over 2000 years ago, cultures in Asia, Africa, and Europe already used simple stencils to create repeatable figurative motives (symbols, labels) on textiles. The easy principle of squeezing ink through a stencil allows a reproduction of patterns on almost all kinds of surfaces (rough, even, plain, bent) and materials (e.g., paper, cardboard, plastic, metal, wood, ceramics, glass, foils), even for nontrained people. In printed electronics, screen printing is mostly used to coat materials, for example, for encapsulation, and for printed electric circuits.

3.4.3.2 Principle of screen printing

A flatbed screen printing press consists of a screen (mesh mounted in a frame (see Fig. 3.16), a squeegee (doctor blade), a flooding blade, and a substrate holder.

Figure 3.16 A screen: mesh mounted into a frame.

The principle of screen printing is shown in Fig. 3.17. At first, the well prepared fluid, *conditioning (0)*, is applied to the screen, *fluid acquisition (1)*. In the second step, the whole mesh is flooded by the fluid using a flooding blade—*predosing (2)*. The texture of the mesh differs for printing- and nonprinting areas. At nonprinting areas, the mesh is covered with an impermeable mask layer, called *stencil material* (see Fig. 3.18).

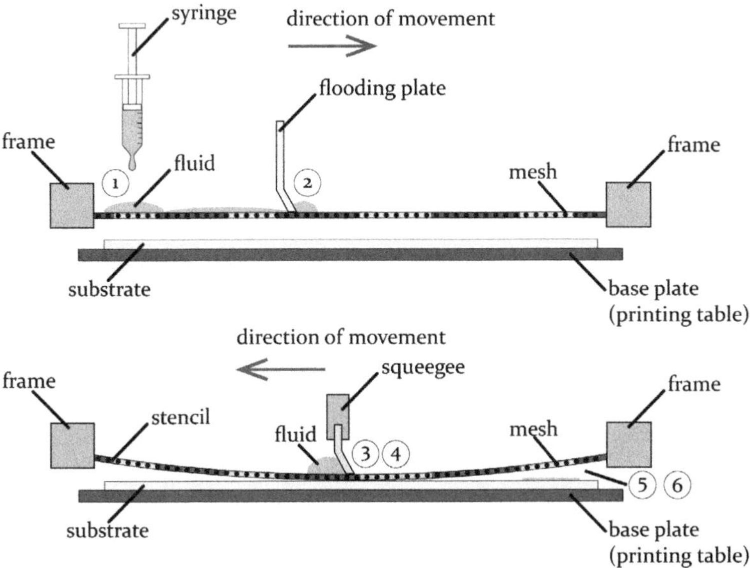

Figure 3.17 Principle of screen printing.

Figure 3.18 Microscopic and topographic image of a screen. The stencil material (purple) covers the fibers.

The squeegee presses the fluid through the mesh at printing areas and brings it in contact with the substrate—*dosing (3)*. When the mesh is lifted, the fluid adheres to the substrate and *transfer (4)* takes place. Beginning with the removal of the mesh, *relaxation (5)* effects appear. At last, solidification and *drying (6)* occur.

3.4.3.3 Characteristics of print equipment

A common screen consists of a mesh that is clamped at a frame, which is mounted on a printing press (see Fig. 3.16). Typical mesh materials are natural silk, plastic, and metal. Another variant is a laser-structured sheet metal mesh. For printed electronics, usually metal meshes are used, which offer chemical stability against solvents and better accuracy than synthetic meshes. Important parameters of a screen are the mesh opening (screen frequency), fiber (respectively thread) diameter, and geometry. Two important parameters in the squeezing process are hardness (shore) and form (round, angular) of the squeegee (see Fig. 3.19). The stencil can be produced directly (photopolymer) or indirectly (transfer film). In flatbed screen printing, the substrate is placed on a table and the flat screen is mounted above. The distance between substrate and screen is called *snap-off distance* and forms one important parameter in screen printing. Mostly in graphical printing, but also for functional materials, cylindrical meshes are used (*rotary screen*). There, the screen is bent and mounted onto a hollow cylinder frame. The squeegee is placed inside, where it squeezes the fluid through the mesh during the rotation.

Figure 3.19 Squeegee in its retainer.

3.4.3.4 Process description

For the printing process, fluid needs to be placed on the screen. This is done by syringes or pipettes in a lab or by ink knives in larger presses. Predosing is realized by a flooding blade, which distributes the fluid on the whole mesh without pressure, not bending it. This is followed by the operation of the squeegee, where the mesh is put under local stress and bent, while the fluid is squeezed through it and brought in contact with the substrate. The squeezed fluid then adheres to the substrate surface and is drawn out of the slots as the elastic mesh lifts off the substrate. Therefore a small snap-off distance (2–3 mm) between screen and substrate is necessary. While pulling the mesh out, fluid also adheres to the fibers and forms ligaments. This means that the wet layer becomes thicker underneath the mesh fibers than in the mesh openings (see Fig. 3.20), which results in rough surfaces. With screen printing it is possible to reach higher layer thicknesses than with any other printing technique. For thin lines, the layer thickness can be controlled by the thickness of the remaining stencil material at the nonprinting areas. Only the fluid that is settled in the cross-sectional area between the top and the bottom of the mesh is transferred to the substrate as the squeegee is pulled over it. Due to the flexibility of the squeegee this does not work for large layers. Here, only the mesh determines the resulting layer thickness. The stencil should be thin to avoid raised edges.

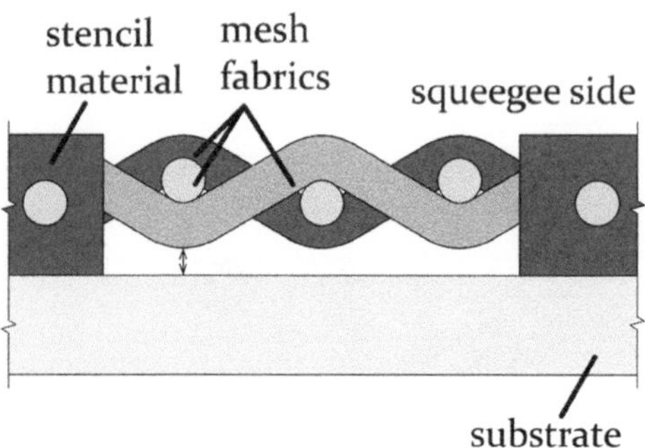

Figure 3.20 Schematic cross section of a mesh.

One important difference between screen printing and other techniques is the usage of highly viscous pastes. Fluids with a viscosity smaller than 200 mPa·s would just run through the open mesh during flooding. Therefore, the paste viscosity always needs to match the mesh opening. This limits the spectrum of materials that can be processed with screen printing. For highly viscous paints (500–5000 mPa·s), graphical layers vary in thickness from 20 to 100 μm. The minimum feature size approximates 70 μm. When integrating screen printing steps into a work flow, one needs to consider that flatbed screen printing is a relatively slow process.

3.4.3.5 Experiences from a practical point of view

First of all, screen printing is a very robust process. This means that it has a large tolerance concerning the parameter settings, for example, squeegee hardness and form or velocity. If the combination of screen and fluid is chosen correctly, even untrained people often succeed in printing.

The main reasons for nonsatisfactory printing results are discussed here: One important factor is matching of the particle size and the mesh opening. The mesh opening should be at least threefold larger than the particle size (rule of thumb). If it is smaller, the particles tend to block the mesh. Blocking of the mesh can also occur if the fluid dries too fast. Both lead to missing bits in the printing result. Additionally, a massive squeegee pressure or a slack mesh can cause slurring effects. Screen printed images can be identified by the characteristic wavy edges. Those inaccurate edges, *sawtoothing* (see Fig. 3.21), can take shape enormously if unsuitable printing parameters are used. This effect is often caused by a false squeegee pressure in combination with thin stencil layers.

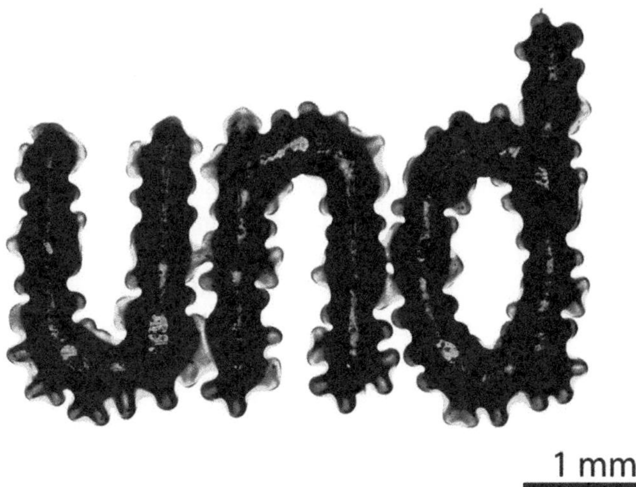

Figure 3.21 Sawtoothing on screen-printed letters.

To sum the facts up, screen printing is a relatively slow technique that is limited concerning the fluid's viscosity. Screen printing cannot reach the register accuracy of other printing techniques and suffers from inaccurate edges of the printed image. But it convinces as a robust technique that can handle particle based fluids (range of micrometers) and that can easily be learned even by untrained people. It also allows large-area printing in the range of square meters and guarantees sufficiently thick layers for, for example, good conductivities.

3.4.3.6 Applications and current research

Screen printing was the first printing technique used for electronic components. Since silver was discovered as an electrical contact for the first time in the 1930s, screen printing of silver conductors has become a popular processing technique for many applications. First researches on printed circuits were documented in 1959. In the automotive section, for example, heating elements for rear windows are produced this way. Furthermore, in printed electronics components like electrodes and wires for solar or fuel cells have been successfully produced.

3.4.4 Inkjet Printing

3.4.4.1 History and facts

The serial production of the first computer systems in the 1950s and the fast technical development toward commonly used devices in the subsequent years forced the need for new ways to reproduce electronically stored information. In 1951 the patent for an inkjet electrocardiogram printer, called Mingograf, built the basis for modern inkjet technology. Further development combined with knowledge in liquid droplet physics created fast, accurate, and affordable inkjet devices for personal use. Today inkjet printers are found in many small offices/home offices (so-called *SOHO sector*) for printing all kinds of documents and images on paper.

3.4.4.2 Principle of inkjet

Inkjet printers differ from conventional printing presses in construction and functionality. The main differences are the contactless ink transfer, because of which it is called a nonimpact printing (NIP) technique, and the absence of a physically existing printing plate.

In inkjet printing, the printing unit is called a *print head* and contains the required circuits and nozzles (see Fig. 3.22). The fluid supply differs for industrial and lab-scale presses: In industrial presses there is usually a fluid tank that is connected to the print head, whereas in smaller presses cartridges are often used. Cartridges, usually known from office inkjet printers, are enclosed reservoirs that are mounted onto the print head.

There are two main concepts of drop formation. One is the *drop-on-demand* (*DOD*) technique, which only produces fluid droplets, if needed, and the other one is the *continuous* technique, where a constant droplet stream is produced and unneeded droplets are removed before striking the substrate surface. Both techniques are subclassified by the actuators that are responsible for the fluid's excitation, for example, piezo or thermal elements.

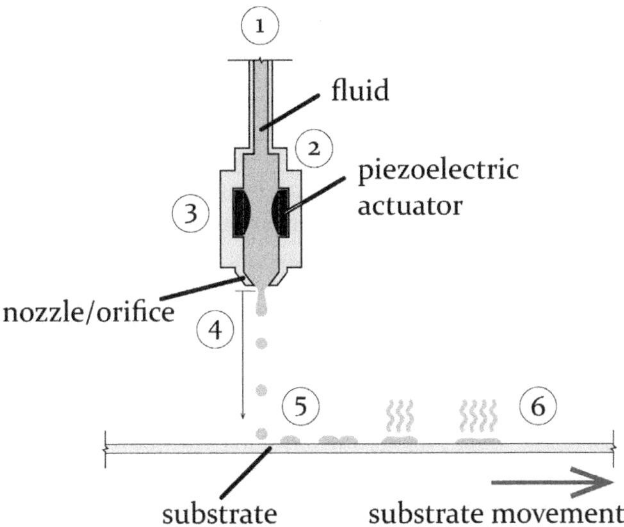

Figure 3.22 Principle of a DOD piezo inkjet.

In the following section, the subprocesses within a DOD piezo inkjet print head are described (see Fig. 3.22). First, the fluid is homogenized and its viscosity is set—*conditioning (0)*. Filled into a cartridge and connected to the print head, the fluid becomes available for the printing process—*fluid acquisition (1)*. In the following step, the fluid is accelerated onto the substrate in several steps that are mostly overlapping in time and therefore not clearly de-limitable. Here, the internal pressure is changed (positive or negative) by the deformation of a piezo element. First, fluid is primed from the fluid reservoir—*predosing (2)*. Pressure waves inside the fluid guarantee the transport of a well-defined amount of fluid through the ink channel and out of the orifice—*dosing (3)*. At the orifice the droplet formation is initiated. If the fluid droplet detaches from the orifice, the *transfer (4)* process begins. The fluid passes a small distance to reach the substrate surface. On the substrate *relaxation (5)* occurs followed by *drying (6)* and solidification of the fluid.

3.4.4.3 Characteristics of print equipment

Print heads consist of an ink supply channel, an electromechanical actuator circuit, an ink channel, and the nozzles. The print head is connected to the fluid reservoir via the ink supply channel. The actuator circuit transforms the digital signals received from the computer into pressure waves that prime, excite, and spread the fluid in the ink channel. For droplet formation, fluid is accelerated toward and ejected out of the nozzle.

The nozzle plate usually consists of nickel or silicon and could be plated with, for example, tantalum or palladium. With etching techniques, several hundred to thousand nozzles and channels are created. Commonly, nozzles have an inner diameter of 10–100 μm. The distances between the nozzles vary from 250 to 500 μm. Each nozzle can be separately controlled during printing. Smaller distances of printed drops are realized by tilting the print head.

3.4.4.4 Process description

Inkjet printing always starts with the conditioning of the fluid. The fluids often need to be mixed again in order to guarantee a homogeneous distribution of, for example, nanoparticles. Due to the small nozzles, clogging by particle agglomerates and gas pockets is always an issue and must be avoided. Therefore, the remixed fluid needs to be de-gassed before it is filled into the cartridge. After connecting the filled cartridge to the print head, the hardware is ready to be used.

Next, the printer's software needs to be fed with the printing layout, appropriate signals for actuator, and the cleaning and printing cycles. Well prepared from both sides (hardware and software), the printing process can start.

In DOD, droplets are forced to pass out of the nozzle by creating an excess pressure inside the ink chamber (due to volume change). This can be realized by local heating and vaporization of the fluid, via thermal elements, or by mechanical distortion of a piezo element. For the piezo technology, the driving signal of the piezo actuator is the most important parameter, as it controls the hydrodynamic process of droplet formation. Finding an adequate pulse shape may be quite complex, as each signal usually comprises a sequence of pulses distinct in amplitude and duration time. Hence, this is termed a *waveform*. The waveform needs to be adapted to the used printing fluid. Thermal inkjet systems locally heat up the fluid until it evaporates. These high temperatures can negatively influence the functional material and even destroy it before applying onto the substrate. Therefore piezo-based DOD can be a better alternative within functional printing.

For the continuous inkjet, a pump in combination with a piezo element is used to create the constant droplet stream. By deflecting certain drops, the printed image is generated. Additionally, a catching and recycling system is necessary to collect the fluid for nonprinting areas or during changeover times, to clean it, and to feed it back into the reservoir.

3.4.4.5 Experiences from a practical point of view

The orifices of the print head are driven in a distance of about 1 to 2 mm to the substrate surface. Drops in DOD form elongated tails or ligaments while flying. Under specific conditions and depending on the fluid properties, the fluid ligament can rupture, causing satellite drops to emerge. Satellite drops can cause inaccurate printing results and noncircular footprints.

The compatibility of the fluid's and the substrate's surface tension is even more important for inkjet printing than it is for conventional printing techniques. Owing to the fact that the ink droplet is accelerated onto the substrate and starts spreading there, fluid dynamics totally differ from those that follow the ink splitting process after the lift-off of any kind of printing plate. This is why surface pretreatment is an often-used tool in inkjet printing. For printing whole areas or straight lines, the distance between two dots needs to be adjusted sensitively. The printed dots should overlap partly in order to create one hydrodynamic system that is able to form a closed layer. Otherwise, the results can be defective, showing individual drops, scalloped or bulging lines, or stacked coins.

One important advantage of the inkjet technology is its compatibility to fluids with a low viscosity (e.g., small-molecule formulations). Therefore, many fluids that are not suitable for conventional printing technologies can be processed with inkjet. Typically, the fluid is enclosed in a cartridge and is therefore protected against solvent evaporation and environmental influences. However, particle-based fluids, even with nanoparticles, or volatile fluids often lead to clogging of the nozzles. Depending on the print head, clogged nozzles can be reinitiated by simple cleaning or in the worst case need to be replaced, which usually means that the whole print head needs to be replaced. Additionally, inkjet fluids have short shelf lives (only a few months) and often fluctuate in quality among different batches of production. Two important advantages are based on the NIP principle: First, the moving height of the print head is easily adjustable to quite a large spectrum of substrate thicknesses and forms, and second, the contactless fluid transfer prevents damage of underlying layers.

Moreover, depending on the substrate dimensions, inkjet printing can be slow because of the limited number of nozzles and print heads. Today, industrial machines reach 5 m/s and provide up to 1000 nozzles per print head. New developments, like the new Memjet print head, are static and nonmoving and include page-wide print heads with up to 70,400 nozzles. This is different in lab-scale printers, where printing often occurs with 5 or fewer nozzles. Here, printing an image of approx. 620 cm^2 (DIN A4) can take hours and is only feasible for research applications. Additionally, advantages like the easy use of inkjet due to digitally adjustable parameters can also quickly turn into disadvantages. If the waveform is not given by the ink supplier, the development of the waveform can become very complicated and time intensive. Also, the parameters cannot be transferred to print heads of other manufactures.

3.4.4.6 Applications and current research

Numerous publications within the last years and the permanently increasing offer of inkjet fluids show the popularity and success of inkjet printing technologies. Compared to conventional printing techniques, inkjet is cheap (initial cost, no printing plate production, small amounts of fluid, less waste of fluid), easy to use (complicated waveforms are often given by ink supplier), and ideal for single sample fabrication (digital printing technique). The range of inkjet goods is therefore widely spread, for instance, silver-based back electrodes for solar cells, organic active layers for OLEDs, and dielectric layers. Common inkjet printers can reach a feature size of 30 μm (~1 picoliter). But the creation of droplets with a diameter of less than 1 μm (subfemtoliter) has already been published.

3.4.5 Coating Techniques

The difference between printing and coating is the ability of printing techniques to create images that are shaped irregularly in two dimensions (layer thickness is the third dimension). Coating techniques are established processes that are used from small-scale, for example, spin coating of silicon wafers, to large-area coatings, for example, easy-release coatings in the square-meter range. The three most important techniques will be introduced in this section.

3.4.5.1 Spin coating

Spin coating is a highly reproducible lab-scale method used as a reference method for producing, for example, thin films of organic semiconductors and verifying the layer formation capabilities of printing formulations. Only small quantities of liquid are required, but the size of the substrates is limited to usually not more than a few centimeters in diameter.

A drop of the liquid is deposited in the center of the substrate, which is mounted on an electrically driven spinning plate. When the spinning plate is accelerated to a specific revolution rate for a specified time, the liquid drop is spread over the substrate surface by centrifugal forces, and abundant liquid is thrown off the substrate from the rim (see Fig. 3.23).

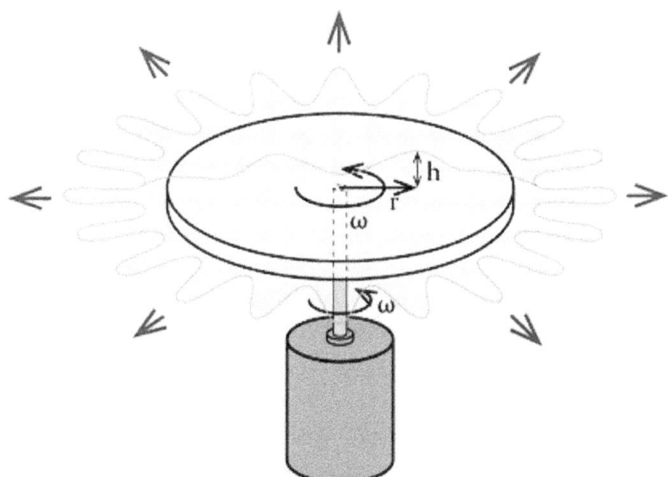

Figure 3.23 Scheme of fluid dynamics on a spin coater.

For the remaining liquid film the balance between centrifugal and viscous forces gives rise to the formation of a smooth and homogeneous layer, and the layer thickness is almost constant over the whole substrate surface. The layer thickness h can be estimated from the rotational speed n (in rpm), and the spinning time t_s as follows:

$$h \propto \frac{60}{2\pi n}\sqrt{\frac{3\eta}{2\rho t_s}} \tag{3.23}$$

where ρ is the mass density of the liquid (in kg/m^3) and η the viscosity. This formula is useful to determine the spinning time or rotational speed that is required to create thin liquid films of well-defined thickness. It applies, however, only in case that evaporation from the liquid films can be neglected. Depending on the fluid's mass density and viscosity, layer thicknesses from tens of nanometers to several micrometers can be fabricated.

If solvent evaporation is substantial, and provided that spinning is continued until the deposited film has dried by solvent evaporation, the thickness can be estimated as

$$h_{\text{evap}} \propto \sqrt{\frac{60}{2\pi n}}\left(\frac{3k\eta}{2\rho}\right)^{1/3} \tag{3.24}$$

where k is a solvent-specific constant proportional to the saturation vapor pressure in the gas atmosphere and to the vapor diffusivity. The result essentially represents the film thickness at which further thinning by liquid flowing to the rim of the substrate is inhibited by viscous friction. It should be noted that the actual thickness of the dry film also depends on the fraction of solute that is contained in the film at this stage.

3.4.5.2 Slot-die coating

Slot-die coating is a well-known technology for roll-to-roll large-format coating with the ability to deposit a micrometer-thin homogeneous layer with a large area. It is famous for very high throughput and cost efficiency. This technology has a long tradition in the production of films for analog cameras. The heart of the coating system is a die with an extraordinary precision. This is essential in order to achieve an even coating in terms of constant thickness and surface smoothness. Dies are very expensive and manufactured up to a width of 5 m for a coating velocity from 0.2 m/min (0.003 m/s) to 2500 m/min (42 m/s). A die is a closed application system with a chamber into which the liquid is pumped. A special pump provides very precise and pulsationless metering of the liquid to the die. Depending on the viscosity of the liquid, layers in the range of 0.5 to 500 μm are achievable. The coating weight is controlled by the pump and the substrate speed. Transverse area coating weight and homogeneity are functions of the manufacturing precision of the die and the lip gap adjustment. The application can be realized with contact of the die to the substrate but can also be contactless.

The advantages of slot-die coating are very high throughput and the usage of a closed system that guarantees cleanness of the application and allows the use of solvent-based liquids. The main disadvantages are the high production costs of the die, the limitation to a roll-to-roll process, and the large amount of substrate that is needed until the process is running stable at the optimal velocity.

3.4.5.3 Blade coating

Blade coating is very similar to slot-die coating. Instead of a die with a closed chamber, a doctor blade (a knife) is used. The working principle of a doctor blade is known from other technologies like gravure printing or flexography: the doctor blade is installed toward a roller. To apply a specified film thickness, the gap between the doctor blade and the roller, carrying the substrate, has to be adjusted very precisely. Alternatively, flatbed blading systems are used.

The advantages of blade coating are the easy construction and operation as well as the low cost of the system. Unlike for slot-die coating there are some lab coaters available, especially for single sheets. Nevertheless, blading also exhibits disadvantages. The doctor blade is an open system, so solvents can evaporate and hence the rheological behavior of the liquid will be changed. The physics of the liquid flow at the doctor blade is very complex and influenced by parameters like viscosity, velocity, predosing of the liquid, and filling level of the fluid reservoir. It is quite challenging to produce a stable, homogeneous film. Even an unsteady thickness of the substrate can lead to an uneven film.

3.4.6 Laser Patterning of Thin Films

Laser process development utilizes the variability of light in terms of wavelength, pulse duration, spatial and temporal pulse shaping, and energy to adapt the process to the material properties. This way, the material can be molten, evaporated, decomposed, and modified, as required. Laser processes, like ablation, cutting, welding, and annealing, can be applied to many applications of organic and printed electronics. For OLEDs, micropatterning of TCO layers allows better control of charge carrier distribution and large-area opening of barrier layers is required for electrical contacting. The integrated series connection of cell strips of organic solar cells benefits from laser patterning, while laser welding can replace lamination with adhesives.

For organic thin-film transistors with high switching speeds, channel widths of <10 μm are required. These small feature sizes cannot be achieved by fast printing processes. Resolution of a few microns or even submicrons can be realized by lithography or microcontact printing (μCP). However, both methods are rather time consuming. Laser patterning is a versatile alternative that can reach the required resolutions. By fine-patterning of preprinted structures, roll-to-roll-compatible throughput is accomplished. Of course, the accuracy is also limited by the registration, which serves as a reference to which the optical system needs to be aligned.

High patterning speeds can be achieved by fast beam deflection, for example, using polygon scanners that use a multifaceted rotating mirror to deflect the beam, or multiparallel processing, which scales the process up linearly by splitting the beam. Laser processes are easily scalable to high throughputs and large formats. In a roll-to-roll process chain, laser processes can be combined with coating, drying, and lamination processes.

Printed electronics materials can be laser patterned by ablation or modification. During the ablation process material is actually removed from the substrate, while the modification leads to a change in chemical and physical material properties, eliminating conductive or semiconductive properties of the thin film. In principal, thin films on every substrate can be patterned by laser ablation. Substrate properties as well as wavelength, pulse duration, and other laser parameters have a strong influence on the ablation process. The absorption spectrum quantifies the amount of laser radiation deposited in the film, providing a first clue on how a certain material reacts on laser irradiation. Especially using ultrashort laser pulses ("ultrashort" denoting a pulse duration of less than 1 picosecond) the intensity is high enough to enable nonlinear absorption phenomena, allowing the patterning of materials with low linear absorption at laser wavelength.

When ablating thin films by laser radiation, four different mechanisms can be observed, depending on material properties and laser parameters. The different ablation mechanisms lead to different patterning results.

For photochemical or cold ablation, Fig. 3.24a, the molecular bonds within the layer are broken, leading to material decomposition in the irradiated area. This mechanism results in steep and clean edges. Due to the mostly gaseous ablation products, little to no debris is observed on the sample. The thermal impact on remaining films and the substrate is negligible. This ablation mechanism occurs when ultrashort laser pulses or deep UV wavelengths ($\lambda < 250$ nm) are used.

During photothermal ablation (see Fig. 3.24b), the layer is molten and evaporated. The ablation mechanism is of thermal nature. Therefore the surrounding material receives a high thermal impact. Around the edges, bulges >100 nm due to resolidified molten material remain after ablation. Laser pulse durations greater than a few hundred picoseconds mostly lead to this thermal process.

Especially when ablating brittle films with little adhesion to the subjacent layer by ultrashort laser pulses, photomechanical ablation occurs (see Fig. 3.24c). During this ablation mechanism, the laser pulse induces thermal and acoustic stress in the film, eventually leading to cracking as the thin film relaxes. The resulting flakes can be found as debris on the sample and as bulges at the edges of the ablated area. Due to the stress-induced nature of this ablation mechanism, thermal impact on the remaining material is minimal, which is beneficial for laser patterning processes. On the other hand, an additional cleaning process is needed to remove debris and bulges from the sample.

Blast-off ablation (see Fig. 3.24d) results in a similar amount of debris and bulging, although the ablation mechanism is different. If the laser radiation is absorbed weakly by the top layer and more strongly by the subjacent material, the subjacent material (thin film or substrate) is ablated. The resulting pressure explosively lifts off the top layer. Due to low absorption of the laser radiation in the top layer, thermal impact on this layer is negligible. The substrate or subjacent layer, however, is damaged by this ablation mechanism.

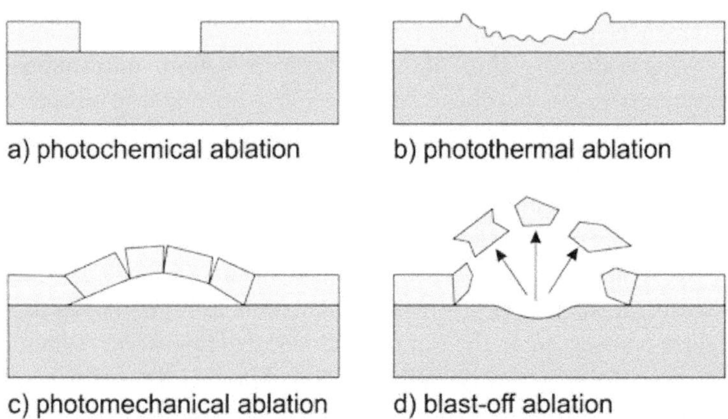

Figure 3.24 Ablation mechanisms that are observable when ablating thin films.

In reality, most often a combination of these four basic ablation mechanisms is observed. The most favorable photochemical ablation mechanism can be achieved by choosing a material- and substrate-specific wavelength and pulse duration.

Patterning via modification by laser irradiation below the ablation threshold can be beneficial compared to ablation. Here, laser radiation changes the electrical properties selectively by chemical or physical modifications of the material, reducing or even suppressing conductivity or semiconductive properties. In organic materials the modification is based on the chemical altering of the π-electron system during the laser process. The modification of transparent conductive oxides is mainly governed by a change of oxygen content inside the film during laser irradiation. Due to the modification below ablation threshold, lower power of the laser pulse is needed compared to ablation. In addition, modification does not change the topography of the film. Therefore debris and bulging are avoided.

In conclusion, laser processes, especially laser patterning, allow versatile and flexible processes, extending the limitations of printing processes.

3.4.7 Additional Coating and Structuring Techniques

Many other printing or coating techniques are used besides the most popular ones, either as substitutions for or as additions to them. Each of them has different advantages and disadvantages and is therefore selected for specific applications, for example, μCP for printing self-assembled monolayers (SAMs). Selected techniques are shortly introduced in the following section.

3.4.7.1 Evaporation deposition

Evaporation deposition is probably the most commonly used family of deposition procedures for thin films in microfabrication processes.

Physical vapor deposition (PVD) implies essentially a transfer by resublimation of the evaporated material on the substrate. The equipment consists of a vacuum chamber in which a source material and the substrate that is to be coated face each other. As the source material is heated to a temperature in the range of its boiling point, particles are evaporated and travel over the distance to the substrate. Here they condensate on the surface, thus forming a homogeneous thin film. Another option is chemical vapor deposition (CVD), where a source material from a gaseous phase is deposited onto the substrate via a chemical reaction. Deposition rates can reach up to several micrometers per second, but also layers of a few nanometers in a high-throughput roll-to-roll process can be achieved. Structuring can be achieved by using shadow masks. The main disadvantage for most of these techniques is the crucial need for a vacuum chamber, since the evaporated particles need a high mean free path to travel without collisions to the substrate.

3.4.7.2 Photolithography

One of the most standardized ways to structure a surface is photolithography. It can, for example, be used to produce 3D structures in the nano- and micrometer range or to create an etching mask that can subsequently be removed from the substrate (stripping).

For this technique, first a photoresist is coated onto the substrate in the desired thickness, normally using spin coating. This photoresist is then precured and dried to evaporate solvents. Afterwards imaging by UV light takes place using a photomask. After a final baking, the pattern is developed by washing away the exposed or unexposed regions of photoresist, depending on the material used.

3.4.7.3 Pad printing

Pad printing can be seen as an indirect gravure printing process, since the printing form is essentially the same.

First, the printing form is filled with fluid using a kind of doctoring chamber that removes all excess fluid. Then an elastomeric silicone-based pad is pressed onto the printing form and lifted off again, taking with it the structured fluid film. Finally, the pad is lowered onto the substrate and lifted off again, thus transferring the fluid. The substrate can be either flat or a 3D shape such as a cup, a pen, etc. The fluid, which is typically solvent based for graphic arts, must be chemically compatible with the high silicone content of the pad, otherwise it will not adhere to it. With this technique structure sizes of 20 μm can be reached.

3.4.7.4 Offset printing

Offset printing is an indirect lithographic process and represents today's backbone in newspaper printing.

On the printing form the printing areas and the nonprinting areas lie essentially in the same surface plane. While the nonprinting areas have hydrophilic surface characteristics, the printing areas are hydrophobic and thus ink accepting. In a dampening unit the plate is covered with an aqueous dampening solution that covers the nonprinting parts. In the inking unit the ink is applied to the printing form and covers only the printing areas. This pattern is then transferred to a cylinder covered by a blanket that finally delivers the ink to the substrate.

Offset printing is highly dependent on many parameters such as ink viscosity or wettability parameters of all fluid–solid contact areas involved.

The viscosity of the ink is very high compared to other printing techniques, that is, 40–100 Pa·s compared to ~0.1 Pa·s in gravure printing. The typical resolution in offset printing is 2540 dpi (i.e., structure sizes of ~10 μm).

3.4.7.5 Nanoimprint lithography

A means of structuring in the nanometer scale is nanoimprint lithography (NIL).

Here a resist is applied to the substrate in a thickness of 100–1000 nm, typically using spin coating. Then a mold with the desired nanostructures is brought into contact with the resist. The final creation and preservation of the structures can take place in several ways, for example, by thermoplastic deformation or UV curing. These different procedures always call for adequate types of resist and mold material. Especially for the latter, a UV-transparent mold is needed. Finally, after removing the mold, the structured resist can be used as an etching mask for the underlying substrate or the resist itself is used as the functional material. Structure sizes of down to 20 nm have so far been created.

3.4.7.6 Microcontact printing

Microcontact printing (μCP) is a combination of standard photolithography and stamp printing used to create patterns of SAMs.

First, a master has to be created using photolithography. The master is then used for the casting of an elastomeric PDMS stamp. This stamp is inked with the desired material, mostly thiol solutions that diffuse into the PDMS, thus creating an ink reservoir. When the stamp is brought into contact with a substrate, a SAM is formed in the pattern of the stamp.

On metallic substrates, such as gold, this SAM can then be used as a wet-etching resist. The resolution of this process is limited by the fabrication of the stamp and the master, but so far structure sizes in the range of 50 nm have been created.

3.4.7.7 Aerosol jet printing

In aerosol jet printing a wide range of materials can be deposited onto a substrate.

Here, a mist generator atomizes the source material, thus creating an aerosol that is refined in a virtual impactor to droplets of 1–5 μm and finally focused onto the substrate via a special deposition head.

The possible feature sizes are about 10 μm, with layer thicknesses of 100 nm and above. The viscosity of the source material must lie between 1 and 1000 mPa·s. Despite a very low process speed the advantages of this technique are that there is no contact of the nozzle to the substrate (nonimpact printing, NIP) and that it is a low temperature process. It is also very well suited for rough substrates.

3.4.7.8 Laser transfer printing

Laser transfer printing is a contactless printing technique like inkjet printing without the difficulties of using printing nozzles.

A continuous transparent belt is permanently coated in an inking station with an ink of metallic content. In an area of the belt where it is situated directly above the substrate, a laser beam actuates droplet formation from the backside of the belt. The droplet then falls onto the substrate and thus a printing pattern is created.

Printing speeds of up to 20 m/min can be achieved here with possible printing resolutions of around 60 μm.

3.4.7.9 Xerography

Xerography, also known as electrophotography, is one of the digital printing techniques and mostly known from office laser printers.

First, a photoconductive cylinder is electrostatically charged by a corona unit. In the following, the surface charges are discharged corresponding to the data input of the image to be printed, using a laser or LED array. In the next step the toner is applied to the cylinder, with the dry toner particles adhering only to the charged areas. Then the cylinder presses onto the substrate and transfers the still dry particles. Finally, the image is fixed in a fusing unit where the particles are melted and permanently secured to the substrate.

Alternatively, instead of the dry toner a fluid toner can be used. This process uses an intermediate blanket, similar to offset printing, to transfer the toner from the photoconductive cylinder to the substrate.

3.4.7.10 Plasma printing

Plasma printing is a process used for the structured pretreatment of mostly polymeric surfaces at ambient pressure conditions.

On the top electrode a structured dielectric is applied. This structure is then pressed onto the substrate that itself is situated on top of a counterelectrode. The structure in the top dielectric forms microcavities with the substrate. In these cavities dielectric barrier discharges are ignited by applying a high voltage between the two electrodes. This plasma modifies the substrate by changing the surface energy properties only in the area of the cavities.

Subsequently, the substrate can be treated, for example, by electroless plating (an autocatalytic deposition of metal layers from an ionic solution) to create printed circuits only on the parts pretreated by the plasma. The resolution of this process is in the submillimeter range and process speeds have so far been reported at a few meters per minute. Nondielectric and even metallic substrates can also be used if the applied high voltage lies in the radio frequency range.

3.4.7.11 Hot embossing

Hot embossing is originally a relief-printing technique for the finishing of brochures, calendars, or other high-quality commercial print jobs with metallic effects.

Basically, instead of a printing fluid a foil is applied to the substrate. This foil consists of several layers, most importantly of the metallic layer, an encapsulation layer, and an adhesive layer. Using a hot (typically 80°C–230°C) structured metal stamp, the foil is pressed onto the substrate. At the elevated areas of the stamp the adhesive is activated and bonds the foil to the substrate.

Current approaches are to use the metal layer, which can, for example, be aluminum, gold, or copper, for conductive printed circuit elements.

3.5 Conclusions

Printing techniques are categorized by the usage of a printing form (conventional vs. digital), by the arrangement of image and nonimage elements (e.g., gravure, letterpress), or by the principle of substrate transport (e.g., R2R, S2S). Different techniques can be identified by specific errors in the printed image, for example, the halo effect in flexography.

Printing can be separated into seven subprocesses: conditioning, fluid acquisition, predosing, dosing, transfer, relaxation, and drying.

Fluid formulation needs to be done for every functional material (polymers, small molecules, or particles). The printing fluids need to be adjusted sensitively in viscosity and surface tension for the given substrate, process, and drying conditions.

As long as printed layers are still wet, a lot of fluid dynamics occurs. In a perfect system, undulations in a printed layer level out before the layer is dried. But often new errors arise in the printed layers, for example, due to the coffee stain effect.

Acknowledgments

The authors would like to thank Felipe Fernandes and Ardeshir Hakimi Tehrani for their support in the creation of the figures.

Glossary of Printing Terms

Cell
Recess in a gravure cylinder or anilox roller that ensures dosing and transportation of the fluid. Within gravure printing the image is created by the arrangement of the cells, whereas in flexographic printing the cells are only used for dosing.

Cell volume
Total volume of the cells; usually used for anilox rollers, not for gravure cylinders. The cell volume is not identical to the amount of fluid that is transferred.

Cell wall
Remaining material between neighboring cells.

Conventional printing techniques
Printing techniques that are based on a printing form.

Doctor blade
Knife-like, often springy unit that is used for wiping off excess ink.

DOD
Drop on demand; Driving principle of an inkjet print head. Alternatively, continuous inkjet systems are used.

Engagement
Changing of the distance between the printing form and the impression cylinder in order to influence printing forces in the nip.

Functional printing
Printing of materials that have a technical function beyond that of graphic prints, for example, layers with certain electronic, magnetic, or morphological properties.

Graphic(al) printing
Replication of structures with the sole purpose of conveying graphic information, for example, via texts or images.

Image element
Contrary to nonimage elements (nonprinting areas), image elements (printing areas) are areas on a printing plate that carry and transfer the printing fluid.

Impression cylinder
The substrate is pressed against the printing roller by the impression cylinder.

Ink *see printing fluid*

Ink splitting
The hydrodynamic process of ink transfer from one surface to another, for example, from the printing roller to the substrate.

Nip
In conventional printing techniques, the area where the ink is transferred, that is, where the substrate touches the printing roller.

NIP
Nonimpact printing; technologies that are based on contactless fluid transfer are called NIP technologies. They generally have no physical printing form.

Prepress
Processes like layouting, making color separations, producing printing forms, etc., that take place before the printing process starts.

Press
The actual production by printing.

Postpress
Summarizes the processes that take place after printing, for example, cutting.

Printability
Qualitative value; describes the properties of the ink and the substrate that are needed for guaranteeing error-free usage on a specific printing press.

Printing
Replication of structures by transferring materials onto surfaces by, for example, mechanical (e.g., gravure printing), hydrodynamic (e.g., inkjet printing), or electromagnetic forces.

> **Direct printing**
> Transfer of the material directly from the printing form onto the substrate.
>
> **Indirect printing**
> Transfer of the material from the printing form onto an intermediate medium and therefrom onto the substrate.

Printing fluid
Liquid matter that will be transferred onto the substrate; solutions or dispersions containing pigments or functional materials, for example, semiconducting polymers.

Printing form
Contains all the information that is needed for the replication of the printing layout, for example, a structured polymer plate or a digital file.

Printed image
Result of printing; printed pattern on the substrate.

Printing layout
Image that should be replicated by printing; the printing layout is created in the prepress process.

Printing machine *see printing press*

Printing parameters
Specific parameters that can be set at the printing press, for example, printing force, engagement, and printing velocity.

Printing plate *see printing form*

Printing press
Machine that consists of at least one printing unit and a substrate transport unit. Printing presses are classified by the substrate transportation system (sheet or web).

Printing pressure
Pressure with which the material is applied on the substrate, for example, via the squeegee in screen printing or the printing plate in flexographic printing; can be expressed as a force per area or length of the nip; sometimes the absolute printing force is used.

Printing roller
Structured cylinder that contains the printing layout and transfers the ink onto the substrate.

Printing result *see printed image*

Printing technique
Classification of the techniques by the printing form: letterpress, planographic, gravure, screen, or masterless.

Printing unit
Part of a printing press where the transfer of the ink from the ink reservoir to the printing plate and in the end onto the substrate occurs.

Register
Accuracy of a printed pattern's lateral positioning. Normally registration marks at the sides of the print are used to control this.

Resolution
Number of producible screen dots or lines per unit of length, usually dots per inch (DPI) or lines per inch (LPI).

Roll-to-roll (R2R) printing *see web-fed printing*

Screen frequency
Resolution of a screen (in the sense of halftones, not webs) or a printing form, respectively, and given in dots or lines per unit of length.

Sheet-fed printing
Printing onto single sheets.

Sheet-to-sheet printing (S2S) *see sheet-fed printing*

SOHO sector
Small offices/home offices sector.

Substrate
The matter/object on which the print is deposited; mostly flat, for example, foils or glass sheets, but in principle it is possible to print on objects of any form.

Tone value
A measure for the optical density of a certain color and is given in %.

Waveform
The description of the sequence of voltage pulses that drive inkjet print heads; waveforms contain information about voltage amplitudes and pulse durations.

Web-fed printing
Printing onto a so-called endless web.

Exercises

3.1 Viscous thin-film flows in lubrication approximation

Consider a thin liquid film of thickness $h(x,y,t)$ on a solid substrate, with viscosity η and surface tension σ. Assume that surface tension $\sigma(x,y)$ is varying in space and that consequently the viscous shear equals the Marangoni drag and equals the gradient $\vec{\nabla}\sigma$. Assume further that the flow velocity $v(x,y,z)$ continuously approaches zero at the solid surface and that viscous shear in the z direction is much larger than that in the x or y direction.

Show that to leading order in $\partial h/\partial x$, $\partial h/\partial y$ (i.e., for sufficiently flat films), the viscous flow

$$\vec{v}(x,y,z)=\frac{1}{2\eta}(z^2-2h(x,y)z)\,\vec{\nabla}(\sigma\nabla^2 h(x,y)-p_{\text{vol}})+\frac{z}{\eta}\vec{\nabla}\sigma$$

solves the Stokes equation. z is normal to the substrate plane.

3.2 Incompressible flows and film thickness dynamics

Using the incompressibility of the liquid, show that film thickness $h(x,y,t)$ in a liquid film is related to the velocity field $v(x,y,z)$ by

$$\frac{\partial h}{\partial t}=-\int_0^{h(x,y)}\left(\frac{\partial v_x}{\partial x}+\frac{\partial v_y}{\partial y}\right)dz$$

3.3 Landau–Levich equation:

Combine the results from Exs. 3.1 and 3.2 to derive the Landau–Levich equation (LLE) [Eq. (3.16)].

3.4 Squeezing liquids between two plates

Consider two parallel plates with infinite length and width b facing each other in the time-dependent distance $D(t)$. Assume that the gap between the plates is filled with some Newtonian liquid and that D changes in time with velocity $\dot{D}=dD/dt$.

(a) Using the Stokes equation, a 2D flow profile ($v_y = 0$) and no-slip boundary conditions on both plates, show that when the plates are moving apart or when they approach each other, the flow velocity is given by

$$v_z = -\frac{6\dot{D}}{D^3}\left(\frac{1}{3}z^3 - \frac{D^2 z}{4}\right)$$

$$v_x = \frac{6\dot{D}x}{D^3}\left(z^2 - \frac{D^2}{4}\right)$$

where the x axis is in the width direction and the z axis is normal to the plates. The coordinate origin (0,0) is assumed to be situated in the center of the gap.

(b) Show that the pressure in the gap is

$$p = p_0 + \frac{6\eta\dot{D}}{D^3}\left(x^2 - z^2 - \frac{b^2}{4}\right)$$

where p_0 is the outside atmospheric pressure.

(c) Consider plates of width $b = 10$ cm, pressed together by a force of 100 N/m of plate length. If the liquid viscosity is 1 mPa·s (water), how long does it take until the distance between the plate becomes smaller than 1 μm (starting at a distance of 1 mm)?

(d) What is the shear rate in the liquid close to the plate surfaces?

3.5 Structural defects

Map the following structural defects to the corresponding printing techniques:

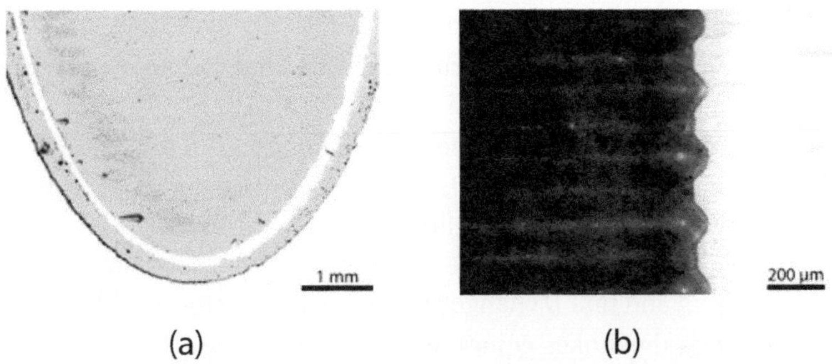

(a) (b)

(c) (d)

Figure E.3.1

Suggested Readings

Fluid Formulation

For a general overview of the theory of polymers and experimental methods we refer to the book by Doi and Edwards. The theory of percolation transitions is presented by Stauffer. For more details on colloidal systems and their stabilization and on the thermodynamics of liquid interfaces we refer the reader to Israelachvili and Takeo. Additional information about the LLE is available in the respective paper.

Doi, M., and Edwards, S. F. (1986). *The Theory of Polymer Dynamics,* Clarendon Press.

Israelachvili, J. N. (2011). *Intermolecular and Surface Forces,* Revised 3rd Ed., Academic Press.

Stauffer, D., and Aharony, A. (1994). *Introduction to Percolation Theory,* Taylor & Francis.

Takeo, M. (1999). *Disperse Systems,* Wiley-VCH.

Printing Techniques

A detailed description of all printing techniques and the aspects of graphical printing is given by Kipphan.

For a deeper understanding of the influence of solvents on flexo printing plates and the characterization for the use in printed electronics we refer to Theopold. More information about the flexo printing process is given by the Flexographic Technical Association (FTA) and Meyer. The Gravure Association of America and the Gravure Education Foundation have comprehensively explained graphical gravure printing. Many aspects and the fundamental principle presented in this book are also valid for printed electronics. Papers by Bornemann et al. and also by Lee et al. and Sankaran et al. deal with the emptying effects of gravure cells. A useful ebook about screen printing and practical advices is edited by Abbott et al. Detailed principles of inkjet technologies are found at H.P. Le.

Abbott, S., et al. (2008). *How to Be a Great Screen Printer: The Theory and Practice of Screen Printing,* MacDermid Autotype.

Bornemann, N., et al. (2012). *Characterization of Gravure Cells Using Confocal Microscopy,* Proceedings of 39th International Research Conference of IARIGAI.

Bornemann, N., Sauer, H. M., and Dörsam, E. (2012). *Experimental Investigation of the Filling and Emptying of Gravure Cells,* Materials Science Engineering (MSE).

Dykes, Y. (1999). *Flexography: Principles & Practices,* Foundation of Flexographic Technical Association.

Foundation, G. E., and America, G. A. O. (2003). *Gravure: Process and Technology,* Gravure Education Foundation and Gravure Association of America.

Kipphan, H. (2001). *Handbook of Print Media - Technologies and Production Methods,* Springer.

Le, H. P. (1998). Progress and trends in ink-jet printing technology, *J. Imaging Sci. Technol.*, **42**(1), 49–62.

Lee, J. A., Rothstein, J. P., and Pasquali, M. (2013). Computational study of viscoelastic effects on liquid transfer during gravure printing. *J. Non-Newtonian Fluid Mech.*, **199**, 1–11.

Meyer, K.-H. (2006). *Technik des Flexodrucks,* Rek und Thomas Medien AG.

Sankaran, A. K., and Rothstein, J. P. (2012). Effect of viscoelasticity on liquid transfer during gravure printing. *J. Non-Newtonian Fluid Mech.*, **175–176**, 64–75.

Theopold, A. (2014). *Charakterisierung von Flexodruckformen hinsichtlich der Eignung für die gedruckte Elektronik,* Books on Demand.

In the following table (Table 3.1) examples of lab printing presses are given.

Table 3.1 Lab printing presses

Flexographic printing press	Printing velocity	Format (width × length)
IGT F1	0.2–1.5 m/s	50 × 190 mm
Saueressig FP 100-300	0.17–0.83 m/s	260 mm
Schlaefli testacolor 171 HighResolution	up to 0.5 m/s	150 × 200 mm
Erichsen Printing Proofer 628	up to 0.67 m/s	95 × 165 mm
GT+W Superproofer 220	up to 5 m/s	50 × 150 mm
Pruefbau FT150	up to 3 m/s	150 × 300 mm

Gravure printing press	Printing velocity	Format (width × length)
IGT G1	0.2–1 m/s	50 × 140 mm
Saueressig CP 90-200	0.17–0.83 m/s	175 mm
Schlaefli Labratester II	up to 1.5 m/s	220 mm
Erichsen Printing Proofer 628	up to 0.67 m/s	160 × 195 mm
GT+W Superproofer 220	up to 5 m/s	50 × 150 mm
Pruefbau FT150	up to 3 m/s	150 × 300 mm

Screen printing press	Printing velocity	Flatbed/ rotary	Format (width × length)
Alraun AT-HTF-760	–	Flatbed	250 × 300 mm
KBA-Kammann K15-QSL	0.1–1 m/s	Flatbed	150 × 150 mm
THIEME LAB 1000	0.05–1 m/s	Flatbed	400 × 400 mm
Coruna Pico23	up to 0.16 m/s	Flatbed	380 × 250 mm

Inkjet printing press	Printing velocity	Print head	Drop volume	Format (width × length) [mm]	Accuracy
Dimatix DMP-2831	–	DOD	1 or 10 pL	210 × 315 mm	± 25 μm
PiXDRO LP50	0.2–0.4 m/s	Various	4–2000 pL	210 × 310 mm	± 5 μm
CeraPrinter L-Series	up to 0.2 m/s	Various	1–80 pL	305 × 305 mm	± 1 μm
Notion n.jet lab	up to 0.9 m/s	Various	Various	305 × 305 mm	± 1 μm

Coating Techniques

For deeper information on planarization during spin coating we refer to the paper by Stillwagon and Larson. Concerning slot-die coating we recommend books by Perez and Kistler, as well as by Pranckh and Sullivan for blade coating. A nice overview is given in Tracton's *Coatings Technology Handbook.*

Kistler, S. F., and Schweizer, P. M. (1997). *Liquid Film Coating,* Springer, Netherlands.

Perez, E. B., and Carvalho, M. S. (2011). Optimization of slot-coating processes: minimizing the amplitude of film-thickness oscillation. *J. Eng. Math.*, **71**(1), 97–108.

Pranckh, F., and Scriven, L. (1988). The physics of blade coating of deformable substrate. *Coating Conference Proceeding*, pp. 217–238, TAPPI Press, Atlanta, GA.

Stillwagon, L. E.,]and Larson, R. G. (1992). Planarization during spin coating. *Phys. Fluids A*, **4**(5), 895.

Sullivan, T. M., and Middleman, S. (1986). Film thickness in blade coating of viscous and viscoelastic liquids. *J. Non-Newtonian Fluid Mech.*, **21**(1), 13–38.

Tracton, A. A. (2005). *Coatings Technology Handbook,* Taylor & Francis.

Laser Patterning and Laser Ablation

A general overview is given in the book by Dowden. For more detailed information we want to refer to the publications by Dyer, Karnakis, Koulikov, Paltauf, Schulz-Ruthenberg, Schaefer, and Srinivasan.

Dowden, J. (2009). *The Theory of Laser Materials Processing,* Springer.

Dyer, P. E. (2003). Excimer laser polymer ablation: twenty years on. *Appl. Phys. A*, **77**, 167–173.

Karnakis, D., et al. (2009). Maskless selective laser patterning of PEDOT: PSS on barrier/foil for organic electronics applications. *JLMN*, **4**, 218–223.

Paltauf, G., and Dyer, P. E. (2003). Photomechanical processes and effects in ablation. *Chem. Rev.*, **103**(2), 487–518.

Schulz-Ruhtenberg, M., et al. (2014). Seminal tools for roll-to-roll manufacturing. *Laser Tech. J.*, **11**(1), 21–25.

Schaefer, M., et al. (2012). Comparison of laser ablation of transparent conductive materials on flexible and rigid substrates. *Proc. LOPE-C*, 39–43.

Srinivasan, R., and Braren, B. (1989). Ultraviolet laser ablation of organic polymers. *Chem. Rev.*, **89**(6), 1303–1316.

Chapter 4

Characterization Techniques for Printed Electronics

Robert Tone and Simone Ganz
Institute of Printing Science and Technology, Technische Universität Darmstadt, Magdalenenstraße 2, Darmstadt, 64289, Germany
tone@idd.tu-darmstadt.de

As a technique for the production of organic electronics, printing has two main advantages, efficient use of the material and high production speed. For these to be truly useful, the quality of the printed layers—their thickness, homogeneity, etc.—must conform to the requirements of the produced devices. Unfortunately, this is not easy to reach. Getting close to the performance of a spin-coated or evaporated device is only possible if the printing process is closely monitored and optimized. This, in turn, necessitates a precise measurement of the properties of the involved materials and the resulting device's structural and electronic characteristics. In this chapter parameters and measurement techniques are presented that are suitable for this.

More precisely, the characterization of substrates and printing fluids, typical structural defects in printed layers, and methods for

Organic and Printed Electronics: Fundamentals and Applications
Edited by Giovanni Nisato, Donald Lupo, and Simone Ganz

ISBN 978-981-4669-74-0 (Hardcover), 978-981-4669-75-7 (eBook)
www.panstanford.com

their detection, as well as techniques for proving the functionality of printed organic semiconductors, are covered.

It is important to note that while this chapter is devoted to printed layers, the characterization techniques and many of the typical defects are the same for other kinds of wet deposition, for example, slot-die coating. Additionally, some literature on coating is listed in the "Suggested Readings" section.

4.1 Viscosities and Surface Tensions

As described in Chapter 3, the printing process can be split into several substeps. For example, the acquisition of printing fluid, its transfer to the substrate, its relaxation on it—these are completely different processes described by quite dissimilar equations. However, two parameters always crop up in wet deposition, the fluid's and adjacent solids' surface tensions as well as the viscosity of the former. These can be complex, changing with strain, time, and the fluid's solid content, and hence need to be thoroughly analyzed. In this section methods for the necessary measurements are presented.

4.1.1 Measuring Viscosity

The viscosity of a fluid moderates any flow within it. To be more precise, *dynamic viscosity* is the ratio of shear stress and shear rate. Dividing this by the fluid's density yields the *kinematic viscosity.* Fluids are categorized by their viscosity's dependence on the shear rate; some examples are shown in Fig. 4.1.

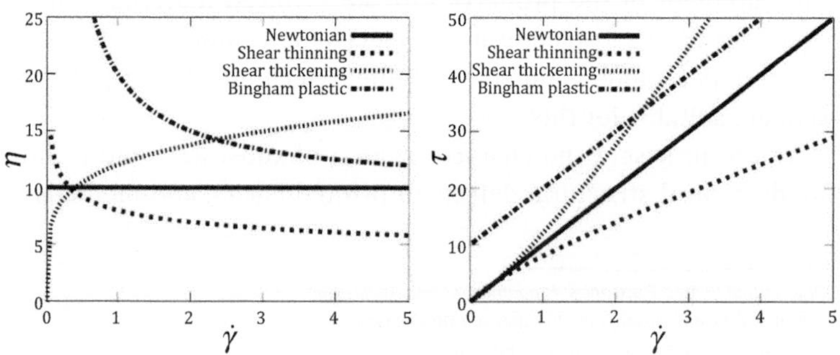

Figure 4.1 Typical viscosities η and shear stresses τ over shear rate $\dot{\gamma}$.

In the case of a shear rate–independent viscosity they are called *Newtonian* (e.g., water). Fluids can also be *shear thickening* (viscosity increases with shear rate, for example, starch in water) or *shear thinning* (viscosity decreases with increasing shear rate, as that of ketchup) and even go from one to the other at different shear rates (e.g., dispersions of hard particles may behave this way). Additionally, these changes are mostly not instantaneous. If the viscosity's time dependency is noticeable, this is called *thixotropy* for shear-thinning fluids and *rheopexy* for shear-thickening ones. Confusingly, multiple definitions for these terms exist, some of which ignore the aspect of time dependency.

Some fluids only start flowing when the shear stress reaches a certain *yield value* and are thus called *plastics*. This behavior is often found in highly concentrated suspensions and dispersions (e.g., mayonnaise) and is normally unwanted, since it inhibits leveling completely.

Viscosity generally decreases with increasing temperature, which means that measurements have to be conducted at a controlled temperature. Also, any measurement technique should allow the correct characterization of all types of fluids. The easiest way to achieve this is to exert the same shear rate throughout the whole volume of the tested fluid. This can be done with a *rotational rheometer*. Its working principle is simple: A fluid is wedged between two probes that are rotated against one another. The fluid's viscosity is then determined by measuring the torque needed to sustain the angular frequency corresponding to a given shear rate or vice versa. Several types of probes can be used, for example, parallel plates or coaxial cups. Not all of these are ideal for non-Newtonian fluids. Suitable and easy to use is a cone-plate geometry. Here, the tangential speed and the gap height are both proportional to the distance from the center, leading to a constant shear rate. By varying the angular velocity, it is possible to plot the viscosity's dependence on the shear rate. The velocity profile for this should not be a slope but a series of steps so that the measurements are conducted in a steady state. It is also possible to use a sinusoidal signal with a varying frequency to detect elastic behavior. To check the time dependency of a fluid's viscosity, one should measure its run after instant changes of the shear rate. A time-dependent viscosity will converge roughly exponentially to its steady-state value after the switches. The time constant derived from this commonly depends on the applied shear rates, a standard procedure can therefore not be defined. Instead, one should choose values that are typical for the analyzed processes, for example, ~1 s^{-1} for leveling and ~100,000 s^{-1} for the blading of the ink.

Not every system can reach such high shear rates. For instance, in the cone-plate setup we introduced before, turbulences often appear at ~1000 s^{-1} causing an apparent change of the viscosity. Even if this does not happen, the maximum applicable shear rate is comparatively low, as the fluid is simply spun out of the gap at some point. Dispersions with particle sizes in the micrometer range are also problematic due to the small gap at the center. Coaxial cups can reduce these problems. They have a constant gap and turbulences appear at much higher shear rates than in a cone-plate geometry. However, they are more difficult to clean and require more fluid. To avoid turbulences and clogging one may also use parallel plates, but here the shear rate is not constant throughout the tested volume. In this setup a non-Newtonian fluid's viscosity will vary with the radial position, requiring a mathematical handling of the data.

A completely different approach is used in *capillary* and *microfluidic rheometers*. Here the viscosity is calculated via the pressure drop and volume flow rate of a fluid that is pressed through a capillary. This can be simple if the *Hagen–Poisseuille equation* can be used. Otherwise, several corrections have to be used to deal with the flow at the ends of the capillary, slip at the walls, non-Newtonian fluids, etc. With these it is possible to get reliable results for extremely high shear rates of more than ~1,000,000 s^{-1}. There are simpler ways to measure viscosities, for example, by letting a sphere sink in a tube filled with the tested liquid or by measuring the time needed to drain the liquid from a cup with a hole in the bottom. These techniques are common in the printing industry and do give precise results, but only for Newtonian fluids. Moreover, comparatively large amounts (ca. 100 mL) of fluid are required.

4.1.2 Measuring Surface Tensions

One of the driving forces of film formation is the surface tension. That of the fluid drives the leveling of the surface, and the relation between the surface tensions of fluid and substrate determines the fluid's *contact angle*, which can—in principle—be calculated.

The surface tension of a substrate is, in turn, calculated from the contact angles of test fluids on it. Several models for all of this exist, taking into account different interactions in different ways. None is suited for every material and all are problematic for printed fluids, which are typically mixtures of materials with different affinities to the substrate. With a well-chosen model approximate contact angles can nevertheless be predicted, reducing the experimental effort.

Normally, the *Owens–Wendt–Rabel–Kaelble* (*OWRK*) model is suited for organic electronics. This assumes that only van der Waals and dipole–dipole interactions can take place, splitting the surface tension σ into a disperse and polar part σ^{d} and σ^{p}. The following calculations are based on this but can be modified to take different interactions into account. To find out the substrate's surface tension σ_{s}, the contact angles of test fluids with different surface tensions σ_{f} are needed.

$$\sigma = \sigma^d + \sigma^p \tag{4.1}$$

Work of adhesion: $$W_{sf} = 2\left(\sqrt{\sigma_s^p \sigma_f^p} + \sqrt{\sigma_s^d \sigma_f^d}\right) \tag{4.2}$$

and *Young–Dupré equation:* $$W_{sf} = (1 + \cos\theta) \cdot \sigma_f \tag{4.3}$$

with contact angle θ lead to

$$\sqrt{\sigma_s^p} \cdot \sqrt{\frac{\sigma_f^p}{\sigma_f^d}} + \sqrt{\sigma_s^d} = \frac{(1+\cos\theta)\cdot\sigma_f}{2\sqrt{\sigma_f^d}} \tag{4.4}$$

The parameters of at least two test fluids—a polar one such as water and a nonpolar one such as diiodomethane—with the corresponding contact angles allow a linear fit, yielding the substrate's surface tension.

To measure the surface tension of a fluid, different test substrates could be used. However, it is easier to measure the fluid's total surface tension and its contact angle on a nonpolar material like polytetrafluoroethylene (PTFE). The surface tension can then be calculated as

$$\sigma_f^d = \frac{\sigma_f^2}{4\sigma_s^d}(\cos\theta + 1) \text{ and } \sigma_f^p = \sigma_f - \sigma_f^d \tag{4.5}$$

In general, only two parameters are needed to calculate the surface tensions of solids and fluids, the total surface tension of fluids and their contact angle on solids. One method for getting the former is to analyze the form of a *hanging (pendant) drop* attached to a hollow needle. The height-dependent pressure within this droplet is counteracted by its surface tension. This can be determined with a fit to the droplet's shape if its density is known. The main advantage of this method is the small amount of fluid needed—a few droplets.

Another way is to pump gas through a submerged, hollow needle to grow bubbles and to measure the pressure within them. This is proportional to the curvature, which is maximized when the droplet forms a half sphere. Its radius is then defined by the needle diameter and allows the surface tension's calculation. With this method the age of the surface can be varied via the air flow, allowing the analysis of dynamic effects.

A fluid's surface tension can also be determined by measuring the force needed to pull a perfectly wetted sample out of it. Typically, rings (*Du Noüy ring*), rods, plates (*Wilhelmy plate*), or tubes of platinum, which has an extremely high surface tension, are chosen for this. This method can also be used to determine contact angles; the force acting on a not-perfectly-wetted substrate will be equal to that on platinum times the cosine of the contact angle.

A more straightforward method for doing this is to put a droplet on a surface, take an image of it at a low angle, and use a suitable fit function to determine the slope at the edge of the droplet.

4.2 Structural Characterization

For any electronic device there are certain demands on the quality of its layers. To meet these, one should know what kinds of surface structures can occur and how to decrease or enhance them. The fluid dynamic effects causing them and the influencing parameters are described in Chapter 3. This section gives a wider but less detailed overview and focuses on how to recognize them. The effects are grouped into those relevant for the lateral extent, the layer thickness and the morphology. After their description follows that of appropriate measurement techniques.

4.2.1 Lateral Extent

The most straightforward characteristic is a printed layer's lateral extent. In evaporated layers the desired pattern is defined by masks. With these it is relatively simple to achieve a high precision, since the movement of deposited molecules is strongly limited. Reproducing printed layers is more complex, since they are not deposited as perfectly even coatings and because they can and will flow before drying. While the latter is needed to compensate the former, it can also lead to a distortion of the layers and their edges. A change of a layer's outline can have disastrous results, e.g. by causing the bottom and top contact of an organic light-emitting diode (OLED) to touch, shorting the device.

To reduce deviations from the wanted form as far as possible and to assess how the layout has to be adapted to take the remaining defects into account, thorough control is necessary. This is problematic because the size of printed structures can vary strongly. For example, pixels for displays can go down to a few micrometers, while lighting devices can reach several square decimeters. Fine structures with a large extent also exist, an example being silver grids that increase the sheet conductivity of contact layers, while retaining their transparency. It is therefore hardly possible to analyze all features of each produced device, especially for quality control in a production line. A combination of different techniques, mostly used on small sections of the devices, is needed to reliably and affordably assess their quality and adjust the production process.

Apart from registration errors, only a few defects can be identified from a layer's outline. One is the separation of very thin lines into droplets, which is caused by the fluid's surface tension and known as the *Plateau–Rayleigh instability*. If no coalescence occurs, printed dots can be seen directly. If it does occur, the dot's arrangement may still cause a visible waviness of the edges. Holes and shrinkage caused by dewetting, as well as spreading, which occurs mostly in thick layers of well-wetting materials, can also be detected.

4.2.2 Layer Thickness

Another obvious structural characteristic of printed layers is their thickness. This can strongly influence a device's performance, not only through the determination of the resistance and capacitance of its layers, but also through more complex interactions. The thickness of layers in OPV cells, for instance, has to be finely tuned; too thin layers cannot efficiently absorb light, while exceedingly thick ones hinder charge extraction. In vapor deposition the material is applied evenly over a large area with a controllable rate, leading to a precise and easily adjustable thickness. For printed layers it is not possible to accurately predict a layer's thickness. For example, for gravure printing it can be roughly estimated using the cell volume and the fluid's solid content, but not all fluid is actually transferred, and especially in simple printing testers the solid content can be changed by an irreproducible drying before the transfer to the substrate.

Additionally, printed layers tend to be inhomogeneous. As in any deposition technique, there is a *roughness* (structures with a very short wavelength, below 1 μm) but there is also a *waviness* (inhomogeneities stemming from fluid dynamic processes or, for example, vibrations of the printing machine, with wavelengths in the millimeter range). When printed layers are characterized, the focus normally lies on the waviness. This is what mostly dominates, cannot be reduced by diffusion on any relevant time scale, and can be comprehensively influenced by the printing process. The roughness is only relevant if the waviness is optimized and can often be reduced by heating.

Several types of defects are common for printing. In a stack each layer can be affected by any of them, making an analysis of all layers very complex. Ideally, each layer should be analyzed separately by printing it on a smooth or already measured layer of the underlying material. The cross section of a stack can also be examined, but for recording all irregularities many cross sections are needed.

In spite of the complex physics behind the defect formation in printed layers, identifying them is mostly quite simple. If the spacing of deposited droplets is too large or the fluid dries too fast for them to coalesce and level, droplets or undulations that are neatly arranged in accordance to the printing form can be seen. For example, in inkjet-printed layers the structure may resemble stacked pancakes. *Viscous fingering* can occur in conventional printing. In this case the undulations form lines parallel to the printing direction (see Fig. 4.2). Raised edges of layers are due to the *coffee stain effect*, shown in Fig. 4.3, which is a result of contact line pinning and can be regulated using mixtures of solvents and by changing the drying conditions.

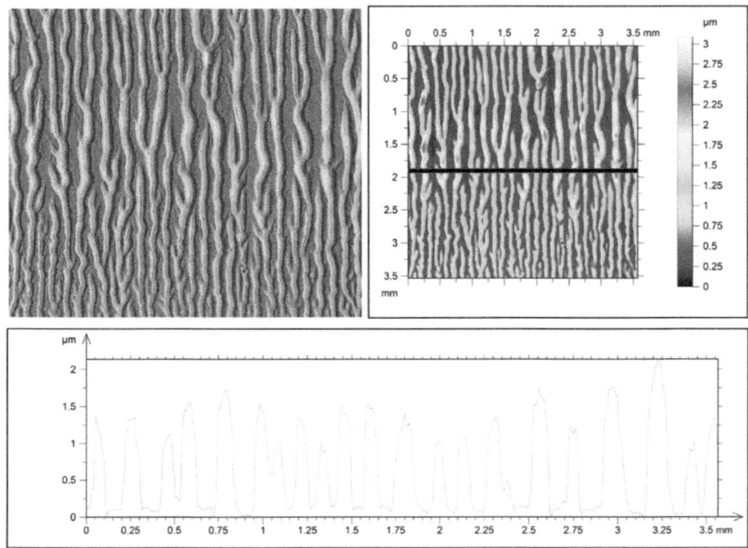

Figure 4.2 Viscous fingering. Microscopic image, topographic image, and height profile.

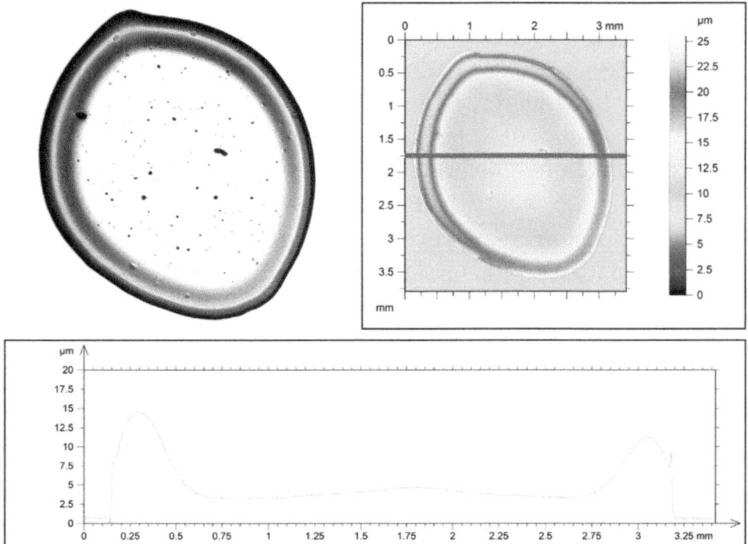

Figure 4.3 Coffee stain effect in a tea stain. Microscopic image, topographic image, and height profile.

Impurities like dust, oil, or structural defects of the substrate, as well as deliberate surface modifications, can lead to pinholes and trigger dewetting (see Fig. 4.4)—this is called *heterogeneous nucleation*. Even small pinholes can cause severe problems by shorting a device. In very thin layers intermolecular forces can cause *spinodal dewetting*. Here, the layer forms undirected and homogeneously distributed undulations, which can culminate in randomly arranged droplets.

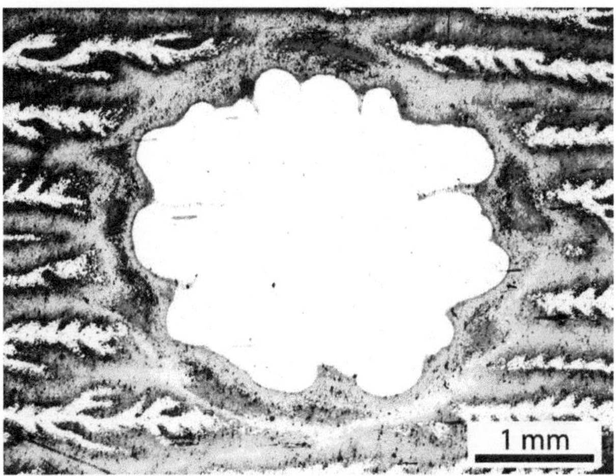

Figure 4.4 Dewetting in a layer with strong viscous fingering.

A combination of defects is also possible. An example for this is shown in Fig. 4.5, where the *Plateau–Rayleigh instability* has caused a bulging of ridges created by viscous fingering in a gravure-printed layer. The pattern of the gravure cylinder can be seen on the right side due to lack of printing fluid.

To optimize a printing process it is important to recognize the appearing defects' causes. To simply decide which of two layers with different defects is better, or when a layer is good enough, a statistical analysis of the surface is more useful. For this it should be clear what is analyzed; all statistical techniques can be used on the *primary* values (P_x), as well as on the *roughness* (R_x) and *waviness* (W_x). For the latter two there is no absolute definition. They can be obtained by simply using a high- or low-pass filter on the primary data, but the cutoff wavelength can be (more or less) freely chosen.

In addition to this, extremely short wavelengths may be cut from the roughness profile to get rid of noise, and for example, extremely long wavelengths may be cut from the waviness profile to separate it from the layer's form. To make things even more ambiguous, the statistics may be applied on 3D data and line scans taken in different directions. Furthermore, there is no standard measurement length or area. This means that it is crucial to note how the analyzed plots were measured and modified to understand what a value means.

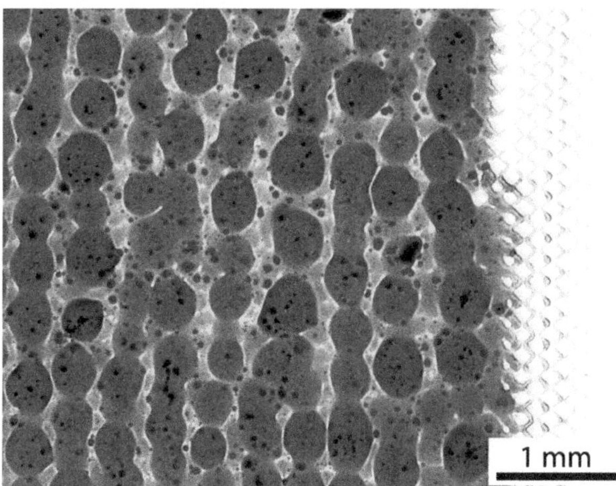

Figure 4.5 Multiple defects in a gravure-printed layer.

Any of a plethora of methods can be used for the characterization of a given profile. The most common are the average distance from the mean line P_a and the root mean square value P_q or P_{rms} here defined for a primary profile of length L with distance to the mean line $\Delta Z(x)$:

$$P_a = \frac{1}{L}\int_0^L |\Delta Z(x)|dx \tag{4.6}$$

$$P_q = P_{rms} = \sqrt{\frac{1}{L}\int_0^L \Delta Z(x)^2 dx} \tag{4.7}$$

Note that a conversion between these values for an arbitrary profile is not possible. They also do not distinguish between many low and a few high peaks and thus more data are needed for a layer's characterization. Shorts can appear if, for example, an insulating layer's maximum valley depth P_v equals its thickness or if the maximum peak height P_p of a conductive layer exceeds the subsequent layer's thickness. Several standards exist that take into account multiple peaks, valleys, or their differences, but their usefulness for organic electronics is limited, since a single peak can be enough for a device not to work. It can also be useful to view the profile's *amplitude distribution function* (*ADF*)—a histogram showing what percentage of the surface lies at which height. Integrating this, beginning at the highest value, yields the more often used *bearing ratio curve*. Simply put, this shows how much material lies above a certain height. Two more often used parameters are skewness P_{sk} and kurtosis P_{ku}:

$$P_{sk} = P_{rms}^{-3} \frac{1}{L} \int_0^L \Delta Z(x)^3 \, dx \tag{4.8}$$

$$P_{ku} = P_{rms}^{-4} \frac{1}{L} \int_0^L \Delta Z(x)^4 \, dx \tag{4.9}$$

P_{sk} shows how asymmetric the profile is; for $P_{sk} < 0$ most of the profile lies above the average value, indicating holes. Smooth layers with a few peaks will have a positive skew. P_{ku} is a measure of the profiles' peakedness. A normal distribution will have $P_{ku} = 3$; a sharper profile will have a higher value.

The challenging part when using these parameters and curves is to determine their weighting and acceptable limits. This must be done for each layer in each device separately, since the demands on these layers can differ strongly. Afterwards rating layers and deciding which ones to cull becomes quite easy—easy enough for an automatic evaluation.

4.2.3 Morphology

The layers' morphology, that is, the arrangement of molecules within them, is another characteristic that strongly influences the functionality of a device, mainly by changing the charge transport. Just like in evaporated samples the molecules can be arranged in partially crystalline domains, they can have a preferential orientation, and the layer can be amorphous, with an arbitrary molecular arrangement. In general, a higher molecular order leads to a narrower distribution of energy levels and thus higher conductivity, while disorder in the structure causes a broader distribution of energy levels. This can split the energy barriers at interfaces into smaller ones and thus ease charge carrier injection.

In evaporated layers a preferential direction is mainly caused by interactions with the substrate. In printed layers printing and drying forces can strongly influence this, for instance, chain molecules can be aligned by the shear flow in the nip. In addition to this, a freshly printed, wet layer can (partially) dissolve the underlying layer(s), leading to intermixing at the interface. This can have various effects, for example, a longer lifetime and changed efficiency in OLEDs, as well as a higher efficiency in OPV cells. In organic field-effect transistors (OFETs) a rough or intermixed interface between the gate insulator and the semiconducting material can strongly decrease the performance. To reduce this, for example, orthogonal solvents or cross-linking—the polymerization of the underlying material—can be used. Among others, the solvents or precursors, air, reaction products and interdiffused molecules from adjacent layers can also remain as impurities in the dried layers. They can strongly influence the device performance, for instance by degradation or the formation of trap states and grain boundaries.

4.3 Structural Measurements

To detect the characteristics presented in the previous section, suitable measurements are needed. Due to the inhomogeneity of printed layers, some techniques that are common for evaporated samples, such as ellipsometry, may be unsuitable for their characterization. In this section methods that can deal with undulating surfaces are listed. We do not give a detailed explanation of their intricacies—for this we refer to the literature cited in the "Suggested Readings" section. Optical techniques and tactile methods are presented separately. Each has different advantages and disadvantages and each can lead to ambiguous results. For the analysis of a new system it is therefore best to use at least one of each and to compare the results. Almost all listed techniques measure a sample's surface profile; therefore edges created by wiping or scratching are necessary to determine layer thicknesses. Otherwise the irregularities of the substrate may distort the results.

4.3.1 Optical Methods

Optical measurements are apt for areas ranging from microns to meters. They can also be nondestructive—for wavelengths in the visible range or longer and if no artificial edges are needed—and fast.

For measuring the outlines of a structure, for example, to determine a layer's size, a "normal" microscope can suffice. These are divided into *simple microscopes*, with a single lens or array of lenses, and *compound microscopes*, containing at least two (*objective* and *eye piece*). Their diffraction-limited resolution rises with the *numerical aperture* (*NA*) of objective and *condenser* (illuminating the sample) and decreases with increasing wavelength. The NA is defined by the acceptance cone's half-angle Θ and the immersing medium's refractive index n:

$$NA = n \sin \Theta \tag{4.10}$$

It typically does not exceed 0.95 (if an immersion fluid cannot be used).

The resolution may also be limited by distortions caused by the objective. Objectives are therefore categorized by how well these distortions have been avoided. The categories are (sorted by increasing image quality): achromats, fluorites (semi-apochromats), and apochromats. As the image quality rises, so does typically the NA. Most objectives are also marked as "Plan," meaning that they project the image on a plane and can therefore be combined with a camera.

The advantages of a compound microscope are its relative simplicity, correct color rendering, and rather low price. With suitable illumination and optical filters fluorescent molecules can be mapped; with polarizing filters birefringence can be found and used to detect crystallites.

Confocal microscopy allows 3D mapping. Here, two pinholes are used to illuminate only one dot in a defined height and to detect only light reflected from this dot. A pixel of an image sensor may serve as the second aperture. By moving the pinholes, an image of one plane can be taken; by varying the plane's height and determining for each point the height at which it is brightest, the contours of a sample can be determined three-dimensionally.

The lateral resolution can be higher than that of a standard microscope. The height of the structures that can be analyzed with this technique is limited by the objective's working distance and can reach several centimeters. For high magnifications with a high NA, the vertical resolution can reach 1 nm. Low-magnification lenses normally allow a resolution in the range of tens to hundreds of nanometers, providing only a rough overview of a printed sample.

A sample's profile can also be determined with *interferometry*. Here, the light is split into two beams, one to a reference mirror and one to the sample, and then combined again, leading to interference. *Vertical scanning interferometry* (*VSI*) uses low-coherence-length white light. The distance between sample and objective is varied and the brightness of each point measured. The interference is strongest for equal optical lengths of both beams and decreases fast with a rising difference, so the height of each point on the surface can be measured by maximizing the fringe contrast between constructive and destructive interference. For *phase shift interferometry* (*PSI*) monochromatic light is used. With three images between which the phase of the reference beam has been shifted by a known amount, the phase relation at each point can be determined. This can be used to determine the sample's height profile.

Like confocal microscopy, the VSI mode can be used for very high structures. The lateral resolution is worse, but the vertical resolution is good—about 1 nm—and independent from the magnification, and the measurement speed is much higher. In the PSI mode the vertical resolution is even better—steps below 0.1 nm can be detected—but the method is only unambiguous if the analyzed structure's height does not exceed a quarter of the light's wavelength. The analyzed material can influence the phase relation, causing errors for all interference-based techniques if the heights of multiple materials are compared.

The maximum slope that can be measured is limited for all optical techniques and dependent on the NA. Often the limit is approximately 70° for confocal and 30° for interference microscopy, meaning that the former is best for rough samples. There are, however, ways to overcome these limits. Some microscopes based on VSI can analyze almost vertical flanks and are thus well suited for rough samples.

Another optical technique is *X-ray photoelectron spectroscopy* (*XPS*), where X-rays are used to eject electrons from the analyzed sample's surface. The surface's composition can then be determined by measuring the electron's energy distribution. This can be combined with sputtering to abrade a smooth sample and thus measure a depth profile—unfortunately, a quite complex and error-prone process.

Instead of light one can use electrons, as in *transmission electron microscopy* (*TEM*), where thin slices of a sample can be analyzed with atomic resolution. By measuring the energy of the detected electrons it is possible to map certain atoms and molecules. With a rotating sample this can be extended into tomography, yielding a detailed 3D image.

In *scanning electron microscopy* (*SEM*) an electron ray probes the surface of a sample. In this technique there is a lot that can be analyzed. *Backscattered electrons*, for instance, stem from the probing *primary* beam and are simply reflected from the sample. The probability of an electron being backscattered rises with the mass of the probed atom, so that different materials can be distinguished by the image's brightness. *Secondary electrons* stem from the sample and are directly caused by the primary ray. Due to their low energy they cannot pass thick layers. This means that only those emitted at the surface can be detected and therefore they can be used to analyze the sample's topography. If an electron has been emitted from an inner shell and the vacancy is filled by an electron from an outer shell of the atom, the energy release can cause an emission of another electron from an outer shell—an *auger electron*. These can be used for the chemical analysis of the surface and, combined with sputtering, to create a 3D map of the sample's composition. Again, this is not easy. In addition to electrons, X-rays and luminescence, the current caused by the electron beam can be used for further analysis. In short, SEM is an extremely powerful method, yielding information about the sample's topography, composition and morphology. Of course, it is also very complex and expensive, especially as it is often coupled to several other analysis tools. Furthermore, insulating samples are problematic, since charges can accumulate on them and disturb the measurement.

4.3.2 Tactile Methods

A fundamentally different approach to measure height profiles is to use tactile methods. They are independent from the analyzed layer's optical properties and can achieve a higher resolution. Unfortunately, they also offer much lower measurement speeds and smaller scanning areas and may damage the sample. This can be checked by measuring twice in the same position, ideally scanning over a larger area in the second go.

The simplest tactile method, called *stylus profilometry*, is to drag a stylus connected to a cantilever over the surface of a sample and to measure its deflection. With this, a line profile can be acquired. By repeating this for an array of lines the 3D surface topography can be determined.

The cantilever is often connected to a ferromagnetic rod that extends into the center of a coil. A change in the rod's position causes an easily measurable change in the coil's inductance, which can then be used to calculate the cantilever's tilt. Other systems use, for example, capacitance or the deflection of a laser beam reflected from the cantilever. The vertical resolution of these systems is often in the range of several angstroms (Å). The lateral resolution is limited by the needle's diameter and normally between tens of nanometers and tens of micrometers. It is also limited by the measurement speed. The measured lines can be relatively long—mostly up to several centimeters—and most machines offer a maximal vertical measurement range of ca. 1 mm. The force pressing the stylus against the substrate can be set, 0.1 mN being a typical value. A major problem is the measurement of soft samples; the stylus tip can engrave a trench into the sample and thus distort the results.

This can be avoided with *atomic force microscopy* (*AFM*). An atomic force microscope also contains a cantilever with a very fine tip. Piezo actuators can drive it in a grid over the sample, while its deflection is measured. An atomic force microscope normally has a higher resolution than a stylus profilometer but can only measure areas with side lengths in the order of a hundred microns and heights in the micron range. It can be used as a stylus profilometer that scans over areas. It is also possible to induce a vibration in the cantilever and measure the resonance frequency. This depends on external forces, which depend on the cantilever's height over a given material, so measuring the height profile that leads to a constant resonance frequency during the surface scan leads to the sample's profile. There are two modes of vibrational operation, a *noncontact* (*NC*) *mode*, where the tip is kept above the surface, and a *tapping mode*, where the tip touches the sample when swinging downward. The NC mode does not cause any degradation of sample and tip and is perfect for soft samples. With a tip that is functionalized with a single molecule it can reach atomic resolution. In ambient air, water can form a film on the sample's surface. It can be penetrated in the tapping mode to measure the actual surface. Degradation is possible, but with correct settings soft samples can still be analyzed.

Making correct AFM images is not easy. For instance, the forces acting on the stylus depend on the sample's composition. This means that if the scan goes over different materials, the measured profile can be wrong. In some cases the control system can even be misled to the point that the tip smashes into the sample, damaging it and breaking the brittle cantilever. In addition to this, the radius of the tip limits the resolution and the measurable slope, so the atomic force microscope probe has to be replaced before the occurrence of any appreciable wear. It is also vital that the tip is kept clean, since small amounts of dirt clinging to it suffice to ruin a measurement.

A stylus profilometer is useful to measure the waviness and height of a printed layer, AFM is better suited to measure the overlying roughness that has a very short wavelength. By using conductive tips in an atomic force microscope, *Kelvin probe* measurements can be conducted, that is, the work function of the sample can be plotted. For this, an alternating current (AC) with the cantilever's resonance frequency and a direct current (DC) signal is applied. When the DC potential equals the difference of the work functions of tip and surface, there is no force acting with the resonance frequency. Thus the work function can be measured by keeping the cantilever's vibration minimal. If a wedge is cut into a sample (with a *focused ion beam, FIB*), different layers within it can be analyzed by the Kelvin probe technique. A nonvibrating conductive tip whose height can be regulated with precise piezo actuators can be used to map tunneling currents (*scanning tunneling microscope, STM*). The tip can be scanned at a constant height over the sample, which leads to a varying current that can be used to assess the sample's profile. The main advantage of this is high speed; the main disadvantages are the risk of crashing into the sample and a comparatively low resolution. For a given material the current depends on its distance to the tip, so keeping this constant by adjusting the height of the tip yields the sample's profile more directly. In this mode the resolution is high enough to resolve molecular orbitals, though this, of course, cannot be achieved easily. One can also scan a tip with an aperture over the sample. This can be used for optical analysis in the nearfield (*near-field scanning optical microscopy, NSOM* or *SNOM*), which can reach molecular resolution.

4.4 Electrical Properties of Printed Layers and Their Characterization

A very important characteristic of organic electronic layers is, obviously, their capability to conduct an electric current. Especially for high-frequency operations the capacitance and inductance are also relevant. This section deals with several methods for the electrical characterization of a sample and with the problems that can arise when using them.

A conductor's most basic parameter is its *conductivity*:

$$\sigma = n\mu q \tag{4.11}$$

with charge carrier density n mobility μ and charge q. Note that there are two types of charge carriers in organic semiconductors—electrons and holes—with separate mobilities and densities. Both can be dependent on the stack used in the device, the occurring voltages, the temperature, etc. If σ is known, the resistance R of a conducting rod with a cross-sectional area A and length l can be calculated as

$$R = \frac{l}{\sigma A} \tag{4.12}$$

With this, for example, the thickness of a conductive layer needed for a desired resistance in any direction can be calculated. While for instance the active layers of an OLED mostly only conduct a current vertically through the layer, contacting layers typically need to conduct along it. Thus, a different parameter is more convenient—sheet resistance:

$$R_s = \frac{1}{\sigma t} \tag{4.13}$$

with thickness t. This makes the calculation of the resistance for a current through a conductor with length L and width W very simple:

$$R = R_s \frac{L}{W} \tag{4.14}$$

Note that the unit of R and R_s is the same—Ω. To differentiate them, the unit of R_s is expressed as ohms per square: Ω/sq. or Ω/□.

Sheet resistance can be measured by a *four-point probe*. This typically consists of four equidistant contacts in a line. The outer contacts are connected to a current source, while the inner contacts are used to measure the induced voltage. Sheet resistance can be calculated by

$$R_s = \frac{\pi}{\ln 2} \frac{V}{I} \tag{4.15}$$

with injected current I and measured voltage V. However, this is only valid for thin, homogeneous, and infinite sheets. For instance, the error for a square whose side length is 10 times the electrode distance is ca. 10%. Correction factors can be calculated.

The contacts can also be placed on the edges of an arbitrarily shaped sheet. A current is induced between two neighboring contacts (I_{12}), while the voltage between the other two (V_{34}) is measured. This is repeated for a different pairing (I_{23} and V_{41}). R_s can then be calculated by

$$\exp\left(-\frac{\pi V_{34}}{I_{12} R_s}\right) + \exp\left(-\frac{\pi V_{41}}{I_{23} R_s}\right) = 1 \tag{4.16}$$

This is known as the *van der Pauw method* and only works correctly for ohmic contacts at the very edge. Repeatability can be checked by switching the contacts providing the current and measuring the voltage. Correction factors for thick layers and nonideal electrodes exist.

Contacting layers is an issue. For hard, inorganic semiconductors these methods can be easily applied. For organic semiconductors that can be more complex because of the material's softness; the measuring tips can penetrate the layer. Bent tips or evaporated contacts can resolve this.

Semiconducting layers that are already integrated in a device are more difficult to characterize. For one, the direction of the current is not the same for all layers in all devices. In addition to this, the conductivity of a semiconducting layer depends on the surrounding materials and fields. It is best to analyze the mobility, e.g. by using the *Hall effect*, and the charge carrier distribution of the layers in the finished device.

A layer's electrical properties do not depend on its resistivity alone. They can be fully described using the *impedance* Z consisting of a real and imaginary part. The real part corresponds to the resistance R, which leads to energy being dissipated, for example, as heat or light. The imaginary part becomes relevant when the voltages and/or currents in a system change over time, for example, for a sinusoidal signal. It stems from energy being stored in an electric or magnetic field and regained later on. This leads to a phase shift between current and AC voltage. An electrical component that stores energy as a magnetic field is called an inductor, and one that uses electric fields is referred to as a capacitor. They can be assigned an inductance L and a capacitance C. These lead to the impedances:

$$Z_L = i\omega L \tag{4.17}$$

and

$$Z_C = \frac{1}{i\omega C} \tag{4.18}$$

The inductance is relevant in elongated structures (antennas, coils) with a low resistivity and can therefore often be neglected in organic electronics. Capacitances are relevant for large, thin layers with low conductivities and thus for organic electronics. The capacitance C of a homogeneous and thin layer can be calculated as follows:

$$C = \varepsilon_0 \, \varepsilon_r \frac{A}{t} \tag{4.19}$$

with area A, thickness t, vacuum permittivity $\varepsilon_0 \approx 8.854 \cdot 10^{-12}$ F/m, and relative permittivity ε_r. The capacitance can strongly influence the performance of many devices. For example, the switching speed of OFETs strongly depends on it, since the gate and the semiconductor form a capacitor that has to be charged for the transistor to switch.

Capacitance can also be used to determine the thicknesses of smooth layers with known area and (frequency-dependent) permittivity. While capacitance is much better suited for thickness measurements than resistance—it is only weakly changed by charge injection—it can still be misleading, for instance if the layer is porous.

The advantage of using capacitance to determine thicknesses is that it is nondestructive. In addition to this, sublayers such as depletion regions in OLEDs, can be analyzed. An appropriate equivalent circuit is needed for this. In an ideal sample each layer can be represented by a resistor and capacitor in parallel (RC element) and multiple layers as several RC elements in series. By measuring the impedance spectrum, the capacitances of all layers and thus their thicknesses can be determined along with their conductance.

To analyze the complex processes in all the layers of a device, special test devices may be needed. To measure the flow of one type of charge carrier, for instance, it can be necessary to use blocking layers for the other type of charge carriers at both ends of a stack. A completely blocking contact is also needed for *charge extraction in a linearly increasing voltage* (*CELIV*). This is exactly what it sounds like: A linearly increasing voltage is applied to a device with one blocking contact and the resulting current is measured. This allows the analysis of charge carriers in equilibrium and yields, for example, their mobility. A modification of this is called *photo-CELIV*, where a photonic pulse is used to excite charge carriers just before the measurement is started. Another way to determine mobilities is to make OFETs from the material of interest and characterize them. In addition to this, the already mentioned Kelvin probe technique can be used for the electrical characterization of a device.

4.5 Conclusions

For reproducible printing results the properties of the printing fluid and substrate have to be controlled. Analyzing these results and understanding the causes for defects is necessary for optimization.

Important parameters are viscosity and surface tension. Both depend on different parameters—for instance, temperature and time. This has to be taken into account for any calculations and for any measurement routine.

Structural characteristics can be grouped into those of the outline, thickness, and morphology. Most problematic for printed layers are those of the outline and thickness, where many more defects can occur than in evaporated layers.

Many different techniques can be used for the structural characterization, all of which are prone to different kinds of errors. It is therefore vital to compare the results of several techniques based on different principles.

A device's look is not the most important thing; its functionality is. This can of course be easily tested, but to find the causes for problems it is necessary to analyze each layer separately. Important parameters are, for example, conductivity and capacitance.

Exercises

4.1 Fluid and substrate characterization
   (a) What is dynamic and kinematic viscosity?
   (b) Define a Newtonian fluid.
   (c) List two techniques for measuring viscosities.
   (d) What is the Owens–Wendt–Rabel–Kaelble (OWRK) model used for? What assumptions are made in it?
   (e) Which fluid characteristic is determined if you put a thin ring into a fluid and measure the force you need to pull it out?

4.2 Structural characterization
   (a) List four typical defects in printed layers.
   (b) What defect is often found when printing thin lines? What is its cause?
   (c) What is the difference between waviness and roughness?
   (d) What is P_a? What is P_{rms}? Can you convert one value into the other? If so, how?
   (e) A layer that you have printed should be approximately 1 μm thick. You could use an atomic force microscope, a confocal microscope, an interferometric microscope in the PSI mode, or a tactile profilometer to measure the exact value. Which techniques are suitable?
   (f) What is the difference between the working principles of AFM and SEM?

4.3 Electrical characterization
   (a) What is sheet resistance and when is it useful?
   (b) What technique allows the measurement of the equilibrium charge mobility in a sample?
   (c) How could you electrically determine a semiconducting layer's thickness?

4.4 Printing a plate capacitor

   You want to print a plate capacitor consisting of the following layer stack: metallic electrode/insulating polymer/metallic electrode.

(a) What important single-layer characteristics should you measure?
(b) How can you do that?
(c) What defects of the conducting layers should you beware of?
(d) What about the insulating layer?

Suggested Readings

Coating was only briefly mentioned in this chapter's introduction. Two recommendable books on this have been written by Gutoff and Cohen and a third one by Tracton.

More information on viscosity and its measurement can be found in the works of Schramm, Irgens, and Macosko. Special consideration on instabilities occurring in rheometers has been given by Larson.

All the important models for the calculation of contact angles are described by Żenkiewicz and Good. Bonn et al. and de Gennes present a host of wetting phenomena. A broader description of interfaces is given by Butt. Arashiro and Demarquette give examples of how pendant drops can be analyzed, while Stalder et al. do the same with sessile drops. Fainerman and Miller give a short description of the maximum bubble pressure method. A rough overview of many types of surface tension measurements is also given in the freely available glossary offered by Krüss, a manufacturer of measurement devices.

Typical instabilities of printed lines have been described by Nguyen et al. and Duineveld. A broader overview of defects of a layer's outline is given by Bonn. Literature on the coffee stain effect has been written by Deegan et al., Berteloot et al., and Majumder et al. Bornemann et al. give approximate formulas for the evolution of viscous fingering and spinodal dewetting, as well as for the leveling time of a thin layer. A more complete description of different types of dewetting has been given by Bäumchen and Jacobs as well as Craster and Matar, the latter also covering a plethora of other instabilities of thin films.

It is much easier to find literature about characterization techniques than about the characteristics of printed layers themselves. Good resources for the basics of optical microscopy are the homepages of microscope manufacturers (e.g., Zeiss, Nikon, Olympus, and Leica), which also offer many interactive tutorials, and the book by Murphy and Davidson. A thorough description of interferometry is given by Malacara, and papers written by Dubois, Yatsenko, and McWaid may help deal with the problem of material dependent phase shift. Tactile methods are covered by the books Morita and Kaupp. Leng and Yao and Wang give a less thorough overview of all kinds of microscopy, while Heath has written a dictionary with extremely brief explanations of thousands of relevant terms.

Schroder goes beyond microscopy, covering all kinds of techniques that are needed for the complete characterization of semiconductors. More about the electrical properties of organic electronics can be found in the works of Brütting and Adachi, Schwoerer and Wolf, and Baranovski. The four-point probe and its pertaining correction factors have been described by the Haldor Topsoe Semiconductor Division and Swartzendruber. The van der Pauw method has been described by van der Pauw himself (introduction), Ramadan (correction of errors caused by shape of sample), Chwang (correction of errors caused by contacts), and Weiss. Lvovich has written a useful book on impedance spectroscopy, covering many aspects of it, while Nowy's paper gives a concise example of how it can be used for organic electronics. For information on the CELIV technique we refer to Juška and Lorrmann.

Arashiro, E. Y., and Demarquette, N. R. (1999). Use of the pendant drop method to measure interfacial tension between molten polymers. *Mater. Res.*, **2**, 23–32.

Baranovski, S. (2006). *Charge Transport in Disordered Solids with Applications in Electronics*, Wiley.

Bäumchen, O., and Jacobs, K. (2010). Slip effects in polymer thin films. *J. Phys. Condens. Matter*, **22**, 033102.

Berteloot, G., Hoang, A., Daerr, A., Kavehpour, H. P., Lequeux, F., and Limat, L. (2012). Evaporation of a sessile droplet: inside the coffee stain. *J. Colloid Interface Sci.*, **370**, 155–161.

Bonn, D., Eggers, J., Indekeu, J., Meunier, J., and Rolley, E. (2009). Wetting and spreading. *Rev. Mod. Phys.*, **81**, 739–805.

Bornemann, N., Sauer, H., and Dörsam, E. (2011). Gravure printed ultrathin layers of small-molecule semiconductors on glass. *J. Imaging Sci. Technol.*, **55**, 40201-1–40201-8.

Brütting, W., Adachi, C., and Holmes, R. J. (2012). *Physics of Organic Semiconductors*, Wiley.

Butt, H.-J., Graf, K., and Kappl, M. (2013). *Physics and Chemistry of Interfaces*, Wiley-*VCH*.

Chwang, R., Smith, B. J., and Crowell, C. R. (1974). Contact size effects on the van der Pauw method for resistivity and Hall coefficient measurement. *Solid-State Electron.*, **17**, 1217–1227.

Cohen, E. D., and Gutoff, E. B. (1992). *Modern Coating and Drying Technology*, Wiley-*VCH*.

Craster, R. V., and Matar, O. K. (2009). Dynamics and stability of thin liquid films. *Rev. Mod. Phys.*, **81**, 1131–1198.

De Gennes, P.-G. (1985). Wetting: statics and dynamics. *Rev. Mod. Phys.*, **57**, 827–863.

Deegan, R. D., Bakajin, O., Dupont, T. F., Huber, G., Nagel, S. R., and Witten, T. A. (1997). Capillary flow as the cause of ring stains from dried liquid drops. *Nature*, **389**, 827–829.

Dubois, A. (2004). Effects of phase change on reflection in phase-measuring interference microscopy. *Appl. Opt.*, **43**, 1503–1507.

Duineveld, P. C. (2003). The stability of ink-jet printed lines of liquid with zero receding contact angle on a homogeneous substrate. *J. Fluid Mech.*, **477**, 175–200.

Fainerman, V., and Miller, R. (2004). Maximum bubble pressure tensiometry: an analysis of experimental constraints. *Emuls. Fundam. Pract. Appl.*, **108–109**, 287–301.

Glossary: KRÜSS GmbH. (2014). Verfügbar unter http://www.kruss.de/services/education-theory/glossary.

Good, R. J. (1992). Contact angle, wetting, and adhesion: a critical review. *J. Adhes. Sci. Technol.*, **6**, 1269–1302.

Gutoff, E. B., and Cohen, E. D. (2006). *Coating and Drying Defects: Troubleshooting Operating Problems*, John Wiley & Sons.

Haldor Topsoe Semiconductor Division (1968). Geometric factors in four point resistivity measurement. Bulletin No. 472-13, 2nd ed., p. 63.

Heath, J. P. (2005). *Dictionary of Microscopy*, John Wiley & Sons.

Irgens, F. (2014). *Rheology and Non-Newtonian Fluids*, Springer.

Juška, G., Arlauskas, K., Viliūnas, M., and Kočka, J. (2000). Extraction current transients: new method of study of charge transport in microcrystalline silicon. *Phys. Rev. Lett.*, **84**, 4946–4949.

Kaupp, G. (2006). *Atomic Force Microscopy, Scanning Nearfield Optical Microscopy and Nanoscratching: Application to Rough and Natural Surfaces*, Springer-Verlag.

Larson, R. G. (1992). Instabilities in viscoelastic flows. *Rheol. Acta*, **31**, 213–263.

Leng, Y. (2013). *Materials Characterization: Introduction to Microscopic and Spectroscopic Methods*, Wiley.

Lorrmann, J., Badada, B. H., Inganäs, O., Dyakonov, V., and Deibel, C. (2010).

Charge carrier extraction by linearly increasing voltage: analytic framework and ambipolar transients. *J. Appl. Phys.*, **108**, 113705.

Lvovich, V. F. (2012). *Impedance Spectroscopy: Applications to Electrochemical and Dielectric Phenomena*, Wiley.

Macosko, C. W. (1994). *Rheology: Principles, Measurements, and Applications*, Wiley-*VCH*.

Majumder, M., Rendall, C. S., Eukel, J. A., Wang, J. Y. L., Behabtu, N., Pint, C. L., Liu, T.-Y., Orbaek, A. W., Mirri, F., Nam, J., Barron, A. R., Hauge, R. H., Schmidt, H. K., and Pasquali, M. (2012). Overcoming the "coffee-stain" effect by compositional Marangoni-flow-assisted drop-drying. *J. Phys. Chem. B*, **116**, 6536–6542.

Malacara, D. (2007). *Optical Shop Testing*, Wiley.

McWaid, T. H., Vorburger, T. V., Song, J.-F., and Chandler-Horowitz, D. (1992). Effects of thin films on interferometric step height measurements. **1776**, 2–13.

Morita, S. (2006). *Roadmap of Scanning Probe Microscopy*, Springer.

Murphy, D. B., and Davidson, M. W. (2012). *Fundamentals of Light Microscopy and Electronic Imaging*, Wiley.

Nguyen, H. A. D., Lee, J., Kim, C. H., Shin, K.-H., and Lee, D. (2013). An approach for controlling printed line-width in high resolution roll-to-roll gravure printing. *J. Micromechan. Microeng.*, **23**, 095010.

Nowy, S., Ren, W., Elschner, A., Lovenich, W., and Brutting, W. (2010). Impedance spectroscopy as a probe for the degradation of organic light-emitting diodes. *J. Appl. Phys.*, **107**, 054501.

Ramadan, A. A., Gould, R. D., and Ashour, A. (1994). On the van der Pauw method of resistivity measurements. *Thin Solid Films*, **239**, 272–275.

Schramm, G. (1994). *A Practical Approach to Rheology and Rheometry*, Haake Karlsruhe.

Schroder, D. K. (2006). *Semiconductor Material and Device Characterization*, Wiley.

Schwoerer, M., and Wolf, H. C. (2008). *Organic Molecular Solids*, Wiley.

Stalder, A. F., Melchior, T., Müller, M., Sage, D., Blu, T., and Unser, M. (2010). Low-bond axisymmetric drop shape analysis for surface tension and contact angle measurements of sessile drops. *Colloids Surf. Physicochem. Eng. Asp.*, **364**, 72–81.

Swartzendruber, L. J. (1964). Correction factor tables for four-point probe resistivity measurements on thin, circular semiconductor samples, Technical Note 199. US National Bureau of Standards.

Tracton, A. A. (2005). *Coatings Technology Handbook*, CRC Press.

van der Pauw, L. J. (1958). A method of measuring the resistivity and hall coefficient on lamellae of arbitrary shape. *Philips Tech. Rev.*, **20**, 220–224.

Weiss, J. D. (2011). Generalization of the van der Pauw relationship derived from electrostatics. *Solid-State Electron.*, **62**, 123–127.

Yao, N., and Wang, Z. L. (2005). *Handbook of Microscopy for Nanotechnology*, Springer.

Yatsenko, L. P., Loeffler, M., Shore, B. W., and Bergmann, K. (2004). Interferometric analysis of nanostructured surface profiles: correcting material-dependent phase shifts. *Appl. Opt.*, **43**, 3241–3250.

Żenkiewicz, M. (2007). Methods for the calculation of surface free energy of solids. *J. Achiev. Mater. Manuf. Eng.*, **24**, 137–145.

Chapter 5

Organic Field-Effect Transistors

Henrique Leonel Gomes
Department of Electronics and Informatics, University of the Algarve, Campus de Gambelas, 8005-139 Faro, Portugal
hgomes@ualg.pt

This chapter aims to provide the reader with practical knowledge about electrical methods of measuring organic thin-film transistor (OTFT) devices. It presents a series of recipes that allow the experimentalist to gain insight into the performance and limitations of the devices and circuits being measured. It also gives guidelines on how to correctly interpret the measurements and to provide feedback to the manufacturing process. Organic-based transistors are particularly suited for sensing and biomedical applications. The operation of these devices is presented, with particular emphasis on the electrical techniques to address them when operating in complex liquid environments.

Organic and Printed Electronics: Fundamentals and Applications
Edited by Giovanni Nisato, Donald Lupo, and Simone Ganz

ISBN 978-981-4669-74-0 (Hardcover), 978-981-4669-75-7 (eBook)
www.panstanford.com

5.1 Introduction

In this chapter various aspects of the characterization and performance of OTFTs are considered. The chapter begins by first addressing individual transistors and afterward addresses the characterization of simple circuits such as logic gates and ring oscillators (ROs). At the transistor level, the emphasis is on how to (i) perform basic parameter extraction and quantify individual transistor performance, (ii) gain insight into the ability of the device to operate over long periods of time by measuring its operational stability, (iii) identify reliability issues that cause device failure, and (iv) identify and circumvent sources of variability in transistor parameters. At the circuit level, the focus is on how the nonideal behavior of individual transistors degrades circuit operation.

This chapter is organized in the following way. A brief history of the thin-film transistor (TFT) is presented on Section 5.2, together with the description of the typical TFT architectures and the basic operation mechanism. The differences between a silicon-based metal-oxide-semiconductor field-effect transistor (MOSFET) and an OTFT device are discussed in Section 5.3. The ideal TFT electrical characteristics are derived. Due to the presence of impurities TFTs show some nonideal characteristics. These are discussed in Section 5.4. Emphasis is given to the physical meaning of the extracted parameters. Recipes to identify nonideal effects caused by extrinsic factors such as impurities, parasitic metal contacts, and imperfect metal contacts are also presented and discussed. Operational stability deserves special attention. Different types of instabilities are briefly explained. It is shown how to quantify and benchmark operational stability. Section 5.4 also addresses variability characterization and draws attention to the importance of design layout that minimizes variability. Electronic active impurities or traps determine transistor performance and stability. Section 5.5 outlines the effects of traps on the electrical properties and describes techniques to study traps using the transistor as a tool. These include the temperature dependence of the field-effect mobility, thermal de-trapping experiments, and small signal impedance spectroscopy. Simple but important circuit types such as the inverter and the RO are presented in Section 5.6. OTFT applications in bioelectronics are discussed in Section 5.7. The major conclusions of this chapter are outlined in Section 5.8. This chapter ends with a list of questions to help the reader apply the concepts learned to some practical examples.

5.2 Field-Effect Transistors

A TFT is a device that uses an electric field to modulate the conduction of a channel located at the interface between a dielectric and a semiconductor. Therefore, it is a field-effect transistor (FET) similar to the well-known MOSFET, which is the basic building block of modern integrated circuits. The development histories of TFTs and MOSFETs are parallel in time. The TFT concept was patented in 1925 by Julius Edger Lilienfeld and in 1934 by Osker Heil, but at the time no practical applications had emerged. In the 1960s several device structures and semiconductor materials like Te, CdSe, Ge, and InSb were explored to fabricate TFTs. However, competition from the MOSFET based on silicon technology forced the TFT to enter a long period of hibernation. In the early 1970s the need for large-area applications in flat-panel displays motivated the search for alternatives to crystalline silicon and the TFT found its niche of application. In 1979, hydrated amorphous silicon (a-Si:H) became a forerunner as a semiconductor to fabricate TFTs. Since the mid-1980s, silicon-based TFTs have successfully dominated the large-area liquid crystal display (LCD) technology and have become the most important devices for active matrix liquid crystal and organic light-emitting diode (OLED) applications. In the meantime, TFTs based on organic semiconductor (OSC) channel layers were introduced in the 1990s with electron mobility equivalent to that of a-Si:H. Nowadays, OTFTs are the candidate for incorporation onto flexible substrates. A representative and particularly important example is full-color, video, flexible OLED displays. Small OLED displays on conventional glass substrates for mobile phone and personal desktop assistant (PDA) applications are rapidly growing and are displacing LCD screens in the small-display sector. OFET technology could be an ideal back plane for this application because of the close material compatibility between OLEDs and OFETs and their excellent mechanical properties. The application niche for

OFETs is not entirely defined. Smart cards, disposable electronics, and electronic skins are also currently under intense research.

5.2.1 Basic Operation of a TFT

A TFT is formed by placing thin films of a dielectric layer as well as an active semiconductor layer and metallic contacts onto a supporting substrate. Figure 5.1 shows a cross-sectional schematic drawing of a MOSFET and a TFT. As evident the substrate of a TFT is an insulating material, whereas the substrate of a MOSFET is a semiconductor material (p-type) of different doping as the source and drain diffusions (n-type). Source and drain contacts to the semiconducting channel material are injecting contacts to the channel in a TFT structure and are a p-n junction in a MOSFET.

The TFT and MOSFET operation is similar in that the current from the source to the drain terminal is modulated by the applied gate electric field. Current modulation in a TFT or in a MOSFET can be explained if the metal-insulator-semiconductor (MIS) part of the TFT is considered as a capacitor. A voltage applied between the metal and the semiconductor causes a charge to build up in the semiconductor and the metal gate.

The energy band diagram of an ideal MIS diode is given in Fig. 5.2 (for a p-type semiconductor). The diode is termed "ideal" because the bands are flat for zero applied voltage. This is the case when Eq. 5.1 is fulfilled.

$$\varphi_{\mathrm{m}} = -\left(\chi + \frac{E_{\mathrm{g}}}{2q} + \varphi_{\mathrm{b}}\right) = 0 \tag{5.1}$$

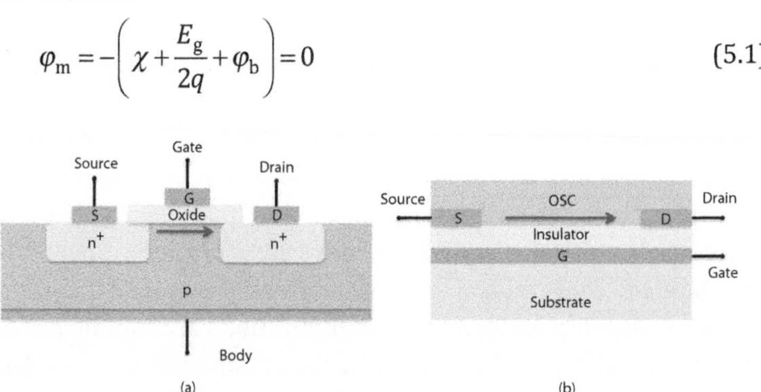

Figure 5.1 Cross-sectional schematic drawing of a MOSFET (a) and an OTFT (b).

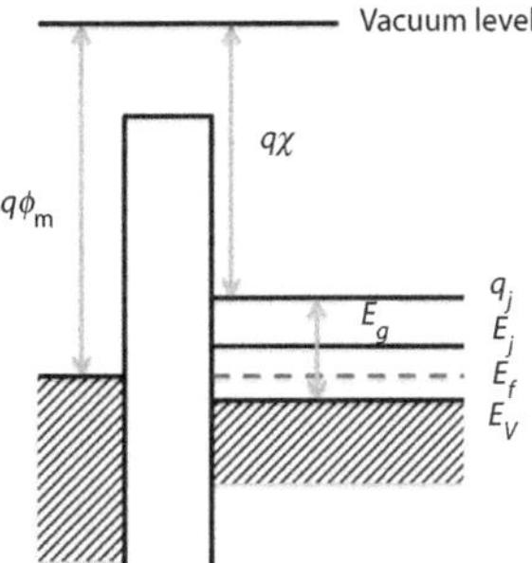

Figure 5.2 Band diagram of an ideal metal-insulator-semiconductor structure at equilibrium.

Here, φ_m is the metal work function, $q\chi$ is the electron affinity measured from the bottom of the conduction band E_C to the vacuum level, E_g is the semiconductor bandgap, q is the absolute electron charge, and φ_b is the potential difference between the Fermi level E_f and the intrinsic Fermi level E_i (which is located very close to the midgap) (in the nonideal case, a small band curvature exists at the insulator–semiconductor interface, and a small potential V_{fb}, the so-called flat-band voltage, must be applied to the metal to get the flat-band conditions.)

When the MIS diode is biased with positive or negative voltages, three different situations may occur at the insulator–semiconductor interface. For a negative voltage (Fig. 5.3a), the bands bend upward and the top of the valence band moves closer to the Fermi level, causing an accumulation of holes near the insulator–semiconductor interface. The interface is thus more conductive than the bulk of the semiconductor. When a small to moderate positive voltage is applied to the gate electrode, majority carrier holes are repelled from the insulator–semiconductor interface so that a depletion layer is formed (Fig. 5.3b). When a larger positive voltage is applied to the metal (Fig. 5.3c), the bands bend even more downward and the intrinsic level eventually crosses the Fermi level. At this point, the density of electrons exceeds that of the holes, and one enters the inversion regime. The inversion mode of operation is usually not observed in OFETs.

Depending on the gate voltage required to form an accumulation layer, a TFT can be classified as either an enhancement-mode or a depletion-mode device.

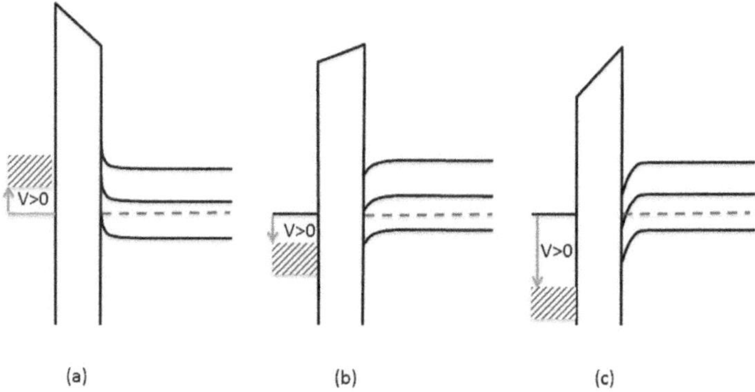

Figure 5.3 Band diagram of an ideal MIS structure under an applied bias. Accumulation (a), depletion (b), and inversion (c) regimes.

In enhancement-mode operation of a p-channel TFT, a negative voltage must be applied to the gate electrode to create an accumulation layer at the insulator–semiconductor interface. For depletion-mode operation, the accumulation layer is already present at zero gate voltage.

Thus, for a p-channel TFT, a positive gate voltage has to be applied to deplete the accumulation channel and turn the device off. Therefore, an enhancement-mode device is a "normally off" device, whereas a depletion-mode device is "normally on." In an ideal TFT model in which traps are neglected, the TFT would be an accumulation-mode device because of the presence of zero-bias carriers (bulk carriers) available for current conduction. However, the presence of doping or traps may give rise to depletion-mode operation of the TFT. From this description we can appreciate that TFTs differ from MOSFET in several aspects. Firstly, a TFT works in the accumulation layer, while a MOSFET works under the inversion layer. Secondly, the semiconductor layer is different from the dielectric material. The lattice mismatch causes the layer to be amorphous with a high density of defects, in particular at the dielectric–semiconductor interface. Thirdly, a TFT is undoped, while a MOSFET is mostly Si doped.

5.2.2 OTFT Architectures

OTFTs can be fabricated on various types of rigid or flexible substrates, that is, there is no need for an (expensive) single crystal wafer. There is no limit to the size or material properties of the substrate as long as it can stand the fabrication process environment. In addition, TFTs can be made from a wide range of semiconductor and dielectric materials.

OTFTs can be configured into four basic structures on the basis of position of the electrodes, as shown in Fig. 5.4. The patterned source/drain electrodes can be deposited prior to the OSC deposition or after it. The former case is a coplanar configuration (popularly called bottom contact) and the latter is staggered configuration (also known as top contact). Both staggered and coplanar configurations are further categorized as bottom-gate and top-gate structures. Different TFT structures can display quite dissimilar device characteristics, while using the exact same materials. In a coplanar configuration, the source–drain contacts and the insulator layer are on the same part of the channel, whereas in a staggered configuration, the source–drain contacts and the insulator layer are on the opposite part of the channel.

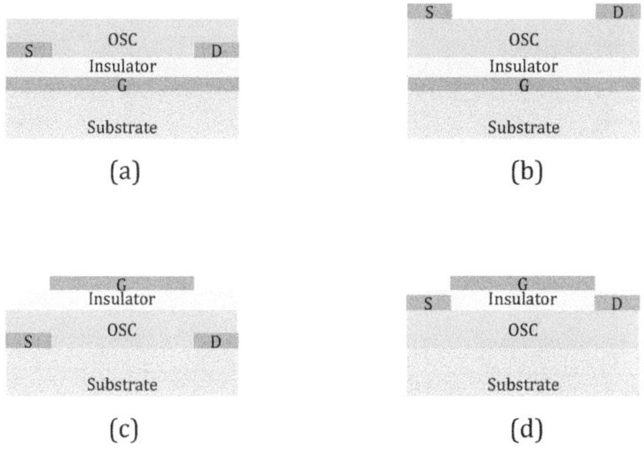

Figure 5.4 Schematic cross section of common OTFT structures: (a) bottom gate and bottom contact, (b) bottom gate and top contact, (c) top gate and bottom contact, and (d) top gate and top contact.

This can be crucial in determining the carrier injection properties of the source–channel interface. In a coplanar structure the metal–semiconductor interface is from a physical point of view a metal–accumulation channel interface (there is an abundance of free carriers in both sides of the interface). Therefore, coplanar devices are expected to be more tolerant to the contact barrier effects. However, the presence of traps may degrade the injection properties of these interfaces. The option for a particular OTFT structure will depend essential on the fabrication technology available (evaporation, spin coating, or printing) and on the best way to achieve clean (trap-free) interfaces.

5.3 The Ideal Thin-Film Transistor

Figure 5.5 shows a cross section of a TFT with the nomenclature used in this chapter. The device consists of a conductor called the gate (made of metal or a highly doped semiconductor), an insulating layer (which we will call the oxide layer, as an inheritance from silicon technology) of thickness d_{ox} (resulting in capacitance density $C_{ox} = \varepsilon_{ox}/d_{ox}$, with ε_{ox} the permittivity of the insulator material), and a semiconducting layer that accommodates the channel of charged carriers and is called the active layer.

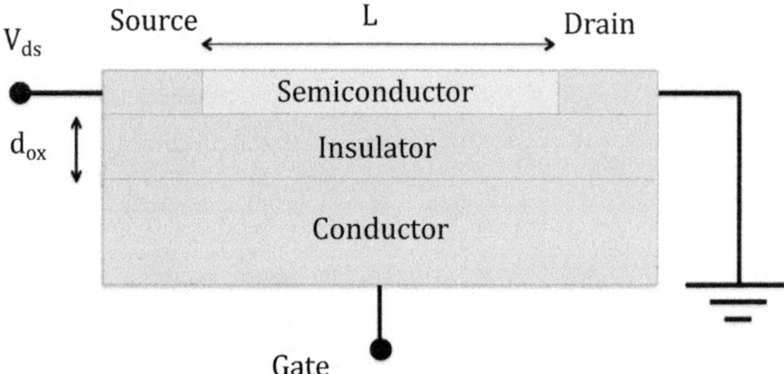

Figure 5.5 Cross section of an OTFT showing the nomenclature used in this chapter.

The basic operation of a FET relies on the charge density modulation in the active layer and thus its conductivity can be modulated by a voltage applied to the gate. The charges are injected and collected by the source and drain electrodes, respectively. Observable external electrical quantities are I_{ds} (the drain–source current), V_{ds} (the drain–source voltage), and V_g (the gate–source voltage). The leakage currents, such as drain–gate or gate–source, are considered zero.

It is common practice in literature to use textbook inversion channel MOSFET theory to describe the behavior of organic transistors. There are two reasons why this might be inappropriate. First, the devices are TFTs and as such do not have a bulk region. Apart from reducing the four-terminal MOSFET devices to a three-terminal TFT from an electronics point of view, the main concern is that a TFT, without a bulk region, cannot accommodate a band bending. Second, OTFTs are all accumulation channel FETs. In this situation, in the absence of localized states (donors) to store immobile positive charge, no band bending can be maintained, even if the active layer is thick. Summarizing, the thick semiconductor in a standard MOSFET can accommodate band bending and will have band bending in inversion mode. Charges induced by the gate are then not all located close to the interface which leads to a complicated charge–voltage which reflects in the current–voltage relation results. In a TFT, or in general an accumulation-type FET, all induced charge is necessarily close to the insulator and the charge–voltage relation is always simply

$$\rho(x) = V(x) - V_g C_{ox} \tag{5.2}$$

with ρ and V the local charge per area and voltage in the channel, respectively. This charge in a TFT might still be either mobile or immobile, though.

To give an idea of how thin the active layer in a TFT can be and still work properly, for a silicon-based device, with −1 V at the gate relative to drain and source and an oxide thickness of d_{ox} = 200 nm, the induced charge is 0.17 mC/m^2. With a density of states (DOS) of N_V of 1.04 × 10^{19} cm^{-3} and assuming continuity, this can fit into 1 Å, which is less than the height of a monolayer. The TFTs have thus effectively two-dimensional charge distributions. This explains why, as has been shown, for OFETs only the quality of the first monolayer matters. At best, the consecutive layers help to stabilize the integrity of the first layer in terms of diffusion of impurities and crystallinity.

For standard MOSFETs, the assumption is made that the induced free charge in the channel is linearly depending on the gate bias. This is because once the channel has been formed, all the charge induced by the gate is free charge. This in turn is caused by the type of semiconductor used in TFTs. For traditional materials, such as Si or GaAs, the acceptors and donors introduce shallow levels, which are consequently all ionized at all operational temperatures. In OSCs, the acceptor and donor states are very deep and abundant. As a result, even at room temperature, not all levels are ionized and temperature and bias can change the degree of occupancy. As we will show, the as-measured mobility does not change because of an increased depth of the acceptor level.

It is easy to show that the equation for currents of a MOSFET is also applicable to TFTs. In the case of a TFT the thickness of the channel is constant, but the density of charges p inside the channel varies from one electrode (source, $x = 0$) to the other (drain, $x = L$). To calculate the currents through the device, we have to understand that, locally, the current $I_x(x)$ at a given point x in the channel is equal to the local induced charge, $C_{\text{ox}}[(V_{\text{g}} - V_{\text{T}}) - V(x)]$, multiplied by the carrier mobility μ, the field felt by the charges, $dV(x)/dx$ (NB: only the drift current is considered), and the channel width W. In other words, we have the following differential equation:

$$I_x(x) = qWp(x)\mu\frac{dV(x)}{dx} \tag{5.3}$$

$$p(x) = \frac{C_{\text{ox}}\left[V(x) - (V_{\text{g}} - V_{\text{T}})\right]}{q} \tag{5.4}$$

V_{T} is the threshold voltage, which will be discussed later in a separate section. However, an important difference with standard MOSFET models is that V_{T} is not related to donor or acceptor concentrations. The threshold voltage can only deviate from zero in the presence of traps. With boundary conditions $V(0) = 0$, $V(L) = V_{\text{ds}}$, and $I_x(x) = I_{\text{ds}}$ for all x, the solution is

$$I_{\text{ds}} = -\frac{W}{L}C_{\text{ox}}\mu\left[(V_{\text{g}} - V_{\text{th}})V_{\text{ds}} - \frac{1}{2}V_{\text{ds}}^2\right] \tag{5.5}$$

V_{ds} and V_{g} are both negative. This equation for TFTs is very similar to the equation for MOSFETs. The only prerequisite is (low) ohmic contacts. The effects of the contacts will be discussed later. The equation is valid up to $V_{\text{ds}} = V_{\text{g}} - V_{\text{T}}$. After that, saturation starts in a region close to the drain is below threshold voltage and is devoid of charges. When the subthreshold conductivity is (close to) zero this region can be infinitely small and still absorb all of the above-saturation voltage $V_{\text{ds}} - (V_{\text{g}} - V_{\text{T}})$. In this way, the charge and voltage distribution across the device (except for an infinitely thin zone) is independent of the drain–source voltage and hence the current is constant at

$$I_{ds} = -\frac{1}{2}\frac{W}{L}\mu C_{ox}(V_T - V_g)^2 \tag{5.6}$$

When the subthreshold conductivity is not zero, the above-saturation voltage can only be supported over a finitely thick zone (l), which depends on the voltage. The remaining voltage drop $V_T - V_g$ then occurs in a region that is not of constant width but shrinks to $L - l$ for increased V_{ds}. The result is that the saturation current is not constant but continues to increase for higher drain voltages.

For low voltages, the quadratic term in V_{ds} disappears from Eq. 5.5 and this is called the linear region. Conventionally, the mobility of a FET is defined via the derivative of a transfer curve ($I_{ds} - V_g$). Using Eq. 5.5 for small V_{ds}

$$\mu_{FET} = -\frac{L}{WC_{ox}V_{ds}}\frac{\partial I_{ds}}{\partial V_g} \tag{5.7}$$

where the subscript "FET" is used to distinguish it from mobilities measured by other techniques. For an OTFT, the as-measured mobility can depend on things such as the temperature and the bias and can substantially deviate from mobilities measured, for instance, by using time-of-flight techniques.

Table 5.1 Simulation parameters used in this work

Parameter	Value	Unit
N_v	1.04×10^{16}	m^{-2}
C_{ox}	160	$\mu F/m^{-2}$
V_{ds}	-0.1	V
W	1	cm
L	10	µm
μ_o	3	$cm^2V^{-1}s^{-1}$
E_g	1.12	eV

5.3.1 Threshold Voltage and Subthreshold Current of Ideal TFTs

It is common practice to use the threshold voltage and the subthreshold current as device evaluation parameters. They are often used to extract information about doping concentrations, traps, and interface states. In the context of the two-dimensional model for the TFT described above, it is important to understand the physical meaning of these device parameters. In this section, we will analyze the trap-free device based on intrinsic or pure materials. Here we call it an ideal TFT. It is shown that the threshold voltage and subthreshold current do have different behavior compared to the conventional MOSFET models.

On the basis of the two-dimensional model for the ideal TFT discussed above the threshold voltage in TFTs is zero because, in the absence of localized states, originating from donors, acceptors, or charged defects, all induced charge is necessarily mobile. Traps can cause a nonzero threshold voltage, which will be discussed later for nonideal devices.

In MOSFETs, the subthreshold current is exponentially depending on the gate bias as well as the drain–source bias. The reason for this is that below threshold the free-carrier density is exponentially depending on the local bias. (The energetic distance between band edge and Fermi level is linearly depending on the voltage drop across the insulator and the free-carrier concentration is depending exponentially on this distance.) In the linear region, the potential at the drain is slightly smaller than at the source. Therefore, p is exponentially smaller at the drain compared to the source. Such a high gradient in density causes the diffusion current to dominate. (Drift currents are still insignificant because the densities are still too small.) The gradient and the current thus depend exponentially on V_{ds} and V_g. The current is proportional to the difference in density at the source and the drain, $I_{ds}\ \mu\ \exp(V_g)-\exp(V_g - V_{ds}) \approx \exp(V_{ds})\ \exp(V_g)$, which leads to the equation normally found in textbooks. Above threshold, the densities depend linearly on the potential and drift currents exceed the diffusion currents.

Because thin films do not have space to accommodate band bending, resulting in the basic Eq. 5.1, the charge density does not depend exponentially on the potential as in MOSFETs but always linearly.

In summary, the similarity of the basic equation, Eq. 5.5 and Eq. 5.6, and the shapes of the curves of Fig. 5.6, with those obtained for inversion channel MOSFETs, explain the persistence in literature of using the MOSFET model to describe OTFTs; empirically, the curves are the same. The complications start when the measured data are analyzed and parameter extraction is attempted. The differences between MOSFET and OTFT parameters are discussed now.

The threshold voltage in an ideal accumulation-type OTFT is zero. This is because, in the absence of localized states, originating from donors, acceptors, or traps, all induced charge is necessarily mobile. Real OTFTs are often nonideal and they have a threshold voltage; however, it is important to keep in mind that the threshold voltage is not an intrinsic device parameter in an OTFT.

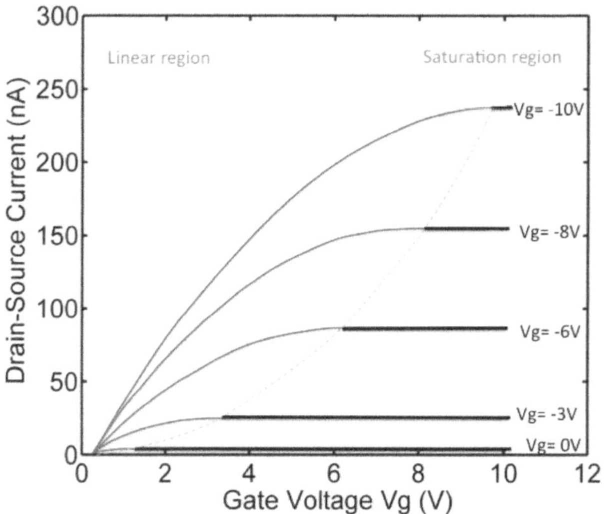

Figure 5.6 I–V curves (I_{ds} vs. V_{ds}) of an ideal thin-film FET resulting from Eq. 5.5 (thin blue lines). The parameters are given in Table 5.1. Thick lines indicate the saturation regime.

5.4 Nonideal Characteristics

In this section we present an overview of nonideal characteristics. Three basic behaviors are commonly reported: (i) nonohmic current–voltage (I–V) characteristics (near the origin), (ii) deviations from linear or from quadratic behavior of the TFT transfer curves, and (iii) large current when V_g = 0 V (off-current). Figure 5.7 shows a typical example of a nonlinear transfer curve measured in the linear region. This behavior prevents a proper extraction of the TFT parameters. It is impossible to unambiguously define a threshold voltage or the field-effect mobility from the slope of the transfer curve. This behavior is caused by the presence of localized charges. The transfer curve follows then a power law. It is possible to quantify how much the curve deviates from the ideal behavior using a γ parameter. This γ parameter can also be related to the density of localized states and its extraction will be discussed later.

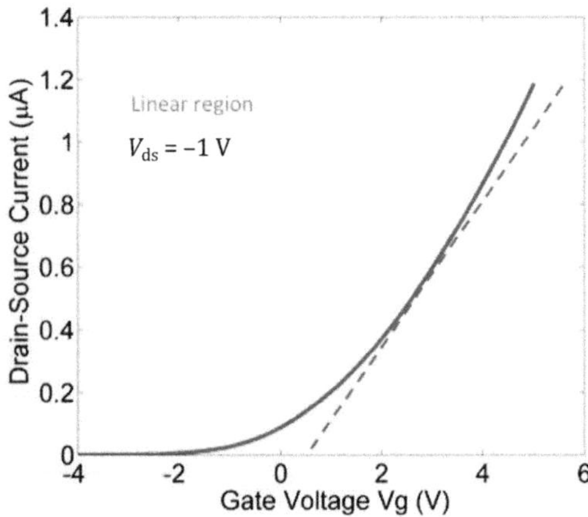

Figure 5.7 Experimental transfer curve for an OTFT showing nonlinear voltage dependence.

Often the metal-semiconductor contacts are nonohmic. This causes nonlinear output characteristics, which are particularly visible when the V_{ds} voltage is low. These nonlinear effects tend to be pronounced for high gate voltages where the TFT channel demands a high current and the metal contact cannot supply it. Figure 5.8 shows two typical examples of this behavior. In Fig. 5.8a the output curves no longer separate from each other. This effect will be visible as a decrease in slope in the in the corresponding transfer curves. Figure 5.8b shows a more complex behavior. These nonlinear behaviors are often caused by the presence of impurities (traps). Their quantification is not simple and it will be discussed later.

Both the output curves in Fig. 5.8 suffer from a large off-current (this is the current when V_{gs} = 0 V). When this current is large it prevents the observation of a well-defined saturation region and degrades the on/off ratio. The off-current may be caused by a parasitic current path through the bulk region or by a built-in channel. The application of a positive gate bias should help to diagnose the physical origin of this current.

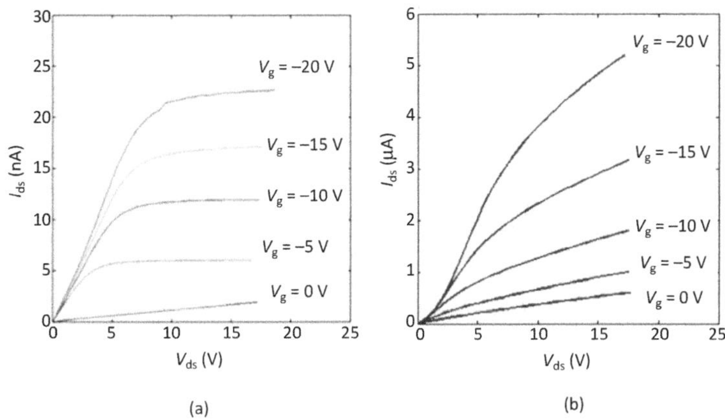

Figure 5.8 Two examples of nonideal *I–V* characteristics. Both TFT characteristics are distorted by the presence of nonohmic carrier injection and a high off-current.

If the current is caused by a built-in channel then it should decrease when a positive gate bias is applied (driving the TFT into depletion). A parasitic leakage path is usually not affected by the gate bias. To minimize the off-current one should make the semiconductor layer undoped and as thin as possible.

Figure 5.9 shows other type of nonlinearity. To simulate this behavior we have to consider a diode element and two TFTs in parallel (see the equivalent circuit in the inset). In this case it appears that the semiconductor region near the metal contact is strongly disturbed. Alterations in the semiconductor morphology, structure, doping, and trap density are likely to occur near the metal contacts giving origin to a complex carrier injection and transport of charges through these interfacial regions. This behavior was observed on an n-type OTFT.

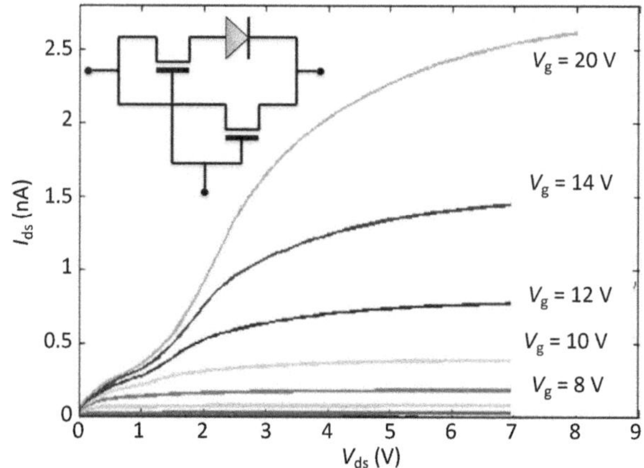

Figure 5.9 *I–V* characteristics observed for an n-type OFET. Severe distortion for low V_{ds} is observed. The inset shows an equivalent circuit to model this behavior.

Table 5.2 summarizes the typical nonideal behavior observed on OTFTs.

Table 5.2 Nonideal TFT characteristics and their physical origin

Nonideal characteristic	Physical origin/Strategies to manage it
Off-current	Possible causes: • The semiconductor is doped. • There is a parallel conducting path through the semiconductor bulk layer. • There is a conducting path through the surface. By applying a reverse gate voltage two behaviors can be observed: • If the TFT accumulation channel shuts down, it confirms that the semiconductor is doped. • If the current remains unchanged, it means that either the semiconductor layer is too thick or there is a leakage path.
Nonlinear transfer curve	Caused by traps.
Contact effects	Caused by traps. Passivation of source and drain electrodes is usually required.

5.4.1 Parameter Extraction

The threshold voltage can be found by fitting a straight line in a transfer curve in the linear region and extrapolating to zero current. However, extracting the threshold voltage in this way is not always easy. Figure 5.7 shows an experimentally obtained transfer curve in the linear region for a FET. From this curve it is clear that the slope (and thus the mobility via Eq. 5.7) and the threshold voltage found by extrapolation to $I_{ds} = 0$ depend on the point of the curve used; both μ and V_T depend on V_g. The awkward situation arises in which both mobility and threshold voltage can depend on the bias point considered. Table 5.3 presents a list of measurements procedures to extract TFT parameters.

Table 5.3 Measurement procedures to extract TFT parameters

Parameter	Measurement procedure	Comments
γ	Plot $I_{ds}^{\frac{1}{1+\gamma}}$ versus V_g.	The linear region is recommended (if there are no contact effects).
Threshold voltage (V_T)	V_T is defined as the intercept of the linear $I_{ds}^{\frac{1}{1+\gamma}}$ versus V_g curve if linear region is used.	The measurements must be carried out in unstressed TFTs and in dark conditions.
Field-effect mobility (μ_{FET})	Once γ is extracted then mobility is estimated from the following equation: $\mu_{FET} = \mu_0(V_g - V_T)^\gamma$	
Parasitic contact resistance (R_C)	It is a voltage-dependent resistance required to fit the I–V curves on the low bias V_{ds} region.	It depends essentially on the trap concentration.
Off-current (I_{OFF})	Measure I_{ds} for $V_g = 0$ V.	It is mostly determined by the organic semiconductor layer thickness and doping.
Gate leakage current (I_g)	Measure the gate current at a particular gate voltage ($V_g = -20$ V).	

5.4.2 Operational Stability

One of the most important stability issues in organic transistors is the shift in the threshold voltage upon applying a prolonged bias to the gate electrode, so-called stressing. The applied negative gate potential for a p-type semiconductor causes a build-up of mobile charges at the semiconductor–insulator interface, which are then trapped. As a consequence, to reach an identical channel current subsequently, a higher gate voltage has to be used. Devices suffering from gate bias stress effects exhibit threshold voltage shifts ΔV_T, current–voltage characteristics with hysteresis and a slow and continuous decrease in the device current. These effects are particular noticeable when the negative gate bias is applied over prolonged time. Therefore, stress effects have a great impact on the application of OSCs in electronic devices, for example, limits the reliability of TFTs.

It is known that the bias stress effect is reversible and that the recovery process can be enhanced by a positive gate bias or by light. A number of studies have also shown that the traps are related to the presence of water. To achieve stable TFTs, it is critical to prevent the incorporation of water, either by introducing hydrophobic capping layers or by using molecular design of new materials less susceptible to water incorporation.

The operational stability of the TFT can be quantified by measuring the time evolution of the threshold voltage shift $\Delta V_T(t)$ during the application of a constant gate voltage. This dependence is described by a stretched exponential function characterized by the parameters τ and β according to

$$V_T(t) = V_0\left\{1-\exp\left[-\left(\frac{t}{\tau}\right)\right]^{\beta}\right\} \tag{5.8}$$

where τ is the relaxation time, the dispersion parameter β equals T/T_0, and $V_0 = V_g - V_{T0}$, where V_g is the applied gate bias and V_{T0} is the threshold voltage at the start of the experiment.

The higher the value of τ, the higher is the transistor operational stability; therefore τ is used as a figure of merit to compare devices fabricated using different technologies.

An example of the use of the stretched exponential formalism to quantify the OTFT stability is illustrated in Fig. 5.10. The time dependence of the threshold voltage shift of transistors having a bare and a pentaflurothiophenol (PFTP)-coated dielectric is compared. By coating both the dielectric and the electrodes surfaces with a PFTP layer, the operational stability improves 1 order of magnitude with respect to uncoated devices. The relaxation time, τ, increases from 2×10^3 s to 2×10^4 s. State-of-the-art TFTs to be used in commercial applications must have a τ in the order of 10^6–10^7 s.

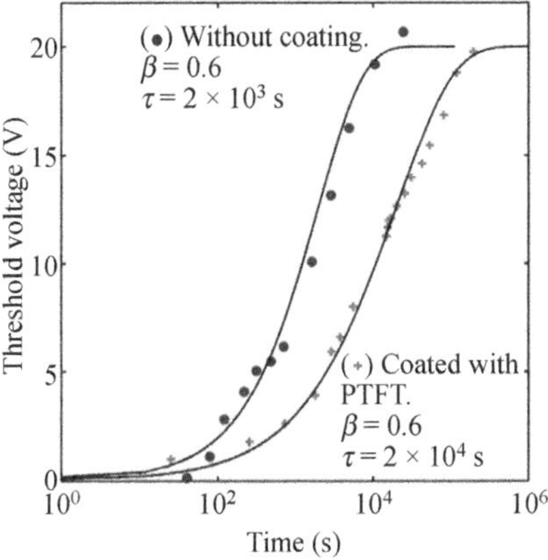

Figure 5.10 Comparison of the stability of two TFTs using the stretched exponential formalism.

5.4.3 Variability

Process variations present during transistor fabrication lead to variability on the resulting transistor parameters. Here we discuss the main sources of variability and the approaches that are generally employed for analyzing and interpreting the mismatch results.

Variations between OTFTs had been a well-known problem more acute than in silicon-based technologies by an inherently higher parameter spread. Reasons for that include

- Irregular morphology of the semiconductor
- Difficulty in controlling the precise dimensions of OTFTs
- Mobile trapped charges in the dielectric

- Uneven material deposition
- Roughness of the semiconductor–gate dielectric interface, which leads to mobility variations between the different transistors

Artifacts, such as bad transistor layout and incorrect data analysis, may also introduce variability.

Interestingly, not all parameters affect the transistor current on the same way. According to Eq. 5.6, variations in the threshold voltage V_T affect the saturation current of the transistor quadratically, while variations in μ, W, L, and d_{ox} influence the saturation current only linearly.

The large transistor variability poses a serious challenge to the cost-effective utilization of organic analogue circuits. It prohibits the use of OFETs in configurations that rely on precisely matched currents, as for instance current-steering D/A converters. Thus, the variability issues are of paramount importance.

Semiconductor foundries run analyses on the variability of attributes of transistors (length, width, dielectric thickness, etc.). This set of files are generally referred to as "model files" in the industry and are used by electronic design-assisted (EDA) tools for simulation of transistor and circuit designs. Designers using this approach run simulations to analyze how the outputs of the circuit will behave according to the measured variability of the transistors for that particular process. These simulations allow estimating the final circuit yield, starting from a basic inverter and stepping up toward the full integrated circuit. Moreover, simulations taking into account the area-dependent variability increase the predicted yield, as expected.

5.4.3.1 Experimental procedures and methodologies to study variability

Variability characterization requires a large number of measurements on a variety of devices, layout styles, and environments. The methodology used in complementary metal–oxide semiconductor (CMOS) technologies to account for local parameter variations and transistor mismatch can be transposed to OTFT technologies. For each TFT, an electrical parameter P is characterized by a continuous distribution with space and specific noise power intensity A_P

independent of the device surface area. Therefore, in the local approach, the variance σ of parameter P for a device with surface WL takes the form

$$\sigma_{\mathrm{p}}^2 = \frac{A_{\mathrm{p}}^2}{WL} \tag{5.9}$$

For instance, the variance σ in the threshold voltage V_{t} will follow

$$\sigma_{Vt}^2 = \frac{1}{C_{\mathrm{OX}}^2} \frac{qQ_{\mathrm{d}}}{4WL} \tag{5.10}$$

where Q_{d} is the channel charge and C_{OX} is the gate oxide capacitance per unit area. Once the extraction has been done, data filtering is usually applied to eliminate erroneous values.

5.4.3.2 Limitations of test structures

Parasitic effects caused by the off-current or parasitic fringe current outside the channel region can introduce errors in parameter extraction and lead to an incorrect variability analysis. Next a specific example is discussed.

Figure 5.11 shows a set of transfer curves for small-area OTFTs produced in identical conditions and located in the same substrate. Apparently, there are variations in the off-current, in the threshold voltage, and on the field-effect mobility. When the off-current is removed from all the curves, the variability on the threshold voltage becomes residual. This is shown in Fig. 5.12.

However, the application of the same correction procedure to a large-area OTFT shows no variations on the mobility. Indeed, all the transfer curves run parallel to each other (see Fig. 5.13). The fact that the dispersion on mobility is dependent on the area of the OTFT, demonstrates conclusively that the variation on mobility arises from the presence of a parasitic source-drain current flowing outside the channel area.

This effect is schematically represented in Fig. 5.14a where the arrows represent the electric field lines. It is expected that this parasitic effect should be more pronounced under a high applied field (saturation regime) and for a smaller-area transistor. To eliminate this parasitic conduction, the semiconductor layer must be restricted to a region inside the interdigitated source and drain electrodes. A new design is shown in Fig. 5.14b. This new design reduces the lateral fringe current.

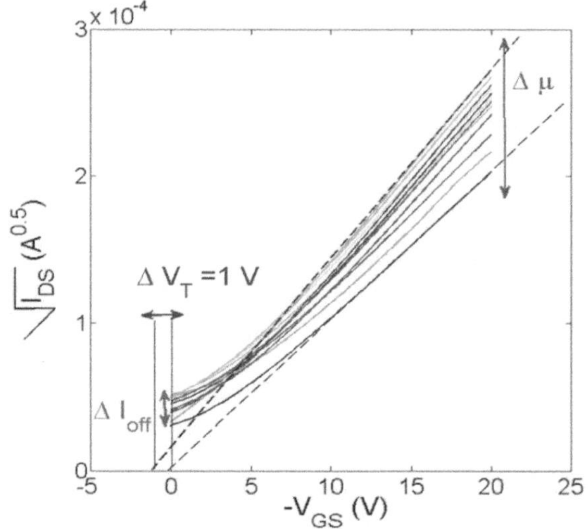

Figure 5.11 Experimental transfer curves measured in the saturation region for a set of identical OTFTs. Dispersion in mobility, threshold voltage, and off-current is observed.

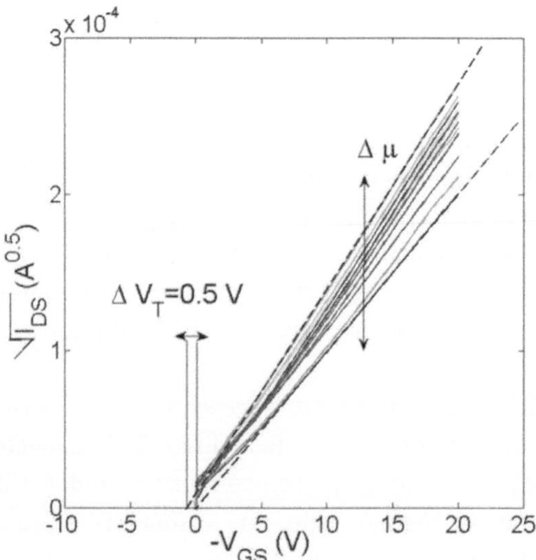

Figure 5.12 Set of transfer curves after removing the off-current. The variability on threshold voltage becomes residual.

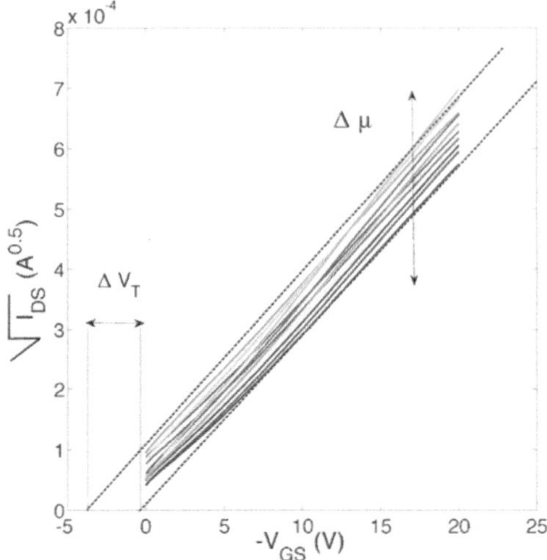

Figure 5.13 Transfer curves measured in an identical set of OTFTs with a large area (W = 40.000 μm; L = 40 μm).

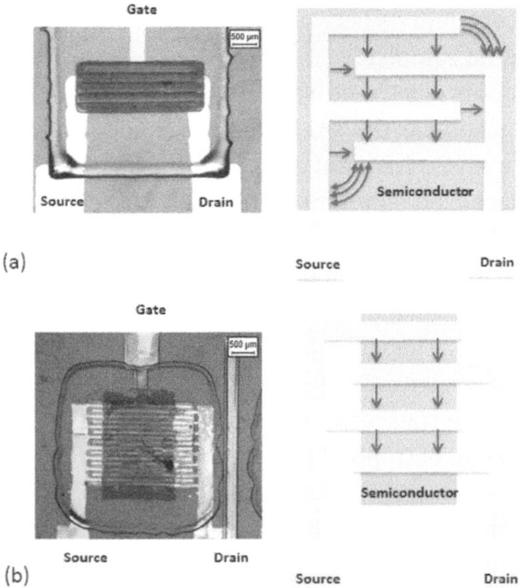

Figure 5.14 (a) Layout design susceptible to parasitic fringing current outside the main channel area. (b) The layout design to reduce lateral fringe current.

Interestingly, all the transfer curves run parallel to each other. Although the transistors have small areas, there are no variations in the mobility.

5.5 Electronic Traps in TFTs

The presence of traps degrades the performance of a TFT. Traps cause hysteresis effects, variations on the threshold voltage, nonideal output or transfer curves, a change in the charge carrier transport, and excess electrical noise. Traps are usually due to the presence of extrinsic impurities introduced during handling or by bad fabrication recipes. In reality the TFT is so sensitive to the presence of traps that is an ideal tool to provide fast feedback of information for adjustment of the fabrication process.

Here we will address some basic techniques to detect the presence of traps and their location on the device geometry. The speed of measurements is of paramount importance. Therefore, these methods are suited to work on a phenomenological basis by looking for correlations between cause and effect and they require a large number of samples.

A trap is an electronic state that can capture a free carrier. Electrical transport is OSCs is by hopping and it is always accompanied by more or less frequent capture of the involved charge carriers in traps of localized states. Such trapped carriers may be released after a specific retention period. The retention period may be short or long. Here we considered deep trap states as an immobile charge that does not participate in the electronic conduction. However, their columbic charge will influence the electric field distribution in a device and therewith the transport, for instance, threshold voltage shifts in TFTs are caused by an immobile trapped charge. However, if the release rate of trapped carriers is sufficiently low, a significant time will be necessary to reach quasi-thermal equilibrium conditions. This causes delay and hysteresis effects in alternating current (AC)-operated devices.

The effects of traps on the electronic properties of TFTs are manifold:

- Reduced mobility
- Threshold voltage instabilities

- Temperature dependence of mobility
- Nonlinearities in *I–V* curves
- Nonlinearities in transfer curves
- Current activation energy and mobility dependency on bias
- Transient response following multiexponentials or stretched exponential decays

A trap is characterized by the following parameters the density (N_t), the energetic depth in respect to the carrier bands (E_t), the thermal emission rate (e_n, e_p), and the capture cross section (σ_n, σ_p).

To study a trap, we have to find a way to control its occupation. This means we must know how to fill and empty it with charge carriers. Then we have to monitor either the filling or the emptying in a controlled way as a function of temperature, bias, frequency, etc. There are a number of recipes in the literature to perform these experiments. However, the majority of these recipes were originally developed for crystalline inorganic devices. Their application to organic-based devices is not always straightforward.

Prior to use trap detection techniques in OTFTs one should take into consideration the following aspects:

- Traps are named "deep" in the sense that the energy required to remove an electron or a hole from the trap to the bands is much larger than the characteristic thermal energy kT. OSCs are relatively wide-band-gap semiconductors, typical around 2 eV. Silicon has a band gap of only 1.1 eV. Therefore, OSCs can accommodate very deep traps. The meaning of a "deep trap" is very different in organic or in silicon technology. A trap of 0.1 eV is considered deep in silicon technology, whereas in organic electronics it is considered a shallow trap. A deep trap in organics may have an energetic depth of 0.5–0.8 eV. This has important consequences. While a deep trap in silicon can be filled and emptied in a matter of seconds at room temperature, in an organic material it may take a day to empty. Some techniques to look at deep traps in silicon operate in timescales of seconds or milliseconds. This means they are not applicable for studying traps in organic electronic devices.
- Traps in OSCs have usually very large cross sections. This means they fill very fast. But because they are deep they empty slowly. Trap-filling times are several orders of

magnitude faster than emptying times. From a practical point of view, in OTFTs it is easier to study traps during the filling process, while in silicon-based devices it is usually standard procedure to study the traps during the emptying process. For silicon-based devices the traps are first filled, usually at low temperatures, to prevent fast escape. This method may be impracticable for organics because of the extremely long times required to monitor a trap release process.

- In inorganic crystalline devices trapped charge carriers can be de-trapped by interaction with light and the resulting current is recorded as a function of the wavelength of the light. In principle, such an optical stimulated current spectrum would directly yield the energy distribution of the trap states if an optical transition from a trap state to the transport states is possible. Unfortunately, in many OSCs a direct transition from the trap state to the transport states is not allowed. Usually, the incident light excites the carrier into an excited state of the same molecule, from where a free carrier is generated by autoionization. Thus, the required optical transition energy is often not related to the energy difference between trap and transport state.
- In organic electronics traps are very abundant. Trap concentration often is higher than the free-carrier concentration.

This section starts by using the nonideal TFT characteristics as a simple tool to detect the presence of shallow localized charge density. These basically provide information about the charges that are not totally free but can participate in the electrical conduction. Techniques to look at deep traps (immobile charge) are discussed in the end.

5.5.1 TFT Nonideal Characteristics as a Tool to Measure Traps

Information about the localized trap density can be extracted by measuring how far a TFT transfer curve deviates from the ideal behavior. Insight into the energetic distribution of these localized charges can be obtained by measuring the temperature dependence of the mobility or the TFT drain–source current. These two cases will be discussed.

5.5.1.1 Extraction of localized charge density from the TFT transfer characteristics

A trap-free TFT should have a linear transfer curve or a quadratic transfer curve on the linear and in the saturation region, respectively. Deviations from this behavior are caused by localized charge. This density of localized charge is then taken into account by the γ parameter as discussed in Section 5.4. The γ parameter is related to a characteristic temperature T_0 of the DOS by

$$\gamma = 2\frac{T}{T_0} - 2 \tag{5.11}$$

T is the absolute temperature and T_0 the characteristic temperature. The energy distribution of the DOS, $g_d(E)$, is expressed as

$$g_d(E) = g_{do}\exp\left(\frac{E_V - E}{kT_0}\right) \tag{5.12}$$

E is the energy of an electron, g_{do} the DOS of traps at the valence band E_V, k Boltzmann's constant, and T_0 a parameter describing the distribution (the slope of a logarithmic plot of the DOS).

γ can be extracted from the transfer curves. This can be done by trial and error or by an integration procedure.

$$H(V_g) = \frac{\int_0^{V_g\max} I_{ds}(V_g)dV_g}{I_{ds}(V_g)} \tag{5.13}$$

Using the measured linear transfer characteristics I_{ds} in the integral function, the slope and intercept of $H(V_g)$ are calculated

$$V_T = \frac{\text{Intercept}}{\text{Slope}} \tag{5.14}$$

and

$$\gamma = \frac{1}{\text{Slope}} - 2 \tag{5.15}$$

The extraction of γ can be made either from the linear or from the saturation region.

Although, the linear region is preferable, bad source and drain metal contacts may cause additional distortion on the linear transfer curve, making it difficult to extract γ.

5.5.1.2 Temperature dependence of the charge transport

The charge carrier mobility of MOSFET devices based on silicon is basically independent of temperature. On the other hand, TFTs based on organic materials do normally not show this characteristic. Complicated temperature dependencies are often observed and reported. These are caused by the presence of traps. Here we will show that Arrhenius plots for different biases may serve as a rapid evaluation tool of the quality of the material. More specifically, they give direct insight into the DOS governing the conduction. Three basic temperature dependences will be briefly described here: (a) abundant discrete trap, (b) low-density discrete trap, and (c) abundant trap that is distributed in energy. A detailed analysis can be found in the references.

5.5.1.2.1 *Abundant discrete trap*

In principle a trap-free TFT should show a charge carrier mobility (μ_{FET}), that is, bias and temperature independent. For this analysis we consider the intrinsic (band) mobility μ_0 to be temperature independent; this may not be necessarily true but is a good approximation. When a discrete trap is assumed, the mobility is lowered significantly by the reduced ratio of free-to-total charge and becomes temperature dependent but remains independent of bias. The reasoning is as follows: Free holes (p) in the conduction band, originally induced by the gate bias, can be captured by the traps, turning these positively charged. At thermal equilibrium, the ratio of densities of holes and charged traps N_T^+ is determined by the energetic distance E_T - E_V between them, the relative abundance of the levels, N_V and N_T, respectively, and the temperature T.

$$\frac{p}{N_T^+} = \frac{N_V}{N_T}\exp\left(\frac{E_V - E_T}{kT}\right) \tag{5.16}$$

The current is only proportional to the free-hole density because the trapped states, by definition, do not contribute to current and the density of electrons is insignificant.

$$p+N_{\mathrm{T}}^{+}=\frac{C_{\mathrm{ox}}V_{\mathrm{g}}}{q} \tag{5.17}$$

$$\mu=\mu_0\frac{N_{\mathrm{V}}}{N_{\mathrm{T}}}\exp\left(-\frac{E_{\mathrm{T}}-E_{\mathrm{V}}}{kT}\right) \tag{5.18}$$

The mobility is thermally activated, and independent of bias, the slope of the Arrhenius plot reveals the activation energy of mobility, which is then equal to the depth of the trap level, $E_{\mathrm{a}} = E_{\mathrm{T}} - E_{\mathrm{V}}$. This behavior is shown in Fig. 5.15.

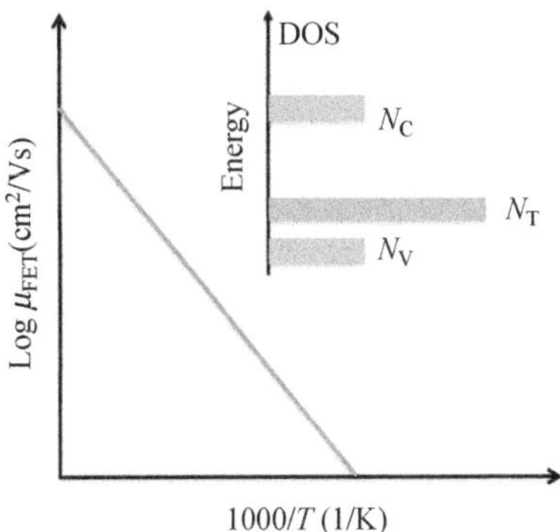

Figure 5.15 Arrhenius plot of the mobility for the case of an abundant discrete trap. The mobility is strongly temperature dependent and the slope reveals the trap depth. The inset shows the schematic DOS.

5.5.1.2.2 *Low-density discrete trap*

If the TFT is relatively clean, the traps can be all filled and exhausted. The induced charge is necessarily free charge (holes) and the mobility returns to the band value μ_0. The traps become exhausted when the induced charge density is comparable to the trap density. This defines the trap-free-limit voltage for the gate bias

$$V_{\mathrm{tfl}}=-q\frac{N_{\mathrm{T}}}{C_{\mathrm{ox}}} \tag{5.19}$$

We can find V_{tfl} by plotting the mobility as a function of the applied gate bias at a constant temperature. When $V_g = V_{tfl}$ a transition is expected to occur from a thermally activated behavior (trap limited) to a temperature-independent behavior (trap-free case). This is schematically represented in Fig. 5.16. Once V_{tfl} is known we can determine N_T.

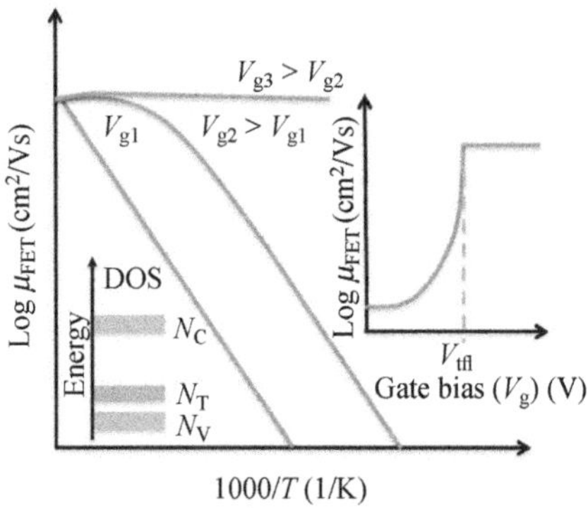

Figure 5.16 Arrhenius plot of the mobility for a discrete trap with density equal to the effective density of valence band states, $N_T = N_V$. The activation energy is equal to the trap depth for biases below the trap-free-limit voltages $V_g < V_{tfl}$. For larger biases, the mobility is no longer thermally activated and the plots resemble those of trap-free devices.

5.5.1.2.3 *Abundant traps that are distributed in energy*

The inclusion of traps distributed in energy makes the analysis of the charge carrier transport complex. The activation energy of drain–source current, I_{ds}, or carrier mobility, μ_{FET}, depends on the gate bias. There exists a temperature, known as the isokinetic temperature, T_{MN}, where the dependence of current or mobility on bias disappears. When presented in an Arrhenius plot, the curves of current or mobility converge to a common point, thus following the Meyer–Neldel rule.

The activation energy of the field mobility (and current alike), as measured via the slope of an Arrhenius plot, depends on the bias in the following way:

$$E_a = -\frac{d\ln(\mu_{FET})}{d(1/kT)} \quad (5.20)$$

$$E_a = kT_2 \ln(N_{T0}) - kT_2 - \ln\left(\frac{C_{ox}V_g}{q}\right) \quad (5.21)$$

Thus, the activation energy of mobility or current does not reveal the depth of an energetic level. Rather, it depends on the parameters of the distribution (T_2 and g_{T0}) and the bias. Figure 5.17 shows this behavior.

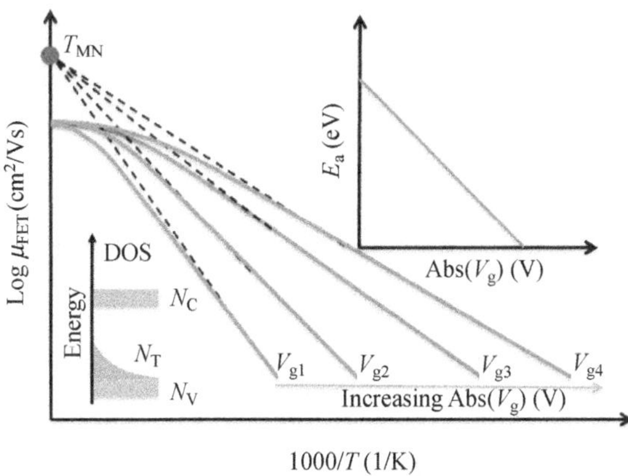

Figure 5.17 Graphical representation of temperature-dependent mobility of Eq. 5.21 of a system with an exponential DOS. The solid circle represents the Meyer–Neldel point (T_{MN}). The inset shows the variation of the activation energy as a function of bias.

In summary, a set of Arrhenius plots for different biases may serve as a rapid evaluation tool of the quality of the material. More specifically, they give direct insight into the DOS governing the electrical conduction. In cases, where a sharp transition in mobility is observed in the transfer curves the density of discrete traps can directly be determined via the trap-free-limit voltage V_{tfl} (see Eq. 5.19). In OTFTs the most common situation observed is the corresponding to traps distributed in energy. Arrhenius plots of the current or mobility are straight lines, but the slope (activation energy) depends on the applied gate bias and they do not provide directly the trap depth. Table 5.4 summarizes the temperature dependence of the mobility typically observed and the information provided about the trap density and energy.

Table 5.4 Summary of different temperature dependence behaviors and the information provided

Temperature dependence of the mobility	Information provided
Arrhenius plots of mobility are straight lines, independent of bias.	Traps are abundant. The slope of the plot reveals the activation energy of mobility, which is then equal to the depth of the trap level, $E_a = E_T - E_V$.
The mobility loses its voltage dependence above a specific gate voltage, the trap-free-limit voltage V_{tfl}.	The number of traps is limited and they are exhausted above a particular voltage. The activation energy is equal to the trap depth for biases below the trap-free-limit voltages $V_g < V_{tfl}$. For larger biases, the mobility is no longer thermally activated and the plots resemble those of trap-free devices.
The mobility is a function of temperature and bias. The so-called Meyer–Neldel rule is usually observed.	The trap states are distributed in energy. The activation energy of mobility or current does not reveal the depth of an energetic level. It depends on the parameters of the distribution.

5.5.2 Electrical Techniques to Study Traps

There are a number of electrical techniques to study traps. Often these methods require the fabrication of dedicated devices such as Schoktty diodes and metal-insulator-semiconductor (MIS) capacitors. However, in some circumstances TFTs may also be used. Techniques that can be applied to TFTs are the following:

- Small-signal impedance spectroscopy (IS)
- Thermally stimulated currents (TSCs)
- Electrical noise

A detailed description of these techniques can be found in the excellent book by Blood and Orton. Here we only briefly describe how to apply them to OTFTs.

5.5.2.1 Small-signal impedance spectroscopy

During TFT manufacturing process it is convenient to make nearby a MIS capacitor structure. A MIS capacitor structure allows the use of small-signal impedance techniques to study in interfacial states. Details of these methods can be found in the papers by Nicollian and Goetzberger. Some TFT configurations, particularly the staggered structure, can also be turned into a MIS capacitor if the drain and source contacts are short-circuited and used as a single terminal, as shown in Fig. 5.18.

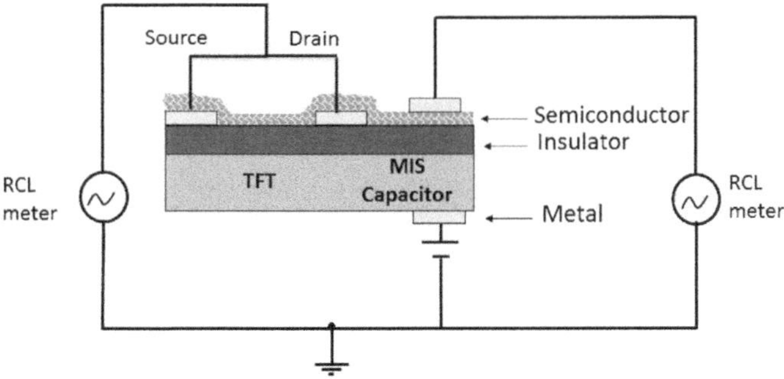

Figure 5.18 Schematic diagram of a TFT wired to be measured as a capacitor. A proper MIS capacitor may also be fabricated in a nearby region.

A MIS capacitor can only provide information about interfacial states if these states can be probed by frequencies below the device relaxation frequency, f_R. A MIS capacitor behaves as a double-layer system. The frequency dependence of the capacitance and resistance shows a dispersion at a particular frequency f_R. This frequency must be within a reasonable observation window (10 Hz–1 MHz) accessible to most of the impedance analyzers. To achieve this condition the OSC layer must have a low resistance.

Once is established that MIS capacitor has a relaxation frequency high enough to perform small-signal measurements, information about interfacial charges can be extracted by measuring the capacitance and the loss for different applied bias and frequencies. The reason for that is because the insulating layer disconnects the interface states from the metal, making them communicate with the semiconductor more readily than with the metal.

The first step is to determine the effect of interface states on the admittance. An AC signal, superimposed on a direct current (DC) bias, will cause the Fermi level to oscillate around a mean position. Any interface states within the modulation depth V_{ac} around this average Fermi level will change their occupancy during an AC cycle and the emitted and captured charges contribute to a capacitance

$$C_{is} = \frac{\Delta Q}{\Delta V} = Aq^2 N_{is} \tag{5.24}$$

with A the area of the device, q the elementary charge, and N_{is} the density of interface states per electron volt per unit area. The charges come out with a characteristic time constant τ. For low frequencies compared to this time constant, the measured capacitance is as given by Eq. 5.24. For increased AC frequencies ω, the response of the states to the signal is diminished; the reduced AC current means a reduced measured capacitance. The slower response also causes a phase lag of the AC current and the capacitance can be measured as a conductance and loss, G and G/ω, respectively. The standard procedure is to model this with an equivalent circuit of a capacitance in series with a resistance (see Fig. 5.19), in which the capacitance is equal to C_{is} and the resistance is such that the time constant RC of this circuit is equal to the relaxation time τ of the levels. Note that whereas the capacitance has real physical meaning in this circuit, the resistance only helps to define the time constant. To find the measured capacitance and loss (C_p and G_p/ω, respectively) of this circuit and hence of the interface states, we have to translate the serial circuit of C_{is} and R_{is} into a parallel circuit.

$$C_p = \frac{C_{is}}{1+\omega^2\tau^2} \tag{5.25}$$

$$\frac{C_p}{\omega} = \frac{\omega\tau C_{is}}{1+\omega^2\tau^2} \tag{5.26}$$

with $\tau = C_{is}R_{is}$. The maximum in loss occurs at $\omega_{max} = 1/\tau$ and is exactly half the low-frequency capacitance

$$\frac{C_p}{\omega_{max}} = Aq^2 N_{is}/2 \tag{5.27}$$

$$\omega_{max} = \frac{1}{\tau} \tag{5.28}$$

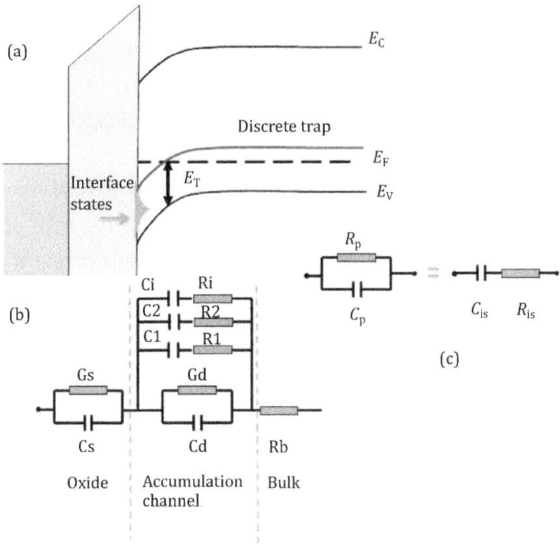

Figure 5.19 Schematic diagram of a discrete level and interface states distributed at the dielectric–semiconductor interface. The AC response of these states can be modeled by equivalent RC networks (b). A parallel R_pC_p circuit can be converted into a series $R_{is}C_{is}$ circuit (c). The converting equations are in the text.

In this way, by measuring the capacitance C_P and loss G_P/ω as a function of frequency, the DOSs at a region qV_{ac} + kT around the Fermi level can be determined.

The second part of the mapping is the determination of energetic position of the states for which we have just found the density. This can be found by determining the movement of the peak in loss with temperature. The trap relaxation time τ follows

$$\tau = \tau_0 T^{-2} \exp\left(\frac{E_T - E_T}{kT}\right) \tag{5.29}$$

with τ_0 still depending on the DOSs at the top of the valence band, the average thermal velocity of the carriers, and the capture cross section but not depending on the temperature. Combining Eqs. 5.27 and 5.28 shows us that an Arrhenius plot of the maximum in the loss as a function of temperature will yield the activation energy, $E_A = E_T - E_V$, of that particular bunch of states:

$$\omega_{\max}(T) = \omega_{\max}^0 T^2 \exp\left(-\frac{E_A}{kT}\right) \tag{5.30}$$

To find the distribution of interface states in energy we can scan the bias and repeat the above procedure at every step.

A simple capacitance voltage plot can provide the information about the dielectric layer capacitance. The presence of shallow traps can introduce hysteresis or humps in the plots.

The presence of interfaces states can also reveal as a peak in a loss–voltage plot. This peak can be due to a discrete state, or to states distributed in energy. A discrete state will cause a peak that is independent of the applied bias. A broad energetic interface state distributed in energy cause a peak structure that moves to higher frequencies for higher applied bias.

5.5.2.2 Thermally stimulated currents

The presence of trapped charge carriers in TFTs can in principle be confirmed using, TSC measurements. The traps are filled by applying a bias while the temperature is such that the trapped carriers cannot be freed by the thermal energy. The temperature is then raised linearly. The liberated carriers contribute to the excess current, that is, the external current minus the leakage current, until they recombine with carriers of opposite type or join the equilibrium charge carrier distribution. This excess current, measured as a function of temperature during heating, is called the TSC.

For a single trap level, a TSC curve has one maximum whose position depends on capture cross section, heating rate, and trap depth. By varying the heating rates the trap depth and capture cross section can also be determined. Because de-trapping currents are extremely small (picoamps), the TSC can only be used with relatively insulating materials. Normally off or fully depleted TFTs satisfy this requirement. When performing TSC experiments the transistor is connected as a MIS capacitor.

The experimental procedure involves the following steps:

1. The trap filling is performed at room temperature. The filling bias conditions are kept until the TFT is brought to temperatures low enough that de-trapping can be disregarded.
2. At low temperatures the TFT is connected as a capacitor. Drain and source terminals are short. A picometer measures the capacitor discharging current between the gate terminal and the drain and source terminals.

3. The TFT is then heated at a constant rate, $\beta = dT/dt$, up to high temperatures. The trapped carriers are released and collected at the grounded source and drain electrodes. The temperature where a current peak occurs is related to the energetic depth of the trap state and the area under the peak is related to the trap density.

Figure 5.20 shows TSC curves recorded using an OTFT. The figure shows also the base line (a curve recorded without trap filling) to confirm that the current is in reality a de-trapping current. For instance, the little peak just below 300 K is not caused by de-trapping.

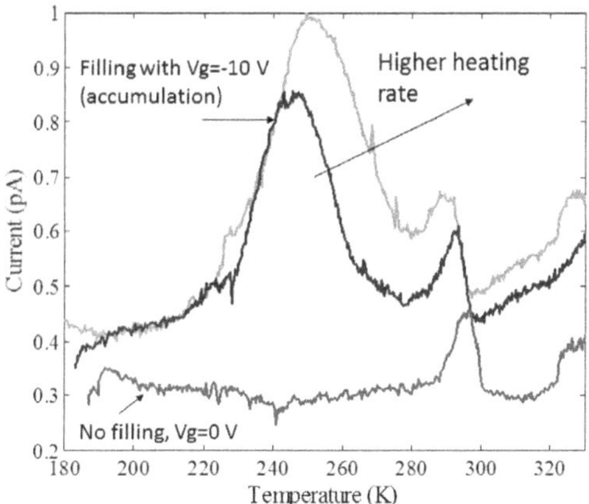

Figure 5.20 TSC curves measured using an organic MIS-capacitor structure.

The density of filled traps, N_t, can be estimated from the time integrated TSC current as

$$Q = eN_tA = \int I dt \tag{5.31}$$

where Q is the integrated total charge and A is the surface area between the electrodes.

The extraction of the trap parameters from TSC measurements is not straightforward. The current temperature profile does not only depend on the density, depth, and distribution of the traps but depends also on the details of the charge transport such as charge carrier mobility and the occurrence of retrapping. Several models can be applied, namely the initial rise time method, the heating rate method, and the curve-fitting method by Cowell and Woods [3].

The initial rise method is valid for all types of recombination kinetics and assumes that the current in the initial part of the curve, when the traps begin to empty, is exponentially dependent on temperature. This method is often used when the full TSC curve cannot be recorded or is distorted by other processes. The method only provides the trap depth and is usually less accurate than the other models.

A more reliable determination of the trap depth is obtained from the relation between the heating rate β and the temperature of the peak maximum, T_m, as described by Blood and Orton [26] in Eq. 5.32:

$$\ln = \left\{ \frac{T_m^4}{\beta(T_m)} \right\} = \frac{E_T}{kT_m} + \ln\left(\frac{E_T}{\sigma\gamma k} \right) \tag{5.32}$$

in which β is the heating rate, E_T is the trap depth, k is the Boltzmann constant, σ is the capture cross section, and γ is a parameter depending on the effective mass. From a series of TSC curves at different heating rates, the peak temperatures, T_m can be determined. The activation energy of the trap can then be obtained from a linear plot $\ln\left(T_m^4/\beta\right)$ versus $1/T_m$. The slope yields a value for the activation energy.

Special care has to be taken into account when using TFTs connected as capacitors to perform TSC experiments. Often the bottom electrode is the entire substrate. When a filling bias is applied, it may charge the entire semiconductor area which is larger than the device area. The charges in the vicinity of the TFT can diffuse to the contacts and the leads, where they are collected and measured in the external circuit. The released charges during the TSC curve are extracted from an area larger than that between the TFT electrodes. This may lead to discrepancies between the total charge expected and the measured charge.

Alternatively, it is possible to fit the complete TSC curves numerically using the classical approach of Cowell and Woods. The underlying assumption is monomolecular recombination of the charge carriers from a discrete set of traps with a single trapping level with a trap depth E_T below the conduction band, with negligible retrapping. The current I then follows from

$$I = \frac{A\exp(-\Theta)}{1 + B\exp(-\Theta(\Theta^{-2}))^2} + I_{\text{off}} \tag{5.33}$$

in which $A = n_t\tau e\mu\nu$, $B = \nu E_T/\beta k = \Theta_m^2\exp(\Theta_m)$, and $\Theta = E_T/kT_m$.

Here, n_t is the initial trap density of traps filled, τ is the average lifetime for a free carrier, μ is the mobility, and ν is the attempt to escape frequency. The TSC curve can be fitted with four fitting parameters, T_m, A, E_T, and I_{off}. We note that at high temperatures, when a significant number of traps is emptied, the conductivity of the TFT channel is partially restored and the associated background current severely distorts the measurements. For temperatures above the TSC peak the current cannot be treated solely as a de-trapping current and the analysis is usually not possible.

5.6 Circuits and Systems

Having studied the electrical performance of individual transistors, it is now important to understand how they behave when connected into circuits. A simple circuit is the inverter. It is based on only two transistors and well suited to prove the capability of OTFTs to build up more complex circuits. The inverter has to show signal amplification, and most importantly, the output signal has to be able to drive a subsequent inverter stage. The last requirement can be tested with a ring oscillator (RO). This device consists of an odd number of inverters connected with each other in series. Figure 5.21 shows the principal electronic circuit of a three-stage RO. An RO is often used to demonstrate a new technology. In silicon microelectronic circuits many wafers include an RO as part of the scribe line test structures. They are used during wafer testing to measure the effects of manufacturing process variations.

It is important to emphasize that one should only embark into the process of making circuits when individual TFTs perform reasonable well. Circuits will not work properly if the threshold voltage of individual TFTs is not stable or if the current modulation ratio is poor. On/off drain current ratios reaching 10^3–10^4 are considered as a minimum to obtain suitable digital circuits.

Inverters and ROs may fulfil several objectives, such as (i) to show that the processing technology is mature enough to fabricate circuits and (ii) to monitor variations caused by systematic and random physical effects. We will present first the electrical characterization of simply inverter circuits and later we will learn how to use ROs to quantify variability.

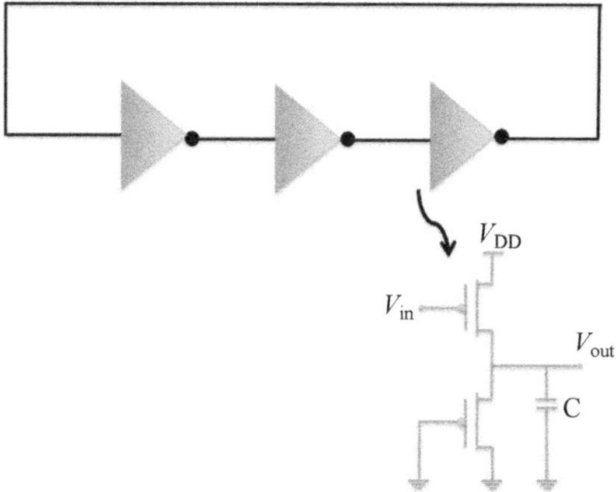

Figure 5.21 Hardware structure of a single-ended inverter-based ring oscillator.

5.6.1 Inverter Circuits

The most basic logic gate is the inverter. It is commonly represented in the form of a pull-down and a pull-up element switching the inverter output to ground (GND) or to the supply voltage (V_{DD}) via a low-resistance path. Figure 5.22 shows the schematics of commonly used configurations of inverters. The selection of a particular configuration depends on the characteristics of the individual TFTs.

Circuits using only one type of semiconductor (either n-type or p-type) require a combination of one pull-up (n) or pull-down (p) load transistor that is always in its on-state and an input-controlled drive transistor that changes the output of the inverter by logically inverting its input. Therefore a low voltage (logic level 0) at the input generated produces a high voltage (logic level 1) at the output and vice versa. The load transistor can be connected to its source (Fig. 5.22a), to its drain (Fig. 5.22b), or to an independent bias voltage source, as shown in Fig. 5.22c. If the gate is connected to its source, a fixed gate–source voltage $V_g = 0$ V results. This is the most popular configuration and will only work if the transistor has a built in channel ($V_T < 0V$) and is referred to as a normally-on TFT. A normally-on transistor in this configuration yields an approximately constant drain current as long as it operates in the saturation region. Hence, the configuration of a load transistor with the gate connected to its source is referred to as current source load (CSL) configuration. If the gate of the load transistor is connected to the drain, a fixed gate–drain voltage $V_g = 0$ V results in a transforming the transistor into a diode for the load, as shown in Fig. 5.22b. Therefore, this configuration is referred to as diode load (DL).

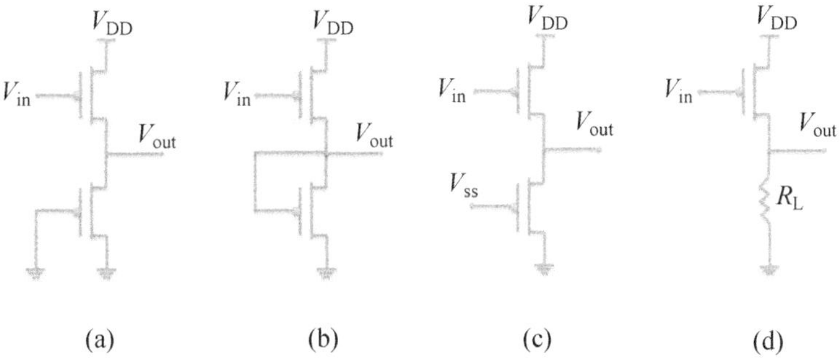

Figure 5.22 Basic configurations of OTFT design styles for basic inverters: (a) current source load (CSL) configuration, (b) load transistor in diode load (DL) configuration, (c) ratioed pMOS logic with two supply voltages, and (d) resistor load (R_L) configuration.

The fact that most OSC materials exhibit weakly normally-on or weakly normally-off behavior (high off-current) leads to poor performance of the input transistor as it can only be switched-off, bringing the transistor into depletion mode, for example, by biasing it in negative voltages for a p-type semiconductor. The configuration in Fig. 5.20c allows the biasing into depletion. By applying a positive V_{SS} we can move the OTFT into depletion and increase the on/off drain current ratio. Although this is a typical manner to illustrate an inverter behavior in the literature, it is a useless approach as those operation conditions does not allow connecting gates in cascade to build up larger circuits.

Figure 5.22d uses a resistor as the pull-up element connecting the output to the supply rail V_{DD}.

Figure 5.23 shows the relationship between output voltages as a function of the input voltage known as the voltage transfer curve (VTC).

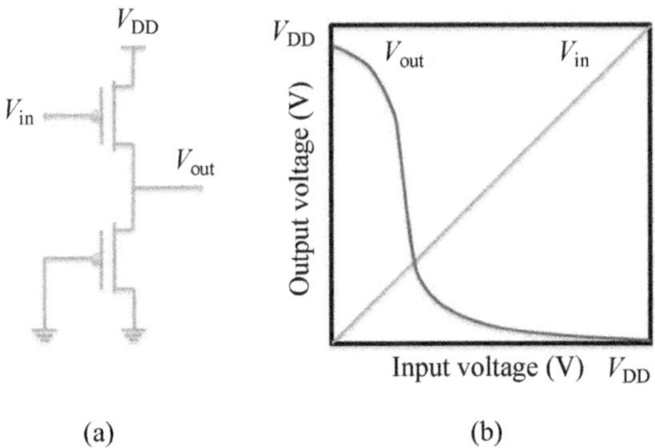

Figure 5.23 (a) Schematic of a CSL-type inverter and (b) voltage transfer characteristics.

The output voltage is determined by the ratio between the channel resistance of the drive transistor at a given input voltage and the resistance of the load element. For this reason these gate topologies are called ratioed logic. This ratioed behavior has also consequences on the circuit speed. A high-resistance element (load) will charge the output capacitance slower than the low-resistance one (drive), which will lead to different propagation delays for propagating logic levels 1 and 0.

5.6.2 Circuit Characterization

A complete characterization of an inverter circuit requires the measurement of the following figures of merit:

- Robustness, that is, valid logic levels and noise margin
- Timing, including rise/fall times, transition times, and clocking requirements
- Driving capability, that is, fan-out characterization
- Input capacitance
- Power consumption

These characteristics and especially the first two (logic levels, noise margin, and timing) can also be used for evaluating the performance of any OTFT implementation. Therefore, it is worth to take a closer look at the characterization process.

One important characteristic of logic circuits is their robustness because most OTFT-based logic circuits suffer from nonideal VTCs. Robustness defines the ability of a logic circuit to detect and issue valid voltage levels for the different logic states in the presence of noise at the input signals. Several possibilities exist to derive information on the robustness of logic circuits. A selection of these will be used in the following sections.

5.6.3 Ring Oscillators

The basic RO circuit consists of an odd number of inverter gates connected in series with a feedback path provided for oscillation. By applying a reset pulse to the circuit, the frequency of oscillation can be measured at the output pin. The frequency is determined by the physical factors of the manufacturing process and can therefore tell the engineer a great deal about those factors. In silicon microelectronics technology, ROs are added to each wafer manufactured, contained in special parametric structures in the scribe lines.

The RO frequency is sensitive to gate length, mobility, threshold voltage, and parasitic capacitances. RO frequency is approximated as

$$f_{\mathrm{RO}} = \frac{I}{CVN_{\mathrm{stages}}} \tag{5.34}$$

where f_{RO} is frequency, I is the transistor current, C is the load capacitance for a single stage, V is the supply voltage, and N_{stages} is the number of RO inverter stages.

Since ROs with a large number of inverter stages can average out the random variations of the individual stages, systematic variation is easily captured by long RO inverter chains. Random variation is easily captured by short RO inverter chains because ROs with a small number of stages experience less averaging at each inverter.

5.6.4 Summary

Electrical properties of the inverter have been presented in the current chapter. The low on/off drain current ratio is a key issue in the design of circuits using only p-type OTFTs, as the load transistor is always on, thus reducing output swing.

OTFT stability plays also an important role. Threshold voltage variations affect the noise margins of the logic gates are and the optimal drive/load ratio. Thus high variability can be transformed into a severe limitation of the number of logic gate stage that can be chained. This limits the amount of circuits that can be implemented. Lack of stability also plays an important role in a circuit that use to work in continuous operation.

Since ROs oscillate at a frequency dependent on the characteristics and dimensions of the devices as well as the loads of each inverter, these circuits can be used to detect device and interconnect process variation through measurements on their oscillating frequencies.

5.7 TFTs for Biomedical Applications

To use OTFTs as sensors we have to disturb or cause a change in one or more of the transistor parameters: (i) threshold voltage, (ii) off-current, and (i) field-effect mobility. Therefore, TFTs are multiparameter and amplifying sensing devices. Chemicals, biological substances, and even living cells may interact with several parts of the device structure and change the TFT parameters. Basically, there are three different strategies to use a TFT as a sensor:

- Floating gate method
- Changes in the TFT channel or at the channel–contact interfaces
- Changes in the dielectric layer

5.7.1 Floating Gate Method

The floating gate method has been substantially explored on inorganic-based TFTs. This method has been used to fabricate neuron–device interfaces, as shown in Fig. 5.24a. Neurons generate small voltage fluctuations on top of the floating gate terminal and modulate the channel current underneath. An identical approach was implemented using organic-based TFTs (see Fig. 5.24b). However, the TFTs used for this purpose have their channels exposed and do not have a built-in dielectric layer or even a gate terminal, they are a resistor. The gate dielectric will be established only when the device is immersed into the electrolyte and the electrolyte itself plays the role of the gate terminal. This occurs because when conductive or semiconductive materials are immersed into electrolytes a double-layer (Helmholtz layer) is established at the metal–electrolyte interface. This layer has a high capacitance and an associated resistance. The double layer is conveniently described by a parallel *RC* network, as shown in Fig. 5.24. This electrochemical layer plays the role of dielectric. The capacitance is usually high (μF/cm^2) and the associated parallel resistance is in the range of a few kΩ/cm^2. Cells generate low-frequency signals ($f < 1$ kHz) with amplitude in the microvolt range. These low-frequency bioelectrical signals are easily coupled trough the double-layer capacitance (C_D) and modulate the TFT channel current.

Bioelectronic organic-based devices can make use of a third metal terminal, which is in contact with the electrolyte to modulate the channel conductance as shown in Fig. 5.25. This channel modulation by a DC voltage in an electrolyte medium can be complex. We may assume that ions brought to near the semiconductor/electrolyte surface electrostatically modulate the channel conductance. Or, alternatively, the ions penetrate into the TFT channel causing a chemical oxidation or reduction of the OSC channel. Independently of the mechanism, these devices show *I–V* characteristics alike a transistor. As expected, the range of DC voltages that can be applied is limited to voltages below |1 V|.

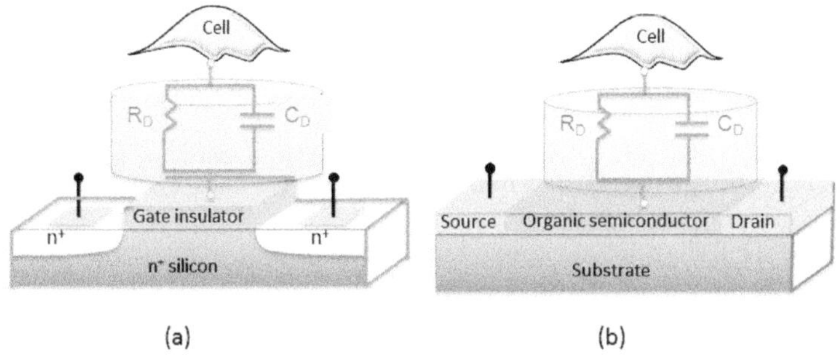

Figure 5.24 Neuron–transistor interfaces: (a) configuration used on inorganic devices and (b) configuration used on organic-based TFTs.

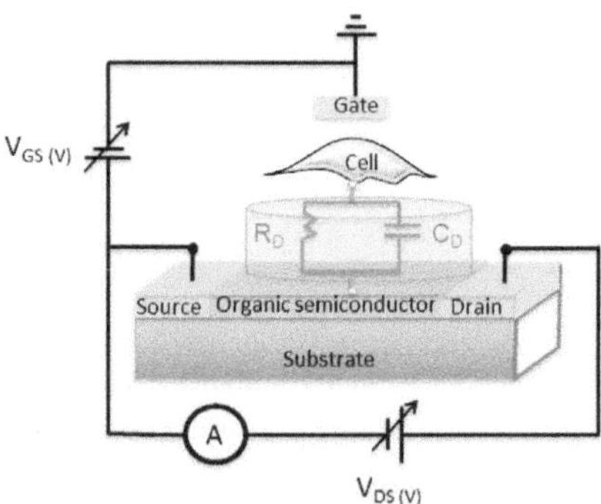

Figure 5.25 Schematic representation of a TFT that uses a gate terminal in contact with the electrolyte.

These sensing TFTs may have also a built-in bottom gate. In this case when immersed in electrolyte solution they become double-gate TFTs. The built-in gate may also be used to select a convenient bias operating point.

These transistors have been successfully explored to measure signals from in neuronal cells. However, before the application of these devices can be successfully established, there are some issues that need to be addressed:

(i) The double-layer capacitance varies with time due to some unknown parameters. Besides, cells also change their surrounding environment and, possibly, change the properties of the double layer.
(ii) The sensing layer is changing continuously with time. These drifts have to be properly understood and quantified prior to establishing a reliable use of the devices in bioelectronic applications.

5.7.2 Organic Thin-Film Transistors as Gas Sensors

If the TFT channel is exposed, the detection of a particular species only depends on the way it interacts with the OSC. For instance oxygen or TNT molecules dope the semiconductor, while water undopes it. This interaction with the semiconductor layer changes the TFT parameters, depending on if the interaction is with the bulk or with the channel layer.

An interaction with the bulk layer causes changes in the transistor off-current. In principle the TFT transfer curve shifts in a parallel fashion in respect to the initial curve. When the change in the off-current is too large may cause apparent changes in the curve shape. Particular care is required in the interpretation of the data. The comparison between curves should only be done for identical current levels.

An interaction with the TFT channel layer can change both mobility and threshold voltage as shown in Fig. 5.26. The transfer curve changes shape. To quantify these changes we have to make use of the gamma parameter explained in Section 5.4.

An interaction with the channel may also affect the carrier injection at the metal/semiconductor contacts. This gives rise to changes in the output I–V curves near the origin. These changes will be easy to pick up by measuring a transfer curve at relatively low bias. Usually a change in slope is observed.

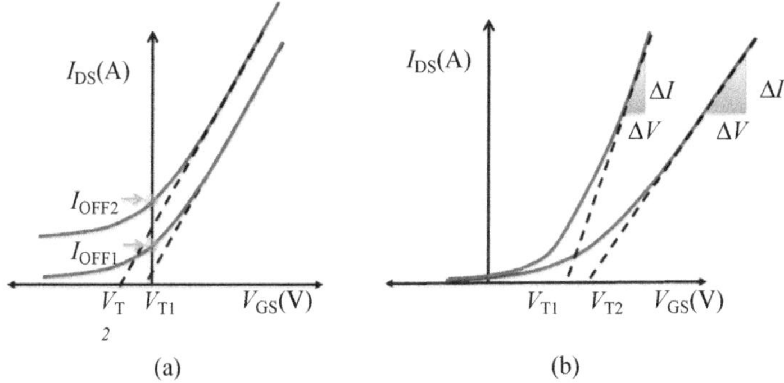

Figure 5.26 Schematic representation of TFT transfer curves and the corresponding changes when the device is exposed to an electrical active substance. Changes on the TFT off-current are expected if the interaction is limited to the semiconductor top surface (a). Changes, both on V_T and on μ_{FET} may occur if the TFT channel is disturbed by the diffusion or electrostatic interact of absorbed species.

5.7.3 Changes in the Dielectric Layer

An interaction with the dielectric layer causes a change in the threshold voltage. This is a well-established sensing method in inorganic-based transistors. Usually, the dielectric layer is exposed and made of a material that can interact with the substance to detect. Often the dielectric is a porous material allowing the diffusion of the analyte.

5.8 Conclusions

The methods and recipes presented in this chapter can be applied to all TFTs, independent of the materials used. In OSCs the carrier transport is slow; traps are abundant and energetically deep. Because of this, the electrical characterization recipes developed to study silicon-based devices are not immediately transposed to OSCs. This chapter pinpoints some of the major differences between the two technologies and how to adapt the characterization methods developed for silicon devices to TFTs based on OSCs. Using the experience gained from the a-Si:H TFT, important progresses have been made in understanding of basic device physics. Influences of the semiconductor band structure, gate dielectric defects, interface DOS, etc., on TFT characteristics, lifetime, failure mechanism, etc., are now understood. This knowledge allows us to use the TFT as a powerful evaluation tool for new materials and new manufacturing technologies. There are constant efforts in searching for new types of organic molecules and polymers to improve performance or to lower the production costs.

Nowadays, the major challenge faced by OTFTs is reliability, especially deterioration of the device characteristics with time. In this context, the methods presented in this chapter to measure traps and interfacial states are a crucial part of the characterization work.

The TFT is a versatile solid-state transistor configuration that can be applied to a wide range of products. It is difficult to predict in which areas or products, OTFTs will have big impact in the future. The key impact areas are assumed to be the following:

- Flexible electronics: OTFTs formed on bendable substrates can be used in lightweight, unbreakable, flexible OLED displays on plastic substrates.
- Simple circuits: With the transistor performance that is available today a wide range of applications can already be addressed, in particular simple logic, memory, and I/O circuits integrated with chemical and biological sensors.

- Sensors, detectors, and bioelectronics interfaces.

TFTs can be easily modified or connected to other devices to detect or generate changes of chemical, biological, optical, magnetic, radioactive, and other properties through controlling the transport of charge carriers, etc.

Still remain many fundamental research challenges that need to be addressed and but it appears likely that OTFTs will find several niche applications.

Exercises

5.1 The definition of V_T from conventional MOSFETs does not apply to OFETs. What is the practical utility of this parameter?

5.2 Charge carrier mobility can be estimated using the Hall effect, time of flight, or space-charge-limited current (SCLC) measurements. How do mobility values extracted from these techniques compare with mobility values extracted by measuring TFT transfer curves?

5.3 The mobility can be extracted from a linear fit of the gate voltage dependence of the current in the linear regime or the square root of the current in the saturation regime. However, questions arise when the device exhibits deviations from this ideal behavior. Summarize the common issues encountered and the strategies to circumvent them.

5.4 TFTs have usually a current in the absence of an external gate voltage. This is a nonideal behavior. Explain how from the shape of this current is possible to gain insight into their physical origin.

5.5 Lateral conduction and fringing effects may cause parasitic effects and lead to wrong TFT parameter extraction. Draw a device structure that can prevent this artifact.

5.6 Explain the physical origin of light-induced effects in OTFTs.

5.7 Outline some strategies to improve TFT operational stability.

5.8 TFTs are multiparameter sensing devices. Outline some strategies as well as some precautions when using TFT parameters in sensing applications.

5.9 OTFTs are low-current and high-impedance devices. Outline some precautions required to perform the dynamic characterization of an inverter circuit.

Suggested Readings

Blood, P., and Orton, J. W. (1992). *The Electrical Characterization of Semiconductors: Majority Carriers and Electron States (Techniques of Physics)*, Academic Press, New York.

Braga, D., and Horowitz, G. (2009). High-performance organic field-effect transistors. *Adv. Mater.*, **21**, 1473–1486.

Campana, A., et al. (2014). Electrocardiographic recording with conformable organic electrochemical transistor fabricated on resorbable bioscaffold. *Adv. Mater.*, **26**, 3874–3878.

Cowell, T. A. T., and Woods, J. (1967). The evaluation of thermally stimulated current curves. *Brit. J. Appl. Phys.*, **18**, 1045–1051.

Cramer, T., et al. (2013). Organic ultra-thin film transistors with a liquid gate for extracellular stimulation and recording of electric activity of stem cell-derived neuronal networks. *Phys. Chem. Chem. Phys.*, **15**, 3897–3905.

Cramer, T., et al. (2013). Water-gated organic field-effect transistors: opportunities for biochemical sensing and extracellular signal transduction. *J. Mater. Chem. B*, **1**, 3728–3741.

Feng, C., Marinov, O., Deen, M. J., Selvaganapathy, P. R., and Wu, Y. (2015). Sensitivity of the threshold voltage of organic thin-film transistors to light and water. *J. Appl. Phys.*, **117**, 185501.

Gomes, H. L., et al. (2004). Bias-induced threshold voltages shifts in thin-film organic transistors. *Appl. Phys. Lett.*, **84**, 3184–3186.

Gomes, H. L., et al. (2005). Electrical characterization of organic based transistors: stability issues. *Polym. Adv. Technol.*, **16**, 227–231.

Gomes, H. L., et al. (2006). Electrical instabilities in organic semiconductors caused by trapped supercooled water. *Appl. Phys. Lett.*, **88**, 08210.

Horowitz, G. (1988). Organic field-effect transistors. *Adv. Mater.*, **10**, 365–377.

Horowitz, G. (2011). The organic transistor: state-of-the-art and outlook. *Eur. Phys. J. Appl. Phys.*, **53**, 33602.

Kim, C. H., et al. (2013). A compact model for organic field-effect transistors with improved output asymptotic behaviors. *IEEE Trans. Electron Devices*, **60**, 1136–1141.

Kim, C. H., et al. (2014). Compact DC modeling of organic field-effect transistors: review and perspectives. *IEEE Trans. Electron Devices*, **61**, 278–287.

Kumar, B., Kaushik, B. K., and Negi, Y. S. (2013). Static and dynamic analysis of organic and hybrid inverter circuits. *J. Comput. Electron.*, **12**, 765–774.

Magliulo, M., et al. (2013). Electrolyte-gated organic field-effect transistor sensors based on supported biotinylated phospholipid bilayer. *Adv. Mater.*, **25**, 2090–2094.

Marinov, O., Deen, M. J., and Iniguez, B. (2005). Charge transport in organic and polymer thin-film transistors: recent issues. *Proc. Inst. Elect. Eng. Circuits Devices Syst.*, **152**, 189–209.

Marinov, O., Deen, M. J., Zschieschang, U., and Klauk, H. (2009). Organic thin-film transistors: part I; compact DC modeling. *IEEE Trans. Electron Devices,* **56**, 2952–2961.

Mathijssen, S. G. J., et al. (2007). Dynamics of threshold voltage shifts in organic and amorphous silicon field-effect transistors. *Adv. Mater.,* **19**, 2785–2789.

Nicollia, E. H., and Goetzberger, A. (1967). The Si-SiO2 interface: electrical properties as determined by the metal-insulator-silicon conductance technique. *Bell Syst. Tech. J.,* **46**, 1055–1133.

Nicollian, E. H., and Brews, J. R. (1982). *MOS (Metal Oxide Semiconductor) Physics and Technology*, Wiley, New York.

Sirringhaus, H. (2009). Reliability of organic field-effect transistors. *Adv. Mater.*, **21**, 3859–3873.

Sirringhaus, H. (2014). 25th anniversary article: organic field-effect transistors: the path beyond amorphous silicon. *Adv. Mater.*, **26**, 1319–1335.

Stallinga, P., and Gomes, H. L. (2006). Modeling electrical characteristics of thin-film field-effect transistors I. Trap-free materials. *Synth. Met.,***156**, 1305–1315.

Stallinga, P., and Gomes, H. L. (2007). Metal contacts in thin-film transistors. *Org. Electron.,* **8,** 300–304.

Stallinga, P., and Gomes, H. L. (2008). Modeling electrical characteristics of thin-film field-effect transistors III. Normally-on devices. *Synth. Met.,* **158**, 473–478.

Chapter 6

The Basics of Organic Light-Emitting Diodes

Mathias Mydlak and Daniel Volz
CYNORA GmbH, Werner-von-Siemens-Straße 2-6,
Gebäude 5110, 76646 Bruchsal, Germany
mydlak@cynora.com, volz@cynora.com

6.1 Introduction

Organic light-emitting diodes (OLEDs) are much like classic light-emitting diodes (LEDs)—able to generate light from electrical energy (see Fig. 6.1). Obvious potential applications are thus similar to what is already known about LEDs, ranging from their use as direct light sources in lamps to backlighting units in displays, when combined with a light guide.

Unlike LEDs, OLEDs consist of several very thin stacked layers and do not rely on small, point-shaped single crystals (see Fig. 6.2). Because of their form factor, they are two-dimensional light sources, which enables new applications. Recent prototypes demonstrated devices with bendable, semitransparent displays using OLED

Organic and Printed Electronics: Fundamentals and Applications
Edited by Giovanni Nisato, Donald Lupo, and Simone Ganz

ISBN 978-981-4669-74-0 (Hardcover), 978-981-4669-75-7 (eBook)
www.panstanford.com

technology. On the other hand, product designers dream of using luminescent foils made from OLEDs on soda cans as smart labels, on semitransparent light-emitting windows, and even on the front of buildings as decorative elements.

Figure 6.1 OLEDs are thin, potentially flexible devices that transform electricity into visible light. The most striking difference between OLEDs and classic LEDs is their form factor; with the latter being small, point-shaped light sources, their organic counterparts may be used for large-area lighting. (Image courtesy: CYNORA)

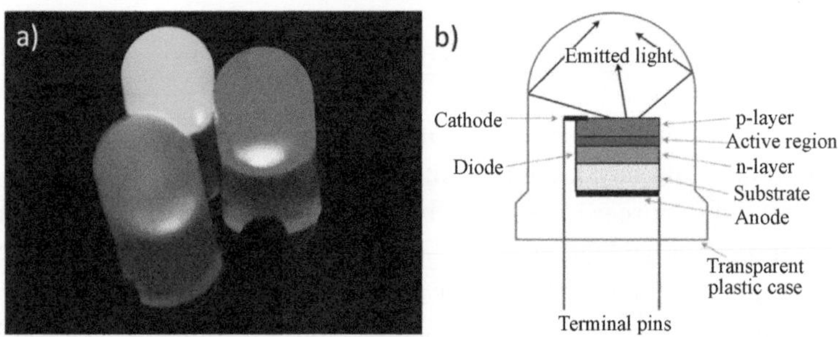

Figure 6.2 (a) Typical shape of RGB LEDs. Published by PiccoloNamek. (b) Principal structure of an LED. Published by J Nava's. Both pictures published under Creative Commons Attribution-Share Alike 3.0 Unsorted.

OLEDs from the ground up can be easily understood following the steps that led to the development of these innovative devices. Therefore, we will give a schematic overview of the research history that led to the original publication of OLEDs by Tang and Van Slyke from Kodak Labs in 1987, before we move on to a detailed description of state-of-the-art OLEDs and the variety of possible modifications that are present in current devices.

6.1.1 History

6.1.1.1 First observation of electroluminescence in the early 20th century

Even though the major parts of OLEDs consist of organic materials, their development has originally been triggered by findings based on inorganic materials. It was in 1907, when the young researcher H. J. Round discovered electrically generated luminescence, electroluminescence, upon application of a voltage of 10 V onto a silicon carbide crystal, also known as *carborundum*. He applied a voltage of approximately 10 V on silicon carbide, a common grinding material that has been known since the 19th century. Round observed yellowish- to orange-colored light, depending on the way the carbide had been synthesized. These findings did not directly lead to a specific application but caused other researchers to study materials under such special conditions. In 1963, Pope and coworkers reported electroluminescence from an anthracene crystal, that is, an organic material. Unfortunately, high voltages above 100 V were required to reach reasonable intensity. This was due to inefficient charge injection into the crystal in combination with poor charge transport within the material due to grain boundaries and defects inherent in the crystal.

In general, the electrical properties of inorganic and organic materials are not alike. Especially in terms of conductivity, things change drastically when moving from inorganic to organic materials. To operate devices with organic materials at low voltages, layer thicknesses of less than 100 nm are common.

6.1.1.2 Inorganic light-emitting diodes

A first triumph in the history of semiconductor physics was the red-emitting GaAs LED developed by Nick Holonyak in 1962. He had been working on diodes with inorganic III–V semiconductors. He used GaAs with different dopings (p- and n-type) to construct a sandwich-type device architecture with an undoped GaAs layer between a p-type doped and a n-type doped layer, which both had been contacted by two electrodes (see Fig. 6.3). Once operating, this diode structure would emit light from the undoped layer, resulting in the first LED. The layers had been vapor-deposited, which is still the process to commercially produce both LEDs and OLEDs today.

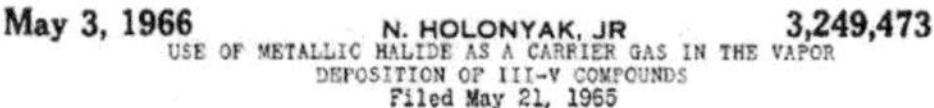

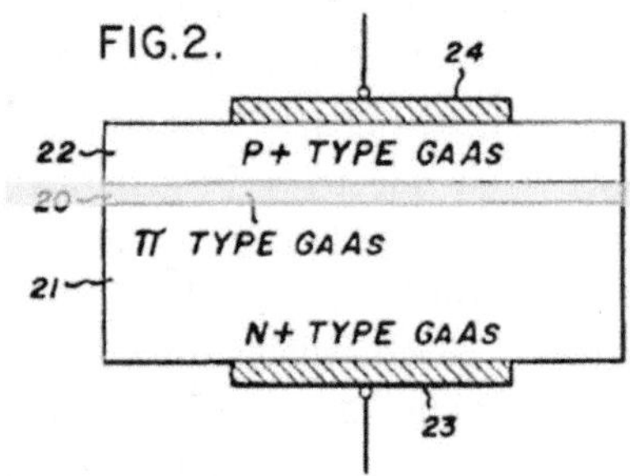

Figure 6.3 Patented three-layer structure of the first inorganic LED by N. Holonyak Jr., May 3, 1966, Pat. 3,249,473.

This discovery started the work on LEDs and resulted in their current potential for (back-)lighting applications across many devices, mostly display and efficient lighting, but also status indicators on many electrical devices. Nevertheless, inorganic LEDs have two limitations, especially for lighting applications.

Since these crystals are small and bright point-shaped light sources, they typically show glare effects and are therefore not suitable for pleasant area lighting. Therefore, small lenses or larger light-guiding systems are used to spread the light over a larger area. Additionally, color tuning in LEDs is complicated because it relies on down-conversion of energy with various phosphors to end up with an emission in the desired part of the visible spectrum.

6.1.1.3 Electroluminescence in organic matter

As already shown before, electroluminescence is not limited to inorganic materials. Using a similar vacuum deposition process as Holonyak Jr. did for GaAs, P. S. Vincett reported relatively efficient electroluminescence from amorphous films of anthracene (and other organics) in 1982. He explained the intense blue emission between 12 and 30 V by homogenous charge transport through the layer.

An important difference between organic and inorganic conductors is the fact that charge is mainly de-localized in so-called energy bands in inorganic materials, whereas charge is transported via hopping between localized states in organic materials. This causes the charge carrier mobility and therefore conductivity of organic matter to be much smaller compared to metals or semiconductors. This is the reason why Pope et al. had to use up to 400 V to generate luminescence. The intrinsically smaller charge carrier mobility of organic substances did not prevent researchers from realizing classical semiconductor applications such as transistors, solar cells, and OLEDs, though.

6.1.1.4 The first OLEDs with Alq_3

A combination of the above findings led to the first report of the first OLED by Tang and Van Slyke in 1987. They had built up a sandwich-type device from vacuum-deposited organic materials, resulting in a sequence of amorphous layers with specific functions, similar to the LED, described earlier. Onto a transparent (inorganic) anode (indium tin oxide, ITO), which was deposited on a glass substrate, they deposited a hole-transporting diamine layer (corresponds to p-type side in LED) and an electron-transporting layer (n-type) of aluminum (III)-quinolate (Alq_3), which also acted as the emitter molecule at the same time. The cathode consisted of coevaporated Mg:Al (see Fig. 6.4).

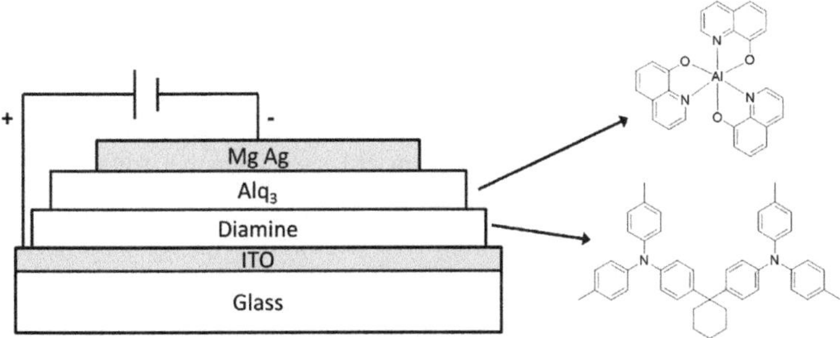

Figure 6.4 The first OLED with a four-layer OLED structure from Tang and Van Slyke, 1987.

This device architecture required only low voltages (<10 V) for efficient light emission above 1000 cd/m^2 (candela per square meter, also called *nits*). For comparison, the light intensity of most display applications usually stays below 300 nits. Even though the efficiency values of 1.5 lm/W (lumen per watt) were relatively high and the external quantum efficiency (EQE) reached a photon:electron ratio of 1%, there was much room for improvement. To discuss these options, we need to take a look at a basic OLED structure with the specific functions of each layers and the material that are used (see Section 6.2).

While the first reported OLED has been based on evaporated small molecules, Burroughes et al. published the first OLED with a polymeric emitter in 1990, using poly(*p*-phenylene vinylene) (PPV). A major improvement of OLEDs in terms of the efficiency step was the use of phosphorescent metal–organic emitters by Forrest and Thompson, which is the foundation of modern OLED emitters in 1998. Further development in highly efficient emitters led to the presentation of thermally activated delayed fluorescence (TADF) in organic emitters by the group of Chihaya Adachi in 2011 and in copper-based emitters by the group of Hartmut Yersin in the same year. The main characteristics of the three emitter concepts will be discussed in more detail in the following section.

6.2 Structure of an OLED

6.2.1 Basic Device Structure

A so-called bottom-emission OLED is shown in Fig. 6.5, that is, with transparent electrode on the substrate side, resulting in light emission into/through the substrate. It consists of a transparent carrier substrate, a transparent anode, followed by a hole injection layer (HIL), which supports efficient charge injection into the organic material. Since the anode withdraws negatively charged electrons from the organic layer, the resulting occupancy can be regarded as positively charged particle moving into a direction opposite to the electrons. This concept is widely used for easier visualization of charge movement within the layers.

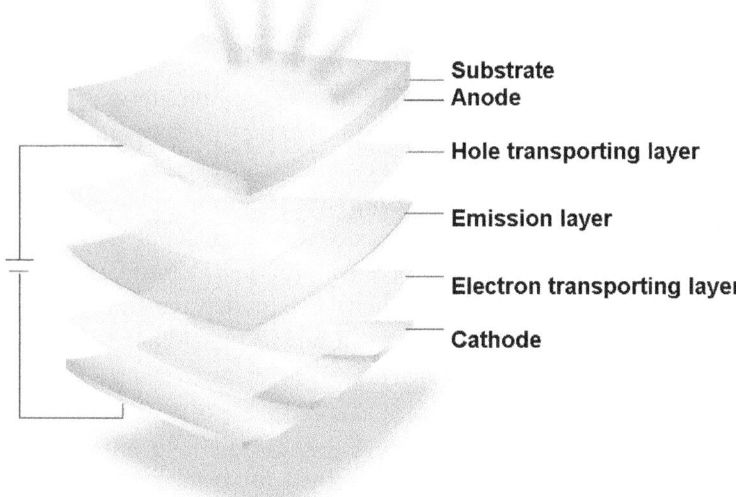

Figure 6.5 Bottom-emission OLED structure: anode/HIL/HTL/EML/ETL/EIL/cathode. OLEDs typically consist of several layers, for example, a supporting substrate made of glass, metal, or plastic foils; electrodes for the generation of charge carriers; layers to enable charge carrier transport; and one or several emission layers, where charge carriers recombine and light is generated. (Image courtesy: CYNORA)

The HIL is followed by a hole transport layer (HTL), which acts as a mediator between the injection layer and the emissive layer (EML), in the center of the device. The EML converts the incoming charges, holes and electrons, into so-called excitons, charge pairs that can be then converted into photons by the emitter molecules. In their original publication, Tang and Van Slyke used a layer of neat Alq_3 in the EML. Further research has led to the adoption of a more sophisticated layout of this layer. It usually consists of a host–guest system, with the emitter doped into a host material at low concentrations. Such design allowed the optimization of charge transport onto the host molecule and the emission properties on the emitter separately from each other. The host molecule was designed to have high stability when participating in the charge transport and high morphological stability to prevent, for example, crystallization and the formation of grain boundaries as well as to transport charge carriers efficiently. The emitting molecule needs to convert energy to light efficiently. Preferably, nonradiative processes are suppressed to achieve internal efficiency values close to 100%. Other parameters to characterize emitters are the emission decay time (synonyms: excited state lifetime or exciton lifetime), which needs to be as short as possible. Ideally, the spectra generated by excitation with UV photons (photoluminescence) and the spectra of OLED devices (electroluminescence) are superimposable. While the spectrum needs to be in the spectral range that can be detected by the human eye, for example, between 400 and 800 nm, the shape depends on the application of the OLED. In displays, the spectrum needs to be narrow to give red, green, and blue in decent quality. For lighting applications, broad spectra can be tolerated to achieve white light.

The electrons are transported into the EML by the electron transport layer (ETL). The ETL is then followed by an electron injection layer (EIL), with the function of enhancing electron injection from the metallic cathode.

When an external direct current (DC) bias is applied, electrons migrate from the cathode through the EIL and the ETL into the EML and holes migrate from the anode via the HIL and the HTL into the EML. The opposite charges attract each other and recombine to an exciton, which then excites an emitter molecule, which then consequently emits light. If a host is present, the charge carrier recombination can take place on the host molecules, which reduces the electrical stress on the emitter and therefore increases the operating lifetime of the OLED device.

6.2.2 Charge and Energy Transfer in OLED Devices

The various processes that lead to light generation in an OLED are complicated, have been described in detail before, and are thus only briefly covered here. The four fundamental steps of light generation in an OLED are depicted in Fig. 6.6. First, charge carriers are generated. Positive and negative charge carriers are called holes and electrons, respectively. Driven by the external electrical field and given the Coulomb attraction between oppositely charged carriers, electrons are moving toward the anode, while holes are moving toward the cathode. Some more details regarding the charge transport are addressed later. After recombination of holes and electrons and various energy transfer processes, the emitting molecules are excited. Due to quantum statistics, only 25% of the excited emitting molecules are in a singlet excited state, while 75% of triplet excited states are yielded. Conventional fluorescent materials such as aluminum quinolate, Alq_3, or polymers such as polyfluorenes are not able to harvest the triplet excited states, thus limiting the theoretical electroluminescence quantum efficiency to 25%. To harvest both singlet and triplet excitons for light emission, either phosphorescent emitters using iridium or platinum or emitters showing TADF are required. In any case, organic materials employed in these layers can be characterized using several parameters. In this section, we will describe these metrics and then look at the specific values that characterize the materials for each layer.

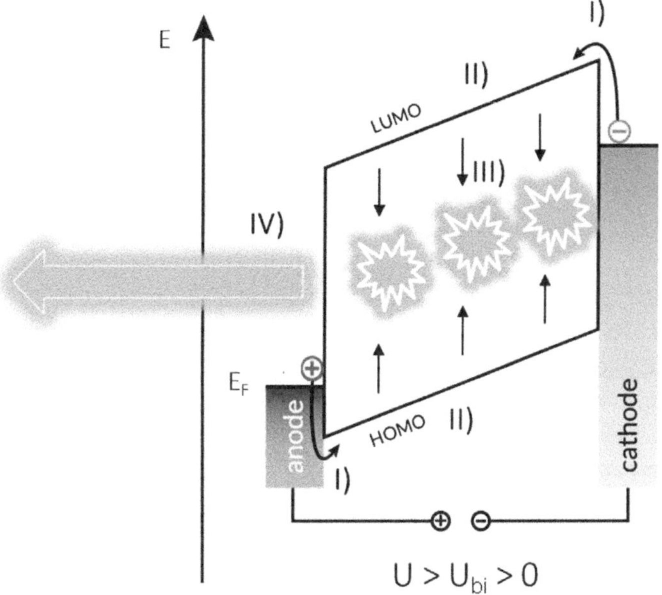

Figure 6.6 Elemental processes during the operation of an OLED. Basic steps are (I) injection of charge carriers into the organic materials; (II) transport of said charge carriers, which is driven by coulomb interaction; (III) formation of excitons; and (IV) emission of light. (Image courtesy: CYNORA)

The organic molecules in OLEDs are semiconductors, in a similar fashion to the III–V semiconductors in LEDs. While inorganic semiconductors are usually crystalline materials with defined conduction bands, their organic counterparts are fused aromatic (i.e., unsaturated) systems that have so-called molecular orbitals. These can transport holes or electrons. The main orbitals that have to be taken into account for charge transport characteristics are the highest occupied molecular orbital (HOMO) and the lowest unoccupied molecular orbital (LUMO). Holes are transported via the HOMO and electrons through the LUMO, which is higher in energy than the HOMO by definition. The energy levels of different materials vary and are the basis for the selection of materials in an OLED stack in order to allow for ideal charge transport. The energy steps between adjacent layers need to be below 0.5 eV in order to reduce barriers for charge carriers that might accumulate at the interface. Therefore, for charge injection purposes hole transport materials are mainly selected by their HOMO values, while electron transport materials are selected with respect to their LUMO values. These are then arranged in a stepwise sequence to guide the charge carriers toward the emitter molecules in the EML.

Depending on the chemical structure of the molecule HOMO/LUMO values can vary not only based on modifications to a given chemical core structure, but can also be influenced by the arrangement of the molecules within the film and therefore also differ between individual molecules of the same chemical structure. This results in heterogeneous HOMO/LUMO distributions throughout the layers. The charge carriers consequently hop from the orbitals of one molecule to the other by a hopping mechanism, vide infra, as to be seen in Fig. 6.7. The parameter that describes how easily this can happen is called charge carrier mobility and is generally investigated in an amorphous film for OLED materials. HTL materials show high hole mobility, while ETL materials show high electron mobility. The charge transport in amorphous organic solids is not yet fully understood. In certain, rare cases, charge carriers may move more freely when strong π-stacking occurs, while most systems may be described more correctly using Marcus's theory.

This so-called hopping model deconstructs the charge transport process into cascades of redox reactions. A schematic description of these processes is given in Fig. 6.7. Through the electrodes, radical cations and anions are generated by chemical oxidation and reduction. These cations and anions react with neighboring neutral molecules and exchange charge. This plethora of redox equilibria is influenced by the external electrical field, resulting in a net flux of charge carriers through the organic layers. A quantitative description of an elemental hopping process can be done with Marcus's basic equation. Without going into too much detail, the conductivity of an organic molecule can be maximized by using a planar, rigid structure.

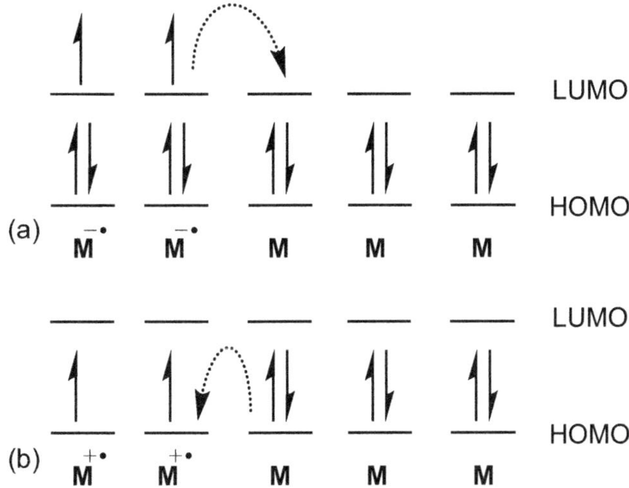

Figure 6.7 Transport of electrons (a) is achieved by chemical reduction, while holes (b) are transported by oxidation. In both cases, there exists a cascade of coupled redox equilibria. The hopping model treats the charge transport as hopping of charge carriers between neighboring molecules. The driving force is the external electrical field.

6.2.3 Relevant Physical Parameters for OLED Materials

OLED materials have to be described with a broad set of properties that are influenced by the frontier orbitals of the molecules. The basics of these can be found in standard chemistry textbooks, both for organic and for inorganic molecules. These frontier orbitals, in combination with the molecular shape and its movement in the excited state, result in different absorption and emission properties, excited state lifetimes (how long the molecule needs to emit light, between nanoseconds in fluorescent materials up to seconds in certain phosphorescent materials), and emission quantum yields (how efficiently it emits light). Furthermore, which potential has to be applied to inject a charge (positive=hole, negative=electron) onto a molecule and how difficult it is for the charge to jump from molecule to molecule, that is, the charge carrier mobility, has to be characterized. On the basis of the combination of these characteristics, the molecules can be classified into emitters, hole transporters, electron transporters, host materials, injection materials, any many more.

6.3 Emitter Concepts

6.3.1 General Differences between the Different Concepts

One major advantage of OLEDs in comparison to LEDs is the possibility to tune the color on the emitter molecule by relatively easy chemical modification. Most of the material classes described in this section can be color-tuned across the visible spectrum, from blue to red. While the color-tuning ability is developed throughout these systems, their efficiency and suitability for commercial application vary to a large extent.

In comparison to the rather small amount of structural motifs that are found in hole- and electron-transporting materials, the spectrum of compounds used as actual emitting materials is much broader. Various concepts are still under evaluation, ranging from fluorescent materials, which have been used in early OLEDs, to conjugated polymers, to phosphorescent materials (so-called triplet emitters), to nanoparticles. There are three main classes of emitter systems that vary in their emission process: fluorescence, phosphorescence, and TADF. To understand the differences, it is relevant to discuss excitons first. To do so, we cannot avoid a few lines of quantum mechanics.

Excitons are formed when two charges recombine (see Fig. 6.8). Electrons and holes have individual spins. Depending on the value of each spin, the resulting exciton can be either a singlet exciton or a triplet exciton, where the latter is lower in energy than the former. Due to spin statistics, roughly 25% of the excitons formed are singlets and roughly 75% are triplets. Once these excitons are formed on a molecule, this molecule is then excited into its singlet or triplet excited state. Only emitter molecules that can transform both kinds of excitons into light can be efficient emitters in OLEDs.

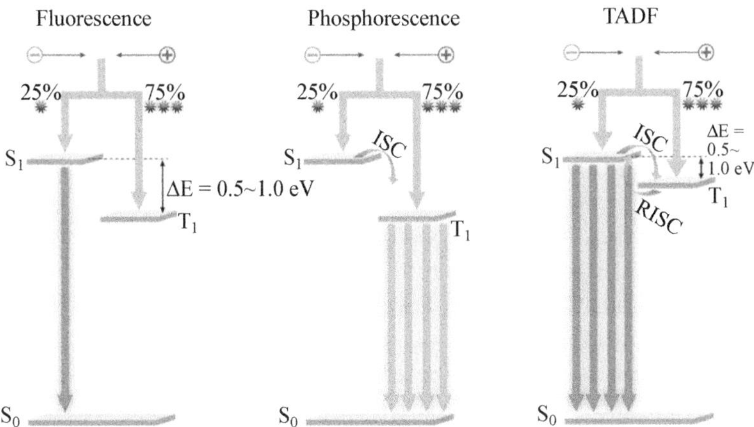

Figure 6.8 Exciton statistics in OLED devices and their emissive decay paths. (Image courtesy: CYNORA)

From the three classes described before, fluorescent emitters can only make use of singlet excitons and therefore have limited performance. In addition, the created triplet excitons, if not used for light emission, can lead to undesired reactions on the molecules or in between adjacent molecules, thus leading to material decomposition and to not emissive charge traps. Fluorescent materials are currently being replaced in commercial products. While there are efficient solutions for red and green emitters, deep-blue materials are still fluorescent, caused by the lack of efficient and hence stable deep-blue emitters. The next two classes of materials, phosphorescent and TADF materials, have one thing in common to separate them from purely fluorescent materials. After the initial excitation via an exciton, they can switch from the singlet state to the triplet state (phosphorescence) or vice versa (TADF) via intersystem crossing (ISC) or reverse intersystem crossing (RISC), which allows them to make use of both exciton types. This ability can either be generated by introducing a heavy metal into the chemical structure or by careful sterical modifications on the molecular level.

Phosphorescent materials, currently state of the art, can convert singlet excitons into triplets and emit light from the triplet state. The absence of heavy atoms and therefore spin–orbit coupling (SOC) in fluorescent molecules have limiting consequences for the device efficiency because they prevent the use of triplet excitons. The earliest strategy to overcome this limitation led to the development of phosphorescent metal–organic emitters. To overcome this issue, metal–organic compounds containing heavy metals such as iridium and platinum were investigated as a different class of emitter materials. These metal ions lead to a strong SOC, which leads to a change in the emission behavior, and these kinds of compounds are theoretically able to reach an internal efficiency of up to 100%. In these cases, a large SOC not only induces fast ISC from higher-lying singlet states to the emitting triplet state but also leads to an emissive transition from the lowest triplet state to the singlet ground state, which is formally forbidden. Therefore, both singlet and triplet excitons can be used for the emission of light. Since all excitons are harvested via the excited triplet state, this emission, which is formally a phosphorescent emission, is called triplet harvesting.

The TADF concept, representing the youngest material class in the focus of OLEDs, can convert triplet excitons into singlet excitons after thermal activation, as becomes clear from the name. Most of such materials have been demonstrated to be based on copper or can be purely organic. While processes for phosphorescent materials are part of commercial production already, TADF materials are being investigated by a growing community for their potential to find a stable deep-blue emitter. In addition, the absence of rare elements is relevant for large-scale manufacturing and the generation of new OLED applications.

In the following section a small selection of different emitters will be presented, sorted by their emission principle.

6.3.2 Fluorescence

Representative examples of fluorescent molecules are given in Fig. 6.9. While the first OLEDs mainly contained known fluorescent dyes such as 4-(dicyanomethylene)-2-methyl-6-(4-dimethylaminostyryl)-4H-pyran (DCM), rubrene, or Alq_3 derivatives, chemists and materials scientists soon developed new emitting compounds that were customized to their intended application in OLEDs, such as $AlMq_3$ or 4,4′-bis(2,2-diphenyl vinyl)-1,1′-biphenyl

(DPVBi). Apart from aluminum, which is a very abundant element, these emitters contained no elements except for carbon, hydrogen, oxygen, nitrogen, and sulfur.

Diverse kinds of small molecules as well as polymer-emitting compounds showing fluorescent emission have been investigated as potential emitting materials since the beginning of OLED research until now, even though these materials suffer from the fact that they can only use up to 25% of the total number of excitons generated in an OLED device. All triplet excitons are lost due to spin-forbidden emissive transitions, leading to radiationless deactivation of excited triplet states, since only the singlet excitons enable spin-allowed emissive transitions from excited singlet states to the singlet ground state.

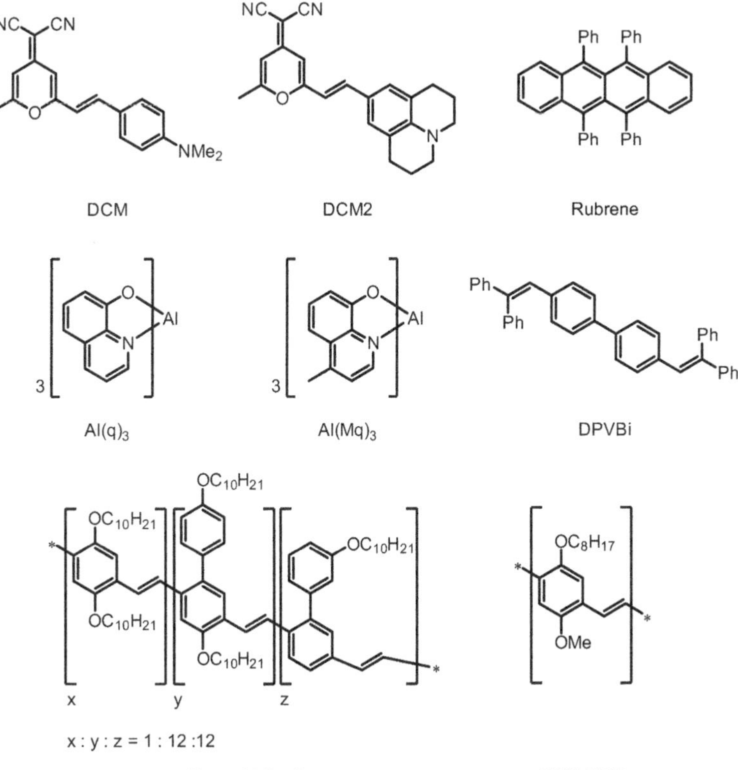

Figure 6.9 Selected examples for classical fluorescent emitter materials. The last two PPV derivatives represent emissive polymers.

6.3.3 Phosphorescence

While many heavy atoms, from osmium, rhenium, and ruthenium to lanthanides, have been tested, two metals have proven very suitable for OLED applications, platinum and iridium (Fig. 6.10). Iridium is considerably more relevant than platinum because of the somewhat easier realization of very high efficiencies and the shorter emission decay time. Although the first triplet-harvesting complex, PtOEP, gave a high efficiency of 4% in 1998, the first device with $Irppy_3$ easily surpassed this efficiency, yielding almost 8% efficiency and a reduced efficiency roll-off at high driving voltages due to the shorter emission decay time of most iridium complexes, in general.

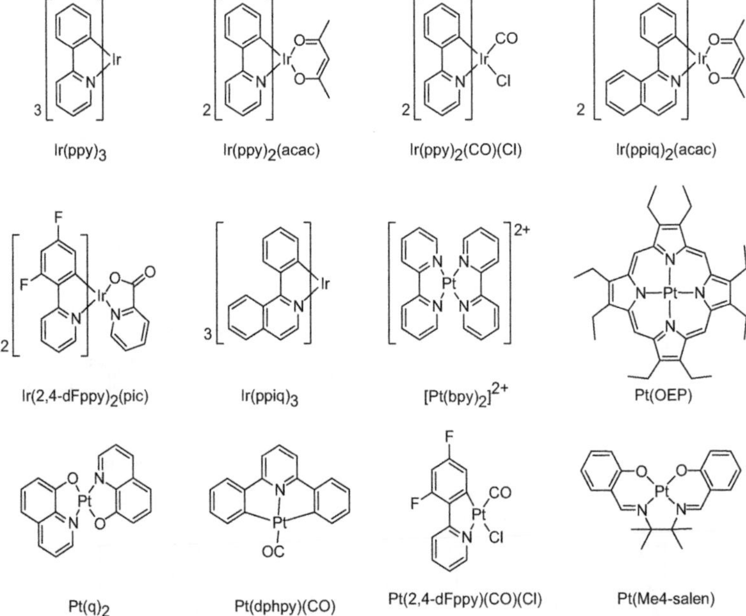

Figure 6.10 Selected examples for platinum- and iridium-based emitter materials.

6.3.4 Singlet Harvesting/TADF

In contrast to triplet-harvesting materials, materials exhibiting TADF are emitting from the singlet states, which is accordingly known as singlet harvesting. This process has already been described before and is another concept to allow efficient emission in OLEDs by including singlet and triplet excitons. Materials representing a small energy gap between the excited singlet state S1 and the triplet state T1 are mostly based on copper or on purely organic core structures. Some representatives of this class of materials are shown in Fig. 6.11.

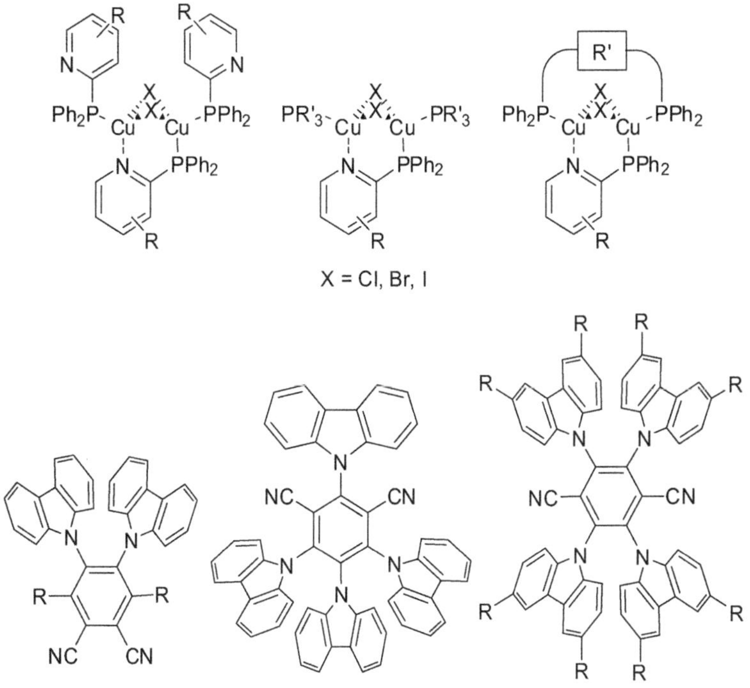

X = Cl, Br, I

4CzPN: R = carbazolyl
2CzPN: R = H

4CzIPN

4CzTPN: R = H
4CzTPN-Me: R = CH3
4CzTPN-Ph: R = phenyl

Figure 6.11 Copper-based and organic structures representing TADF emission.

6.3.5 Host Materials

Due to charge transport reasons and to avoid quenching effects that can come up when many molecules in excited states are in close distance, the emitting materials are diluted in matrix materials, often with doping concentrations around 10 wt%. These host materials often share structural motifs with hole- and electron-transporting materials. So-called ambipolar host materials often contain both hole- and electron-transporting functional groups fused in one molecule.

Host materials have several functions, as already indicated before. Since both charge carriers recombine on host molecules within the EML, they need ideal mobilities for both electrons and holes in order to avoid charge accumulation at one of the interfaces (to the ETL or the HTL). Additionally, the exciton needs to quickly diffuse to an emitter molecule and not to any of the neighboring layers. Depending on the general stack architecture, materials with higher hole or electron mobilities can be favored. Hosts with similar mobilities for both charges are called ambipolar. An ambipolar host system can also be emulated by mixing hole- and electron-transporting materials. This approach can lead to fine-tuned transport capabilities but adds a layer of complexity to the device, with a three-component mixture in the EML.

Once an exciton is formed, it also has to be efficiently onto an emitter molecule; therefore efficient electronic coupling between both molecules is required. If that is not possible, the exciton can lead to light emission from the host molecule, which is generally not an efficient emitter, leading to efficiency losses. An efficient electronic coupling includes matching HOMO and LUMO values, that is, higher LUMO than the emitter and lower HOMO than the emitter, and matching excited-state energies, which will be described in more detail in the emitter section.

6.4 Additional Aspects

6.4.1 Encapsulation

Most OLED materials are sensitive to moisture or oxygen. Cathode materials are generally highly reactive metals that strongly react with water and oxygen and lead to device malfunction. Organic materials, on the other hand, can react with oxygen or water, not just while they are electrically excited and lead to defect areas or enhanced device degradation. The stability of the device and materials versus undesired reactions is represented in the shelf life, while degradation during operation is represented in the operational device lifetime. In any case, with the aim of avoiding such negative effects, OLEDs are generally encapsulated for protection.

The classical concept is glass encapsulation on glass substrates. This is a rigid encapsulation that is glued on top of the substrate, with the glue applied around the active device area. Glass encapsulation is generally very long lived and influenced strongly by the barrier properties of the glue. Since the encapsulation is performed under inert atmosphere, changes in external pressure can lead to cracks in the glue and a consequent loss of barrier function. Major pressure changes like in plane take-offs or landings can even lead to cracks in the glass substrates. Desiccant pads are often incorporated into the barrier glass in order to extend the device lifetime by absorbing diffusing water.

Alternatives are barrier films, which consist of a plastic-type substrate (mostly polyethylene terephthalate [PET]) and a sequence of alternating organic/inorganic layers to avoid diffusion. Some of these barriers can be flexible, while it has to be noted that bending can cause cracks in the inorganic layers and limit the barrier properties. In any case, these barriers are becoming the favorites because they are easy to apply, are lightweight, and can be made flexible.

6.4.2 Flexible Substrates

Similar to encapsulation materials, substrates can be made from glass, PET, PIN, or others. The selection of the substrate depends on several aspects. In the first place, glass substrates can only be used for rigid, flat OLED applications. This is typically the case for current OLED displays (smartphones, tablets, TV) and lighting applications. For conformable applications, thin glass can still be used like all plastic-based substrates, but it has cannot be employed for completely flexible applications.

Since plastic-based substrates are more rugged than glass-based ones, they are also employed in first smartwatches, even without the requirement of a curved or flexible application in the final device.

Similarly, the production process has to be adapted to the substrate. Since flexible and even conformable substrates have to be rigid during the deposition process of the individual layers, they are often laminated onto another carrier substrate like glass and released from it after the production process. In contrast, a roll-to-roll inline deposition process would be perfectly suited for a flexible rollable substrate without a carrier backbone. While such processes are still in pilot production stages, processing of flexible films on top of carrier substrates is already taking place on the medium scale.

6.4.3 Multilayer Architectures

6.4.3.1 Substrate and electrode materials

The current standard material for transparent anodes is ITO. It is being deposited onto the substrate using a sputtering process. Generally, it is possible to purchase substrates like glass or PET foils that already carry a layer of ITO. This layer is characterized by its roughness and its conductivity (or resistivity), which corresponds to its thickness. Even though ITO is today one of the standard materials in used in displays, the sheet conductivity is still quite low, especially for large-area applications like lighting. Therefore, many alternatives are being evaluated, like combinations with metallic grids, to reduce sheet resistivity over large areas.

While plastic substrates are becoming more important for applications with flexible/bendable devices, the rigid structure of ITO becomes an even bigger problem, even in combination with metallic grids that can compensate the formation of islands upon fractures of ITO layers. A very popular alternative is graphene, which can form transparent layers with high conductivity, in principle. But due to the planar structure of graphene, it usually aligns flat on the substrate, resulting in high in-plane conductivity but low conductivity normal to the device plane. As processing generally affects the performance of organic materials to a very large extent, this issue might be overcome in the future.

A promising alternative for flexible anode structures is modifications of a material that has formerly only been used as a solution-processed HIL. PEDOT:PSS is an ionic mixed polymer with good conductivity and reasonable transparency. While there will be more details in the next paragraph, it has to be mentioned that many successful attempts have been made using special formulations of this material as the anode, with and without supportive metallic grids.

In most OLEDs, cathodes are made from more or less abundant metals such magnesium, aluminum, barium, silver, or copper. When bulk films are used, the cathode is not transparent, whereas some degree of transparency may be regained when using nanowires, for example, made from silver. Common anode materials are optically transparent and ought to have high conductivity. Because of this, transparent conductive oxides (TCOs) are commonly used. A well-established TCO is ITO, $(In_2O_3)_{0.9}(SnO_2)_{0.1}$. Other feasible TCOs are discussed as alternatives.

On top of the TCOs, injection layers are often employed, first to compensate the high roughness of the oxides and second to reduce the barrier between the work function of the TCO and the HOMO energy of the subsequent layers. A popular material for this purpose is PEDOT:PSS (see Fig. 6.12), a water-soluble composite material made from a conjugated polythiophene and deprotonated polystyrene sulfonic acid. Beside PEDOT:PSS, materials such as molybdenum oxide, vanadium oxide, or copper(I) salts may be used.

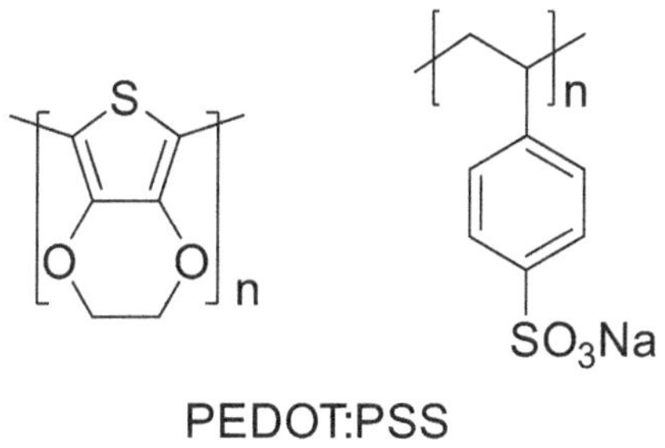

Figure 6.12 PEDOT:PSS is a mixture of a conjugated polythiophene and polystyrene sulfonic acid.

6.4.3.2 Charge transport materials

Most hole injection and hole transport materials consist of electron-rich aromatic systems, that is, six-membered rings. Many variations of the structure, size, amount, and position of heteroatoms are possible and lead to a whole variety of energy levels, film-forming properties, and optical characteristics.

Apart from the molecular design, HTL materials can also be modified by doping during the deposition process. Similar to inorganics, this is possible during sublimation processes and already in commercial application. Examples of hole-transporting materials are given in Fig. 6.13. Commonly, electron-rich aromatic systems such as triarylamine or carbazole derivatives are used.

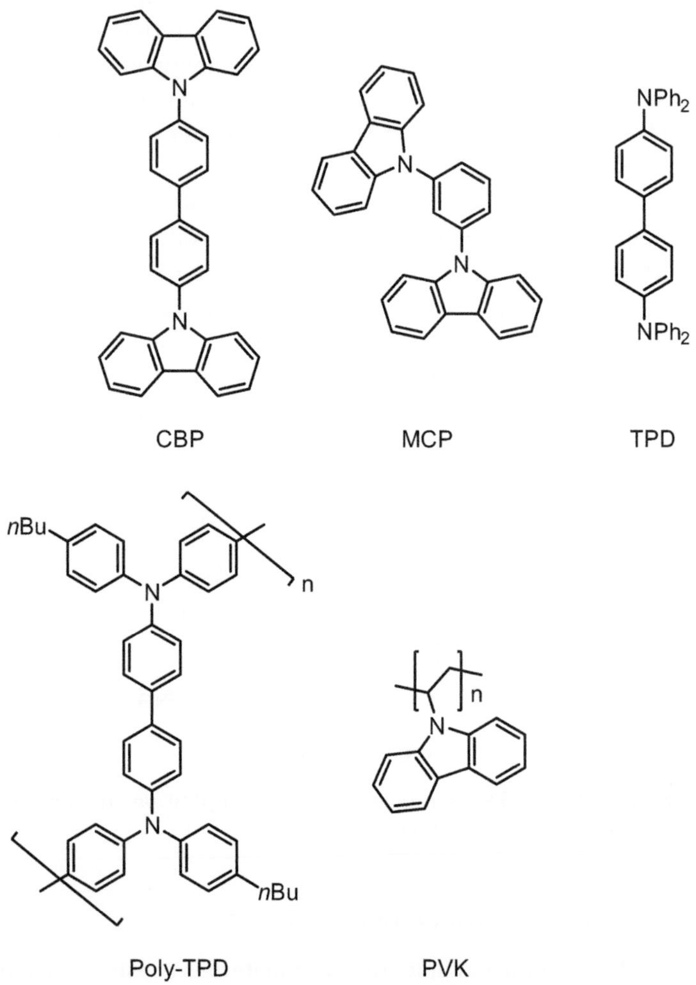

CBP MCP TPD

Poly-TPD PVK

Figure 6.13 Selected examples for hole-transporting materials. Often, electron-rich aromatic compounds such as carbazole derivatives or triaryl amines are used.

Electron transport materials are characterized by low-lying LUMO values, making it easy to inject electrons into these orbitals. Furthermore, they are optimized for high electron mobilities. This is generally achieved with electron-poor aromatic systems, which are often five-membered heteroaromatics, for example, benzimidazoles, pyridine, and triazole, but also phosphinoxide and sulfones as functional groups (see Fig. 6.14).

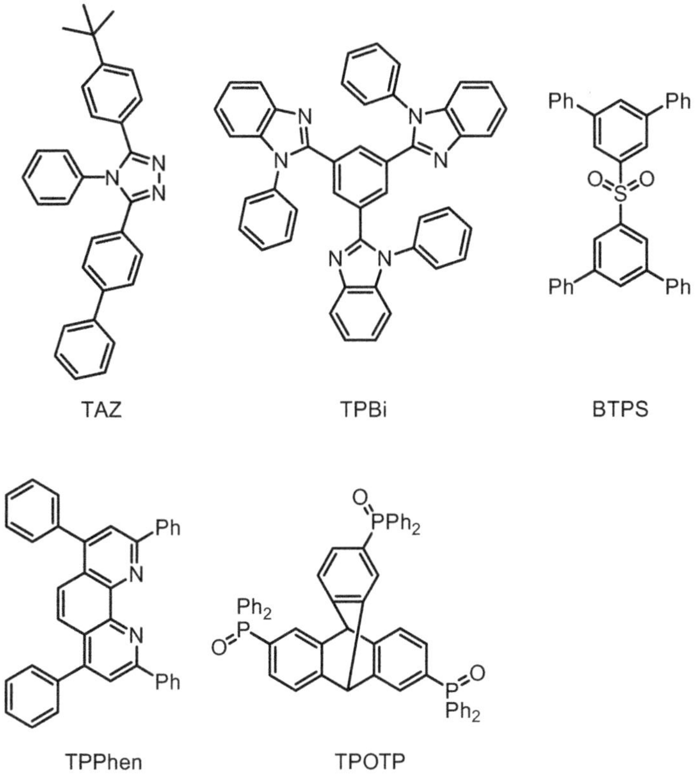

TAZ TPBi BTPS

TPPhen TPOTP

Figure 6.14 Selected examples for electron-transporting materials. Often, electron-poor aromatic heterocycles such as phenanthroline, pyridine, triazole, and benzimidazole are used. Further relevant functional groups are sulfones, phosphinoxides, and triaryl boranes.

6.4.3.3 Other layers for improved device performance

In addition to all the layers described before, there can be many other layers incorporated in OLED devices to improve charge transport, control the charge balance in the EML, or protect certain materials from charge carriers. There are some HTL materials that decompose when reduced or ETL materials that decompose when they are oxidized. To avoid this from happening, blocking layers (hole-blocking/electron-blocking layers, HBLs/EBLs) can be used between the EML and the respective layers with energy levels that block holes or electrons from moving through these layers. Similarly, there are also exciton-blocking layers that avoid exciton diffusion from the EML to adjacent layers.

6.4.4.4 White-light OLEDs

To generate white light from OLEDs with different colors, there are several concepts. Since white light is composed of several individual colors, as represented in the CIE diagram below (Fig. 6.15), once can either use two or three colors to create white light. The quality of the light can be described by the color point CIE x and CIE y and the color-rendering index (CRI). These parameters are mainly affected by the shape of the emission spectrum of the emitting compound. The CIE coordinates describe the color composition, that is, the relative intensities of the included colors, while the CRI describes the impression of colors when illuminated by the given light source. The lower the CRI, the worse is the color impression, as shown in Fig. 6.15. Instead of using the CIE coordinates, it is common in the lighting industry to use the color temperature. It is represented by the black-body curve and has higher values towards the blue side of the CIE diagram. While 2000 K represents a warm light impression, 6000 K is a value for cold blueish light, like known from fluorescent tubes.

The modification of the CRI is also crucial for a nice color perception. It is desirable to have a very broad emission spectrum to have a high CRI. Therefore, a three-component while OLED (WOLED) is often superior to two-component WOLEDs.

The stack design of WOLEDs is a very important topic because it has to take into account the aforementioned points, while maximizing the device efficiency to reduce the cost per light output. Consequently, WOLEDs mostly consist of so-called tandem devices, where two or three OLEDs are directly stacked on top of each other and then connected to an external circuit. This maximizes the device efficiency but makes the production more complex because tens of layers have to be deposited on top of each other with perfect alignment and no defects. In between such stacked OLEDs charge generation layers (CGLs) are used to generate holes for one diode and electrons for the other one.

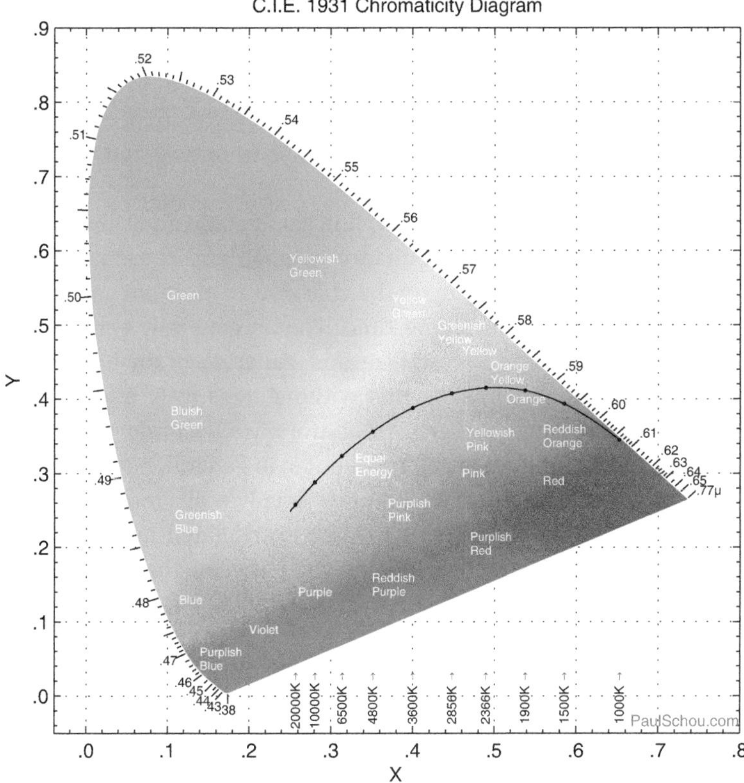

Figure 6.15 CIE diagram with the black-body curve indicated. (Image courtesy: PaulSchou)

Process stability during device production is crucial because small variations in layer thicknesses or morphology can cause color shifts in between different devices. The human eye is very sensitive and can spot such variations quickly in OLED assemblies, which are very common when creating large OLED luminaires.

6.4.4.5 Outcoupling

To maximize the light that leaves the device there are many optimization techniques available that can be divided into two groups, internal and external outcoupling.

Internal outcoupling techniques are all measures within the stack architecture that maximize the outcoupled light. A very popular technique is optical fine-tuning of charge transport layers. This means the thickness variation of layers in order to generate a cavity between the electrodes for constructive interference of the desired wavelength, that is, the emission wavelength of the dopant. Such variations can lead to trade-offs in electrical performance; therefore there is generally a compromise in optical and electrical properties of the device.

Furthermore, internal outcoupling can include additional layers within the OLED stack that enhance outcoupling. These layers can be materials that participate in the charge transport but can scatter the emitted light at the same time. This is generally achieved by materials that tend to crystallize and generate grain boundaries within the layers. Achieving this without negatively affecting the device performance is not straightforward; therefore external outcoupling techniques are more popular, in general.

External outcoupling techniques involve all measures that can be applied on the outside of the encapsulated device. Due to the difference of refractive indices (η) between the substrate or encapsulation of the device and the surrounding air, a large fraction of light cannot leave the device through the emission plane but leaves the device on the edges or is completely lost in so-called waveguide modes. To adjust for this problem, there are plastic films that can be laminated on top of the device and support the outcoupling due to refractive index matching and a special surface structure. A very common representative is the microlens array structure (MLA). Films with variations of this structure are very common, especially in lighting applications since they can also act as diffusors to homogenize the light impression on larger OLED panels. This function can also be used to hide conductive structures like metal grids that are used in large-area OLED panels.

6.5 Outlook and Challenges

We hope that this chapter gave an insight into many aspects of the very broad field of OLEDs. While the applications in displays are slowly growing along with the OLED lighting market, there is still fundamental research ongoing in order to achieve highly efficient deep-blue materials and the production of flexible backplanes for flexible OLEDs, and also the selection of the production process for many kinds of OLED devices is still not finished. Therefore, there are many ways to contribute, and we strongly encourage the reader to participate in this exciting field of applied research and development.

Exercises

6.1 Why is it easier to realize a thin, large-area light-emitting device with OLEDs than with LEDs?

6.2 Describe the charge transport mechanism in noncrystalline organic matter. Name the main differences between this mechanism and the processes occurring in conductive metals.

6.3 In Section 6.1.1.1, it was mentioned that thin layers of anthracene, an organic fluorescence emitter, show electroluminescence when being contacted with 30 V, while larger crystals of the same material required up to 400 V. What are possible reasons for this behavior?

6.4 Given are the CIE coordinates of colored OLEDs. To what color do those coordinates correspond?
- (a) CIE $x = 0.20$, CIE $y = 0.15$
- (b) CIE $x = 0.60$, CIE $y = 0.25$
- (c) CIE $x = 0.50$, CIE $y = 0.40$
- (d) CIE $x = 0.33$, CIE $y = 0.33$
- (e) CIE $x = 0.20$, CIE $y = 0.70$

6.5 For materials a–d, the following characteristics have been measured. Which one is likely to be fluorescent, phosphorescent, and TADF?
- (a) Energetic difference ΔE(S1–T1) between S1 and T1 is 0.1 eV, the excited-state lifetime τ at room temperature is 50 μs, and the excited-state lifetime τ at 77 K is much larger.
- (b) ΔE(S1–T1) is 1 eV, τ at room temperature is 5 μs, and τ at 77 K does not change much.
- (c) ΔE(S1–T1) is 1 eV, τ at room temperature is 25 ns, and τ at 77 K is much larger.
- (d) ΔE(S1–T1) is 0.04 eV, τ at room temperature is 2 μs, and τ at 77 K is much larger.

6.6 Which of the three following OLED stack architectures is more likely to give efficient green Alq_3-based electroluminescence and why?
- (a) ITO/PEDOT:PSS/Alq_3/TPBi/Al
- (b) ITO/TPBi/Alq_3/PEDOT:PSS/Al
- (c) ITO/Alq_3/PEDOT:PSS/TPBi/Al

6.7 ITO is used as an anode material in OLEDs and has a working function of −4.9 eV. Which of the following materials would you choose as hole transport material to assemble a working OLED architecture in the following OLED stack: ITO(WF= −4.9 eV)/*X*/$Irppy_3$ doped in CBP (WF = −5.9 eV)/TPBi/LiF/Al?
   (a) *X* = CDBP (WF = −6.3eV)
   (b) *X* = TAZ (WF = −6.6 eV)
   (c) *X* = MCP (WF = −5.9 eV)

6.8 Why does it often pay off to make more sophisticated multilayer OLED architectures?

6.9 Which kinds of charge-conducting materials are used in OLEDs?

6.10 Which quasi-particles in OLEDs are converted into light?

6.11 Name the main emission principles available to convert excitons into light.

6.12 Describe the general structure of an OLED.

6.13 Why is charge balance in an OLED so crucial?

6.14 How can the charge balance be controlled?

6.15 Which kinds of OLED setups rely on CGLs?

6.16 Name the different types of outcoupling techniques.

6.17 What are the reasons for device encapsulation?

6.18 Which kinds of OLED applications do you know, apart from rigid devices?

Suggested Readings

We recommend the following textbooks for more details on the following topics:

- **Photophysics/emissive materials**

  Turro, N. J. (1991). *Modern Molecular Photochemistry*, University Science Books.

  Yersin, H. (2004). *Triplet Emitters for OLED Application. Mechanisms of Exciton Trapping and Control of Emission Properties*, Springer-Verlag.

- **Spectroscopy**

  Lakowicz, J. R. (2006). *Principles of Fluorescence Spectroscopy*, 3rd ed., Springer: New York.

- **OLEDs in general**

  Yersin, H., et al. (2008). *Highly Efficient OLEDs with Phosphorescent Materials*, 1st ed., Wiley-VCH, Weinheim.

  Shinar, J. (2003). *Organic Light-Emitting Devices: A Survey*, Springer.

Chapter 7

Organic Solar Cells

Claudia Hoth,[a] Achilleas Savva,[b] and Stelios Choulis[b]
[a]*BELECTRIC OPV GmbH, Landgrabenstraße 94 Nürnberg, 90443, Germany*
[b]*Molecular Electronics and Photonics Research Unit, Department of Mechanical Engineering and Materials Science and Engineering, Cyprus University of Technology, 45 Kitou Kiprianou Street, Limassol 3041, Cyprus*
claudia.hoth@belectric.com

During the last few years solution-processed organic photovoltaics (OPVs) have gained huge attention and are one of the future photovoltaic technologies for low-cost production. The first OPV products entered the market after the power conversion efficiency increased from 1% in 1999 to almost 7% in 2009 and beyond 12% nowadays. This significant progress shows the potential of this technology to be one of the key technologies for future power generation. Most of the scientific literature in the field of OPV is very focused on the fundamental science, the materials, and their optical and electrical properties, as well as device physics. These disciplines are essential to continue driving the OPV technology toward high efficiencies and high device stability. This chapter addresses the basic state-of-the-art technologies around the OPV science. Section

Organic and Printed Electronics: Fundamentals and Applications
Edited by Giovanni Nisato, Donald Lupo, and Simone Ganz

ISBN 978-981-4669-74-0 (Hardcover), 978-981-4669-75-7 (eBook)
www.panstanford.com

7.1 will explain the basic principles of OPV. Manufacturing and production technologies will be described in Section 7.2, including solution-processed technologies such as printing and coating and finally vacuum deposition. The various aspects of material choices will be discussed in Section 7.3 by addressing materials for interfacial layers; photoactive materials, including small molecules and polymers; and design and criteria for electrode materials and encapsulation materials. Device architectures and concepts of multilayer solar cells (tandem solar cells) are demonstrated in Section 7.4. The metrology and electrical characterization methods of organic solar cells are demonstrated in Section 7.5. Solar cell device stability, degradation, and device lifetime will be addressed in Section 7.6. Moreover, examples of applications of OPV are shown in Section 7.7. A summary and outlook are given in Section 7.8.

7.1 Basic Principles of Organic Solar Cells

The first research activities on organic solar cells concentrated on pure layers of conjugated polymers. The main chain of these polymers can be represented by an alternating arrangement of single and double bonds along the polymer backbone. The double bonds are in fact only a simplified representation of the state of the p_z electrons in the polymer. They are highly delocalized along the polymer backbone, forming a molecular orbital. In monomers, the energy levels of all molecular orbitals are clearly defined. For longer polymer chains, the orbitals of all included monomers combine to yield a band structure. The upper energy limit of the bonding molecular orbital, the π-band, is known as the level of the highest occupied molecular orbital (HOMO). In the classical theory of semiconductors, this corresponds to the valence band. At the same time, the antibonding π^* orbitals form a π^* band, which is known as the lowest unoccupied molecular orbital (LUMO). It is comparable to the classical conduction band. The band gap between the energy levels of the HOMO and the LUMO is small enough to allow semiconducting behavior. As a result, conjugated polymers are sensitive toward excitation, which can be caused by the incident light.

A breakthrough for polymer solar cells was the introduction of the bulk heterojunction as the photoactive layer. The bulk heterojunction is based on the photo-induced charge transfer between an electron donor (conjugated polymer, for example, poly(3-hexylthiophene) [P3HT]) and an electron acceptor (fullerene or fullerene derivative, for example, $PC_{61}BM$). The photo-induced charge transfer proceeds at the interface of the conjugated polymer donor and the fullerene acceptor on a femtosecond timescale. The bulk heterojunction is a mixture of the donor and acceptor materials on the nanometer scale. Mixing of conjugated polymers and fullerenes leads to a three-dimensional heterojunction and therefore, efficient charge generation within the entire bulk occurs. The morphology of the bulk, consisting of the donor and acceptor domains, affects the charge separation, the photogenerated current, and the total power conversion efficiency (PCE) significantly. Too large domains result in insufficient charge separation, whereas too small domains favor recombination of free charges and charge transportation is hindered.

The conversion of light into electricity occurs in several successive steps, which is demonstrated in Fig. 7.1:

- Absorption of photons
- Exciton generation
- Charge separation
- Selective transportation and extraction of the charges to the opposite contacts

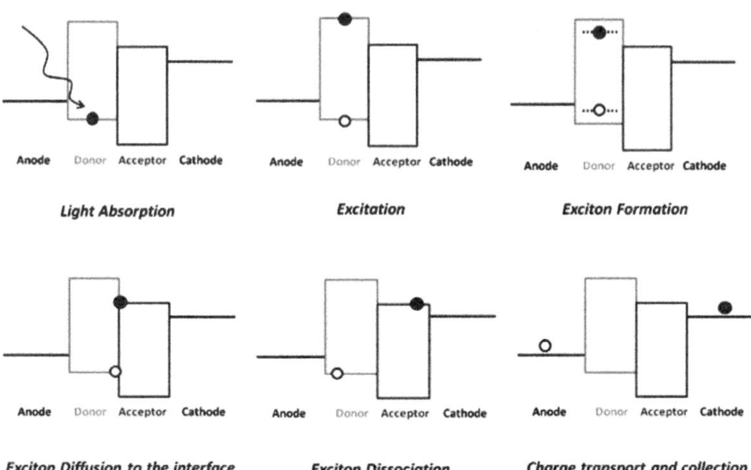

Figure 7.1 Photon-to-electron conversion in an organic solar cell device.

Photons with energy smaller than the band gap, E_g, will not be absorbed by the polymer, leading to reduced charge generation. With the absorption of photons by the polymer, having an energy larger than the band gap, E_g, electrons can be excited over the band gap from the valence band to the conduction band. In a conjugated polymer, this excitations are the π–π^* or HOMO–LUMO excitations. The so-called excitons are generated. An exciton is a bounded electron–hole pair. For efficient charge generation, an exciton generated anywhere in the blend has to reach, by drift and diffusion, a donor–acceptor interface within its lifetime. The separation is determined by the diffusion length of the excited electron, which is in the range of 10 nm. The diffusion length is the distance that the electron–hole pair travels prior to recombination occurs. For organic materials the binding energy is in the excess of 0.5 eV, resulting in metastable excitons with lifetimes up to several microseconds at room temperature. The excitons have to be separated at the interface between the donor and the acceptor prior to recombination within the active layer. The electric field across the polymer layer is the driving force for the separation. In the bulk heterojunction the interface between the donor and the acceptor is distributed throughout the photoactive layer. With incident light, the charge transfer on the interface takes place and the electron is transferred from the LUMO of the conjugated polymer to the LUMO of the fullerene. This charge separation occurs due to an energetic difference of the LUMO levels of the polymer and the fullerene. Through this step, the exciton is separated into a free hole on the polymer and a free electron on the fullerene. The charge transport competes with recombination of charge carriers. The charges may encounter an opposite charged carrier and recombine with this carrier. Conjugated polymers show high mobilities along the polymer chain. However, the mobility is limited by the hopping process between the chains. The mobility is a dimension for the velocity of the charge carriers in the electric field. For a conversion to current the generated charges have to be transported selectively to the metal contacts. The holes are transported through a hopping process by the polymer to the cathode and, respectively, the electrons to the anode by the fullerene. The charge carriers are collected at the metal contacts. The amount of absorbed light is directly related to the short-circuit current.

7.2 Manufacturing and Production Technologies

Organic materials have the potential to be processed at low temperatures from solution. The application of low-cost and scalable manufacturing processes such as printing or coating from roll to roll can be realized with this class of materials. Printing as a high-throughput fabrication method has evolved to one of the most prominent manufacturing technologies for organic electronics. In these elegant processes the materials are deposited from solution directly or indirectly in defined spots onto flexible plastic substrates or rigid sheets. Herein, a series of production technologies for organic photovoltaics (OPV) such as blade coating, spray coating, inkjet printing, gravure printing, slot-die coating, and screen printing are briefly mentioned. A more detailed description of manufacturing technology can be found in Chapter 3 of this book on printing processes. For successful printing and coating each of these technologies requires distinct ink properties that differ in viscosity, surface tension, vapor pressure, and boiling point. The solid concentration and selection of appropriate solvents define not only the ink characteristics during processing but also the drying kinetics and film formation of the printed or coated layer.

7.2.1 Blade Coating

In principle, this technique uses a coating knife to apply an ink evenly over the substrate and is a common technique in research and development. Film thicknesses strongly depend on the applied ink volume; therefore, a defined ink volume is deposited either directly on the substrate or, in the case of low viscous inks, into the gap between the coating knife and the substrate. The film formation occurs with vaporization of the organic solvent.

The blade can be located as close as a few microns above the surface of the substrate. The narrower the gap, the thinner the wet film. Apart from ink volume, viscosity, and solid concentration, this gap between the substrate and coating knife determines the wet film thickness. The target is to receive a wet film with a designated homogeneous thickness over the entire area. Dry film thicknesses can be calculated over solid concentrations of the ink. The vapor pressure, boiling point, and surface tension of the solvents affect the drying kinetics of the wet film and determine the uniformity of the dry film. Additionally, the film uniformity is influenced by the blade-coating parameters, such as temperature of the substrate holder and the coating knife, the coating velocity, the gap between the substrate and the coating knife, and the geometry (angle and shape) of the blade. Blade coating has evolved to a state-of-the-art coating technique on the laboratory scale and this method is compatible to a large area and can be easily transferred to a roll-to-roll fabrication.

7.2.2 Spray Coating

In the spray-coating technique, the films are generated stepwise. Due to the transfer of nitrogen gas pressure, single droplets are deposited at high velocities onto the surface of a substrate. As a result the spray-coated films are formed via droplets and have therefore distinct film morphologies compared to all other film preparation methods. The principle of spray coating is demonstrated in Fig. 7.2.

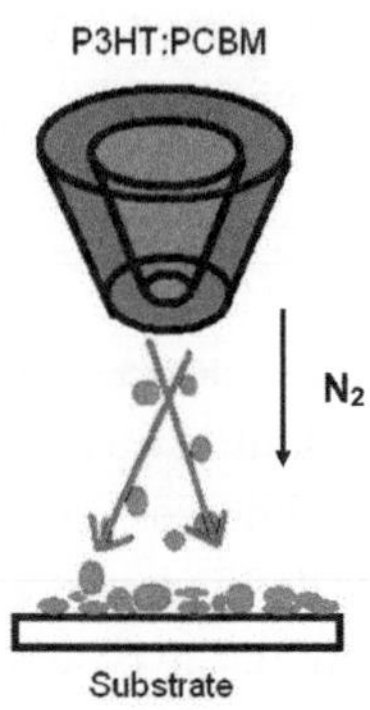

Figure 7.2 Principle of spray coating.

The ink and spraying parameters are adapted such that droplets dry immediately at their impact on the substrate. Hereby, organic solvents with relatively high vapor pressures are preferred to guarantee the drying of droplets at their impact. Too low vapor pressures result in low drying rates and the liquid droplets are pushed sideward by the pressure gas of the airbrush. Too high vapor pressures lead to droplets that will not adhere to the substrate prior to drying of the droplets or vaporization of the organic solvent. The surface tension of the organic solvent is another important aspect for decent wetting properties of spray-coating inks on substrates. In this technique the ink viscosities are on the lower edge in the range of 10 mPa·s; however, the viscosity can be modified by heating systems of the airbrush setup. The surface roughness of spray-coated films is rather high, in the tens of nanometers, compared to other film deposition methods, which is related to the transfer of rather large droplets. However, the implementation of this large-area technique in a roll-to-roll fabrication is rather straightforward and promising results have been demonstrated.

7.2.3 Inkjet Printing

This technique allows an accurate deposition of inks in defined spots. A postpatterning procedure of printed films is needless due to the drop-on-demand (DOD) technology. Inkjet-printed films are formed by the coalescence of single adjacent droplets.

For excellent droplet formation and jetting behavior, the drive waveforms of the printing head have to be modified for each ink. The quality of droplet formation is very dependent on both the rheology of the printing ink and the driving conditions of the print head. Inkjet printing requires significant efforts to properly design inks in combination with optimized waveforms of the print head. However, with optimized inks and printing parameters, promising results to deposit functional layers by inkjet printing have been published.

7.2.4 Gravure Printing

Gravure printing is based on an engraved printing roll, an ink bath, a coating knife, and a counterpressure roll. The print pattern is engraved into the printing cylinder and filled with ink by rotating through the ink bath. Excessive ink is peeled off by the coating knife and the ink remains in the engraved cells. During imprint, ink is transferred to the substrate by touching the surface of the substrate.

Hereby, the viscosity of the ink is very low. This high-resolution technology is a promising candidate for high-throughput fabrication from roll to roll.

7.2.5 Slot-Die Coating

Slot-die coating has evolved to a promising printing method for the manufacturing of organic electronics from roll to roll. This technology is rather comparable to doctor blading, which is often used on the laboratory scale. Slot-die coating allows the use of inks over a wide range of velocities. Due to the high homogeneity of coated films and high-throughput potential slot-die coating is a promising production technology for flexible organic electronics. The biggest drawback of slot-die coating is its incompatibility with patterning. However, the implementation of concepts of postpatterning methods allows the use of a large-area slot-die coating technology for roll-to-roll production.

7.2.6 Screen Printing

Screen printing is based on a patterned screen with an image by closing selected areas. Screen printing ink viscosities are 10 to 100 times higher than inks used for doctor blading, etc. Inks should show thixotropical behavior, high viscosities in the resting state that attenuate with increasing shear stress, and a rapid viscosity recovery time interval when shear stress discontinues. Commercially available screen-printing pastes therefore contain, among others thixotropic and viscosity modifiers, binders and thickeners to allow a reproducible and reliable screen-printing process. Recently, research studies on screen-printed anode and cathode materials with high resolutions have been published. These promising results prove the potential of this technology for the high-throughput fabrication of OPV devices.

7.2.7 Roll-to-Roll Manufacturing

Attractive features of OPV are the possibilities for thin flexible devices that can be fabricated using high-throughput and low-temperature approaches that employ well-established printing techniques in a roll-to-roll process. Such printing technologies which are capable of being integrated in high-throughput roll-to-roll manufacturing (Fig. 7.3) need to be qualified for the variety of deposited materials. The possibility of using flexible plastic substrates in an easily scalable high-speed printing process can reduce the balance of system cost for OPV devices, resulting in a shorter energetic payback time.

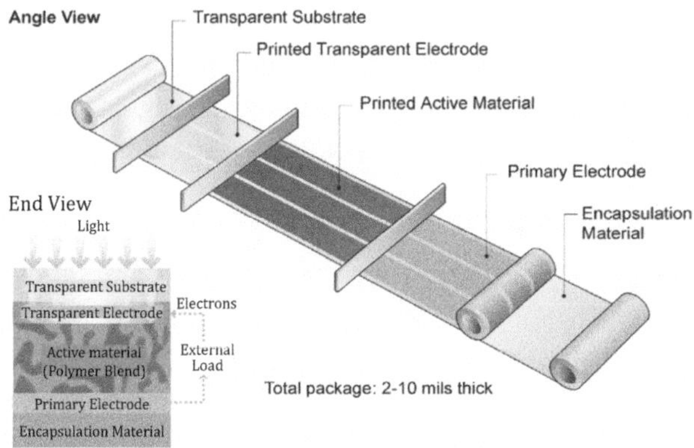

Figure 7.3 Roll-to-roll manufacturing of OPVs.

7.3 Materials of Organic Solar Cells

7.3.1 Interfacial and Intermediate Layers

7.3.1.1 PEDOT:PSS and other highly conductive polymers

The most common highly conductive polymer in the polymer electronics is poly(3,4-ethylenedioxythiophene):poly(styrenesulfonate) (PEDOT:PSS), a macromolecular salt, which is a mixture of two ionomers. The sulfonyl groups are deprotonated and negatively charged, whereas PEDOT is a conjugated polymer and is positively charged. PEDOT:PSS shows a slight blue color. It is used in almost all bulk heterojunction solar cells as a charge carrier–sensitive layer, which results in enhanced injection. The acidity of PEDOT:PSS leads to an etching of the transparent conducting oxide (TCO) under environmental conditions. Therefore, this material is often used as an intermediate layer in a grid electrode constitution due to its high conductivity and high transmission in the visible range. PEDOT:PSS can be easily modified in its conductivity, morphology, and wetting behavior by the PEDOT:PSS ratio, the use of a surfactant, solvent formulations, etc. The deposition via different coating and printing technologies is rather straightforward. Transmission of 85% and sheet resistances in the range of 50–100 Ω/sq. are possible. Another highly conductive polymer in polymer electronics is polyaniline (PANI), which can be used as an interlayer and nowadays this material is mostly used as a corrosion layer. The high conductivity is reached by a protonic doping of PANI. A solution-processed layer shows a conductivity of around 10–200 S/cm. PANI offers reasonable ambient stability properties, is a nonacid, and can be processed from solution.

7.3.2 Nanoparticles and Nanotubes

Nonpolymeric materials as the intermediate layer for a grid electrode offer also a huge innovative potential. So far most of the investigations were done on carbon nanotubes, which provide a wide range of transmission (80%–85%) and sheet resistance between 20 and 1000 Ω/sq. The conductivity remains unchanged when the layers are bent. The intrinsic work function ranges from 4.5 to 5.1eV and can be tuned by n- or p-type doping. The processing can be performed in ambient conditions. However, single-walled carbon nanotubes (SWCNTs) are expensive and the air stability of SWCNT films is only several 100 hours. An alternative for bottom and top electrodes are nanoparticles based on indium tin oxide (ITO). The films can be deposited from solution in ambient conditions. However, high-temperature annealing is necessary for high conductivities, which is unfavorable for polymer substrates. In contrast silver needles or nanoparticles as electrodes can also be deposited from solution but high-temperature annealing steps seem to be redundant for optimized coating fluids. Coated films based on silver needles show a transparency in the range of 80%–90% in the visible spectrum and a sheet resistance in the range of 10–200 Ω/sq.

Previous studies confirm higher short-circuit currents compared to ITO reference devices due to higher transparency and an improved optical path length coming from the scattering of incident light.

7.3.3 Photoactive Materials

At present, bulk heterojunction structures based on blends of a conjugated polymer as donor and a soluble fullerene derivative as acceptor represent the material system with the highest PCE reported until now. In organic solar cells the photoactive layer constitution is a homogeneous mixture of an electron donor (p type) and an electron acceptor (n type). Donor and acceptor form morphologies with phase separation in the nanometer scale. This leads to a larger interface between the p- and the n-type semiconductor. The donor material is based on a conjugated polymer with semiconducting properties, whereas the acceptor materials are fullerene derivatives.

The first research activities were based on poly[2-methoxy,5-(2′-ethyl-hexyloxy)-p-phenylene vinylene)] (MEH-PPV)/C_{60} composites, which were later substituted by the combination of poly[2-methoxy-5-(3′,7′-dimethyloctyloxy)-1,4-phenylene vinylene] (MDMO-PPV)/ 1-(3-methoxycarbonyl) propyl-1-phenyl[6,6]C_{61} (PCBM) because these materials showed improved processing conditions. However, the highest PCEs of OPV based on these materials were limited to a 3% PCE, which is related to the rather large band gap and low carrier mobility of the PPV-type polymers. Later, the research focused on the promising poly-alkyl-thiophenes. With this class of materials, the researchers demonstrated encouraging results with P3HT:PCBM mixtures with a weight ratio of 1:3, leading to high efficiencies in the range of 5% PCE. Hereby, the improved short-circuit current density could be related to an outstanding high external quantum efficiency (EQE) showing a maximum of 76% at a wavelength of 550 nm. Further efficiency enhancement of P3HT:PCBM devices was reported due to thermal annealing of the photoactive layer. Subsequently, for high efficiencies of this material system the research was mainly focused on the optimization of device-annealing conditions and processing of the photoactive layer. The optimum ratio for P3HT:PCBM blends is in the range of 1:1 by weight and the most suitable organic solvents are chlorobenzene, dichlorobenzene, or xylene. Figure 7.4 demonstrates the chemical structures of the p-type donors MDMO-PPV and P3HT.

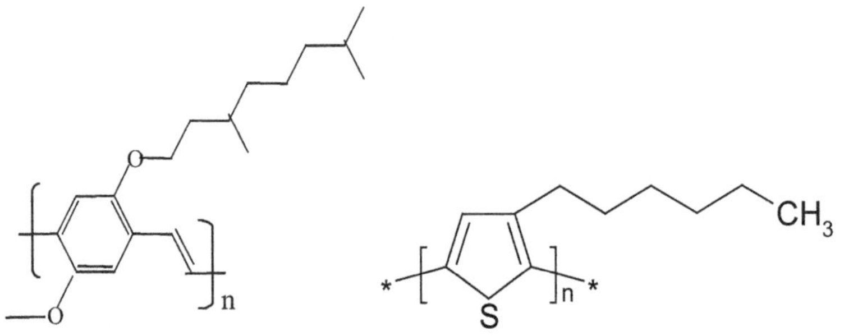

Figure 7.4 Chemical structures of poly[2-methoxy-5-(3′,7′-dimethyloctyloxy)-1,4-phenylene vinylene] (MDMO-PPV) and poly(3-hexylthiophene) (P3HT).

The chemistry of conjugated polymers offers powerful methods to tune the HOMO and LUMO levels and to modify the band gap of the material. The donor–acceptor theory where alternating electron-rich (donor) and electron-poor (acceptor) units are present and coupled together to form the polymer backbone is an efficient approach to tune the HOMO and LUMO levels and the band gap of the material. This modification is related to the consequent reduction of the bond length alternation by resonance structures of the donor–acceptor polymer. Several promising semiconductor materials based on alternating donor–acceptor groups have been synthesized. A variety of promising structures are copolymers based on thiophenes, fluorines, carbazoles, cyclopentadithiophenes, and metallated conjugated polymers having optimized band gaps. In bulk heterojunctions these compounds have an efficiency potential between 7% and 10% PCE in combination with the fullerene acceptor PCBM.

Polyfluorenes quickly became a prominent donor–acceptor structure and were tested for OPV applications. APFO polymers are a successful demonstration of the donor–acceptor approach and OPV solar cell hero efficiencies beyond 5% PCE (AM1.5 corrected for the spectral mismatch) were demonstrated together with the acceptor PCBM. This high performance is mainly attributed to the high open-circuit voltage, which is a typical feature of fluorine-based polymer solar cells due to a lower-lying HOMO level. The use of carbazole copolymers in organic solar cell devices was reported by different groups with record efficiencies of over 7% PCE. Cyclopentadithiophene-based polymers have attracted enormous attention in the last years. The most prominent structure is poly[2,6-(4,4-bis-(2-ethylhexyl)-4H-cyclopenta[2,1-b;3,4-b]-dithiophene)-alt-4,7-(2,1,3-benzothiadiazole)] (PCPDTBT) showing a low band gap ($E_g \approx 1.45$ eV) and high hole mobility, which are necessary for excellent charge transport (Fig. 7.5). On the basis of this material high OPV performance beyond 5% PCE has been reported due to an optimization of the donor–acceptor blend nanomorphology.

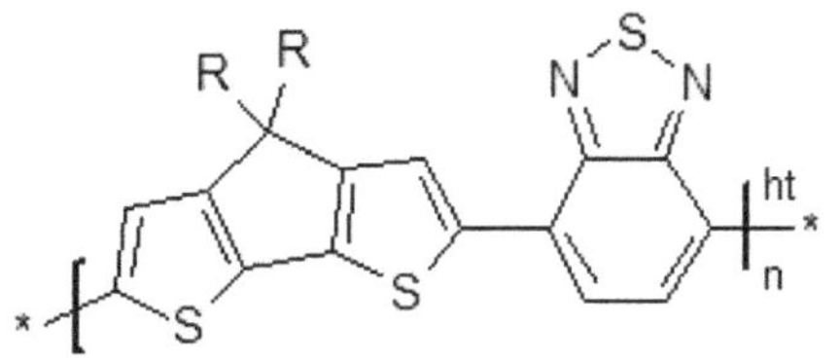

Figure 7.5 Chemical structure of poly[2,6-(4,4-bis-(2-ethylhexyl)-4H-cyclopenta[2,1-b;3,4-b]-dithiophene)-alt-4,7-(2,1,3-benzothiadiazole)] (PCPDTBT).

Modifications of the polymer backbone, namely by substituting the C atom by an Si atom to bridge the two thiophene units, lead to an optimization of the processing and, thus, the blend morphology. This results in solar cell device performances of above 5%. The only drawback of this material class is the rather high HOMO level, which limits the open-circuit voltages to values of 700 mV when mixed with PCBM. The current research is therefore focused on two strategies to overcome this limitation by bithiophene copolymers with low HOMO levels or novel acceptor materials with high LUMO levels. Metallated conjugated polymers have attracted as emitter materials in polymer light-emitting diodes (PLEDs) but have been demonstrated also in OPV devices as donor materials with PCEs in the range of 5%. PCBM as an acceptor material in solar cells was first reported in 1995. Up to now, PCBM is a suitable acceptor material for OPV devices.

The ideal acceptor requires strong absorption complementary to the absorption profile of the polymer donor. Moreover, the LUMO level offset of the donor to the acceptor has to be optimized to guarantee not only efficient charge transfer but also high open-circuit voltages. Furthermore, in donor–acceptor blends the electron mobility of the acceptor should be high enough. In OPV devices a variety of acceptor materials have been investigated. Among them fullerene derivatives, carbon nanotubes, perylenes, and inorganic semiconducting nanoparticles. However, only derivatives of C_{60} and C_{70} fullerenes resulted in highly efficient organic solar cell devices. Research studies focused also on improved processability, on HOMO–LUMO level variations, and on affecting the morphology in solution-processed donor–acceptor composites. Apart from that, the biggest boost in PCE is reached with the optimization of the acceptor LUMO level, whereas a reduction of the LUMO offset is directly translated in an increased open-circuit voltage. Researchers investigated multi-adducts of fullerenes with ~200 mV higher lying LUMO values compared to C_{60} fullerenes and demonstrated record efficiencies beyond 6% PCE for P3HT solar cells with bis-indene C_{60} and C_{70} adduct as acceptor material.

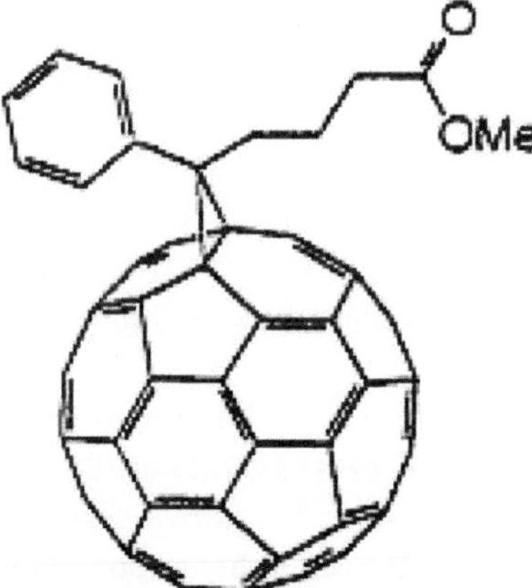

Figure 7.6 Chemical structure of [6,6]-phenyl C_{61} butyric acid methyl ester ($PC_{61}BM$).

However, a favorable arrangement of the HOMO and LUMO levels of the donor and acceptor materials is a prerequisite for highly efficient solar cell device performance. The synthetic efforts to create new electron donor polymers were guided by the following desired attributes of the material: (1) broad absorption covering most of the visible and extending to near-IR; (2) high hole mobilities for efficient charge transport; and (3) optimum leveling of energy states of donor, acceptor, and electrode materials, allowing efficient charge separation with minimum losses to thermal energy, while minimizing the energy barrier to the collecting electrodes. A recently reported model has demonstrated that the maximum PCE of a bulk heterojunction solar cell can be predicted by the aforementioned properties, namely the energy band gap and the LUMO level of the polymer donor. Hereby, an optimized morphology of the donor–acceptor composites in the nanometer scale to control charge transport properties (charge carrier mobilities in the range of 0.001 $cm^2/V{\cdot}s$) and recombination within the blend has to be taken into account to yield high PCEs.

7.3.4 Electrodes

7.3.4.1 Physical Vapor Deposition of Top Electrodes

Apart from other techniques, the top electrode can be deposited by a physical evaporation process under high vacuum in the range of 10^{-6} mbar. The vacuum chamber lowers the boiling point of the metals, but the intention of the high vacuum is a direct and homogeneous vaporization from the source to the sample. The samples are placed with shadow masks in a rotating sample holder on top of the chamber. The device area and layout is defined by a shadow mask, which is placed on the surface of the sample. The energy for the evaporation is provided by a current source. The flowing current heats the so-called evaporation boat filled with metal pellets. The material of the boat is usually tungsten or molybdenum. When the temperature of the boat exceeds the boiling point of the metal, the metal becomes gaseous and deposits on the sample. The evaporation is controlled via a deposition controller. It measures the deposited layer thickness by an oscillating crystal, whose frequency varies with the amount of deposited metal. With respect to the deposition rate, the current delivered to the source is adjusted.

Typical electrode combinations deposited by vacuum evaporation are LiF/Al or Ca/Ag. The low-work function (LWF) material (LiF or Ca) is used to fine-tune the work function of, for example, Ag and for optimized electron injection and to prevent diffusion of the high-work function (HWF) material (Al, Ag, or Au) into the photoactive layer. However, these electrode combinations are not environmentally stable and need to be handled under inert atmosphere excluding oxygen and water. Moreover, Au and Ag build a nonrectifying contact with the photoactive layer resulting in significant fill factor (FF) and open-circuit voltage losses of the device caused by surface recombination. Hereby, the power loss might be related to the work functions of Au (5.1 eV) and Ag (4.6 eV) in contrast to Al (4.3 eV). Another approach to fine-tune the work function and to suppress the diffusion of those evaporated metals is to deposit intermediate layers such as PEDOT:PSS or ZnO between the metal and the active layer or by introducing a buffer layer based on MoO_3, WO_3, or V_2O_5. However, the transfer of evaporated electrodes to roll-to-roll compatible processing is rather complex. The quality of these layers as well as the challenge to structure the electrodes in register to the rest of the module stack remain to be solved. Vacuum processing and high-temperature processing are rather unfavorable due to their considerably high processing cost compared to printing technologies. Apart from that, the compact top electrode can also be printed as an opaque metal layer. General requirements for OPV electrodes are defined by the material properties such as conductivity, transmission, reflectance, and interaction with semiconductor materials but also need to take into account the compatibility to roll-to-roll processing, sufficient flexibility, and good environmental stability—all in combination with low material costs.

7.3.4.2 Printable Materials for Top Electrodes

The requirements for nontransparent electrodes are high conductive materials with surface conductivities in the order of 1 Ω/sq. or lower. Printable electrode materials such as metal pastes; inks comprising of nanoparticles; metal inks based on silver, copper, or gold; or organometallic metal inks are highly compatible to roll-to-roll production. Film thicknesses are in the range of 100 nm to 1 μm, depending on the printing or coating technology. The suitability of such inks and pastes for OPV is the chemical compatibility with the OPV cell stack, the processing with printing or coating technologies, and a low drying temperature to gain the required conductivity with an upper limit of 150°C due to the utilization of polymer substrates and the sensitivity of the functional layers. Another approach is semitransparent electrodes comprising of a metal grid deposited onto a transparent conductive intermediate layer such as PEDOT:PSS, which has high conductivity and rather high transmission in the visible range. The high qualitative deposition of PEDOT:PSS on top of the photoactive layer is realized by the use of surfactants or the modification of the solvent formulation. This material can be easily printed or coated by a wide range of fabrication technologies. The conductivity and transmission of PEDOT:PSS vary strongly with the deposition method and the used solvent formulation. In general, transmissions around 85% with a sheet resistance of 50–100 Ω/sq. are possible.

7.3.4.3 Bottom Electrodes

Transparency is a key parameter to keep the optical losses to a reasonable minimum. For enough light absorption a transparency of more than 80%–90% over the whole solar spectra is required. The second key parameter is the surface conductivity, which should be as high as 10 Ω/sq. to prevent electrical power losses. Due to a very thin solar cell layer stack of only a few hundred nanometers, the surface roughness of the bottom electrode should be in the range of only a few nanometers root mean square (rms) to prevent local shunting. Moreover, the chemical stability against organic and halogenated solvents, like toluene, xylene, chlorobenzene, etc., as well as excellent adhesion of the bottom electrode to the substrate are of fundamental interest for the selection of the bottom-electrode materials. The most common materials for transparent electrodes are TCOs such as ITO, doped zinc oxide (ZnO), or doped tin oxide (SnO). Most of these oxides are deposited by vacuum processing technologies such as sputtering or by sol–gel deposition techniques. ITO is a composite based on two binary oxides: 90% of indium oxide (In_2O_3) is doped with 10% tin oxide (SnO_2). The intrinsic work function of ITO ranges between 4.4 and 4.9 eV. ITO shows reasonable chemical stability, which can be modified by doping with other materials. The disadvantage of highly conductive ITO is its brittleness, which is of fundamental interest for the use of flexible substrates; its high cost (1000 USD/kg); and scarcity of indium. Therefore, alternative transparent electrode solutions for OPV are multilayer electrodes, which are sputtered on flexible substrates. Here, layer stacks with a dielectric/metal/dielectric (DMD) structure are used. Silver or silver alloys are frequently used as the metal layer, whereas the dielectric often consists of TCOs, which are responsible for the sheet resistance in this stack. Advantages of multilayer systems are high flexibility due to a lower dielectric thickness, high transparency, and conductivity. However, transmission, conductivity, and surface topography, for example, rms roughness and wetting behavior of multilayer systems, are crucial and depend on metal layer thicknesses and insulators. An alternative to TCOs are grid electrode composites, which are a combination of a highly conductive transparent layer, from conducting polymers like PEDOT:PSS, nanoparticles, or nanorods, vertically stacked with a nontransparent metal grid. The metal grid serves a carrier collection and transport and is necessary since most of the transparent conductive coatings have insufficient conductivity at the required high transparency. The 10 Ω/sq. value might be a regime where metallic grids will not be necessary any more. Both electrode types can be used in the bottom- and top-electrode configuration. However, for processing reasons, most solar cell configurations use TCO electrodes as transparent bottom electrodes, whereas grid electrodes are frequently used as top electrodes. It is important to note that the layout and constitution of the metal fingers or metal grids are of fundamental interest for efficient charge carrier collection and transportation.

7.3.5 Encapsulation Materials

Due to the sensitivity of the materials to air exposure, the devices need to be covered by a hermetic sealing. In particular, organic solar cell materials require protection from oxygen and moisture in the atmosphere and employ therefore any form of encapsulation for high stability of the materials and long lifetimes of the product. One common roll-to-roll compatible procedure to encapsulate the organic electronic device is to sandwich it between barrier films sealed by a UV-curable or thermal curable adhesive (e.g., on an epoxy resin base), which is applied by lamination processes. There is a variety of multilayer stack barrier materials available, which can be combinations of polymers as well as inorganic and organic layers. Glass encapsulations show excellent barrier properties but are only applicable in rigid products. The fundamental requirements for flexible barrier films are among others excellent barrier properties against oxygen and moisture, good processability and adhesion, high transparency, chemical stability, aging durability, and high flexibility. Another approach to extend device lifetimes is to implement dry seal getters into the encapsulation that capture water and/or oxygen. The quality of the sealing can be determined by permeability measurements of water vapor (water vapor transmission rate [WVTR]) and oxygen (oxygen transmission rate [OTR]). The WVTR is given by the water mass transported through a given area (1 m^2) per defined time (1 day). The permeation rates are measured on samples exposed to climate chambers under accelerated conditions, for example, 38°C with 90% relative humidity. The lower the values for the WVTR and the OTR, the better the barrier quality.

7.4 Solar Cell Device Architectures

The earliest version of OPV devices was just a single-layer device structure with an organic photosensitive semiconductor sandwiched between the transparent bottom electrode and the top electrode. In 1994, this structure was created by Marks et al. using 50–320 nm thick PPV sandwiched between an ITO and an LWF cathode. The reported quantum efficiencies for this device were around 0.1% under 0.1 mW/cm^2 intensity. The bilayer OPV cell structure includes an additional electron-transporting layer (ETL) than is found in the single-layer OPV structure, which is demonstrated in Fig. 7.7. This structure was first realized by C. W. Tang in 1985 and the device structure comprises ITO/copper phthalocyanine (CuPc)/perylene tetracarboxylic derivative (PV)/silver (Ag). The reported PCE was 1% under simulated conditions and resulted from improving exciton dissociation efficiency by adding electron-transporting material that forms an offset energy band with hole-transporting material.

In contrast to classical bilayer junction devices, a bulk heterojunction device consists of a mixture of donor and acceptor material, as shown in Fig. 7.8. Most often, the donor is a conjugate polymer and the acceptor is a fullerene, soluble in common organic solvents. On the point that bilayer OPV cells collect very small amounts of excitons created near the donor–acceptor interface, bulk heterojunction OPV cells, which have an intermixed composite of donor and acceptor, have an advantage in terms of their having a much larger interface area between donor and acceptor. The efficient charge separation results from photo-induced electron transfer from the donor to the acceptor at the large interface, and the high collection efficiency results from a bicontinuous network of internal donor–acceptor heterojunctions.

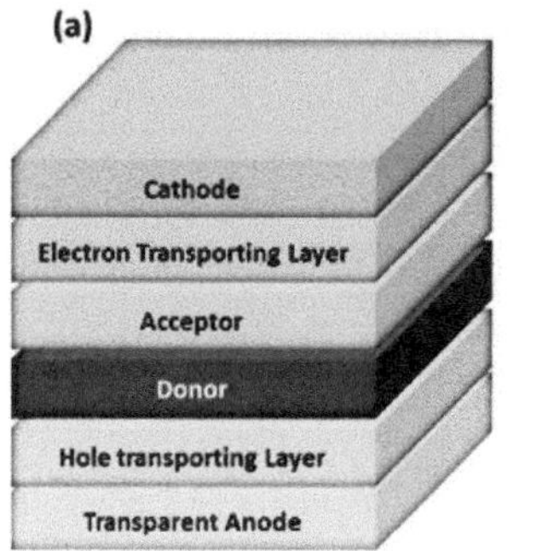

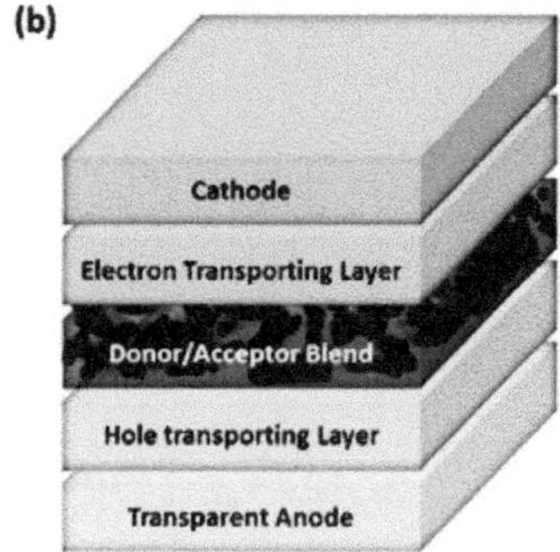

Figure 7.7 Device architectures: (a) bilayer or planar heterojunction organic solar cells and (b) bulk heterojunction organic solar cells.

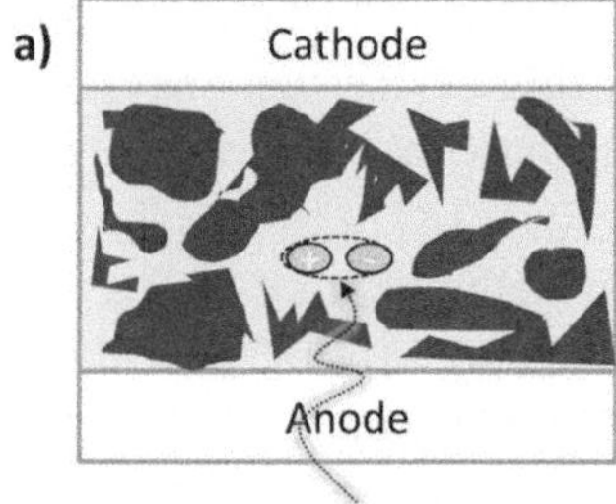

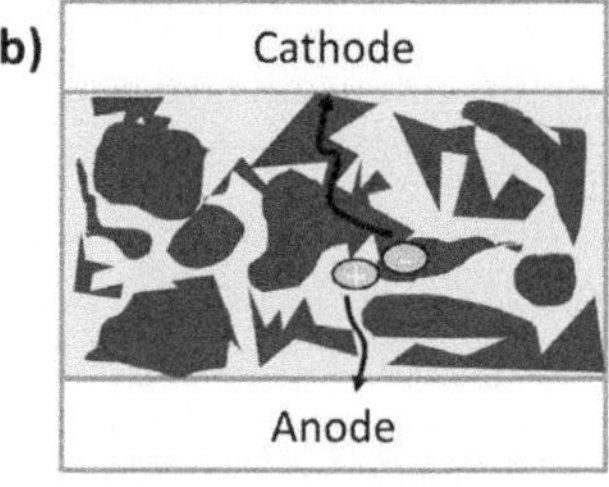

Figure 7.8 Schematic of a polymer–fullerene bulk heterojunction layer. (a) Absorption of a photon from the donor and strongly bounded exciton formation and (b) exciton split at the donor–acceptor interface and transport of the charges to the corresponding electrodes.

The cathode interface in normal structured OPV devices is the upper electrode consisting of an LWF metal, such as Al. On the other hand, in the so-called inverted structured OPV devices, the cathode is the bottom electrode, consisting of a TCO, because the electrons are forced to move down by using an ETL between the active layer and the TCO. The normal structure usually comprises an ITO/PEDOT:PSS/photoactive layer (usually used P3HT:PCBM)/LWF metal, i.e., Al). In the inverted structured OPV devices the current flow is reversed by reversing the polarity of the electrodes and is based on ITO/ETL (i.e., TiO_x or ZnO or n-doped metal oxides)/photoactive layer/HWF metal (i.e., Ag). Both device structures are shown in Fig. 7.9.

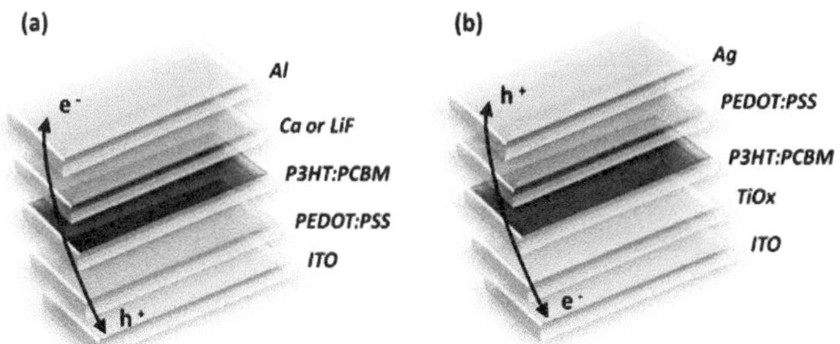

Figure 7.9 Device structures: (a) normal device structure and (b) inverted device structure. Commonly used materials are indicated for each of the device functional layers.

An inverted device architecture where the nature of the charge collection is reversed is a good normal device alternative. This architecture has recently gained considerable research attention due to the device stability and processing advantages compared to the conventional architecture. In this inverted architecture, the polarity of charge collection is the opposite of the conventional architecture, allowing the use of higher-work function (Au, Ag) and less air-sensitive electrodes as the top electrode for hole collection. The use of higher-work function metals offer better ambient interface device stability and the possibility for using nonvacuum coating techniques to deposit the top electrode, helping to reduce fabrication complexity and costs. Electron-selective buffer layers (TiO_x, ZnO) have been successfully inserted between the bottom metal electrode and the organic active layer to efficiently transport electrons and block holes.

The realization of tandem solar cells covers efficiently the emission spectrum of the sun. A typical organic tandem cell comprises two distinct devices stacked on top of each other, each of them being based on a donor–acceptor composite. The device structure of such a tandem cell is depicted in Fig. 7.10.

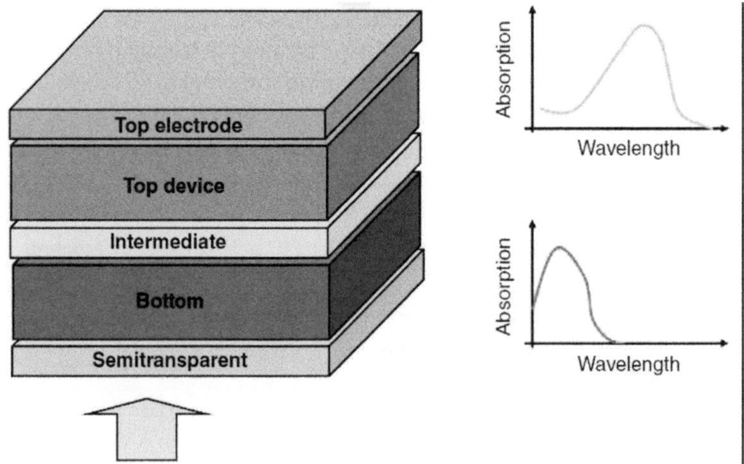

Figure 7.10 Device architecture of a tandem solar cell comprising two subcells having different, complementary absorption spectra.

The light that is not absorbed by the bottom solar cell can further impinge on the top solar cell. The different band gaps of materials reduce the losses due to thermalization. The two solar cells can be connected either in series or in parallel, which is defined by the intermediate layer and the two electrodes, whereas in most reports the series connection is preferred. The intermediate layer ensures the alignment of the quasi-Fermi level of the bottom cell acceptor with the quasi-Fermi level of the top cell donor, which can be directly related to the recombination of holes and electrons coming from each other. According to Kirchhoff's law, the voltage of the tandem solar cell is equal to the sum of the voltage of each subcell.

$$V_{oc_1} + V_{oc_2} + V_{oc_3} \ldots = V_{oc_{tandem}} \tag{7.1}$$

The short-circuit current density of a tandem solar cell depends on the current-matching conditions and therefore on the FF of each subcell. The most efficient tandem cells require materials with a band gap energy difference of about 0.3 eV, which leads to a maximum value of 15% PCE. This corresponds to a bottom donor having a band gap of 1.6 eV and a top donor having a band gap of 1.3 eV.

7.5 Electrical Characterization of Solar Cells and Modules

The cell can be considered as a two-terminal device that conducts like a diode in the dark and generates a photovoltage when charged by the sun. The operating regime of the solar cell is the range of bias, from 0 to V_{oc}, in which the cell delivers power. The cell power density is given by P_{max} and reaches a maximum at the cell's operating point or maximum power point. This occurs at some voltage V_m, with a corresponding current density J_m. The performance characteristics for a typical organic solar cell device under illumination are demonstrated in Fig. 7.11.

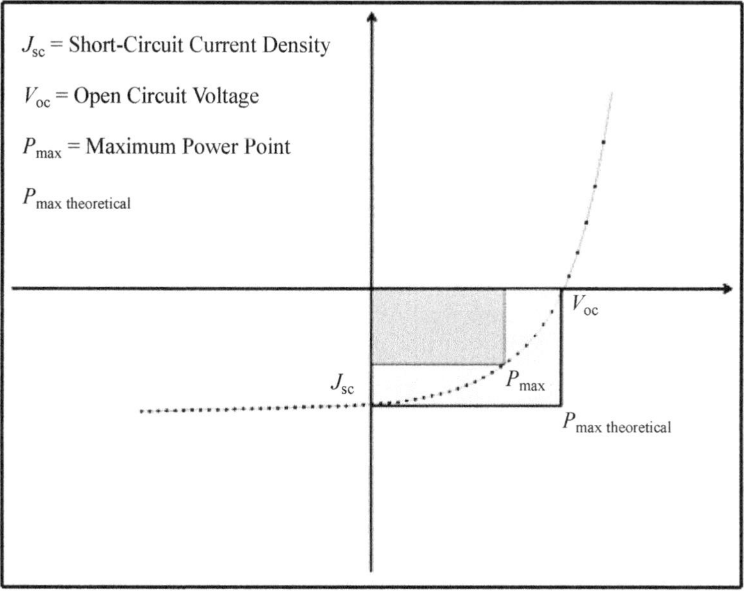

Figure 7.11 Typical current density vs. voltage characteristics (J/V) under illumination of an organic solar cell.

Four quantities— short-circuit current density (J_{sc}), open-circuit voltage (V_{oc}), FF, and PCE—are the key performance characteristics of a solar cell. Additional information for the device performance can be extracted with the use of the dark J/V characteristics. Regions R_p, n, and R_s illustrate how the different components of the solar cell equivalent circuit dominate the J/V response of the cell at different voltages. At low voltages (region R_p), the J/V characteristics are primarily determined leakage currents, at intermediate voltages (region n) by the diode recombination currents (morphology of the active layer), and at high voltages (region R_s) by series resistance. A steep slope in the R_s region generally means a low-R_s device.

Figure 7.12 depicts the widely used solar cell equivalent circuit model. It deconstructs the solar cell J/V behavior into four constituent parts: a photocurrent source, a diode, a series resistor, and a shunt resistor. The photocurrent source is simply the result of converting absorbed photons to free charge by the solar cell, the diode represents electron–hole recombination at the p-n junction, the series resistor accounts for the internal resistance of the cell to current flow, and the shunt resistor models leakage current through the cells.

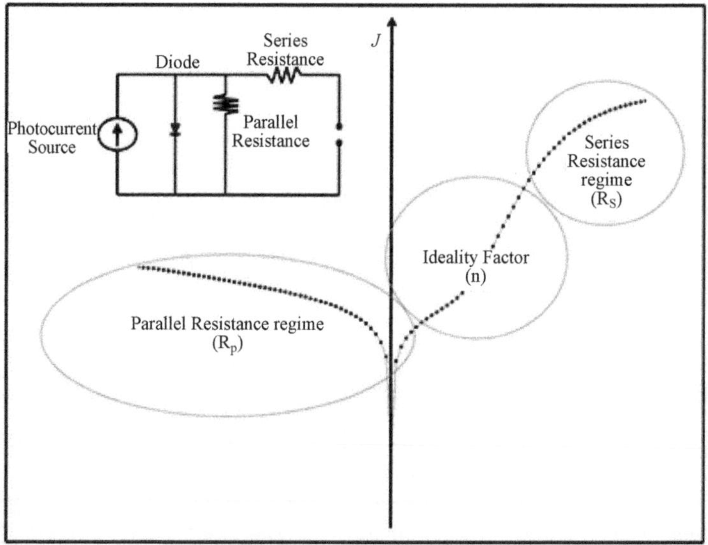

Figure 7.12 Dark current density vs. voltage characteristics and equivalent circuit model of organic solar cells.

The open-circuit voltage V_{oc} is the voltage across the solar cell when $J = 0$, which is the same as the device being open-circuited. Because $J = 0$ and power is the product of current and voltage, no power is actually produced at this voltage. However, the V_{oc} marks the boundary for voltages at which power can be produced. The open-circuit voltage can also be thought of as the point at which the photocurrent generation and dark current processes compensate one another. Although it is evident that V_{oc} is strongly affected by the difference of the HOMO of the donor and the LUMO of the acceptor, the origin of the open-circuit voltage is a conflict point among the scientific community and it has to be examined in future.

Similar to V_{oc}, the short-circuit current density J_{sc} is the current density when $V = 0$, which is the same conditions as the two electrodes of the cell being short-circuited together. Again, there is no power produced at this point, but the J_{sc} does mark the onset of power generation. In ideal devices, the J_{sc} will be the same as the photocurrent density (J_{ph}). However, several effects can lower the J_{sc} from this ideal value like the dark current:

$$J_{nh} = J_{sc} - J_{d} \tag{7.2}$$

While V_{oc} and J_{sc} mark the boundaries of power production in a solar cell, the maximum power density produced, P_{max}, occurs at the voltage V_{max} and current density J_{max} where the product of J and V is at a maximum in absolute value. Because of the diode behavior and additional resistance and recombination losses, J_{max} and V_{max} are always less than J_{sc} and V_{oc}, respectively. The FF describes these differences and is defined as

$$\mathrm{FF} = \frac{J_{max}V_{max}}{J_{sc}V_{oc}} \tag{7.3}$$

The FF is an indication of how close J_{max} and V_{max} come to the boundaries of power production of J_{sc} and V_{oc} and also an indication of the sharpness of the bend in the exponential J–V curve that connects J_{sc} and V_{oc}. Since a higher FF is related to a higher maximum power, a high FF is desired; however, the diode-like behavior of solar cells results in an FF always being less than 1. Devices with a high J_{sc} and V_{oc} can still have a low FF, suggesting that something has to be done to improve device quality.

The most discussed performance parameter of a solar cell is the PCE, η, and is defined as the percentage of incident irradiance (I_L = light power per unit area) that is converted into output power. Because the point where the cell operates on the J–V curve changes depending on the load, the output power depends on the load. For consistency, the maximum output power is used for calculating efficiency. In equation form, efficiency is written as

$$\eta = \frac{[J_{\mathrm{max}}] \times V_{\mathrm{max}}}{I_{\mathrm{L}}} \times 100\% = \frac{\mathrm{FF} \times [J_{\mathrm{SC}}] \times V_{\mathrm{OC}}}{I_{\mathrm{L}}} \times 100\% \tag{7.4}$$

This form clearly shows that FF, J_{sc}, and V_{oc} all have direct effects on η. Furthermore, the area used to calculate J can affect η and should include inactive areas that are integral to the solar cell, such as grids and interconnects, when calculating efficiency for large-area devices or modules. PCE is important since it determines how effectively the space occupied by a solar cell is being used and how much area must be covered with solar cells to produce a given amount of power. Since larger areas require more resources to cover with solar cells, a higher η is often desirable.

The EQE of a device is the fraction of incident photons converted into current and depends on wavelength. One reason for the wavelength dependence is that the absorption in the active layers is a function of wavelength:

$$\mathrm{EQE}(\%) = \frac{\text{Number of collected electrons}}{\text{Number of incident photons}} \tag{7.5}$$

To fully understand the performance of OPV devices, measuring only the EQE is not sufficient. The internal quantum efficiency (IQE) calculation is a useful tool to examine the electronic processes subsequent to the light absorption in the photoactive layer. By definition, the ratio of the number of extracted charge carriers to the number of absorbed photons in the active layer is the IQE:

$$\mathrm{IQE}(\%) = \frac{\text{Number of collected electrons}}{\text{Number of photons absorbed in the A.L.}} \tag{7.6}$$

$$\mathrm{IQE}(\%) = \frac{\text{Number of collected electrons}}{\text{Number of incident photons}} \times \frac{\text{Number of incident photons}}{\text{Number of photons absorbed in the A.L.}}$$

$$\mathrm{IQE}(\%) = \frac{\mathrm{EQE}(\%)}{\text{Absorption of the A.L.}} \tag{7.7}$$

A high IQE in fact confirms efficient exciton separation, charge carrier transport, and collection. Several reports have published IQE numbers higher than 90% for OPV devices, with a PCE above 6%.

The short-circuit current density expected under a light source can be estimated from the EQE and the spectral irradiance of the light source by integrating the product of the EQE and the photon flux density. For the standard AM1.5 G spectrum, the calculation is

$$J_{\mathrm{SC}} = \int_{0}^{\infty} e\mathrm{EQE}(\lambda)\frac{\lambda}{hc}E_{\lambda}^{\mathrm{AM1.5G}}(\lambda)d\lambda \tag{7.8}$$

where $E_{\lambda}^{\mathrm{AM1.5G}}$ is the spectral irradiance of the AM1.5 G spectrum, λ is the wavelength, h is Planck's constant, c is the speed of light, and e is the elementary charge.

7.6 Lifetime and Degradation of Organic Solar Cells

The competitive position in the energy market and the field of application are a function of efficiency, lifetime, and, in particular, the cost per kilowatt-hour or watt peak. The potential of organic solar modules is assessed by these key parameters. Besides PCEs, a remarkable recent development in lifetime and fabrication cost reduction has been made. The degradation of solar modules is an aging effect that results in lifetime-limiting effects in organic solar cells. The lifetime and the understanding of the degradation mechanisms are of huge interest for any successful product application. Qualitative experiments have demonstrated that parameters such as temperature, humidity, irradiation, and electrical and mechanical stress limit long-term stability. The lifetime of OPV modules is studied under accelerated conditions. Hereby, OPV devices are exposed to stress conditions such as damp heat climate conditions, such as 38°C, 65°C, and 85°C with, for example, 85% relative humidity and/or 1 sunlight soaking or outdoor exposure. An adequate instrument to predict lifetimes of organic solar cells is the accelerated lifetime model to determine acceleration factors. The mechanism of the accelerated lifetime degradation at elevated temperatures and stress conditions can be described by the Arrhenius model.

Environmental lifetime factors such as water, oxygen, high temperature, and light exposure have a strong influence on the stability of the structural layers in OPV devices. For instance, conjugated polymers that comprise the active blend of OPV devices are unstable in air and react through photolytic and photochemical reactions when exposed to sunlight, causing photodegradation of the polymer. However, such mechanisms can be much slower than degradation occurring within the electrode layers of the device, such as oxidation of the metal contact. OPV devices rapidly degrade due to the susceptibility of metals, usually used for the back electrode, to oxidation in the presence of absorbed oxygen molecules. The degradation of the metal electrode leads to the formation of thin insulating oxide barriers, hindering electric conduction and collection of the charge carriers. Certain metals such as Al, Ca, and Ag are commonly used as electrodes in OPV devices because of their high electrical conductivity and suitable work function. Interestingly, two basic degradation mechanisms of the metal electrode have been identified. These are, primarily, oxidation and, secondarily, chemical interaction with polymers within the active layer interface.

The degradation due to oxidation at the electrode–polymer interface can result in the formation of an oxide layer on the top metal surface as well as within the metal–polymer interface. This oxidation layer hinders the charge selectivity of the electrode, thus reducing device performance. For example, in Ca/Al electrodes, the oxide formation is faster due to Ca presence; it has been reported that Ca/Al electrode degradation in air is due to considerable changes at the metal–organic interface (Fig. 7.13).

In addition, PEDOT:PSS is an essential hole-selective contact for achieving efficient normal and inverted OPV devices. This particular buffer layer is extremely sensitive to moisture and oxygen. The detrimental effects of atmospheric air on the electrical properties of this material have been studied by Vitoratos et al., who showed that when the PEDOT:PSS layer absorbs water, it promotes irreversible structural modification of its networks due to its highly hygroscopic nature, resulting in conductivity reduction, which may consequently result in shortage of device lifetime. Moreover, the PEDOT:PSS layer can influence the degradation of the active layer. It has also been reported that the PEDOT:PSS layer can induce degradation of P3HT:PCBM OPV, which is demonstrated by a decrease in the absorbance and the formation of aggregates in the active layer.

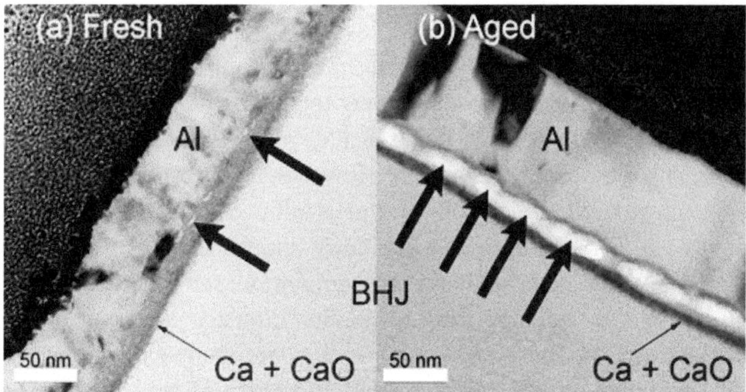

Figure 7.13 Cross-sectional transmission electron microscopy (TEM) images have revealed the formation of void structures to be the primary degradation mechanism for Ca/Al contacts. These structures grow as the electrode ages and becomes oxidized (b). Reprinted from Lloyd, M. T., Olson, D. C., Lu, P., Fang, E., Moore, D. L., White, M. S., Reese, M. O., Ginley, D. S., and Hsu, J. W. P. Impact of contact evolution on the shelf life of organic solar cells. *J. Mat. Chem.*, 19, pp. 7638–7642, with permission from Royal Society of Chemistry, Copyright 2009.

Recent studies are reporting durability of 1 year under real environmental conditions and accelerated lifetime testing indicates even longer lifetimes. Recently, researchers demonstrated semitransparent OPV modules laminated between flexible barrier films and glass sheets that meet the IEC-61646 environmental chamber tests, as required for thin-film solar modules. The investigated flexible and rigid OPV modules, respectively, demonstrated less than 4% and 8% efficiency changes after damp heat, thermal cycling, and a sequence test. This remarkable development confirms that OPV modules packaged between flexible barriers or rigid glass sheets can withstand critical IEC stress tests. The main degradation mechanisms of OPV are of intrinsic or extrinsic nature. Among others, the device lifetime limitation is related to chemical degradation and oxidation reactions of functional layers and electrode materials, loss of barrier, substrate and adhesive quality, delamination between layers, and electrical stress in or between OPV cells.

7.7 Application Examples

There has been growing interest in the field of OPV in the recent years. The ability to produce photovoltaic systems on flexible plastic carriers utilizing organic materials has clear benefits in terms of cost and production. As a relatively new technology, OPV does not have the target to compete with classical established solar technologies concerning efficiency or lifetime. However, improvements are occurring all the time. Integrated solar cells allow organic semiconductor devices to achieve their full flexibility and potential. OPV modules have many potential applications from remote sensors to large-scale building integrated application. Leading industry players from various market segments approach joint development activities regarding technology and custom-shaped and custom-sized applications. OPV is already playing an important role in a number of areas. The option to customize OPV is a major benefit of this technology compared to classical photovoltaics (PV). Key products can be semi-transparent film or glass with integrated OPV.

The short-term vision for OPV module applications is to encourage market adoption by portable power sources and off-grid applications. Typically, portable electronic devices require around 2–5 watts of power and 2–3 years of lifetime. Among others, the products in this category can be chargers for devices like mobile phones, tablets, and electronic readers, or the OPV module is integrated to the back surface of the electronic devices or bags. Examples are shown in Fig. 7.14. Areas with limited access to electricity have an increasing market for off-grid applications. OPV technology is a perfect fit for off-grid power in developing countries. Lighting in terms of solar power lamps or foldable and rollable solar chargers are only a few examples for off-grid applications. OPV modules for off-grid power require efficiencies in the range of 3%–5% and a lifetime of more than 5 years. In the long term, as the performance of OPV modules improves, an important commercial application is building integrated photovoltaics (BIPVs). Potential OPV applications in BIPVs include facades, roofing, curtain walls, windows, awnings, shadings, and other architecture applications, as demonstrated in Fig. 7.15. Hereby, in the foreground BIPVs serve as an aesthetic building material, which is attractive to architects and is colored, semitransparent, low cost, lightweight, flexible, and easy to install. Due to the excellent low-light performance of OPV, it is an attractive source for power generation in such BIPV applications. Konarka has commercialized numerous products such as the integration of OPV systems in San Francisco's bus shelters.

Figure 7.14 Portable application examples of OPV modules.

Figure 7.15 Roof of a San Francisco bus shelter, shading element, BIPV, semitransparent, multicolored, and custom-shaped applications.

OPV devices are lightweight, customizable, and flexible and ultimately have a lower cost per unit area. In addition, the energy required to manufacture OPV devices is much less than conventional PV, with energy payback on the order of two months, whereas traditional PV is on the order of two years to generate the energy equivalent required to manufacture the product. Another key difference is that OPV is not a finished product ready for the end user. The freedom to customize OPV in size and layout is a major benefit of this technology compared to classical PV.

The possibilities for OPV are huge and the industry leaders recognized the potential for OPV. For a successful and competitive OPV product, the cost has to be low. The feedback from the market so far has been very encouraging. OPV is on a strong upward path and will be commercially viable within the next few years. On a short-term basis, the OPV market will develop gradually toward a mature technology. In the long term, the technology has the potential to provide truly green energy to everybody, wherever it is needed. Power is generated at the point of use and it can be integrated seamlessly and become almost invisible. This could really transform the PV market fundamentally. The vision is that OPV becomes ubiquitous, a part of everyday life and not just only on rooftops.

7.8 Summary and Outlook

The knowledge of fundamentals of OPV was gained between 2005 and 2010. During this period, the third generation of OPV materials was established, tandem concepts were successfully developed, the electrode and interface materials were deeply understood and reported, and key production methods for large-area volume processing were found and implemented. However, PCEs are in the range of 10% level for single-junction cells. The basis in terms of know-how and knowledge are established to reach record efficiencies of over 12%, as demonstrated recently. A module PCE of around 10% would be cost efficient to transfer the OPV technology from the laboratory scale to large-scale production. The state of the art is very close to that goal. In parallel, the progress in stability and lifetime is encouraging. Indoor, outdoor, and accelerated lifetimes of organic solar modules are high enough for first applications. OPV technology was developed outstandingly rapidly, specifically in the field of functional materials, electrode materials, packaging materials, and production processes.

Exercises

7.1 Describe the main light-to-electron conversion steps using organic solar cells.

7.2 Why does the bulk heterojunction structure lead to more efficient organic solar cells compared to the bilayer structure?

7.3 An organic solar cell was measured using a solar simulator under 1.5 G and 100 mW/cm^2. The V_{oc}, J_{sc}, and FF were found to be 0.59 V, 9 mA/cm^2, and 60%, respectively. Calculate the PCE (%) of the aforementioned organic solar cell.

7.4 What are the main environmental factors that affect the lifetime of OPV devices?

7.5 Give three reasons to use organic solar cells instead of other solar cell technologies.

Suggested Readings

Brabec, C. J., and Sariciftci, N. S. (2000). *Semiconducting Polymers,* Chapter 15, eds., Hadziiannon, G., and van Hutten, P. F., pp. 515–560.

Brabec, C. J., Cravino, A., Meissner, D., Sariciftci, N. S., Fromherz, T., Rispens, M. T., Sanchez, L., and Hummelen, J. C. (2001). Origin of the open circuit voltage of plastic solar cells. *Adv. Funct. Mater.,* **11**, pp. 374–380.

Brabec, C. J., Dyakonov, V., and Scherf, U. (2003). *Organic Photovoltaics,* Wiley Online Library.

Brabec, C. J., Sariciftci, N. S., and Hummelen, J. C. (2001). Plastic solar cells. *Adv. Funct. Mater.,* **11**, pp. 15–26.

Bredas, J. L. (1986). *Handbook of Conducting Polymers,* Chapter 25, Vol. 2, ed., Skotheim, T. A., p. 859.

Brinker, C. J., and Scherer, G. W. (1990). *Sol-Gel Science: The Physics and Chemistry of Sol-Gel Processing,* 1st ed., Academic Press.

Bunshah, R. F. (1994). *Handbook of Deposition Technologies for Films and Coatings: Science, Technology and Applications,* 2nd ed., Materials Science and Process Technology Series, Park Ridge, NJ, Noyes.

Dennler, G., Scharber, M. C., Ameri, T., Denk, P., Forberich, K., Waldauf, C., and Brabec, C. J. (2008). Design rules for donors in bulk-heterojunction tandem solar cells-towards 15% energy-conversion efficiency. *Adv. Mater.,* **220**, pp. 579–583.

Gilot, J., Wienk, M. M., and Janssen, R. A. J. (2007). On the efficiency of polymer solar cells. *Nat. Mater.,* **6**, pp. 704.

Hauch, J. A., Schilinsky, P., Choulis, S. A., Childers, R., Biele, M., and Brabec, C. J. (2008). Flexible organic P3HT:PCBM bulk-heterojunction modules with more than 1 year outdoor lifetime. *Sol. Energy Mater. Sol. Cells,* **92**(7), pp. 727–731.

Hoppe, H., and Sariciftci, N. S. (2004). Organic solar cells: an overview. *J. Mater. Res.,* **19**(7), pp. 1924–1945.

Hoth, C. N., Choulis, S. A., Schilinsky, P., and Brabec, C. J. (2007). High photovoltaic performance of inkjet printed polymer: fullerene blends. *Adv. Mater.,* **19**, pp. 3973–3978.

Hoth, C. N., Choulis, S. A., Schilinsky, P., and Brabec, C. J. (2009). On the effect of poly(3-hexylthiophene) regioregularity on inkjet printed organic solar cells. *J. Mater. Chem.,* **19**, pp. 5398–5404.

Knupfer, M. (2003). Exciton binding energies in organic semiconductors. *Appl. Phys. A,* **77**, pp. 623–626.

Koster, L., Shaheen, S. E., and Hummelen, J. C. (2012). Pathways to a new efficiency regime for organic solar cells. *Adv. Energy Mater.,* **2**, pp. 1246–1253.

Krebs, F. C. (2012). *Stability and Degradation of Organic and Polymer Solar Cells,* John Wiley & Sons.

Lenes, M., Wetzelaer, G.-J. A. H., Kooistra, F. B., Veenstra, S. C., Hummelen, J. C., and Blom, P. W. M. (2008). Fullerene bisadducts for enhanced open-circuit voltages and efficiencies in polymer solar cells. *Adv. Mater.,* **20**, pp. 2116–2119.

Lloyd, M. T., Olson, D. C., Lu, P., Fang, E., Moore, D. L., White, M. S., Reese, M. O., Ginley, D. S., and Hsu, J. W. P. (2009). Impact of contact evolution on the shelf life of organic solar cells. *J. Mat. Chem.,* **19**, pp. 7638–7642.

Nelson, J. (2003). *The Physics of Solar Cells,* Imperial College Press.

Park, S. H., Roy, A., Beaupré, S., Cho, S., Coates, N., Moon, J. S., Moses, D., Leclerc, M., Lee, K., and Heeger, A. J. (2009). Bulk heterojunction solar cells with internal quantum efficiency approaching 100%. *Nat. Photon.,* **3**(5), pp. 297–302.

Peters, C. H., Sachs-Quintana, I. T., Kastrop, J. P., Beaupré, S., Leclerc, M., and McGehee, M. D. (2011). High efficiency polymer solar cells with long operating lifetimes. *Adv. Energy Mater.,* **1**(4), pp. 491–494.

Sargent, E. H. (2009). Infrared photovoltaics made by solution processing. *Nat. Photon.,* **3**, pp. 325–331.

Sariciftci, N. S., Smilowitz, L., Heeger, A. J., and Wudl, F. (1992). Photoinduced electron transfer from a conducting polymer to buckminsterfullerene. *Science,* **258**, pp. 1474–1476.

Savva, A., Petraki, F., Eleftheriou, P., Sygellou, L., Voigt, M., Giannouli, M., Kennou, S., Nelson, J., Bradley, D. D. C., Brabec, C. J., and Choulis, S. A. (2013). The effect of organic and metal oxide interfacial layers on the performance of inverted organic photovoltaics. *Adv. Energy Mater.*, **3**, pp. 391–398.

Schaer, M., Nuesch, F., Berner, D., Leo, W., and Zuppiroli, L. (2011). Water vapor and oxygen degradation mechanisms in organic light emitting diodes. *Adv. Fun. Mater.,* **11**(2), pp. 116–121.

Scharber, M. C., Koppe, M., Gao, J., Cordella, F., Loi, M. A., Denk, P., Morana, M., Egelhaaf, H.-J., Forberich, K., Dennler, G., Gaudiana, R., Waller, D., Zhu, Z., Shi, X., and Brabec, C. J. (2010). Influence of the bridging atom on the performance of a low-bandgap bulk heterojunction solar cell. *Adv. Mater.,* **22**, pp. 367–370.

Scharber, M. C., Mühlbacher, D., Koppe, M., Denk, P., Waldauf, C., Heeger, A. J., and Brabec, C. J. (2006). Design rules for donors in bulk-heterojunction solar cells: towards 10% energy conversion efficiency. *Adv. Mater.*, **18**, pp. 789–794.

Schilinsky, P., Waldauf, C., and Brabec, C. J. (2002). Recombination and loss analysis in polythiophene based bulk heterojunction photodetectors. *Appl. Phys. Lett.*, **81**, p. 3885.

Servaites, J. D., Ratner, M. A., and Marks, T. J. (2011). Organic solar cells: a new look at traditional models. *Energy Environ. Sci.*, **4**(11), pp. 4410–4422.

Slooff, L. H., Veenstra, S. C., Kroon, J. M., Moet, D. J. D., Sweelssen, J., and Koetse, M. M. (2007). Determining the internal quantum efficiency of highly efficient polymer solar cells through optical modeling. *Appl. Phys. Lett.*, **90**, p. 143506.

Tanaka, J., and Tanaka, M. (1986). *Handbook of Conducting Polymers,* Chapter 35, Vol. 2, ed., Skotheim, T. A., p. 1269.

Wang, Y. (1992). Photoconductivity of fullerene-doped polymers. *Nature,* **356**, pp. 585–587.

Wong, W.-Y., Wang, X.-Z., He, Z., Djurisic, A. B., Yip, C.-T., Cheung, K-Y., Wang, H., Mak, C. S. K., and Chan, W.-K. (2007). Metallated conjugated polymers as a new avenue towards high-efficiency polymer solar cells. *Nat. Mater.*, **6**, pp. 521–527.

Yu, G., Gao, J., Hummelen, J. C., Wudl, F., and Heeger, A. J. (1995). Polymer photovoltaic cells: enhanced efficiencies via a network of internal donor-acceptor heterojunctions. *Science,* **270**, pp. 1789–1791.

Zagorska, M., Pron, A., and Lefrant, S. (1991). *Handbook of Organic Conductive Molecules and Polymers.* Chapter 4, Vol. 3, ed., Nalwa, H. S., p. 183.

Chapter 8

Printable Power Storage: Batteries and Supercapacitors

Sampo Tuukkanen[a] **and Martin Krebs**[b]
[a]*Department of Automation Science and Engineering, Tampere University of Technology, P.O. Box 692, Korkeakoulunkatu 3, FI-33101 Tampere, Finland*
[b]*VARTA Microbattery GmbH, Daimlerstraße 1, 73479 Ellwangen, Germany*
sampo.tuukkanen@gmail.com, martin.krebs@varta-microbattery.com

8.1 Introduction

Most electric and electronic equipment needs energy storage. The purpose of an energy storage device is to provide electrical energy when the main source of power does not deliver the needed energy. This is valid especially for portable devices.

In electrochemical energy storage devices the energy can be stored in two different ways, in the volume of the electro-active electrode materials (batteries) and on the surface of the electrode materials (supercapacitors). The physical structures and phenomena are quite similar in batteries and supercapacitors. However, the principal

Organic and Printed Electronics: Fundamentals and Applications
Edited by Giovanni Nisato, Donald Lupo, and Simone Ganz

ISBN 978-981-4669-74-0 (Hardcover), 978-981-4669-75-7 (eBook)
www.panstanford.com

difference in operation causes some difference in their performance. A slower charge-capturing mechanism in the case of batteries results in lower power output when compared to supercapacitors, where chemical reactions do not take place. On the other hand, batteries can have higher energy densities than supercapacitors because the ions can penetrate into the electrode material in large quantities. Further, the lifetimes of batteries are lower because the volume changes caused by the penetration of charges into the electrode material induce structural fatigue to the device structure.

The following parameters are commonly used to describe the performance and functionality of electrochemical energy storage devices:

- Storage capacity or charge density (C/l or C/kg)
- Energy density (J/kg or Wh/kg)
- Power density (W/kg)
- Voltage efficiency, ratio of output (discharging) voltage and charging voltage
- Lifetime, shelf-life (time of becoming unusable), or cycle lifetime (charge/discharge cycles)

8.2 Principles of a Cell/Battery

In **battery technology**, different electrochemical systems are available. An electrochemical system consists of an anode material (negative terminal), cathode material (positive terminal), and an electrolyte. Because of these combinations and the structure of the cell, the electrical properties vary over a wide range. The most important properties are:

- Cell voltage
- Charge contents (capacity)
- Load capability
- Peak load capability
- Cycle ability for rechargeable (secondary) cells

8.2.1 Voltage

Voltage is the difference of the electrical potential between the two terminals of an electrochemical cell. It depends on the electrochemical system, consisting of the electro-active materials of both the electrodes and the electrolyte, open-circuit voltage U_{oc}, and the current flowing in the external circuit, closed-circuit voltage U_{cc}. The lower voltage limit where the application stops operation is called cutoff voltage or end-point voltage.

The discharge voltage curve of the electrochemical cell is presented in Fig. 8.1. The discharge stops when the end-of-discharge voltage U_{eod} is reached, either at a pulse at t_p or at the base load at t_b.

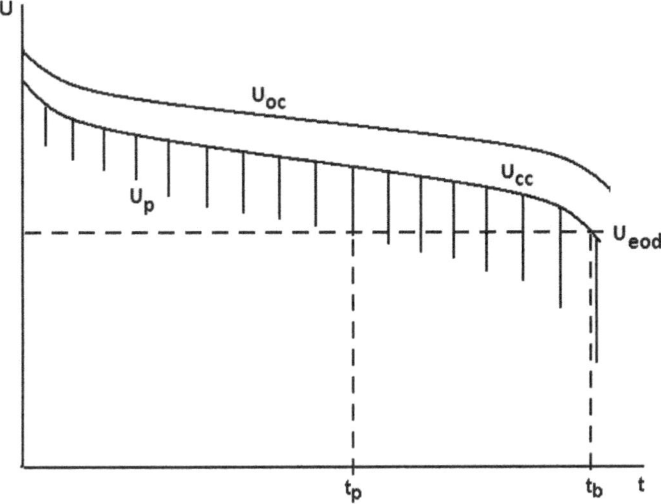

Figure 8.1 Discharge voltage curve under continuous and pulse load conditions.

8.2.2 Charge "Capacity"

Charge capacity is the charge that is stored in the electro-active materials of the electrodes. The real value is defined by the electrode having the lower charge. The charge is also reduced when the electro-active material cannot be discharged completely, for example, because of kinetic reasons.

8.2.3 Internal Resistance/Impedance

Impedance is the sum of all impedances of the parts in the electrochemical cell, which are connected in parallel or series.

Impedance depends of the frequency of the electric load. It is important to ensure that the voltage drop, which occurs because of the impedance, is not too high to make the electronic device work. For users' convenience the direct current (DC) resistance is measured after longer load times (0.2 s up to 2 s). This value is often given in component specification sheets.

8.2.4 Pulse Load Capability

Usually the impedance of a battery cell decreases strongly with higher frequencies. Therefore the impedance for short pulses (up to 0.1 s) is much lower than the DC resistance. This is important for applications were a high current is drawn for a short time, for example, measurement taking in a sensor or wireless data transmission. This capability can be derived from the impedance spectrum or directly measured.

8.2.5 Cycle Life

During cycling the capacity of a secondary battery decreases. This effect is due to degeneration processes of the electrodes and/or the electrolyte. It can be reduced by constructive measures. Typically the cycle life is defined by the number of cycles, when the residual capacity is 80% of the initial value.

8.2.6 Cell versus Battery

One storage element is called a **cell**. It consists of two electrodes, a separator, an electrolyte, and a housing that closes the element hermetically. If these cells are connected together, they are called a **battery**. For batteries two different ways of connections are possible, serial and parallel. A serial connection is used to increase the battery voltage; a parallel connection is used to increase the capacity.

8.2.7 Battery as an Energy Source

The smart objects that need electrical power usually have a certain demand of voltage. It depends on the parts that shall be powered. For most parts, like organic light-emitting diodes (OLEDs), ASICs, and electrochromic displays (ECDs) a voltage of 3 V is enough, but there are also parts that require a higher voltage. For example, organic field-effect transistors (OFETs) may need voltages up to 50 V.

Also the capacity of printed batteries is very important. It defines the energy content and insofar the duration for the use of the smart object. It can be distinguished between applications for primary batteries (single use) and for secondary batteries (rechargeable).

A primary battery is applicable if the power consumption is so low that the charge stored in the battery is enough for the whole lifetime. In this case there is no need to replace the battery or recharge it.

If the power consumption is higher a secondary battery is applicable. In this case it is important to provide the energy for recharging, whenever it is needed. That may be from an external power source by wire or wireless through some energy-harvesting means.

The power demand of a device is characterized by its current consumption. This can be low over a long period of time, a base current. For example, is it possible to power an electronic device in sleep mode or supply a backup current for a memory device.

Some other features cause higher currents for a short period of time. For example, current can be needed for taking a measurement by an analog-to-digital converter (ADC) or sending a protocol by wireless transmission. The charge needed is the integral of the current over the time ($Q = I \times t$).

For high peak currents it has to be considered whether the internal resistance allows to take the respective currents. It causes a voltage drop and leads to a lower cell voltage. If the cell voltage drops below a certain level, defined by the consumer, proper function is no more guaranteed.

If the duration of the peak current is below 1 s the impedance of the cell has to be regarded. The impedance can be measured as a function of the frequency. It shows that each battery has a certain capacity that allows for a short time (some milliseconds) to draw a current that is much higher than the DC current.

8.3 Principles of a Supercapacitor

8.3.1 Conventional Capacitors

A capacitor is defined as a device where opposite charges are accumulated with a certain distance. The amount of charge and the distance give the resulting capacity. The capacitance of a conventional capacitor is defined by

$$C = \frac{1}{\varepsilon_r \varepsilon_0} \frac{A}{d}, \tag{8.1}$$

where A is the area of capacitor electrode and d is the distance of the electrodes. For typical capacitors that have the areas of some square meters and dielectric thicknesses of down to 1 μm the capacitance values can by up to several μFs, at usual geometries. A thinner dielectric thickness can be achieved by using oxide layers on aluminium foils in the electrolyte (Elco) where capacitances up to 1 mF can be achieved.

8.3.2 Supercapacitor Structure

During the recent 10 years, capacitors that use an electrochemical double layer (EDL) instead of a dielectric layer were developed. In the EDL the ions in the electrolyte accumulate to the close vicinity of the opposite electrode surfaces. The separation of charges in the EDL (also known as the Helmholtz layer) is very small, in the nanometer range, which leads to very high capacitances. These types of energy storage devices are so-called supercapacitors but also known as electric double-layer capacitors (EDLCs) or ultracapacitors, depending on the context. There is no consensus about the terminology of electrochemical capacitors (ECs). However, in several recent papers it has been proposed that bare EDLCs would be called ultracapacitors, whereas pseudocapacitors would be called supercapacitors. In this chapter, we will solely talk about EDLCs and call them supercapacitors.

In general, ECs can be divided into supercapacitors and pseudocapacitors. The difference between these is that a supercapacitor is a physical charge storage device where the charging is based on the accumulation of charges on the electrode surfaces, whereas in the pseudocapacitor the charges undergo fast surface redox (oxidation–reduction) reactions with the electrode surface material.

Thus, a pseudocapacitor aligns somewhere between batteries and supercapacitors. In this chapter, we solely focus on the comparison of supercapacitors and batteries.

The supercapacitor structure is described in Fig. 8.2. Supercapacitors consist of two electrodes, an electrolyte between them, and usually a porous separator layer. The ions in the electrolyte are able to move in the electrolyte and through the separator while the supercapacitor is charging. The specific capacitance (C/A) of an EDL is usually of the order of 10 μF/cm^2. Commercial supercapacitors available today have capacitances up to 5000 F.

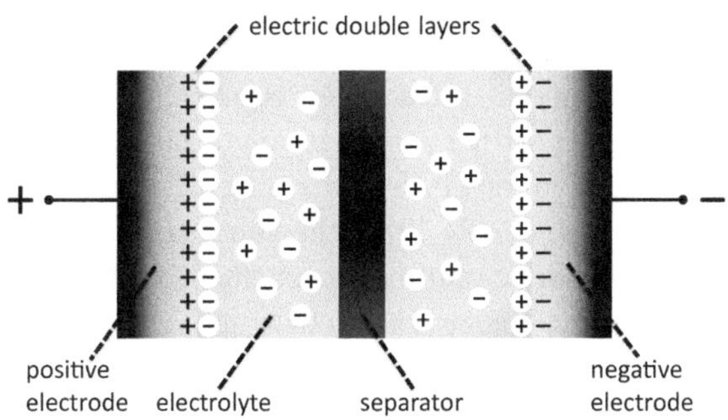

Figure 8.2 Structure of a supercapacitor.

8.3.3 Supercapacitor Parameters

The supercapacitor device is composed of two individual capacitors, C_1 and C_2, which are located on the electrode–electrolyte interface. The total capacitance of the device if then defined by

$$C = \frac{C_1 C_2}{C_1 + C_2} \tag{8.2}$$

If the capacitors C_1 and C_2 are identical the total capacitance is $C = C_1/2$. The capacitance of a supercapacitor is usually reported per mass of electrode material and is called the specific capacitance (in units F/g).

The maximum energy stored in such a capacitor is given by

$$E_{\max} = \frac{1}{2} C V^2 \tag{8.3}$$

where V is the voltage over the capacitor. This energy for supercapacitors is usually reported per mass of electrode material and is called the energy density (in units Wh/kg). This maximum energy E_{max} results from the definition of electric power $P = IV$ and the definition of capacitance given by

$$Q = CV \tag{8.4}$$

Because of this capacitive behavior the charge/discharge curve of the supercapacitor shows linear voltage versus time dependency. This is a disadvantage of supercapacitors when compared with batteries. For a lot of applications, for example, OLEDs, a fairly constant voltage plateau is required and thus in the case of supercapacitor DC/DC converter is needed.

Another important parameter in addition to capacitance is an internal resistance R_S (also called equivalent series resistance [ESR]). RS is caused mostly by the electrolyte and separator through which the ions have to travel when charging/discharging the device.

The maximum power P_{max} that the supercapacitor can produce is determined by short-circuiting a capacitor C and a resistor RS in series. The maximum power is given by

$$P_{max} = \frac{V^2}{4R_S} \tag{8.5}$$

This power is usually reported per mass of electrode material and is called the specific power or the power density (in units W/kg).

8.3.4 Electrical Characterization of Supercapacitors

A conventional way to characterize the electrochemical behavior of a supercapacitor is cyclic voltammetry, which can be performed using an electrochemical workstation equipped with a potentiostat. A cyclic voltammogram (CV) curve shows the current that develops in the electrochemical cell while the potential difference of electrodes is linearly varied. CV measurement can be done in either three-electrode or two-electrode configuration, which have some significant differences. Three-electrode electrochemical cells, containing a working electrode, a reference electrode, and a counterelectrode, are generally used for electrochemical material analysis. The potential is measured between so-called reference and working electrodes, while the current is measured between the working and counterelectrodes. The purpose of the counterelectrode is to allow accurate measurements to be made between the working and reference electrodes. The role of the counterelectrode is to ensure that current does not run through the reference electrode, since such a flow would change the reference electrode's potential. However, highlighted sensitivity of three-electrode configuration analysis can give raise to large errors when aiming at analysis of the material's energy storage capability.

An example of a CV curve measured for a solution-processed carbon nanotube (CNT) supercapacitor in the two-electrode configuration is presented in Fig. 8.3a. The charge storage capacitance of the device can be, in principle, obtained from the area closed by the CV curve. In a symmetrical two-electrode cell, the counterelectrode and the reference electrode are connected together and the potential differences applied to each electrode are equal to each other, that is, one-half of the values than in the case of three-electrode cell. The twice-larger voltage range in a three-electrode cell results in doubling of the capacitance calculated from the CV curve area. Even when using a two-electrode configuration for measuring the device capacitance, there are other sources of errors. Due to the resistive losses in the system, the voltage-scanning rate changes the shape and size of the CV curve, which then results in large errors in the capacitance calculated from the CV.

A galvanostatic measurement, which is based on keeping a constant current over the electrolytic cell, can be used to overcome the drawbacks of CV measurement. On the basis of the industrial standard IEC 62391-1 the supercapacitor capacitance can be measured reliably using the constant current discharge method. This galvanostatic discharge measurement procedure is presented in Fig. 8.4. The supercapacitor is expected to be fully charged after 30 min in constant voltage VR. After this the device is discharged with a constant current $I_{\text{discharge}}$. Capacitance can then be determined from the galvanostatic curve with the equation

$$C = I_{\text{discharge}} \cdot \frac{t_2 - t_1}{V_1 - V_2} \tag{8.6}$$

The RS is defined from the IR drop voltage change in the beginning of the galvanostatic discharge curve by the equation

$$R_S = \frac{IR\,\text{drop}}{I_{\text{discharge}}} \tag{8.7}$$

A high performance supercapacitor must have a large capacitance *C*, high cell operating voltage *V*, and minimum equivalent series resistance RS. Examples of galvanostatic discharge curves measured for a solution-processed CNT supercapacitor are presented in the Fig. 8.3b–c.

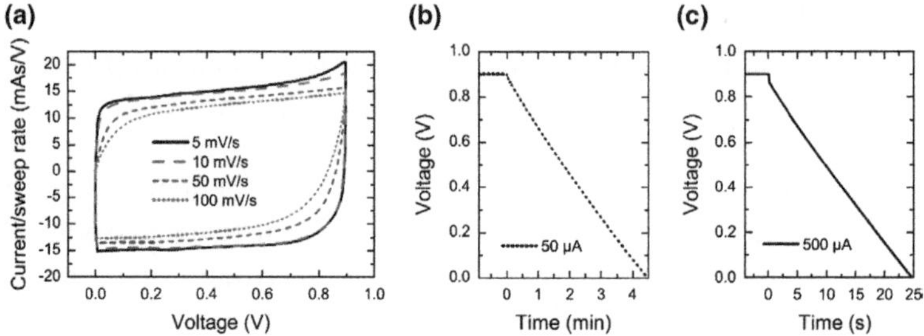

Figure 8.3 (a) A cyclic voltammogram (CV) curve and galvanostatic discharge curves for measuring for (b) capacitance *C* and (c) series resistance R_S for the carbon nanotube-based supercapacitor. Reprinted with kind permission from Springer Science+Business Media: Lehtimäki, S., Tuukkanen, S., Pörhönen, J., Moilanen, P., Virtanen, J., Honkanen, M., and Lupo, D. (2014). Low-cost, solution processable carbon nanotube supercapacitors and their characterization. *Appl. Phys.* A, **117**(3), p. 1329.

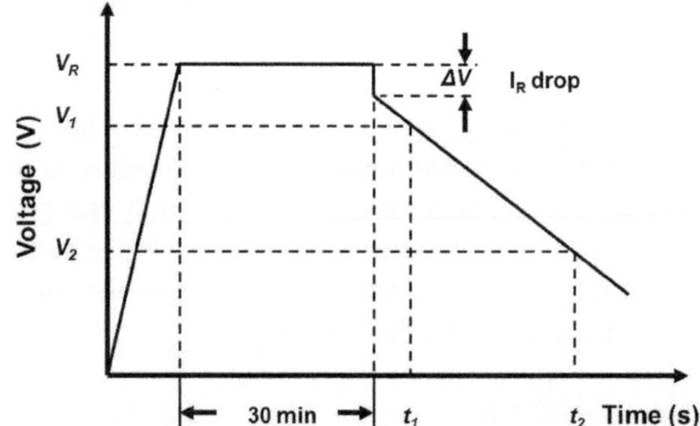

Figure 8.4 Principle of a galvanostatic measurement of supercapacitor parameters. Reprinted from *International Standard: Fixed Electric Double Layer Capacitors for Use in Electronic Equipment, IEC 62391-1*, with permission from International Electrotechnical Commission, Copyright © 2006 IEC Geneva, Switzerland. www.iec.ch

8.3.5 Comparison of Supercapacitor and Battery

Supercapacitors are energy storage devices that have a large power density and a long cycle life. Compared with Li ion batteries, supercapacitors have hundreds of times longer cycle lives and over 10 times larger power densities, leading also to much shorter charging times. Supercapacitors have approximately 10 times lower energy density compared to batteries. A comparison of different types of electrochemical devices is shown in Fig. 8.5.

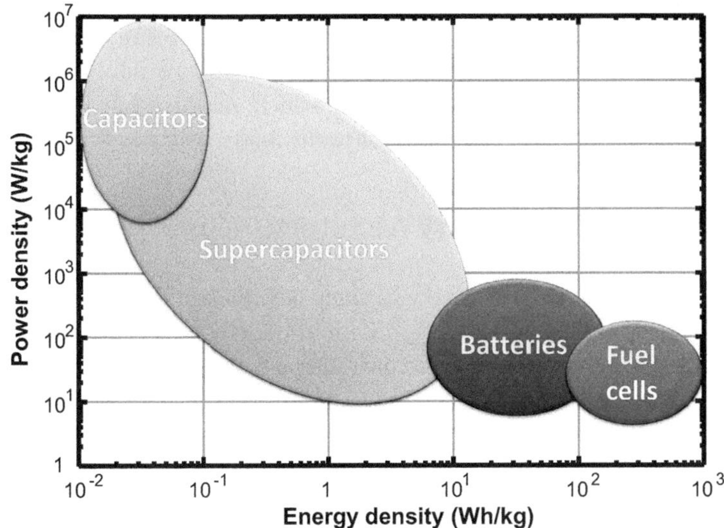

Figure 8.5 A Ragone plot of different energy conversion and storage devices.

In a battery the charge is stored in the volume of the electro-active materials. The electrochemical process takes place on the surface in contact with the electrolyte. So the electrode needs a high ionic conductivity. The required electrons shall be conducted to or away from the **point of electrochemistry**. This requires a suitable percollative network. Especially in the cathode, which typically consists of oxide materials, conducting agents are necessary. Therefore the charges cannot be stored in a high speed, but the amount of charges that can be stored in the lattice is much higher. Because of these fundamental considerations the application range of both elements can be defined.

- Capacitors are useful if a charge shall be stored or discharged with a high rate in a short time. Typical applications are the flash of a mobile phone, which is realized by a light-emitting diode (LED) that can stand a very high current for a very short time.
- Batteries are used if a high amount of charge is needed over a long time. This is the case for the most smart objects.

The more recent development of electrochemical energy storage devices has led to battery/supercapacitor hybrid constructions that are aiming at combining the advantages of both devices at once. The battery provides a high-energy density to the storage system, whereas an active converter controls the energy flow from the battery to a coupled supercapacitor, which, on the other hand, can provide a high output peak power to the load.

8.4 Printed Battery Architectures

The performance of a battery strongly depends on the construction. Especially the length of the ion path in the electrolyte is important for the DC resistance. In principle there are two different structures, which are described next.

8.4.1 Stacked Assembly

In a stacked assembly the components of the printed battery are stacked face to face, as presented in Fig. 8.6. Usually both electrodes are built up on both substrates, one for the positive and one for the negative electrode.

In the first step, the current collectors are printed on both sides and printed layers are dried (or baked/cured). Next, the electrode materials are printed on the current collectors, the anode material to the negative terminal and the cathode material to the positive terminal, and then dried.

The electrode/separator combination can be applied in two ways. In the case of a liquid electrolyte, similar to state-of-the-art cells, a separator sheet has to be cut and placed on the electrode. It should be fixed by a little drop of glue for further processing. Then the electrolyte is injected on top of the separator and soaked during a waiting step.

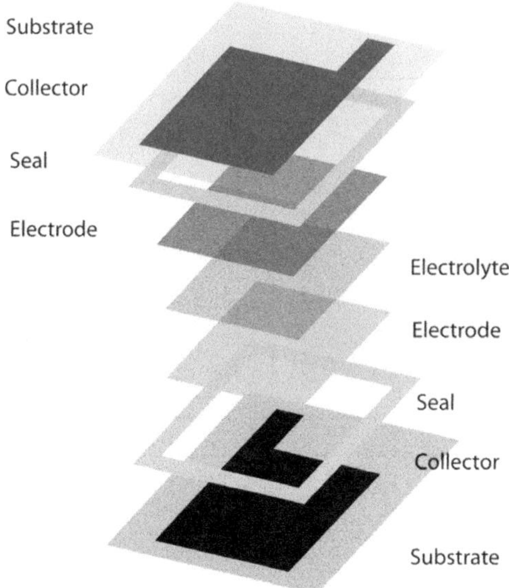

Figure 8.6 Components of a stacked printed battery assembly.

In the case of a solid electrolyte the paste is printed directly on both electrodes. After a drying step the cell can be further processed. Then the electrolyte is not completely dried but has a rubber-like consistency.

The last step is the printing of a glue frame to seal the battery hermetically. After the glue has dried, both parts are assembled and closed by lamination, hot embossing, or ultrasonic welding.

In Fig. 8.7, the deposition of the components of the printed battery is done on the same substrate. This substrate has a perforation in the middle axis to fold it easily. In another construction both electrodes are printed on different substrate strips and led together directly before the lamination. In this case the registration (alignment) is very important.

In the stacked configuration presented in Fig. 8.8, the way of the ions through the electrolyte is very short. It can be in the order of 10–20 μm. Also the area of the electrodes face to face is very large. Therefore the internal resistance of this construction is very low. An example of the printed stacked battery is shown in Fig. 8.9.

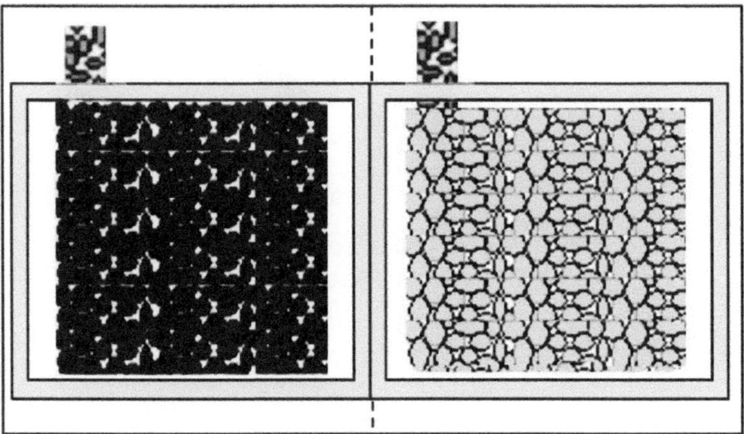

Figure 8.7 Deposition of the components of a printed battery on the same substrate.

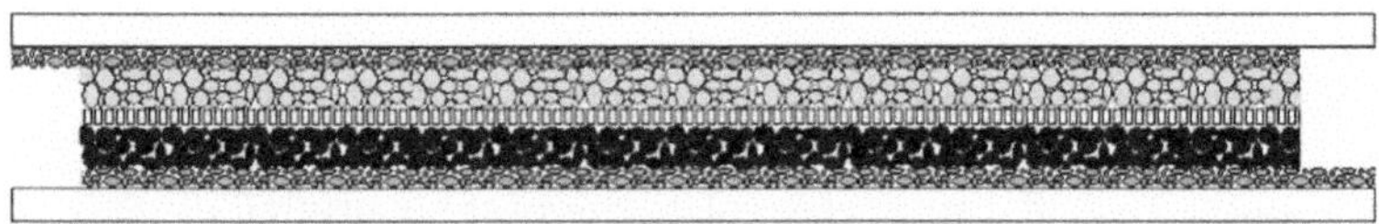

Figure 8.8 Assembly of the electrodes and separator/electrolyte in a stacked configuration.

Figure 8.9 Example of a printed stacked nickel/metal hydride battery. The device was fabricated in Hochschule der Medien (HdM), Stuttgart, Germany.

8.4.2 Co-Planar Assembly

In the co-planar assembly presented in Fig. 8.10, the electrodes are printed next to each other on the same substrate, having a small gap between them. Then the separator/electrolyte is placed on both electrodes, including the gap. Also a solid electrolyte can be applied.

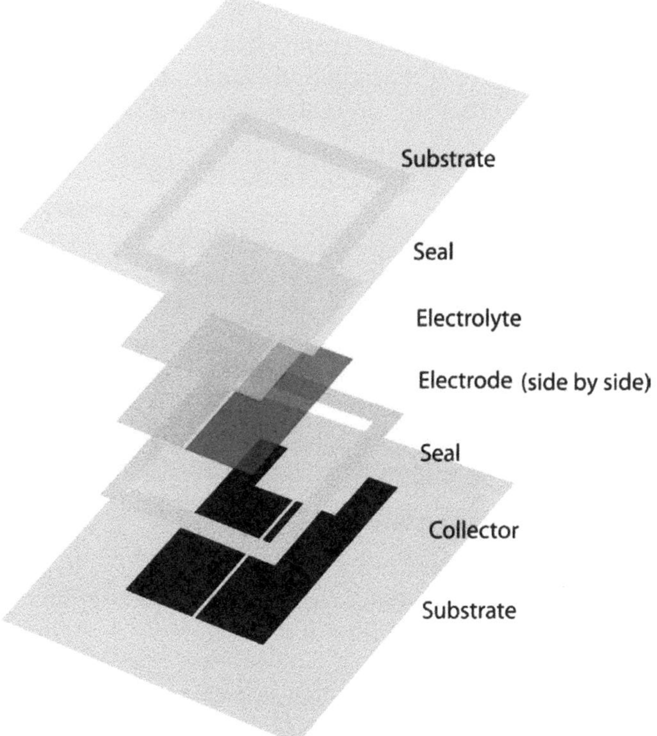

Figure 8.10 Components of a co-planar printed battery.

The ions move through the electrolyte from one electrode to the other. In the beginning of the discharging sequence, the ions travel the shortest bath, which is the size of the gap between two electrodes, which can be seen in Fig. 8.11. The internal resistance is then low when the electrode material close to the gap is discharged. Later during the discharging progress, the bath whose ions have to move gets longer and longer, which continuously increases the internal resistance.

Figure 8.11 Assembly of the electrodes and separator/electrolyte in a co-planar configuration.

An example of a co-planar nickel/metal hydride battery is presented in Fig. 8.12. It shows both electrodes, of which the positive is the green one. They are printed on a current collector of silver covered with a thin layer of carbon.

Figure 8.12 Example of a printed co-planer nickel/metal hydride battery. The device was fabricated in Hochschule der Medien (HdM), Stuttgart, Germany.

The advantages and disadvantages of stacked and co-planar battery configurations are compared in Table 8.1. A serial connection is easily possible for both cases.

Table 8.1 Comparison of stacked and co-planar constructions

	Stacked	**Co-planar**
Advantage	Low internal resistance, high rate capability	Easy to print, thin construction, flexible
Disadvantage	Thick and unflexible	High internal resistance, strong decay in the discharge curve

The serial connection of a stacked battery is presented in Fig. 8.13. The current collector is printed on top of the step to connect the negative electrode of one cell with the positive electrode of the next cell.

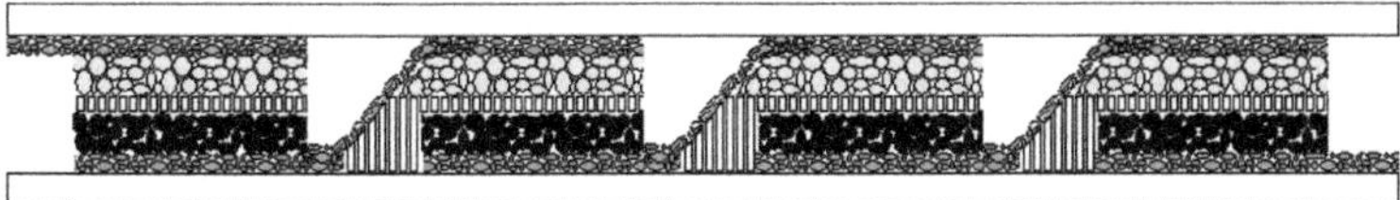

Figure 8.13 Serial connection of stacked cells.

The serial connection of a co-planar battery is presented in Fig. 8.14. The positive electrode and the negative electrode of the cell beneath are printed on the same current collector. Then the electrolyte is printed on the both electrodes of each cell, including the gap. It has to be made sure that the electrolytes of neighboring cells do not contact each other.

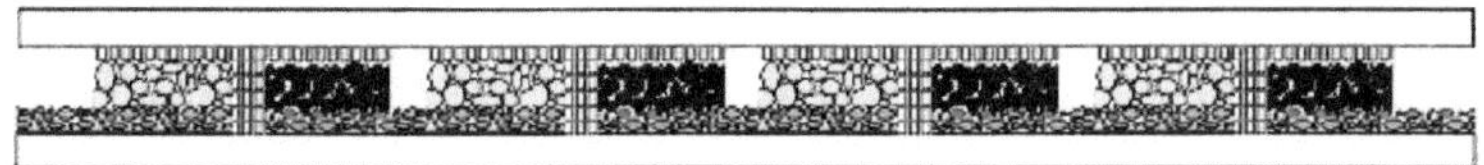

Figure 8.14 Serial connection of co-planar cells.

8.4.3 Liquid Electrolyte

Usually batteries use liquid or slightly gel-type electrolytes. These provide a high ionic conductivity. They also migrate into the pores of the electrodes in a short time. So the ions can be transported to or from the point of the electrochemical reaction easily. On the other hand, the liquid electrolyte tends to spread along the surfaces of the casing and may generate leakage. Especially the KOH electrolyte shows this type of behavior, and thus it cannot easily be encapsulated.

One more aspect is the mechanical property. The liquid electrolyte cannot transfer forces. So the electrodes are not mechanically connected with each other. They can move along each other and create damages in this way. If inside the packaging a small amount of gas is generated, it causes an increase in the thickness. There is a bubble between the electrodes, which interrupts the ionic contact between the electrodes. This defect is called **pillowing**.

8.4.4 Solid Electrolyte

These problems are widely avoided by a solid electrolyte. This is an electrolyte/separator paste that has a high conductivity and a high amount of gelling agents. So it can be printed easily.

After a short drying period the paste shows a high viscous, rubber-like form. The electrolyte is partly soaked by the electrodes. Then both the half-cells are assembled and laminated by a certain pressure/temperature influence. After this procedure both half-cells stick together very strongly. The interface between the electrodes is thin and has a high adherence. This way a high ionic conductivity can be reached. The printed battery forms a solid body that can stand many mechanical impacts.

8.4.5 Battery Materials

In principle there is a large variety of materials that can be used in batteries. However, because of a set of requirements some typical materials have shown to be best suitable for the cells. The selection of anode, cathode, and electrolyte materials determines the electrical behavior of the whole cell.

The purpose of the anode is to deliver electrons to the negative terminals when the electro-active material is oxidized. While recharging the direction of the processes are vice versa. For high-energy efficiency it is important that the electro-active materials be completely consumed at the end of discharge. During the charging process the reduced materials shall deposit at the same places where they were before the discharge, indicating dimensional stability and absence of deformation. This is important for cyclability. This can be achieved by using host lattices where the reduced atoms are intercalated. Typical anode materials are listed in Table 8.2.

The purpose of the cathode is to take electrons from the positive terminals when the electro-active material is reduced. While recharging the direction of the processes are vice versa. The cathode materials typically are metal oxides where the metal atoms change their oxidation states. In this case, deformations are not expected. The oxides form the host lattice for hydrogen or lithium atoms. For all rechargeable electrodes the volume effect has to be taken into account. Typically the lattice of the host expands when the atoms are intercalated. This can lead to a mechanical destruction of the electrode and insofar for capacity loss. Typical cathode materials are listed in Table 8.3.

Table 8.2 Typical battery anode materials

Electrolyte system	Primary	Secondary (rechargeable)
Water-based electrolyte	Zinc	Metal hydride (MH)
Organic electrolyte	Lithium (metal)	Lithium (metal)
		Lithium in carbon
		Lithium titanate
		Lithium in silicon

Table 8.3 Typical battery cathode materials

Electrolyte system	Primary	Secondary (rechargeable)
Water-based electrolyte	Manganese dioxide	Nickel hydroxide
	Silver oxide	
	Nickel oxy Hydroxide	
	Oxygen (air)	
Organic electrolyte	Manganese dioxide	Lithium cobalt oxide (LCO)
		Lithium nickel mangan cobalt oxide (NCM)
		Lithium iron phosphate (LFP)

The electrolyte consists of a Li ion containing salt and a solvent. The solvent can be organic carbonates. The amount of the salt should be in the order of 1 M. Typical electrolytes are listed in Table 8.4.

Table 8.4 Typical battery electrolyte materials

Electrolyte system	Primary	Secondary (rechargeable)
Water-based electrolyte	KOH solution (alkaline) Zinc chloride solution (neutral)	KOH solution
Organic electrolyte	Lithium perchlorate in EC/DME	Lithium phosphor fluoride in EC/DEC

8.5 Printed Supercapacitor Architectures

8.5.1 Printable High–Surface Area Materials

The most common material used in supercapacitor electrodes is activated carbon, which is very porous and relatively inexpensive, although its conductivity is so low that it has to be blended with more conductive materials and a metallic current collector must be used.

A very promising new direction has been found in carbon nanomaterials, mostly CNTs and graphene, whose high electrical conductivity and nanoporous structure allow for the electrode to function as a current collector itself. This opens the door to flexible, thin-film structures, which can be fabricated, for example, using low-cost and high-throughput printing techniques. An example of CNT-based solution-processable material is shown in Fig. 8.15. Potentially interesting future solutions are composite materials where the high surface area of activated carbon is combined with the high conductivity of CNTs or graphene.

8.5.2 Printable Supercapacitors

There are several studies where supercapacitor electrodes are made from solution-processed materials. Various solution-processing methods, such as inkjet printing, screen printing, spray coating, doctor blading, bar coating, dip coating, painting, and brush-on coating, have been demonstrated. These solution-processing methods can be (at least in principle) scaled up to roll-to-roll mass production where actual printing techniques are used.

Printable supercapacitors have been recently fabricated, for example, from activated carbon, CNTs, graphene carbon black, and carbon fibers, as well as their composites with conducting polymers. An example of a printable CNT-based supercapacitor is shown in Fig. 8.16. This type of disposable, low-cost energy storage device is suitable for energy-harvesting applications where energy is gathered, for example, from radio-frequency field and mechanical movement.

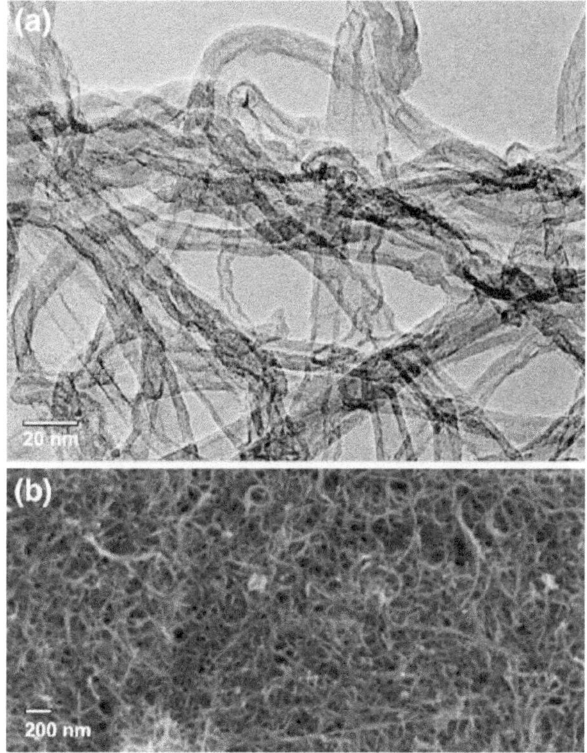

Figure 8.15 An example of new printable nanocarbon-based high-surface-area materials used in supercapacitor electrodes. (a) TEM and (b) SEM images of CNT/xylan nanocomposite ink. Reprinted with kind permission from Springer Science+Business Media: Lehtimäki, S., Tuukkanen, S., Pörhönen, J., Moilanen, P., Virtanen, J., Honkanen, M., and Lupo, D. (2014). Low-cost, solution processable carbon nanotube supercapacitors and their characterization. *Appl. Phys. A*, **117**(3), p. 1329.

There are several demonstrations of printed supercapacitor electrodes, but to have a fully printable supercapacitor, the separator film also should be printable, except in the case where electrodes are printed directly on the separator film. Further, application of the electrolyte and device encapsulation has to be done using scalable methods. One promising approach toward fully printed supercapacitors was recently demonstrated by Tuukkanen et al., who used nanocellulose film as a printable separator in a CNT-based supercapacitor.

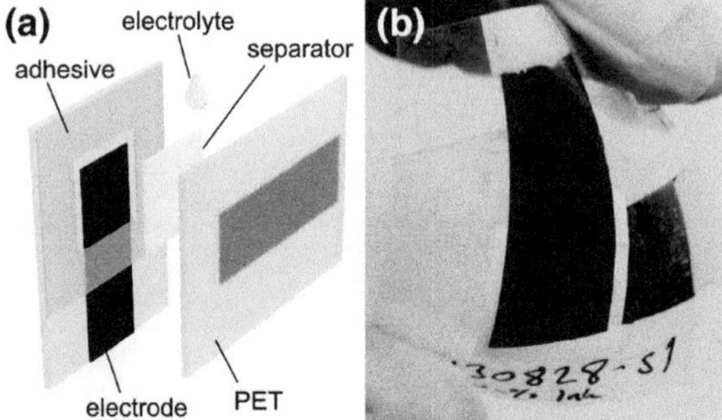

Figure 8.16 An example of printable and flexible CNT-based supercapacitors. The high-surface-area electrodes, which simultaneously act as current collectors, are fabricated from solution-processable CNT nanocomposite ink. (a) The assembly of the device is done by sandwiching two CNT electrodes on a separator paper soaked with aqueous NaCl electrolyte. (b) Photograph of a bent CNT supercapacitor. A commercial paper separator was used. Reprinted with kind permission from Springer Science+Business Media: Lehtimäki, S., Tuukkanen, S., Pörhönen, J., Moilanen, P., Virtanen, J., Honkanen, M., and Lupo, D. (2014). Low-cost, solution processable carbon nanotube supercapacitors and their characterization. *Appl. Phys. A*, **117**(3), p. 1329.

8.6 Challenges of Printed Electrochemical Systems

The printability of different electrochemical systems depends on a selected set of materials and the architecture. Here we summarize some aspects that determine whether the energy storage device is printable or not. Table 8.5 summarizes different battery chemistries and their suitability for printing.

In general, porous electrodes can be easily printed. After drying the electrodes made of electro-active materials, binders, and solvents form a porous body that can act as an electrode. Air electrodes have very complicated structures and insofar are considered not to be printable with today's skills.

Cells using an organic electrolyte are difficult to print because they require an inert atmosphere. Organic electrolytes should not be exposed to oxygen during the whole manufacturing chain before the closing of the cell. Printing production lines with a special atmosphere are quite complicated to build. Further, especially the encapsulation should have barrier properties in the order of 10^{-6} g/m^{2}day H_2O permeation.

Table 8.5 Overview of battery technologies and their printability

Primary batteries	V_{max}	**Printability issues**
Zinc/manganese dioxide	1.5 V	Easy to print, open system
Zinc/air	1.4 V	Complicated cathode, alkaline electrolyte
Zinc/silver oxide	1.5 V	Alkaline electrolyte
Lithium/manganese dioxide	3.0 V	Affected by water
Secondary batteries (rechargeable)	V_{max}	Printability issues
Nickel/metal hydride	1.2 V	Alkaline electrolyte
Lithium ion	3.7 V	Affected by water

Obtaining a fully printable power storage device, where the electrodes and the separator are monolithically printed on top of each other, is difficult due to the need for a printed separator. The challenge in printed separators is the probability of having pinholes that can cause short circuiting of the electrodes.

Acknowledgments

Many thanks to Prof. G. Hübner, Hochschule der Medien, Stuttgart, for providing outstanding support related to printed batteries and their components.

Exercises

8.1 What is the principle difference between batteries and supercapacitors?

8.2 What kinds of properties are required from supercapacitor electrode materials?

8.3 Calculate the capacitance from the CV curve in Fig. 8.3a.

8.4 Calculate the capacitance from the galvanostatic curve in Fig. 8.3b.

8.5 Calculate the ESR from the galvanostatic curve in Fig. 8.3c.

8.6 A battery has a capacity of 20 mAh. A current is drawn by a data logger including wireless communication. The base current is 10 μA. For measurement a current of 800 μA is drawn for 1 ms every 15 minutes. Once a day the data are transferred to the base station, which takes 10 mA for 0.5 s. How long does the battery last?

8.7 A device demands an operational voltage of 7.5 V. How many primary cells of zinc/manganese dioxide type have to be connected in series? How many cells have to be connected if a nickel/metal hydride secondary system and a lithium ion system is used?

Suggested Readings

Candelaria, S. L., et al. (2012). Nanostructured carbon for energy storage and conversion. *Nano Energy*, **1**(2), pp. 195–220.

Chen, P., Chen, H., Qiu, J., and Zhou, C. (2010). Inkjet printing of single-walled carbon nanotube/RuO2 nanowire supercapacitors on cloth fabrics and flexible substrates. *Nano Res.*, **3**(8), pp. 594–603.

Chen, T., and Dai, L. (2013). Carbon nanomaterials for high-performance supercapacitors. *Mater. Today*, **16**(7), pp. 272–280.

Chen, T., and Dai, L. (2014). Flexible supercapacitors based on carbon nanomaterials. *J. Mater. Chem. A*, **2**, pp. 10756–10775.

Choi, H.-J., et al. (2012). Graphene for energy conversion and storage in fuel cells and supercapacitors. *Nano Energy*, **1**(4), pp. 534–551.

Conway, B. E. (1999). *Electrochemical Supercapacitors*, Kluwer, Norwell, MA.

Gao, L., Dougal, R. A., and Liu, S. (2005). Power enhancement of an actively controlled battery/ultracapacitor hybrid. *IEEE Trans. Power Electron.*, **20**(1), pp. 236–243.

Halper, M. S., and Ellenbogen, J. C. (2006). *Supercapacitors: A Brief Overview*. Report No. MP 05W0000272, MITRE Corporation, McLean, Virginia.

Hu, L., Choi, J. W., Yang, Y., Jeong, S., La Mantia, F., Cui, L. F., and Cui, Y. (2009). Highly conductive paper for energy-storage devices. *Proc. Natl. Acad. Sci.*, **106**(51), pp. 21490–21494.

Hu, L., Wu, H., and Cui, Y. (2010). Printed energy storage devices by integration of electrodes and separators into single sheets of paper. *Appl. Phys. Lett.*, **96**(18), p. 183502.

Hu, S., Rajamani, R., and Yu, X. (2012). Flexible solid-state paper based carbon nanotube supercapacitor. *Appl. Phys. Lett.*, **100**(10), p. 104103.

International Electrotechnical Comission (2006). International standard: fixed electric double layer capacitors for use in electronic equipment, IEC 62391-1.

International Electrotechnical Comission (2006). International standard: primary batteries, IEC 60086-1.

Jayalakshmi, M., and Balasubramanian, K. (2008). Simple capacitors to supercapacitors-an overview. *Int. J. Electrochem. Sci.*, **3**(11), pp. 1196–1217.

Jost, K., Stenger, D., Perez, C. R., McDonough, J. K., Lian, K., Gogotsi, Y., and Dion, G. (2013). Knitted and screen printed carbon-fiber supercapacitors for applications in wearable electronics. *Energy Environ. Sci.*, **6**(9), pp. 2698–2705.

Kaempgen, M., Chan, C. K., Ma, J., Cui, Y., and Gruner, G. (2009). Printable thin film supercapacitors using single-walled carbon nanotubes. *Nano Lett.*, **9**(5), p. 1872.

Keskinen, J., Sivonen, E., Jussila, S., Bergelin, M., Johansson, M., Vaari, A., and Smolander, M. (2012). Printed supercapacitors on paperboard substrate. *Electrochim. Acta*, **85**, pp. 302–306.

Kossyrev, P. (2012). Carbon black supercapacitors employing thin electrodes. *J. Power Sources*, **201**, pp. 347–352.

Kötz, R., and Carlen, M. (2000). Principles and applications of electrochemical capacitors. *Electrochim. Acta*, **45**(15), pp. 2483–2498.

Le, L. T., Ervin, M. H., Qiu, H., Fuchs, B. E., and Lee, W. Y. (2011). Graphene supercapacitor electrodes fabricated by inkjet printing and thermal reduction of graphene oxide. *Electrochem. Commun.*, **13**(4), pp. 355–358.

Lehtimäki, S., Li, M., Salomaa, J., Pörhönen, J., Kalanti, A., Tuukkanen, S., Heljo, P., Halonen, K., and Lupo, D. (2014). Performance of printable supercapacitors in an RF energy harvesting circuit. *Elec. Power Energy Syst.*, **58** p. 42.

Lehtimäki, S., Tuukkanen, S., Pörhönen, J., Moilanen, P., Virtanen, J., Honkanen, M., and Lupo, D. (2014). Low-cost, solution processable carbon nanotube supercapacitors and their characterization. *Appl. Phys. A*, **117**(3), p. 1329.

Li, X., and Bingqing W. (2013). Supercapacitors based on nanostructured carbon. *Nano Energy*, **2**(2), pp. 159–173.

Linden, D., and Reddy, T. B. (1865). *Handbook of Batteries*, McGraw-Hill Professional.

Liu, C., et al. (2010). Advanced materials for energy storage. *Adv. Mater.*, **22**(8), pp. E28–E62.

Liu, Q., Nayfeh, M. H., and Yau, S.T. (2010). Brushed-on flexible supercapacitor sheets using a nanocomposite of polyaniline and carbon nanotubes. *J. Power Sources*, **195**, pp. 7480–7483.

Lu, M., Beguin, F., and Frackowiak, E. (2013). *Supercapacitors: Materials, Systems and Applications*, John Wiley & Sons.

Pandolfo, A. G., and Hollenkamp, A. F. (2006). Carbon properties and their role in supercapacitors. *J. Power Sources*, **157**(1), pp. 11–27.

Pech, D., et al. (2010). Elaboration of a microstructured inkjet-printed carbon electrochemical capacitor. *J. Power Sources*, **195**(4), pp. 1266–1269.

Pörhönen, J., Rajala, S., Lehtimäki, S., and Tuukkanen, S. (2014). Flexible piezoelectric energy harvesting circuit with printable supercapacitor and diodes. *IEEE Trans. Electron Devices*, **61**(9), pp. 3303–3308.

Stoller, M. D., and Ruoff, R. S. (2010). Best practice methods for determining an electrode material's performance for ultracapacitors. *Energy Environ. Sci.*, **3**(9), pp. 1294–1301.

Tuukkanen, S., Lehtimäki, S., Jahangir, F., Eskelinen, A.-P., Lupo, D., and Franssila, S. (2014). Printable and disposable supercapacitor from nanocellulose and carbon nanotubes. *5th Electronics System-Integration Technology Conference* (ESTC 2014).

Wendler, M., Hübner, G., and Krebs, M. (2011). Development of printed thin and flexible batteries. *IC International Circular of Graphic Education and Research. Hsgb: The International Circle of Educational Institutes for Graphic Arts, Technology and Management*, **4**, pp. 32–41.

Xu, Y., Schwab, M. G., Strudwick, A. J., Hennig, I., Feng, X., Wu, Z., and Müllen, K. (2013). Screen-printable thin film supercapacitor device utilizing graphene/polyaniline inks. *Adv. Energy Mater.*, **3**(8), pp. 1035–1040.

Yu, G., Xie, X., Pan, L., Bao, Z., and Cui, Y. (2012). Hybrid nanostructured materials for high-performance electrochemical capacitors. *Nano Energy*, **2**, pp. 213–234.

Zhang, J., and Zhao, X. S. (2012). On the configuration of supercapacitors for maximizing electrochemical performance. *ChemSusChem*, **5**(5), pp. 818–841.

Chapter 9

Encapsulation of Organic Electronics

John Fahlteich,[a] **Andrea Glawe,**[b] **and Paolo Vacca**[c]
[a]*Fraunhofer Institute for Organic Electronics, Electron Beam and Plasma Technology, Winterbergstraße 28, 01277 Dresden, Germany*
[b]*KROENERT GmbH & Co. KG, Schützenstraße 105, 22761 Hamburg, Germany*
[c]*SAES Getters S.p.A, Viale Italia, 77, 20020 Lainate MI, Italy*
john.fahlteich@fep.fraunhofer.de

9.1 Introduction to Encapsulation

Encapsulation of an organic electronic device describes the packaging of devices to protect them against damage caused by the environment (extrinsic degradations factors). These degradation factors include mechanical damage (scratches, kinks, pressure), reactions with harmful gases (moisture, oxygen) or liquid chemicals (e.g., lipids from fingerprints, tensides, or acids), and light-induced degradation (e.g., exposure to direct sunlight/UV wavelengths). The term "encapsulation" thereby refers to both the packaging materials (e.g., glass) and the packaging processes (e.g., lamination).

This chapter gives an introduction to commonly used materials and processes for the encapsulation of organic electronics. One

Organic and Printed Electronics: Fundamentals and Applications
Edited by Giovanni Nisato, Donald Lupo, and Simone Ganz

ISBN 978-981-4669-74-0 (Hardcover), 978-981-4669-75-7 (eBook)
www.panstanford.com

major cause for degradation of organic devices is the reaction of the electrically active materials with ambient water vapor and/or oxygen. Therefore a major task of the encapsulation is to prevent water vapor and oxygen from reaching the device. Two major mechanisms of device degradation are observed in devices when they come in contact with water: (1) pixel shrinkage caused by water penetration from the side and (2) local degradation and dead-spot formation due to defects in the encapsulation. Figure 9.1 illustrates the damage to an organic light-emitting diode (OLED) caused by reaction of the OLED layers with water.

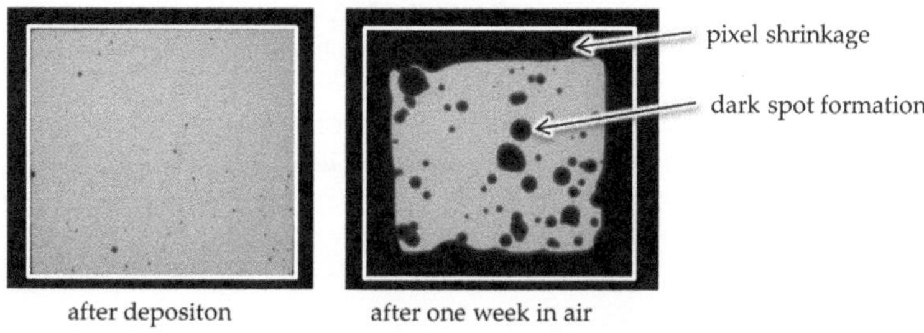

Figure 9.1 Aging of an orange OLED with insufficient encapsulation.

Organic electronic devices show the strongest requirements in comparison to other products that require protection against these gases. These requirements depend on the technology, materials used, and desired product properties and expected product lifetime. In literature, often the water vapor transmission rate (WVTR) and oxygen transmission rate (OTR) are used as quantities to describe on the one hand the performance of an encapsulation system and on the other hand the maximum allowed amount of water to reach an organic device before device failure. These values (WVTR and OTR) describe a mass of water or volume of oxygen permeating through the encapsulation system per time unit (commonly one day) and area (commonly 1 m^2). The units of the values are given as [WVTR] = $g/(m^2 day)$ and [OTR] = $cm^3/(m^2 day\ bar)$. As we will see below both WVTR and OTR are **no** material constants but depend on the environmental conditions (mainly temperature and relative humidity) at which they are measured. Therefore also WVTR and OTR requirements given for organic electronic devices depend on their desired application environment. There is a big difference between indoor use (23°C and about 50% relative humidity) and, for example, outdoor use in tropical conditions (up to 38°C and 90% relative humidity).

Browsing the literature about encapsulation of organic electronic devices, the requirement for an encapsulation system with a WVTR of 10^{-6} g/(m^{2}day) can be found in many papers. This requirement is usually derived for OLEDs, which sometimes include highly reactive low-work function metals as the cathode. The assumption of primary metal oxidation, causing injection/extraction barriers for charge carriers, is well justified and indeed requires very strong encapsulation in the range below 10^{-5} g/(m^{2}day) for reasonable lifetimes. While OLEDs strongly suffer from local degradation at defects in the encapsulation (see Fig. 9.1), other devices such as organic solar cells are more sensitive to a total amount of water reaching the device over time on a certain surface area. Only 20 mg of water is sufficient to reduce the power efficiency of a small-molecule organic solar cell with a size of 1 m^2 by 50%. Assuming an expected lifetime of 10,000 hours, that corresponds to a WVTR of 2 × 10^{-5} g/m^{2}day. This is still less demanding than the 10^{-6} g/m^{2}day given for OLED. The easiest way to reach such low WVTR is to use glass to encapsulate the device. Glass is impermeable to both water and oxygen and has no local defects that could lead to dead-spot formation. However, it has limited resistance against mechanical impacts and damage and is not flexible. To allow flexible (meaning bendable and formable) devices, coated polymer webs can be used as encapsulation, as shown in the next sections below.

9.2 Types of Encapsulation

Depending on the desired lifetime, the inherent stability of the system and the target market for the organic device, different encapsulation approaches may be chosen, as shown in Fig. 9.2. Among these a glass/glass encapsulation Fig. 9.2a), consisting of a glass substrate with the device and a second glass sheet mounted on top, sealing the device, is still the strongest and most durable encapsulation. However, this type of encapsulation only allows rigid and flat devices. Mostly thin-film barrier-coated polymer webs (e.g., polyethylene terephthalate [PET]) are used to manufacture flexible devices. These barrier films are laminated on a device that was processed on a rigid glass substrate or metal foil (Fig. 9.2b) or they can be used as a substrate for the device with an additional barrier film laminated on top (Fig. 9.2c). The technologically most challenging option is shown in Fig. 9.2d: thin-film encapsulation (TFE) applied directly on the organic device. This TFE is done by vacuum coating techniques such as plasma-enhanced chemical vapor deposition (PECVD) or atomic layer deposition (ALD) and is often combined with subsequent barrier film lamination (similar to Fig. 9.2c). In any case, it is important to know which side of the device is the active side. The active side thereby means the side of the device to which the light is emitted by an OLED or from which an organic solar cell or sensor is illuminated. On this side the encapsulation has to be optically transparent. Metal foils or thick metal layers cannot be used on the active side. Glass, many oxide barrier layers, and often used polymer webs fulfill the transparency requirement.

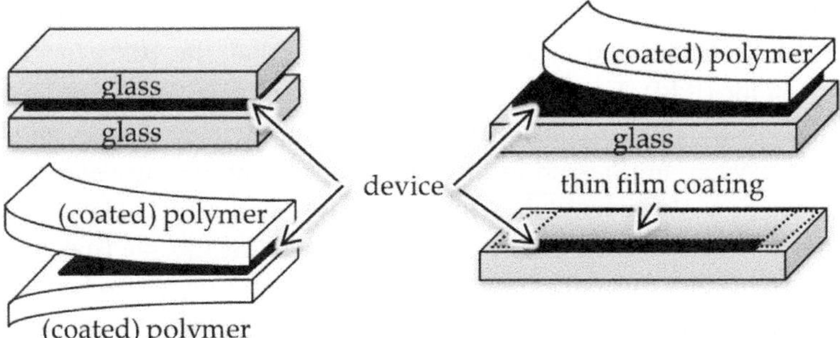

Figure 9.2 Different approaches for encapsulation. (a) Top-left image: glass/glass encapsulation; (b) top-right image: lamination of a flexible barrier film on top of an intrinsic barrier substrate like glass or metal; (c) bottom-left image: processing on a flexible barrier film and lamination of another film on top; and (d) bottom-right image: direct thin-film encapsulation of the device.

9.2.1 Glass/Glass Encapsulation

Macroscopically thick glass sheets are practically a perfect barrier against oxygen and moisture. In addition to that, glass is durable against other chemicals, fingerprints, and to some degree also mechanical damage. Usually, the device is processed on a glass substrate to prepare a glass/glass encapsulation. Subsequently the device is encapsulated by mounting a second glass sheet on top. Thereby an edge seal or a full-area adhesive is used to stick the top glass sheet onto the device. Of course it is also possible to mount an organic device that has been processed on a flexible film between two glass slides. The weak point of this encapsulation approach is the permeation through the connection of the two glass slides. Whereas it is certainly possible to cover the full area of the device with adhesive or bonding material, usually only an edge sealing of the assembly is performed, since the area is protected by the glass and chemical or mechanical influence of the adhesive on the device can be avoided. Therefore two main mechanisms determine the remaining permeation of water into the sealed cavity (Fig. 9.3).

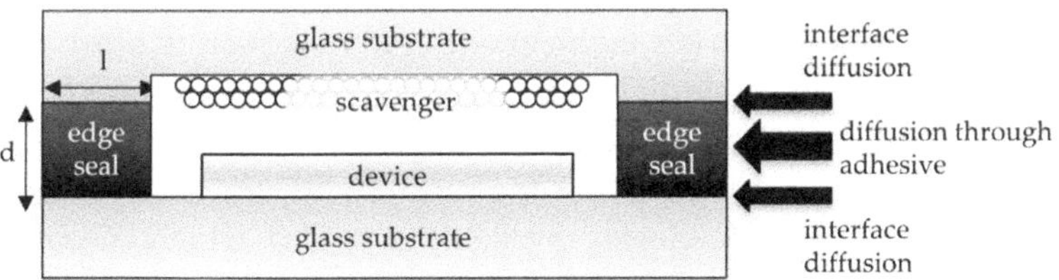

Figure 9.3 Schematic drawing of glass/glass encapsulation. The device is located between two glass slides and can be equipped with a getter material to enhance the device lifetime. Parasitic diffusion of water and oxygen takes place through the edge seal.

The first is the permeation of water or oxygen through the adhesive or bonding material itself, which depends on the used material, the adhesive thickness (d) and the length of the edge seal (l). Given that the edge-sealing process has to be applied as last processing step, the overall substrate temperature that can be applied to cure or modify the edge seal is limited by the thermal stability of the device. Thus most edge-sealing systems rely on polymer materials, which do not have very high barrier properties. In this case the parasitic permeation can be decreased by changing the geometries of the seal, that is, use as thin as possible adhesive systems with a large lateral extension to create long diffusion pathways.

The second mechanism for parasitic permeation in an edge-sealed glass/glass system is the permeation at the interface between the glass surface and the edge seal material and through any coated thin layers below the edge. These layers can be, for example, indium tin oxide (ITO) electrode structures, screen-printed porous passivation or metal structures, or even organic materials. The second effect results often in lower water quantity reaching the device compared to the first but cannot be neglected. By smart design and optimized adhesives the permeation through the edge seal in glass/glass encapsulations can be lowered into the range of 10^{-6}. Glass/glass encapsulations can also be scaled up, which requires some engineering of the edge-sealing process to avoid defects. Another big advantage of glass/glass encapsulations is the equal thermal expansion coefficient of the top and the bottom seal, as well as the mechanical rigidity minimizing the mechanical strain in the sealing layer. Also the edge sealing can be combined with scavenger materials in the device cavity. The role of scavenger materials is to react with the permeating oxygen or moisture faster than other materials involved in the organic device. Therefore glass/glass encapsulations are the most stable and most reliable encapsulations technologies available and are widely used as reference for permeation and device measurements.

9.2.2 Lamination of Barrier Films

Glass/glass encapsulation does not allow bendable and flexible devices. For that, so-called permeation barrier films are used as encapsulation material. Permeation barrier films are usually polymer webs that are coated with inorganic layers or multilayer stacks that prevent water and oxygen from permeating through the film. A huge advantage of this technology is the possibility to manufacture the barrier films in advance, so their production conditions are not limited by the sensible organic device. Therefore higher temperatures, radiation, chemical baths, and gases can be applied to the barrier film, which might destroy an organic device instantly. Due to the separated processing of the barrier films and device, the complete range of coating processes, as described below, is available and can be used, limited only by substrate compatibility. However, compatibility of the device is still needed for lamination step itself. In this step the barrier film is stuck onto the devices (usually on the full area instead of only using an edge seal) using an adhesive polymer. Depending on the type of adhesive—the lamination process requires also high temperature (hot-melt adhesives), pressure (pressure-sensitive adhesives [PSAs]), or radiation (UV-curing adhesives). The types of adhesives and the processing for lamination are described below in Section 9.4.3. With barrier films, also roll-to-roll processing of flexible devices is possible.

The same parasitic permeation paths for water vapor and oxygen as shown for the glass/glass encapsulation above also apply for the encapsulation with a barrier film. In addition to that, moisture permeation takes place also at defects in the barrier film, leading to, for example, dead-spot formation. Also, mechanical stress in the system has a higher impact to the device compared to glass/glass encapsulation, causing, for example, bending or in worst-case delamination, and has to be minimized. Depending on the type of adhesive and the desired application, and depending on buffer layers, the distance between the actual barrier layers and the devices might be larger than for glass/glass encapsulations, leading to higher lateral permeation of water into the device. Sometimes an additional edge sealing is applied to lower this parasitic permeation.

If the device is processed on top of the barrier film, the barrier must not be damaged during the following processes, which is especially critical for pattering steps by laser ablation or wet etching, for example, for lateral electrode structuring in thin-film modules.

Barrier films are available in many different qualities and permeation barrier strengths, from food packaging to ultrahigh barrier systems according to the short overview in Section 9.4.2.

9.2.3 (Direct) Thin-Film Encapsulation

The term "thin-film encapsulation" (TFE) describes direct coating of the device with a thin single or multilayer stack using common deposition techniques such as ALD, PECVD, sputtering, or also printing techniques. The module is thereby processed on glass, metal foil, or barrier film. Ideally, such a process can be integrated into either sheet-to-sheet or roll-to-roll device manufacturing. Using TFE, parasitic effects and device contamination can be reduced and/or damage due to a separate mechanical lamination process is avoided. The main issue of TFE is the need to process on the organic device, which severely limits the processing conditions. The exact limits depend on the device technology used, but higher substrate temperatures exceeding 80°C–100°C as usual for many deposition processes, UV radiation, or solvent exposure are often not allowed because they would damage the device. Therefore many technologies, using plasma, reactive gases, or high temperatures to achieve dense inorganic barrier layers are difficult to use in TFE processes.

Just like in barrier films, multiple layers can be stacked to improve barrier properties. TFE can also be combined with other encapsulation technologies, for example, lamination between glass slides or barrier films after TFE. Another significant advantage of combining TFE with suitable production layouts is the direct contact between the bottom and top barriers, rendering additional perimeter sealing obsolete. It is in principle possible to combine the barrier, electrode, and antireflection coating as multifunctional TFE.

9.3 Introduction to Permeation in Solids and Thin Films

The most important task of the encapsulation is to prevent water vapor and oxygen from reaching the device. Therefore permeation barrier performance of the encapsulation materials is a critical parameter for the device application. Understanding gas permeation in solids (glass), polymers (e.g., adhesives), and thin films (inorganic oxide barrier layers) is therefore crucial to optimize the encapsulation of the device.

Permeation is described as mass transport of gas or liquid through a solid material. It is a compound process based on the fundamental physical processes sorption, solution, and diffusion. Figure 9.4 shows the subprocesses that define the permeation through a solid or thin film. These processes are:

1. Adsorption and solution: The permeating gas is adsorbed at the surface of the solid, and if the gas is soluble in the material, it is dissolved in the solid.

2. Diffusion: Depending on the concentration gradient between the two solid surfaces, the gas diffuses through the solid material.
3. Desorption: The gas particles are released from the backside solid surface.

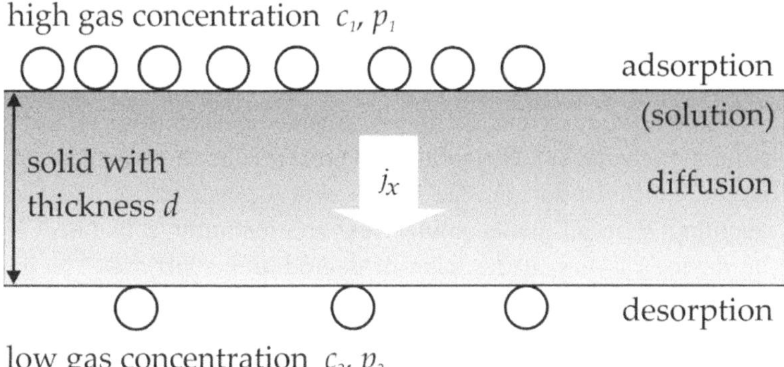

Figure 9.4 Fundamental processes involved in gas permeation in solids.

The physical quantity related to the gas permeation is the permeation coefficient (*P*), being the product of the diffusion coefficient (*D*) and the sorption coefficient (*S*).

$$P = D \cdot S$$

9.3.1 Fick's First Law of Diffusion

Diffusion in solids can be described in general by Fick's laws (commonly referred to as **Fickian diffusion**). Fick's first law covers gas and liquid diffusion through a solid material under equilibrium condition. It defines the amount of gas diffusing through a solid within a defined time interval (*t*) with the particle current density (*J*) that can be derived from the diffusion coefficient and the concentration gradient of the permeating gas in the solid material:

$$\vec{J} = -D\Delta\vec{c}$$

When talking about permeation in thin films or even in polymer webs on large areas, the lateral dimensions of the solid are orders of magnitude larger than its thickness *d*. Under that boundary condition and with the assumption that the gas concentration on the solid surface is homogeneous, Fick's first law can be simplified to a 1D equation with a concentration gradient only perpendicular to the solid surface. With the additional assumption that the solid is isotropic with a constant diffusion coefficient, the particle current density can be described as

$$J_x = \frac{\partial n}{\partial t}\frac{1}{A} = -D\frac{c_1 - c_2}{d}$$

There by n describes the quantity of the permeating gas and A is the surface area of the solid. In typical setups the moisture and oxygen concentration on the device side (c_2) is zero at all times, assuming that all water molecules are instantly consumed by the device's layers in the form of degradation reactions. The gas concentration on the outer surface of the encapsulation (c_1) is determined by (ad)sorption of the gas to the surface according to the surface properties and the environmental conditions in which the device is stored. Thereby, c_1 may not be measured directly in an experiment. Instead several models exist to describe adsorption and desorption of gases on solid surfaces and solution of these gases within the solid material in dependence of the gas's partial pressure at the surface. Sorption in general can be from physical nature (physisorption, dipole or van der Waals force) or chemical nature (chemisorption, chemical covalent bonding). Thereby it is well known that chemisorption is much stronger than physisorption. The simplest model to describe sorption processes is Henry's law, which proportionally links concentration (c) of a permeating gas on/in a solid material to its partial pressure at the surface (p_g):

$$c = S \cdot p_g$$

The proportionality factor is the sorption coefficient S. The gas particle flow through a solid can be now described with

$$J_x = -DS\frac{p_1 - p_2}{d} = -P\frac{p_1 - p_2}{d}$$

Henry's law can be applied only under specific conditions that are:

- Only physisorption takes place (e.g., no gas molecule dissociation at the surface).

- The permeating gas cannot condensate on the solid surface at the given temperature.

Therewith sorption of oxygen on and in many polymers at room temperature may be well described using Henry's law. However, this is not the case for water vapor and even CO_2 as these gases may condensate and dissociate at room temperature. Although Henry's law still gives a satisfying approximation, more refined models such as the Langmuir sorption model or Brunauer–Emett–Teller sorption model are needed to correctly describe water sorption on typical encapsulation materials and organic devices.

The available partial pressure for oxygen permeation is easy to determine and equals 20% of the total air pressure according to the composition of air. For water, the determination of the partial pressure and concentration on the air side is more difficult. It depends on temperature and relative humidity and is discussed in a separate section of this chapter. However, the main consequences of diffusion physics for the encapsulation of organic electronic devices are:

1. The permeation through a glass encapsulation can be described by Fick's permeation laws. However, the glass thickness is very high and the related diffusion coefficient of the material for water vapor or oxygen is very low (e.g., oxygen diffusion in different glasses at room temperature: between 10^{-25} cm^2/s and 10^{-18} cm^2/s). This means that a glass encapsulation can be treated as impermeable.
2. Permeation through polymer webs, organic layers, and adhesives can be described with the Fickian diffusion model, assuming that the materials are free of local irregularities (defects) and have a homogeneous thickness and material composition. The steady-state permeation rate is inverse proportional to the material/layer thickness with a constant diffusion coefficient. Because of high diffusion coefficients (e.g., $D = 10^{-9}$ cm^2/s for O_2 in polyesters) the materials are not sufficient for encapsulation of organic devices alone.
3. The permeation through thin metallic and oxide thin films cannot be described with the Fickian model as the layer properties are not homogeneous and the layers contain

defects that provide a much higher permeation than predicted from the Fickian model based on bulk material. Different defect permeation models are used to describe water vapor and oxygen diffusion through these layers.

9.3.2 Defect Models for Permeation in Thin-Film Encapsulation

9.3.2.1 Classification of defects

Inorganic thin films such as metal or oxide layers are used both as TFE deposited directly onto the device and as a barrier layer that is deposited onto a polymer web that is laminated onto the device. The solid-state (Fickian) gas diffusion through such barrier layers is several orders of magnitude lower than experimentally determined permeation results. For example, the permeation through a 100 nm thick "bulk" SiO_2 layer would be 13 orders of magnitude lower than measured for 100 nm thick physical vapor deposition (PVD)-coated SiO_2 layer.

It is well known that thin films contain defects of different sizes. A common understanding is that these defects are almost solely responsible for the permeation of gases through the thin films, while the Fickian solid-state diffusion is negligible. Defects in thin films are classified with respect to their size and permeation mechanism according to Fig. 9.5. Lattice defects include defects within the solid-state structure, such as dislocations or atomic holes. Gas permeation through these defects is very low and can be neglected. Nanodefects have a size in a range between 0.3 nm and approximately one to a few nanometers. The number of these defects in a typical thin oxide barrier layer can be up to 4 orders of magnitude higher than the number of macrodefects. Even though the permeation through a single nanodefect is very low, it is believed that these defects are the driving force for water vapor permeation just due to their large number. In addition to that, water condensation can take place at these defects accelerating the permeation further. Due to their small size as well as the fact that oxygen does not condensate at ambient conditions, it is believed that these defects don't play an important role for oxygen permeation.

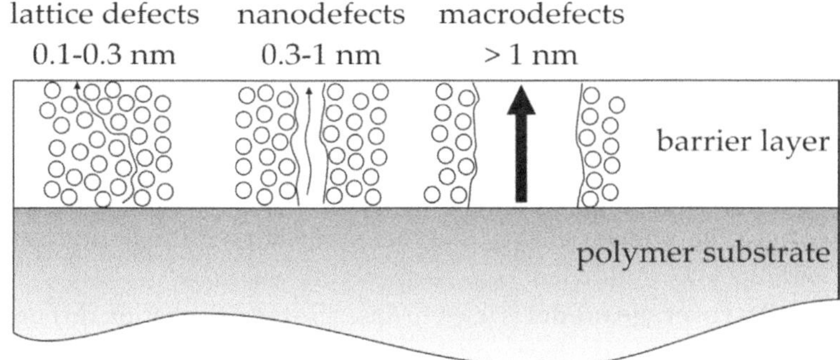

Figure 9.5 Oxford classification of defects.

Macrodefects with a size larger than 1 nm up to several micrometers are caused for example by substrate surface defects, particles, and dust, as well as damage due to mechanical cracking. Macrodefects can be classified further regarding their size and the dominant diffusion mechanism. Defects that have a size considerably larger than the mean free path of the permeating gas in air (>>66 nm on ambient conditions) allow unhindered gas diffusion. They are the driving force for oxygen diffusion through thin films and play an important role also for water vapor permeation and local degradation in encapsulated organic electronic devices.

9.3.2.2 Models to describe gas permeation at layer defects

Several models exist to estimate gas permeation in thin-film single layers and layer stacks. Often the encapsulation system contains more than one layer (e.g., a polymer film, an inorganic coating, and an adhesive). The resulting permeation in such a layer stack can be described by the **ideal laminate model** (sometimes also called the **electrical analogy model**). This model is based on the assumptions that

- All layers are permeable with a permeation coefficient other than zero ($P_1 ... P_n \neq 0$).
- All layers are free of accelerated diffusion paths such as layer defects.

The ideal laminate model describes the gas permeation through the layer stack in the same way as the electrical current in a series of electrical resistances. The total water vapor permeation may be calculated according to the following equation:

$$\frac{d_{\text{total}}}{P_{\text{total}}} = \frac{d_1}{P_1} + \frac{d_2}{P_2} + \cdots + \frac{d_n}{P_n}$$

This finally means that in a multilayer system, the total gas permeation is always lower than the permeation through the layer with the lowest individual gas permeability would be.

The **coverage model** is used to describe the permeation through an inorganic single layer on any permeable substrate with the assumption that the permeation through the barrier layer only takes place at defects, while the layer material itself is impermeable. In that case, the total gas permeation through the system is determined by the permeability of the substrate and the total defect area in the layer. However, the coverage model does not take into account that the gas spreads in all directions within the permeable substrate after passing through the inorganic layer.

The coverage model has been extended by Prins and Hermans to the well-known **pinhole model** to include this behavior. If the average size and the density of defects are known, the total gas permeation though the system layer + substrate can be calculated using this equation:

$$J_x = -D\frac{A_{\text{defect}}}{A_{\text{total}}} \cdot \frac{c_1 - c_2}{d_{\text{substrate}}}\left(1 + 1.18\frac{d_{\text{substrate}}}{r_0}\right)$$

The equation is valid with the boundary conditions that the defect area is low compared to the substrate's total area ($A_{\text{defect}} << A_{\text{total}}$) and that the thickness of the substrate is larger than about one third of the average defect radius ($d_{\text{substrate}} > 0.3 \cdot r_0$). The model assumes that an average defect radius (r_0) can be used for calculation and the influence of different defect sizes on the gas spreading in the polymer is negligible. There are several enhancements of the pinhole model that try to take into account different sizes of defects. However, in most cases the pinhole model gives already a good approximation, although both the coverage model and the pinhole model only describe the permeation through macrodefects without taking into account possible chemical interactions of the gas with the layer material (e.g., dissociation) and without including permeation through nanodefects.

9.3.2.3 Defects in multilayer systems

Stacking thin layers to a multilayer system is a common approach to reduce the effect of defects in the individual constituents of the system on permeation through the whole stack. Commonly, polymer layers and oxide barrier layers are stacked alternately on the substrate (see Section 9.4.3). While the oxide layers have low gas permeability but contain layer defects, the polymer layers are permeable but have the capability to interrupt defect growth and cover particles on the substrate or in the layers below. On assumption behind stacking dyads (one oxide layer plus one polymer layer) to a multilayer stack is that—because of a different distribution of defects in the first and subsequent oxide layers—these defects are decoupled so that water or oxygen have to diffuse laterally first and saturate the polymer layer before reaching the next defect in the subsequent oxide layer and finally the device. This so-called tortuous path of diffusion model allows a significant delay of water permeation through the encapsulation system (see the next section). Also the probability of defect growth caused by imperfections in the substrate, the device electrode or first barrier layer is reduced by introducing an interlayer between the first and the second barrier layer.

9.3.3 Time Dependence of Permeation

All explanations up to now include only steady-state permeation after reaching a constant concentration gradient between the air side and the device side of the encapsulation. However, usually encapsulation of organic devices is done in an inert (dry) atmosphere prior to exposing it to ambient condition with water vapor and oxygen (permeates) in the atmosphere. That means that the initial device conditions and the concentration gradient in the barrier film/encapsulation are in a nonequilibrium condition having—idealized—no permeate molecules solved within the encapsulation film. A certain "time lag" is needed to reach a concentration gradient within the sample according to the ambient condition on the outer surface of the encapsulation (e.g., 50% relative humidity at 20°C) and the time-constant zero concentration of the permeate on the device side (assuming that the degrading device binds all incoming permeate molecules). The time dependence of permeation under the nonequilibrium condition can be described using Fick's second law:

$$\frac{\partial \vec{c}}{\partial t} = -\nabla\left(-D\nabla\vec{c}\right) \qquad \frac{\partial c_x}{\partial t} = D\frac{\partial^2 c_x}{\partial x^2}$$

Of course this is only true in case solid-state diffusion is the dominating mechanism without taking into account permeation through defects. Also a position- and concentration-independent diffusion coefficient D is assumed. As stated above, the concentration of the permeate at the outer side of the sample is assumed to be constant at all times (c_1 = const.). Within the encapsulation film, the permeate concentration is zero at all positions initially, continuously increasing until the steady state is reached. The concentration on the device side of the encapsulation film remains zero at all times ($c_2 = 0$).

Fick's second law can be used together with these boundary conditions to calculate the total amount of gas permeating through the solid ($Q(t)$):

$$Q(t) = \frac{Dtc_1}{d} - \frac{dc_1}{6} - \frac{2dc_1}{\pi^2}\sum_{n=1}^{\infty}\frac{(-1)^n}{n^2}e^{-Dn^2\pi^2t/d^2}$$

The total amount of water depends on the diffusion coefficient D, the thickness of the encapsulation film d, and the water vapor concentration on the outer side of the encapsulation. The time needed to reach a constant concentration profile in the encapsulation profile is the so-called lag time of diffusion, which is defined for large t ($t >> 0$) as the intercept of the linear terms in the equation above:

$$t_0 = \frac{d^2}{6D}$$

It is thereby important to keep in mind that the lag time does not equal the time needed to reach a constant WVTR through the encapsulation system and also does not equal the required minimum measurement time. However, increasing the lag time of a TFE system by alternately stacking inorganic and organic barrier layers is a common concept to reach sufficient lifetime of organic electronic devices. Organic devices usually tolerate a maximum quantity of water vapor molecules reaching the device's surface before failure regardless of whether that water amount reached the device with a constant WVTR over the time or with an initially low WVTR that increases after the time lag is over. Figure 9.6 illustrates this behavior in detail. The figure shows the time dependence of the total amount of water reaching the device and the WVTR for two different encapsulation systems. One has a low WVTR but a short lag time; the other one has a high lag time but only a lousy WVTR after reaching that time. This difference is important to know when deciding for an encapsulation solution with respect to the desired lifetime of the device. Some groups reported simulated results for multilayer stacks showing a lag time of water permeation of several years, which would by far be sufficient for encapsulating, for example, OLED displays for mobile phone application, regardless of whether the steady-state permeation rate would be far too high for a reliable device encapsulation.

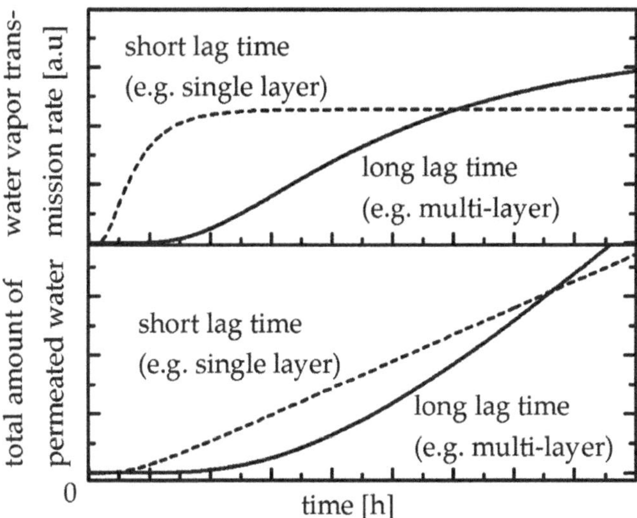

Figure 9.6 Schematic representation of time-dependent gas permeation through two different initially dry encapsulation films.

9.3.4 Influence of Ambient Conditions on the Gas Permeation Rates in an Encapsulation System

The gas permeation rates not only depend on the material properties of the encapsulation but also on the ambient conditions that determine the gas concentration on the outer surface of the encapsulation layers. These conditions are the partial pressure of the gas at the surface and the ambient temperature.

9.3.4.1 Partial pressure of the permeate gas

Calculation of the surface partial pressure is easy for oxygen as it equals the fraction of oxygen in air multiplied by the total air pressure under normal ambient conditions. Therefore, in typical application environments of organic electronic devices the partial pressure of oxygen is 0.2 bar remaining constant even at different temperatures or different climates (e.g., tropical climate versus indoor climate in Northern Europe). Therefore only temperature dependence of permeation according to the section below has to be considered.

The water vapor partial pressure, however, is more difficult to determine. Usually WVTRs are given for a certain relative humidity and temperature. Relative humidity (r.h.) is thereby defined as the quotient of the actual water vapor partial pressure and the saturation vapor pressure of water at a given temperature and air pressure:

$$r.h.(\%) = \frac{p_{\text{vapor}}}{p_{\text{saturation}}} \cdot 100$$

However, the saturation vapor pressure of water in air is temperature dependent (exponentially increasing in a temperature range of 0°C–60°C), which means that 90% relative humidity corresponds to a much lower absolute water vapor partial pressure at 20°C compared to, for example, 38°C. Therefore, the WVTR through an encapsulation film is also much lower at 23°C than at 38°C. The temperature dependence of the water vapor saturation pressure is given by the Magnus formula in a range between −45°C and 60°C:

$$\ln p_s = \ln(611.2\,Pa) + \frac{17.62 \cdot \vartheta}{243.12°\text{C} + \vartheta}$$

Thereby the temperature ϑ is given in degrees Celsius. The factors and constants in the equation have been determined experimentally and vary for different temperature ranges. Table 9.1 summarizes the water vapor saturation pressure at common testing conditions for aging of organic electronic devices. To obtain the available amount of water, these values have to be multiplied with the relative humidity percentage at the given measurement condition.

Table 9.1 Saturation pressure of water vapor at different typical testing conditions for organic electronic devices

Description of the condition	Temperature (°C)/ Typical relative humidity (%)	Saturation vapor pressure of water at given temperature (Pa)
Ambient	20/50	2333
Tropical	38/90	6616
Accelerated aging	60/90	19,993
Damp-heat test	85/85	~58,680

9.3.4.2 Temperature dependence of permeation

Fickian diffusion is a thermally activated process following an Arrhenius equation for temperature dependence. This results in an exponential increase of gas permeation with increasing temperature. As both diffusion and sorption depend exponentially on temperature the permeation coefficient can be determined by

$$P = P_0 \cdot e^{-\frac{E_A}{RT}}$$

The activation energy of permeation is defined as the sum of the activation energy of the gas diffusion (E_D) and the heat of solution or activation energy of sorption ($E_A = E_D + \Delta H_S$). Temperature dependence of permeation through defects in thin films differs from those shown for Fickian diffusion. In large macrodefects the permeating gas and the thin film material do not interact but the gas is permeating virtually unhindered through the defect. There is no contribution of the thin film to the activation energy of permeation in the full stack. If many smaller defects dominate gas permeation, the permeation mechanism changes depending on the defect size to

- Molecular flow for a defect size in the range of the mean free path of the gas molecules in air (around 60 nm)
- Interface diffusion and, if applicable, condensation in capillaries for small defects a few nanometers in size and less

In both cases, the activation energy of the encapsulation system is influenced by the defects in any barrier layers, leading to mostly stronger increase of the gas permeation with increasing temperature.

Oxygen permeation is dominated by macrodefects for many oxide barrier layers independently from the coating technology. The activation energy of oxygen diffusion through coated polymers is almost the same as of the uncoated polymer substrate being, e.g., for PET: $E_A = 29 \pm 3$ kJ/mol.

Water vapor permeation, instead, is not only dominated by macrodefects but also by an interaction of the water with the coating material and/or by condensation at nanodefects. The activation energy of water vapor permeation strongly depends on the layer material and also the coating technology. Al_2O_3 layers, for example, show a higher activation energy of water vapor permeation if they are coated by sputtering compared to evaporation. This result points to a more macrodefect-dominated permeation in evaporated layers and a more nanodefect-dominated diffusion in sputtered layers. The values range from $E_A = 3$ kJ/mol for uncoated PET and evaporated layers to $E_A = 14$ kJ/mol for some sputtered Al_2O_3 layers.

9.4 Encapsulation Substrates and Materials

The most often used encapsulation material is glass with a thickness in the range of less than 1 mm to a few 10 mm. Glass encapsulation is well established for large devices and a module size of up to several 100 cm^2 and even square meters. Glass encapsulation is processed in sheet-to-sheet processing machines at industrial substrate sizes from generation 1 (Gen 1 = 300 × 400 mm^2) up to generation 6 (Gen 6 = 1500 × 1800 mm^2) or rarely higher. This section deals with alternative encapsulation substrates and materials that can be used for flexible devices that are currently not available with glass encapsulation.

9.4.1 Encapsulation Substrates

9.4.1.1 Polymer substrates for encapsulation

Flexible organic electronics devices and encapsulation systems can be processed on polymer webs such as polyethylene terephthalate (PET) or polyethylene naphthalene (PEN). However, most polymers are highly permeable for water vapor and oxygen and require a thin-film coating to reach a sufficiently low WVTR for organic device encapsulation. Basic requirements to the polymer webs being used as base substrates for encapsulation of organic devices are:

- Uniform surface and low intrinsic surface roughness without spikes to support high quality of any coatings on the substrate (including active organic layers, transparent electrodes, and barrier layers).
- Low density of particles and surface defects: Typically used PVD and PECVD techniques suffer from shadowing effects during layer deposition. Through these shadowing, nodular defects are formed at particle and surface defect sites (providing accelerated diffusion paths for water vapor and oxygen).
- High optical transparency and low absorption for front-side encapsulation: High optical transparency is needed to maximize the efficiency of organic solar cells and also OLEDs.
- Good adhesion of printed and/or vacuum-coated thin films on the polymer. A certain level of layer adhesion is required to ensure that the thin films can fulfil their desired functionality such as permeation barrier performance.
- Mechanical stability versus bending, strain, and also scratching: A defined level of tensile stress must be applied during roll-to-roll processing (web tension). The polymer must not strain higher than a certain percentage (<0.3%) to avoid damage to, for example, barrier layers. Also, often the polymer is facing the environment requiring a scratch and abrasion resistance of the material.
- Long-term stability against moisture and UV light: Especially in outdoor applications, the device is exposed to UV light and high ambient moisture. The encapsulation must not only protect the device but also be intrinsically stable against UV and moisture, which is, for example, not true for PET substrates.
- Temperature stability for higher processing temperatures during thin-film deposition: Some processes in organic electronics manufacturing require processing temperatures of 120°C and above. These are, for example, curing of lacquers, lamination, or PVD deposition of conductive layers. Standard-

grade PET films exhibit significant shrinkage at 150°C and change their properties when exposed to temperatures higher than their glass transition temperature.

Polymer film manufacturers are currently working on the development of suitable polymer substrates for manufacturing flexible electronic devices as up to today no product is available that fulfils all abovementioned requirements, especially the low defect and particle density on a large area for a reasonable price per square meter. Very often a high-quality, expensive, planarized polymer film (e.g., TEONEX® Q65FA PEN by DuPont Teijin Films) is used as substrate for the encapsulation film and device manufacturing. This film is thermally stabilized to allow a processing temperature of up to 150°C and has a very smooth surface with minimum particle contamination. In contrast to that, low-cost substrates such as a MELINEX® 400 PET film (also from DuPont Teijin Films) show significantly higher particle and surface defect density, have lower thermal stability (shrinkage at 150°C), and degrade under the influence of moisture during a longer time of use.

Beside PET and PEN substrates, high-performance polymers are used if a specific application demands superior properties, for example,

- High thermal stability, which is addressed with polyimide (PI) or polyether etherketone (PEEK) films that are stable up to 250°C (PEEK) or even >300°C (PI)
- UV and moisture stability, which can be addressed with fluoropolymers such as ethyletetetrafluorethylene (ETFE)

However, these polymers have other drawbacks preventing them from being used for a broad range of applications. Mostly, material cost (several hundred dollars per kilogram), low adhesion of thin films on the material (PI and fluoropolymers), low optical transparency (especially PI), and minor surface quality in terms of roughness prevent a wider use of these polymer substrates for organic device encapsulation.

Polymer substrates may be pretreated to improve the substrate's quality prior to their use in organic electronics or for encapsulation. The three most common pretreatment methods are:

1. Surface cleaning to remove particles and dust by, for example, mechanical cleaning using rollers with a sticky surface (tacky

rollers), ultrasonic bath cleaning, and air jet or CO_2 snow cleaning

2. Surface planarization by applying either a wet coating (using slot-die, reverse gravure, or other coating methods) or special vacuum-coating techniques (e.g., a liquid–vapor–liquid–solid [LVLS] process, as described later in Section 9.4.3)
3. Adhesion promotion by plasma or corona pretreatment of the surface directly prior deposition of the active layers

Especially applying a planarization layer may provide a significant difference of an order of magnitude or even more in permeation barrier performance of an oxide layer on the substrate.

Figure 9.7 shows the WVTRs of reactively sputtered zinc tin oxide (ZTO) permeation barrier layers (see Section 9.4.2.2) that have been deposited onto different polymer substrate products under the same process conditions. That example shows a difference of up to 2 orders of magnitude between the lowest WVTR on a planarized PEN substrate and the highest WVTR on a thermally stabilized PET film with standard-grade surface quality.

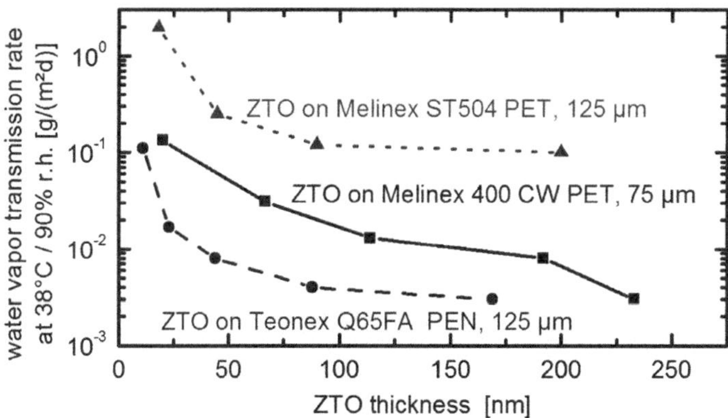

Figure 9.7 Comparison of permeation barrier performance of ZTO layers on different substrates.

9.4.1.2 Ultrathin glass

Within the last few years, ultrathin glass was introduced as a new substrate type for organic electronics by the large glass manufacturers Corning, Asahi Glass, Nippon Electric Glass, and

Schott. Ultrathin glass is defined as glass with a thickness of 200 μm and less going down to even 25 μm. Because of its low thickness and because of chemical additives the glass is bendable with a certain radius (7.5 cm radius has been demonstrated to work for 100 μm glass thickness). In spite of its low thickness, ultrathin glass is still virtually impermeable for water vapor and oxygen, having a WVTR and an OTR below the limit of detection of most available permeation measurement devices (< 10^{-5} g/m^{2}day). In contrast to inorganic coatings on polymer substrates, ultrathin glass does not suffer from large local defects that would lead to dead spots in an organic device. Also ultrathin glass can be laminated both in sheet-to-sheet and in roll-to-roll processes by using well-established encapsulation processes and adhesives. Another important advantage of ultrathin glass is the temperature stability allowing us to deposit high-performing transparent electrodes on the substrate at typical processing temperatures of 250°C or even higher.

However, a disadvantage of ultrathin glass is its fragility and tendency to crack even at fairly low mechanical stress. It is therefore difficult to handle and often needs to be laminated onto a carrier film (mostly a polymer web) for convenient processing. Several research groups currently work actively on methods for improved handling and processing of ultrathin glass to make roll-to-roll processing on a larger roll width and for longer rolls available to the industry.

9.4.2 Thin-Film Encapsulation Materials and Processes

9.4.2.1 Thermal evaporation

Thermal evaporation of aluminum for metallization of polymer substrates for food packaging applications has been a common method already for many years. For manufacturing transparent barrier layers either reactive thermal or electron beam evaporation of aluminum (using oxygen as reactive gas) or thermal evaporation of understoichiometric SiO or electron beam evaporation of SiO_x is used.

All these processes have in common a very high deposition rate and very low deposition costs for producing large areas of barrier films with a medium barrier performance. For thermal evaporation of aluminum usually ceramic boats are used that are heated by applying an electrical voltage. A wire feed guides an aluminum wire into the boat, which melts when coming into contact with the boat surface. Evaporation rates of several grams per minute lead to a dynamic deposition rate of up to 5 μm·m/min in a continuous roll-to-roll process. Oxygen is injected into the process chamber to achieve transparent coatings. Thermal evaporation of understoichiometric silicon oxide is done by filling a crucible with SiO granulate and heating the crucible. For electron beam evaporation also a crucible is filled with a solid coating material. A high-power electron beam (usually several 10 kV and up to a few 100 kW power) periodically scans the material surface and locally melts and evaporates the deposition material.

Adding plasma (e.g., by using hollow cathode plasma sources) to the evaporation process potentially leads to a better barrier film quality, reducing the gas transmission rates and improving process stability and reproducibility. The plasma assistance can be used both for thermal and for electron beam evaporation. Using evaporation processes, WVTRs of about 1 g/m^2day or even less have been reported on several substrates such as PET, polypropylene (PP), or cast polypropylene (CPP). Up to now, no sufficiently low WVTRs for the encapsulation of most organic electronic devices have been reported. Also direct TFE of devices using this high-rate evaporation has not been done yet, although it is possible if the thermal load on the substrate can be kept low. However, with reactive evaporation a highly productive large-scale roll-to-roll process is available for medium permeation barrier performance. Industrial roll-to-roll coating machines are operating with a coating width up to 3 m and even higher at a web speed up to 10 m/s to produce low-cost packaging films.

9.4.2.2 Plasma-enhanced chemical vapor deposition

Up to now, one of the most often used TFE methods is PECVD. A plasma discharge is used to provide the activation energy for a chemical reaction between different reactants (precursor gases) that finally will precipitate on the substrate (device or, e.g., polymer web) as a thin layer. A lower substrate temperature is possible with using PECVD in comparison to other CVD processes such as hot-wire CVD or thermal CVD. As the thermal stability of organic devices and typical polymers (e.g., PET webs) is limited to often less than 150°C, PECVD is the favorable CVD process.

Figure 9.8a shows a well known setup for PECVD deposition of thin layers: The system is based on a parallel plate capacitor in which a radio frequency (from 13.56 MHz up to 80 MHz) AC voltage is applied to the plates. The monomer is injected into the chamber either from the side or through holes in the second plate. Several modifications of this setup exist to for example, add a substrate bias voltage or to use noneven formed electrodes to vary ion bombardment on the substrate surface. Radio-frequency (RF) PECVD is mostly used to deposit SiO_2 or SiN_x barrier layers in a batch-coating process. The process pressure typically lies in a range between 5 Pa and 100 Pa. Sometimes an inductive coil is used to apply the RF voltage, leading to higher ionization compared to the capacitive plasma. RF PECVD sources require an impedance-matching network to maximize power output to the plasma process. Also scaling of the RF PECVD process to larger coating areas is difficult because of the required impedance matching.

Microwave sources, operating at 2.45 GHz, are available also for large coating areas and are used to deposit SiO_x and SiN_x barrier layers on both small and large substrates. Figure 9.8b shows a cross section of a linear microwave source. The source design is similar to a coaxial waveguide and consists of a copper rod as the inner conductor surrounded by a quartz tube with a microwave magnetron at each end of the tube. While atmospheric pressure is present within the tube, the plasma chamber can be evacuated to a pressure in the range between 10 Pa and 1000 Pa. Microwaves with a frequency of 2.45 GHz are generated at both ends of the and fed into the copper conductor. The microwave PECVD process can be enhanced by overlaying a magnetic field with a magnetic flux of 0.087 T to the electric field. This magnetic flux corresponds to an electron cyclotron resonance (ECR) frequency of 2.45 GHz, increasing the energy of the electrons and leading to a higher plasma density. ECR plasma sources are widely used for the deposition of microwave-PECVD barrier layers. However, the deposition of conducting layers is not possible using microwave PECVD. Microwave and ECR PECVD can also be done at atmospheric pressure allowing much simpler process equipment without having a vacuum vessel.

Also magnetron plasma sources may be used for generating dense PECVD plasmas. Yasuda described the magnetron PECVD process already in 1985. However, large-area magnetron PECVD coatings and the application of this process for the deposition permeation barrier layers have been propagated first in the 2000 s. Magnetron PECVD operates at a lower process pressure (0.2 Pa to 5 Pa) compared to other PECVD processes. Often a dual magnetron is used as plasma source. The two magnetron targets are switched alternately as cathode and anode with a frequency of 50 kHz. Mostly low-sputtering yield materials are used as the target material. As the target will be covered with an oxide layer during the deposition, the amount of target material measured in the PECVD layer is very low. Other variants of linear mid-frequency plasma sources such as the plasma beam source (PBS) exist with the same goal to scale the PECVD process to large coating areas. Very high deposition rates compared to microwave of RF PECVD are achieved for the deposition of plasma-polymer layers using either dual-magnetron PECVD or the dual PBS.

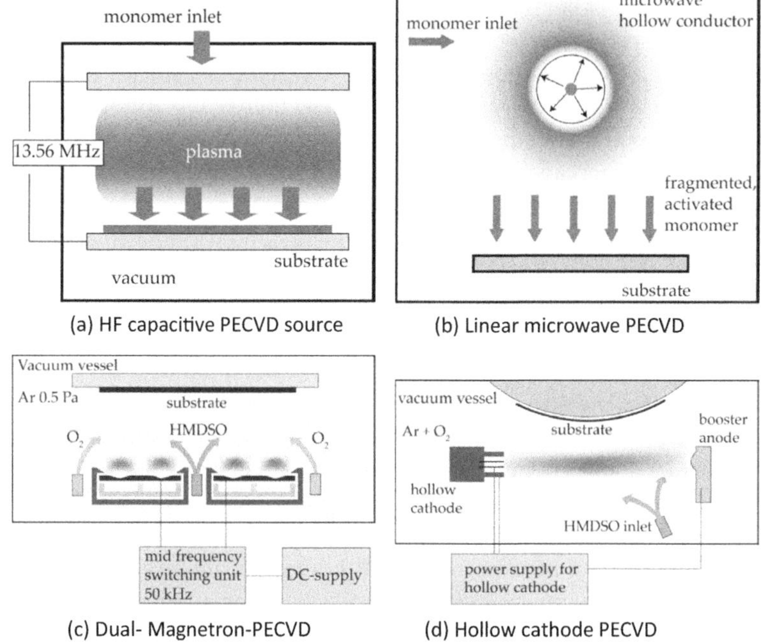

Figure 9.8 Examples for different PECVD plasma sources and deposition techniques.

Finally a hollow cathode that is normally used for plasma-assisted reactive evaporation can be used as PECVD source for TFE or barrier film. High-density plasma can be generated in the process chamber to allow high fragmentation of the monomer in the plasma.

Typical barrier layer materials deposited by PECVD are silicon oxide (SiO_2) and silicon nitride (SiN_x) or a composite SiO_xN_y. Widely used monomers are

- tetramethylsilane (TMS): $Si(CH_3)_4$
- hexamethyldisiloxane (HMDSO): $(CH_3)_3$-Si-O-Si-$(CH_3)_3$
- tetraethylorthosilicate (TEOS): $(CH_3$-CH_2-$O)_4$-Si
- silane: SiH_4

for the deposition of silicon oxide as well as

- hexamethyldisilazane (HMDSN): (CH3)3-Si-NH-Si-(CH3)3
- silane: SiH_4

for the deposition of silicon nitride barrier layers. PECVD processes thereby have the great advantage that the composition of the oxide layer can be adjusted by changing the amount of monomer and reactive gas being injected in-line during the process. With using a low amount of monomer and a high oxygen flow, silicon oxide or nitride barrier layers can be produced. Increasing the monomer flow while at the same time reducing the oxygen flow leads to a plasma polymer layer, which can be used as interlayer in a multilayer barrier. Several groups use a PECVD process to deposit multilayer stacks with alternating oxide barrier layers and polymer interlayers for low-stress encapsulation systems by alternately changing the monomer and reactive gas flow.

WVTRs of PECVD based SiO_x or SiN_x layers on polymers stongly depend on the actual process conditions, the monomer, the plasma source, and the substrate material. The lowest WVTR in the 10^{-5} g/m^2day range at 23°C/50% relative humidity have been reported by using a microwave PECVD for SiN_x films on PEN substrates using the monomer silane with the reactive gas ammonia (NH_3). PECVD has been used to successfully encapsulate OLED devices demonstrating a lifetime of 10,000 hours and more either with a several-micrometer-thick silicon oxide layer or a multilayer based on two 100 nm thick silicon nitride layers separated by a printed polymer interlayer.

9.4.2.3 Magnetron sputtering

Besides PECVD, sputtering is the most often used technology for the deposition of barrier layers for the encapsulation of flexible electronic devices. Permeation barrier properties of sputtered layers are superior to evaporated layers, even for the same layer material.

Often rectangular magnetrons and recently also cylindrical/rotatable magnetrons are used to deposit sputtered oxide or nitride barrier layers. The magnetron consists of a sputtering target that is bonded onto a cooling plate. The cooling plate is water-cooled on the backside to avoid thermal damage to the target. The sputtering target itself acts as the cathode and a negative voltage is applied to it accelerating inert gas ions (mainly argon) to the target surface after the plasma has been ignited. A permanent magnet is installed behind the target. The magnetic field overlaps the electric field in a way that the electrons are forced onto a cycloid path above the target surface. This area is the zone with the highest plasma density, leading to the typical sputtering racetrack on the surface of a rectangular target. For a single magnetron typically a direct current (DC) or a pulsed DC power supply is used. Pulsing the plasma voltage leads to improved process stability.

Sputtering oxide or nitride layers from metallic targets, however, requires special attention regarding the coverage of the anode and the cathode with electrically isolating oxide/nitride material. The electrodes need to be kept metallic (conductive) either by hiding the anode behind special shielding or by using a dual-magnetron system. In a dual-magnetron setup, two cathodes are switched alternately as cathode and anode in a bipolar process using a midfrequency (10–50 kHz) power supply. In case, nonconductive oxide targets must be used for barrier layer deposition (e.g., intrinsic zinc oxide or even silicon oxide), DC or bipolar sputtering cannot be used. In this case an RF power supply is needed. RF processes typically have a lower deposition rate compared to DC or pulsed DC sputtering.

Much attention is paid to the way the reactive gas is injected into the process chamber in a reactive sputtering process. Injecting a fixed amount of reactive into the process chamber often leads to one of the two following situations:

1. The reactive gas flow (O_2 and/or N_2) is too low: In that case substoichiometic layers are formed on the substrate having insufficient optical transparency to be used as encapsulation layer on the active side of the device.

2. The reactive gas flow is too high and is not fully consumed by the sputtered metal atoms. Free reactive gas flow molecules in the plasma tend to damage the substrate and—in case of direct TFE of the device—lead to instant corrosion of the device's cathode. Also deposition rates as well as the permeation barrier layer properties of the oxide layers are lower in that case.

The best results have been achieved when the reactive gas flow is actively controlled in a proportional-differential-integral control loop. Thereby the control variable can be the discharge voltage, the optical emission of excited metal atoms in the plasma or other parameters such as the reactive gas partial pressure. In such a controlled process, only the required amount of reactive gas is injected to form transparent oxide coatings. The target coverage with oxide material is maintained constant and the partial pressure of the reactive gas is kept low.

Many different materials may be deposited as permeation barrier/encapsulation coatings with using a sputtering process. Figure 9.9 shows scanning electron microscopy (SEM) cross-sectional images of different sputtered oxide layers on a PET substrate. Figure 9.10 shows a comparison of the WVTRs of these materials at different layer thicknesses. It is clearly visible that a dense amorphous structure of the layers is favorable for a low WVTRs. Al_2O_3 and ZTO have a very dense structure, as shown in Fig. 9.9a for ZTO and also a very low WVTR even below 10^{-2} g/m^{2}day. Other widely used materials are SiO_2 (also having a dense structure but with a slightly increased surface roughness), indium tin oxide (ITO) which at the same time may serve as transparent electrode and rarely TiO_2, which however has a very columnar structure and shows a higher WVTR and OTR. Depending on the requirements regarding the optical layer properties (refractive index and absorption), Al_2O_3 (low refractive index) or ZTO (higher refractive index) are a good choice for manufacturing a barrier film to encapsulate organic electronic devices. ZTO thereby has a higher deposition rate, which allows faster coatings and a higher throughput.

Other widely used materials are SiO_2 (also having a dense structure but with a slightly increased surface roughness), indium tin oxide (ITO), which at the same time may serve as a transparent electrode and rarely TiO_2, which, however, has a very columnar structure and shows a higher WVTR and OTR. Depending on the requirements regarding the optical layer properties (refractive index and absorption), Al_2O_3 (low refractive index), or ZTO (higher refractive index) is a good choice for manufacturing a barrier film to encapsulate organic electronic devices. ZTO thereby has a higher deposition rate, which allows faster coatings and a higher throughput.

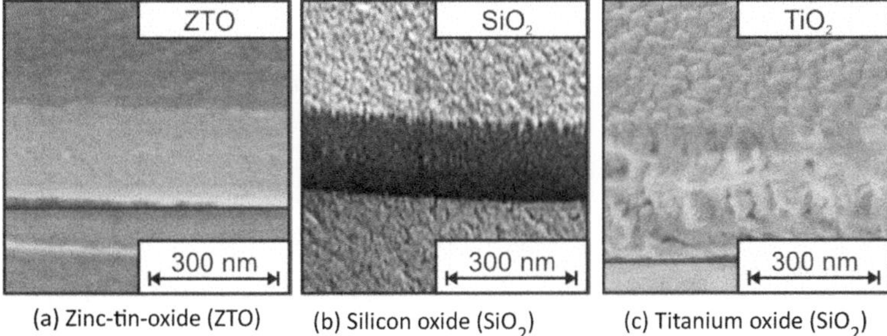

Figure 9.9 Cross-sectional images of different sputtered oxide layers with a thickness of 300 ± 40 nm.

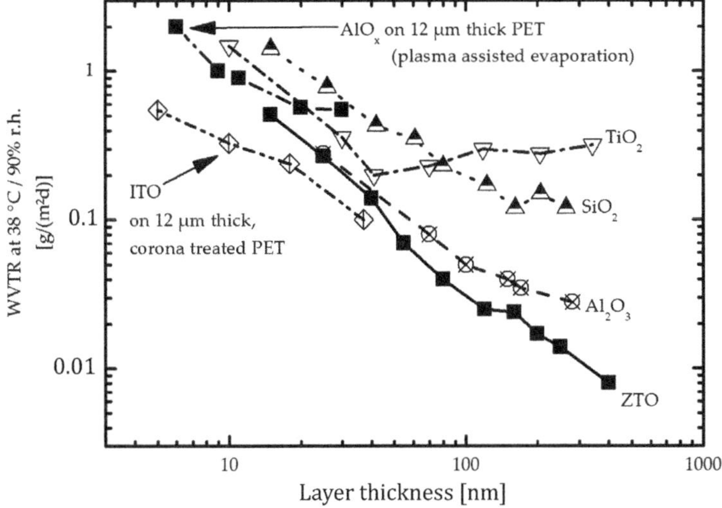

Figure 9.10 Water vapor transmission rates of different single-barrier layers on PET Melinex 400 CW at 38°C/90% r.h.

9.4.2.4 Atomic layer deposition

The atomic layer deposition (ALD) process is a special process that can be used to produce very dense, almost defect-free thin films on different substrates. Sometimes ALD is also referred to as atomic layer epitaxy (ALE). The ALD process is based on the cyclic alternating injection of two different reactants into a processing/coating chamber with intermediate purging steps. Figure 9.11 illustrates schematically the different steps of a typical ALD cycle.

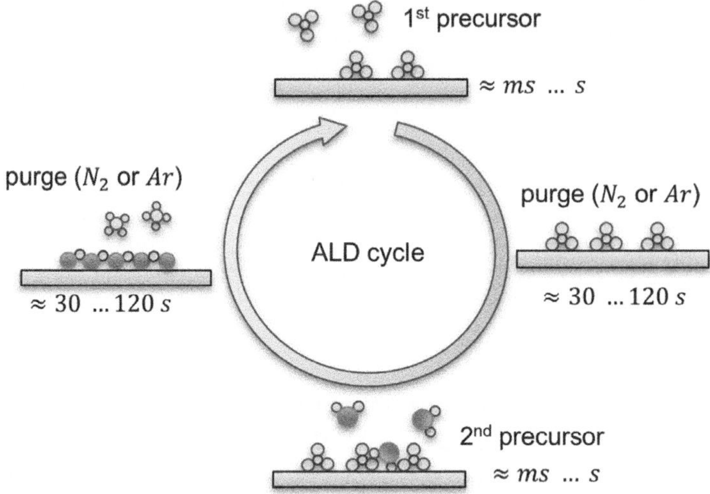

Figure 9.11 Scheme of an ALD cycle.

One cycle begins with injecting first a highly reactive precursor (e.g., trimethylaluminium [TMA], $Al(CH_3)_3$, for Al_2O_3 layers). The reactant will precipitate onto the substrate and the coating chamber walls. In the second step the coating chamber is purged with an inert gas for a certain (longer) time to remove residual precursor material, only leaving a monolayer on the substrate and the chamber walls. The third step is the injection of a second reactant (e.g., water) that reacts on the substrate surface with the first reactant, forming usually less than a monolayer of solid oxide material. Finally another purging step is needed to clean the process chamber from the second reactant before starting the next cycle.

To deposit a 12 nm Al_2O_3 layer using TMA and water approx. 100 cycles are required. The deposition rate per cycle is about 0.12 nm at a substrate temperature between 100°C and 150°C. The most time consuming steps within an ALD cycle are the purging steps as it must be made sure that all residual precursor gas is removed from the substrate and the process chamber before the second precursor is injected. Typical cycle times are in the range of less than one minute to a few minutes, whereas the reactants are injected for less than one second to a few seconds and purging is done for about 10 to 60 seconds at an elevated substrate temperature between 100°C and 175°C for Al_2O_3 coatings. At a lower substrate temperature higher purging times are required to achieve good layer quality.

One of the most important properties of the ALD technology is the self-limitation of the reaction, which means that the deposition rate is equal or less to a monolayer for each cycle, leading to an ideal layer by layer growth. Because of that ALD layers have a very high density and a very low number of defects. As the precursor gases can enter large trenches or shadowed areas it can precipitate on every surface in the chamber. This means that ALD allows the isotropic coating of noneven surfaces, including trenches, particle surfaces, and shadowed areas. Therewith very good surface defect coverage is expected when coating for example, polymer webs with residual particles on it. This is a great advantage over PVD processes such as evaporation or sputtering. These processes suffer from shadowing effects and do not allow the complete coverage of particles on a substrate. Consequently ALD layers reach lower WVTRs compared to the other processes, even on substrates with some particle contamination.

The most often ALD-deposited barrier material is Al_2O_3 using the precursors TMA and water (H_2O) or ozone (O_3). A WVTR in the range of 10^{-4} g/m^2day or even less has been achieved using a 10 to 25 nm thick Al_2O_3 layer on different substrates such as PET, PEN, or PI. However, the quality of the substrate surface still plays an important role also for the ALD process. Proper cleaning and handling of the substrate is required to achieve very low WVTR values, less than 10^{-4} g/m^2day. For other applications also titanium oxide (TiO_2) layers are deposited using ALD. However, TiO_2 is due to its (nano)crystalline structure not as good as aluminum oxide for permeation barrier application. WVTRs below 3×10^{-3} g/m^2day can be achieved, even with TiO_2.

The major drawback of the ALD process is its very low deposition rate and therewith productivity. The deposition of a 25 nm thick Al_2O_3 layer requires 200 cycles, meaning a deposition time of 200 minutes in a batch-coating process. To improve the deposition rate several groups use plasma activation or a special precursor gases such as ozone. With using a plasma-assisted ALD process, the purging time can be reduced by a factor of 2 or more, leading to a lower cycle time of only a few seconds without losing the favorable layer properties. Another promising option for increasing productivity is switching from a time-separated injection of the precursor gases to a spatial distribution of the gas inlet with a moving substrate. Several groups introduced spatial and also roll-to-roll ALD setups to improve the layer deposition drastically. One concept is called the **showerhead** principle, separating the gas injection by using separated gas nozzles to inject the different process gases. Thereby, the substrate is moving along a row of gas injection nozzles: every other nozzle is used to purge the chamber with an inert gas, whereas the nozzles in between carry alternately the two different precursors. This concept allows also roll-to-roll processing. Another concept is separating the precursor gases to different chambers of a vacuum vessel with an inert-gas-purged separation chamber in between. The substrate is thereby wound repeatedly through each of the chambers to deposit a barrier layer.

Using roll-to-roll ALD, a WVTR of 10^{-4} g/m^{2}day has been reported. However, winding the coated polymer web is still a problematic issue in terms of damaging the layer by contact to rollers or mechanical strain.

ALD is one favorable process for direct TFE of organic electronic devices, as it allows hermetic sealing of patterns (e.g., pixels/ patterned electrodes) and particles on the substrate surface and is less prone to local layer defects, leading to dead spots in the device. Also ALD may operate at device-friendly conditions without UV irradiation (which often occurs in plasma processes) and low substrate temperatures.

9.4.2.5 Summary of single-layer encapsulation materials

Table 9.2 summarizes again important properties of different single-layer barrier technologies, taking into account not only important process properties and the productivity but also achievable WVTRs and their capability for direct TFE of organic devices.

Table 9.2 Comparison of different barrier layers done with different technologies, including the typical process pressure *p*; a typical dynamic deposition rate determined with continuous substrate movement, R_D; a typical layer thickness *d*; and typically reported WVTR values

Property	Uncoated polymers	(Reactive) evaporation	PECVD	Reactive sputtering	ALD
Materials	PET, PEN	AlO_x, SiO_x	SiO_x, SiN_x	Al_2O_3, SiO_2, Zn_2SnO_4	Al_2O_3, TiO_2
p (Pa)	–	<1	1–1000	0.3–2	20 Pa–air pressure
R_D (nm·m/min)	–	4800	10–1000	50–100	0.1 nm/cycle roll-to-roll up to 10 nm·m/min
d (nm)	100,000	10–20	100–>1000	100	10–25
WVTR on PET (g/m^{2}d)	2–5 (at 100 μm)	1	0.001–0.1	0.01	<0.001
Roll-to-roll/Dynamic processing	–	Yes	Yes	Yes	Yes
Direct thin-film encapsulation of devices	–	No	Yes	Yes	Yes

9.4.3 Multilayer Thin-Film Encapsulation Systems

As discussed before, single layers show very substrate-dependent gas permeation and are sensitive to particles and defects in underlying layers or the substrate itself. Also, increasing single-layer thickness does not mean an improvement of the barrier performance in many cases. Shadowing effects just lead to further growth of defects instead of covering particles. With increasing single-layer thickness, also the mechanical stability of the system is reduced due to increasing intrinsic stress in the layer. This results in higher brittleness as well as reduced flexibility (bendability).

These issues can be addressed by depositing several thin inorganic barrier layers that are separated by a polymer interlayer. Depending on the substrate, at first a base polymer may be applied to planarize the substrate surface and cover particles on it. After the base polymer an oxide barrier layer is deposited using one of the single-layer technologies described above. Mostly sputtering and PECVD, and sometimes also ALD, are used for that. After that a polymer layer is deposited on top of the barrier layer to protect it from mechanical damage and to cover and planarize defects and particles in the barrier layer (see Fig. 9.12, center image, for an example). Sometimes the dual layer (barrier + polymer top coat) is repeated several times to form multilayer stacks with alternating barrier and polymer layers (see Fig. 9.12, right image). Different methods are known for the polymer layer deposition that can be commonly divided into vacuum technologies and nonvacuum technologies.

Among the vacuum technologies the most well-known process is the vacuum LVLS process. An acrylate material is evaporated from a liquid precursor within the vacuum, and it precipitates on the substrate as liquid viscose material that needs to be cured afterward using for example, UV light. On the basis of this technology, VITEX Technologies Inc. developed a multilayer barrier that can be used as substrate or TFE for OLED or organic photovoltaic (OPV) devices. Commonly sputtered Al_2O_3 or SiO_2 layers are used as barrier material in the layer stack. WVTRs down to 10^{-5} g/m^{2}day have been reported. The PML process has been adapted to a roll-to-roll coating machine and commercialized by the company 3M to manufacture permeation barrier film rolls based on alternating sputtered SiO_2 barrier layers and acrylate polymer layers.

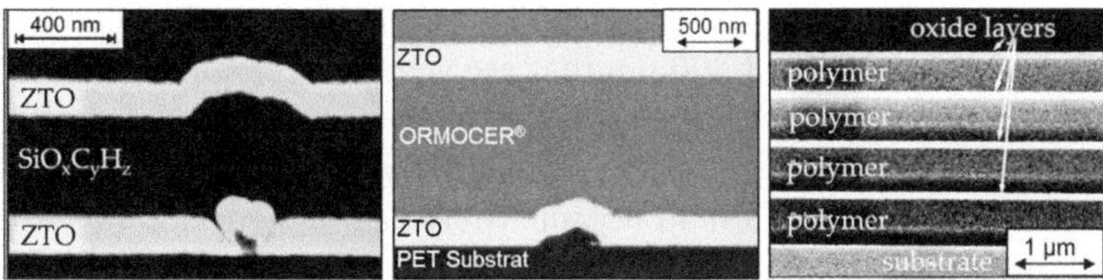

Figure 9.12 Examples for multilayer barriers for encapsulation of organic electronic devices. Left: Combination of sputtering and PECVD done in a roll-to-roll vacuum coater (source: Fraunhofer FEP, 2009). Center: Combination of sputtering and reverse-gravure coating of an ORMOCER®, showing good defect planarization (source: Fraunhofer POLO®, 2012). Right: Four dyads of sputtered oxide and in-vacuum coated acrylate layers (source: VITEX Systems, 2006).

Another way to produce multilayer all-in-vacuum is using a PECVD process to deposit the interlayer. Some of the groups thereby use a PECVD process for making both the barrier layer and the interlayer. The processes are based on an adjustment of the layer composition by changing the precursor gas flow, for example, adding oxygen to deposit silicon oxide layers or only use a monomer such as HMDSO without or with little amount of oxygen to deposit $SiO_xC_yH_z$ plasma-polymer layers. WVTRs in the range of 10^{-5} g/m^{2}day have been achieved using a graded-layer stack with three inorganic and two organic zones. With a three-layer stack based on a combination of reactively sputtered barrier layers with interlayers grown by magnetron PECVD WVTR in the range of 1×10^{-3} g/m^{2}day (38°C/90% relative humidity) are obtained on a PET substrate. However, as visible in the left image of Fig. 9.12, the PECVD process is not able to fully planarize a surface particle/defect due to shadowing effects that are typical for vacuum deposition techniques with a directed particle flow. A surface irregularity still remains after PECVD coating, although the accelerated path for water diffusion has been closed.

Improved surface planarization can be achieved using a multilayer barrier that uses wet-coated polymer interlayers. Hereby many of the well-known printing and coating processes can be used that are also used for adhesive application. One example for wet deposition of the interlayer is the use of a reverse gravure coating technology to apply an ORMOCER® layer to a substrate or precoated barrier film. The center image in Fig. 9.12 shows a typical three-layer stack that consists of two ZTO barrier layers with a thickness between 150 nm and 200 nm that are separated by an ORMOCER® hybrid polymer layer with 1 μm thickness. An ORMOCER® thereby is an organic-inorganic hybrid polymer in which functional clusters are cross-linked within an organic (polymer) network. A WVTR of $<2 \times 10^{-4}$ g/m^{2}day, measured using a calcium test at 38°C and 90% relative humidity ($<5 \times 10^{-5}$ at 23°C/50% relative humidity) has been reported using this three-layer stack on a standard, commercially available PET substrate (Melinex 400 CW with 75 μm thickness).

A combination of PECVD-coated SiN barrier layers and printed planarization polymer layers has been proven to provide superior encapsulation performance for OLED display and lighting modules using just a three-layer stack that consists of two SiN barrier layers and one polymer interlayer. The stack is deposited directly on the device showing a WVTR of 10^{-6} g/m^{2}day and an OLED device lifetime >10,000 hours with virtually no dead spot on 10×10 cm^2. Table 9.3 summarizes the properties of different multilayer encapsulation and permeation barrier systems showing that low WVTR of less than 10^{-4} g/m^{2}day are possible even in roll-to-roll processing. These are already near the requirements from organic devices for encapsulation systems. However, independent from the specific thin-film technique used, it is important to keep in mind that device encapsulation is not just based on a permeation barrier performance and low water vapor or oxygen transmission rate (OTR). Many different properties of the encapsulation systems need to be considered when selecting one for the device encapsulation. These properties comprise, for example:

1. Optical properties of the layer stack, including transparency, absorption, and reflection, as well as haze to optimize light-incoupling to organic solar cells and optical sensors and light-outcoupling from OLED devices.
2. UV stability of the encapsulation and UV absorption/reflection to protect the organic device in an outdoor application.
3. Thermal stability: Laminating the barrier film onto the solar cell is often done at an elevated temperature higher than 100°C. The barrier film as well as the device must not be damaged during this process. Devices for outdoor or automotive applications must be stable in a temperature range from –40°C to 85°C over a long time.

Table 9.3 Summary of different multilayer barrier technologies (poly: polymer)

Multilayer type	All-in-vacuum	Reported WVTR (g/m^2day)	Productivity/ Roll-to-roll	Typical layer stack and thickness
Sputtering + Magnetron-PECVD	Yes	1×10^{-3} at (38°C, 90%)	Roll-to-roll, single run, 1 m/min	ZTO, 100 nm poly, 400 nm ZTO, 100 nm
Sputtering + ORMOCER®	No	2×10^{-4} (38°C, 90%) 5×10^{-5} (23°C, 50%)	Roll-to-roll, three runs 1 m/min + 3 m/min + 1 m/min	ZTO, 150 nm poly,1000 nm ZTO, 150 nm
Sputtering + PML	Yes	$<1 \times 10^{-3}$ single dyad, at 50°C $<10^{-4}$ multidyad (38°C, 100%)	Roll-to-roll	Base polymer $n \times$ (barrier + polymer)
PECVD graded layers	Yes	10^{-5} (23°C, 50%)	Batch deposition, graded layers	SiO_x + $n \times$ (SiO_xC_y/ SiO_x)
PECVD $SiN_x + SiO_x$	Yes	3.2×10^{-6} (25°C, 40%) with SiN_x + 6·(SiO_x/SiN_x)	Subsequent batch deposition	SiN, 50 nm $n \times$ (SiO_x, 70 nm SiN_x, 50 nm)
PECVD + Planarizing polymer	No	$\sim 10^{-6}$	Batch deposition, direct encapsulation of organic devices roll-to-roll in development scale (2014)	SiN, 100 nm $n \times$ (polymer + SiN, 100 nm)
Al_2O_3 ALD + SiN PECVD	Yes	$<5 \times 10^{-5}$ (38°C / 85% r. h.)	Subsequent batch deposition	Al_2O_3, 5 nm SiN, 100 nm

4. Mechanical stability: Processing the encapsulation, for example, in a roll coater must not negatively influence the barrier performance. Also, the encapsulation must survive subsequent handling and mechanical loads or impacts (e.g., hail) in application.
5. Moisture stability: Even if the encapsulation is used to protect the organic device against moisture, often used barrier substrates are not stable against moisture in long-term use (such as PET). If the substrate directly faces the environment, it must be protected against the ingress of moisture or a stable substrate must be used.

9.4.4 Encapsulation Sealants and Barrier Adhesives

Transport of gases and vapors through polymers is a complex phenomenon and its comprehension needs to establish a correlation between different mechanism and laws, relating solubility and transport in polymer layers to their molecular properties, the nature of the permeates, and the polymer morphology. As reported in Section 9.3, the permeability of a gas or a vapor through a dense polymeric barrier can be described by the Fickian diffusion model where the diffusion coefficient D determines how fast the permeant can move in the barrier and the solubility coefficient S determines how much of the permeate can be dissolved in in the barrier. It is important to underline that the diffusion coefficient is a kinetic parameter, while the solubility coefficient is a thermodynamic parameter, and they show opposite trends in combination with thermal effects. Moreover, as described in Section 9.3.3, usually encapsulation of organic devices is done in an inert atmosphere where the initial device conditions and the concentration gradient in the barrier film/encapsulation are in a nonequilibrium condition. In this process, three different phases can be identified: (1) a first phase in the transient regime, the lag time, during which the barrier starts to be filled and an increasing of permeant flux inside the encapsulated system is observed; (2) a second phase characterized by a constant distribution of the permeant inside the barrier and a resulting constant flux, where the steady-state regime can identify the barrier WVTR; and (3) a third period where a flux decreasing is observed due to a reduced concentration gradient of the penetrant.

The competition between transient and steady-state regimes is strongly affected by the type of encapsulation system and by the adopted adhesive polymers. An overview about different types of encapsulation is given in Section 9.2. Here, we analyze the properties of adhesive polymers by considering functional and composition characteristics. From a functional point of view, adhesives can be classified in **passive** systems if they work like a physical barrier to the permeate or **active** systems when a physical or chemical interaction between adhesive and permeate takes place.

9.4.5 Passive Adhesives

In the case of passive systems, the steady-state regime is strongly prevalent and permeation can be reduced through two main approaches, (i) to employ a polymer with intrinsically low permeability and (ii) to increase the tortuosity of the diffusion path by dispersing a defined amount of impermeable particles in the polymer matrix.

Regarding the first approach, transport properties strongly depend on the free volume within the polymer and on the mobility of the polymer chains. For instance, polymers with low glass transition temperatures show greater segmental mobility and a higher diffusivity. Moreover, matrix cross-linking level and polymer molecular weight can influence the transport process. With increasing molecular weight, the number of chain ends, representing material discontinuity, decreases and the diffusivity can be reduced up to a factor of 10. Other polymer structure modifications can be considered, like the introduction of bulky or reactive substituents able to induce an increasing rigidity of the polymer backbone and a decreasing free volume available for the diffusion of the permeant molecules. In the opposite direction, the addition of plasticizers to a polymer matrix, including glycidyl ethers of aliphatic, arylaliphatic glycols, di- and polycarboxylic acids, and polyester polyols, results in increased segmental mobility and generally in increased permeate transport. To perform a correct evaluation about polymer chain mobility, **dynamic mechanical thermal analysis** (DMTA) can be applied in determining critical parameters like the glass transition temperature (T_g), the storage modulus (E), the loss factor (tan δ), and the coefficient of thermal expansion (CTE).

The second approach to reduce the steady-state permeation is represented by the introduction of a filler material. Nanometric silica filler particles are generally adopted and finely dispersed in epoxy resins. For specific applications, also zirconia and titania particles can be introduced. The goal is to reduce the free volume and to increase the diffusion tortuosity according to the volume fraction of the filler, the shape, and the distribution of particles. When the filler is finely distributed into the matrix, the barrier rigidity is also increased due to lower polymer chain mobility. Moreover, as a function of nanoparticle surface properties, a physical interaction to the polymer chains can be also induced through dipole interactions to polymer functional groups resulting in a physical cross-linking and a subsequent lower chain mobility. Some drawbacks have also to be mentioned: If the filler is incompatible with the polymer, if its content is excessively high, or if a suitable interaction is not obtained at the filler–polymer interface, voids tend to occur at the interface with opposite results—an increase of free volume of the system and higher permeability. Inorganic filler content can be evaluated by submitting the sealant to a thermogravimetric analysis (TGA) where the degradation of the epoxy resin is generally promoted in an oxygen gas flowing up to 600°C–650°C and the remaining inert filler can be easily detected. Filler particles distribution can be detected by SEM analysis where filler–polymer interfaces can be also investigated.

So far we have discussed the steady-state regime for barrier adhesives by focusing on permeation through an adhesive. However, as described in Section 9.2.1, the second mechanism for parasitic permeation is the permeation at the interface between the substrate surface and the edge seal material and this effect can become relevant. For this reason, it is mandatory to adopt barrier adhesives able to ensure suitable adhesion strength and to produce defect-free interfaces. In this regard, the adhesive chemical structure with its functional groups, the employment of polar flexibilizers like polyols and acids, and the selection of an appropriate photoinitiator with a counterion able to balance the cure speed and the nucleophilic behavior are critical factors to modulate the chain mobility in determining a regular interface. However, the viscoelastic properties of the barrier have to be carefully investigated. This may be done, for example, using lap shear tests, because if, from a point of view of sealant properties, increased chain mobility induces a higher adhesion, on the other hand it is synonymous of a higher permeation through the polymer.

9.4.6 Active Adhesives and Getters

The employment of a polymer with intrinsically low permeability or the introduction of an inert filler material in producing a barrier adhesive are typical approaches to reducing the steady-state permeation. This phenomenon is regulated by the concentration gradient between inside and outside parts of the barrier and permeation works in equalizing the external and internal permeant concentration, but available technologies do not ensure making permeation across the sealing material negligible. However, to limit dark-spot area growth and pixel shrinkage in organic electronics devices below values significantly affecting the device uniformity, maximum pressure for some dangerous permeants (H_2O and O_2) has to be maintained at a very low level, for example, H_2O pressure has to be limited inside at $\sim 1.1 \times 10^{-4}$ Torr. The more efficient way to control the H_2O and O_2 partial pressure is to adopt a combination of a **getter** material with the adhesive to make the barrier an active barrier.

Usually a **getter** is a material, a composite, or an alloy able to capture gaseous impurities according to the following key characteristics: (i) very fast kinetics of the capture process (adsorption, chemical reaction), (ii) suitable capacity in terms of weight of specific chemical species captured per unit weight of the getter, and (iii) high activity also for low partial pressure of a specific chemical species in equilibrium with the getter.

Different getter systems can be considered (pure metals, metal alloys, inorganic materials, or organic-inorganic composites) but due to their peculiar characteristics, getter materials based on organic-inorganic polymer composites are the ideal systems for integration in organic electronics devices. Proper assessment of different getter materials for organic electronic devices is essential for the choice of a more suitable solution as well as for the accurate design of the dryer accommodation inside the device. In this regard, the characterization of getter characteristics and performances is extremely important: measurements of sorption performances in conditions as close as possible to the operating conditions expected to occur in the final application have to be addressed. Specific methods to measure theoretical total capacity, initial pumping speed, and dependence of pumping speed on capacity have to be developed. Three different methods are continually optimized in this regard: (1) volumetric test, (2) microgravimetric test, and (3) climatic cell exposure test. In a volumetric test, sorption performances are measured by exposing the sample, under vacuum, to a constant pressure of a specific gas. The gas pressure drop occurring in a known sampling volume directly connected to the test volume is measured and weight gain versus time can be calculated. The main information that can be obtained from such a test is the initial pumping speed of the getter. Only limited information is given on the effective capacity of the getter. In a microgravimetric test, sorption performances are measured by means of a test with the sample exposed to a defined concentration of gas in a nitrogen backfilling at atmospheric pressure. Weight change versus time is gained and sorption speed versus capacity can be calculated. The gas level can be adjusted and it is set close to the typical concentration expected to occur inside an organic electronic device (10 ppmv for moisture). The main information that can be obtained is the initial pumping speed but also the dependence of pumping speed on capacity. A climatic cell exposure test is the simplest and most widely used measurement method for moisture evaluation, especially as an incoming inspection control. Samples are exposed to a defined atmosphere inside a climatic cell. Temperature and relative humidity are controlled (usually 25°C, 55% relative humidity; 85°C, 85% relative humidity; or 60°C, 90% relative humidity). Weight increase is measured periodically.

The main information that can be obtained from such a test is the getter total capacity for water, which is directly related to the content of the active absorbing material in the getter system. Information on sorption speed can be obtained only if the weight gain is frequently measured during early stages of the test. Figure 9.13 shows an example of a gravimetric test result for a polymer composite–based getter material. The sorption capacity is determined by saturation of the weight increase after a longer time. Sorption speed can be characterized from the slope of the weight increase during the initial stage of the test (in the example within the first 3 minutes).

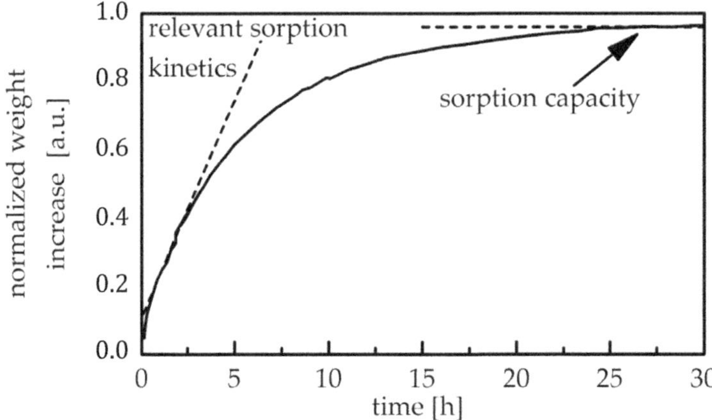

Figure 9.13 Weight increase of a polymer composite getter material over time.

Particular interest has been devoted to the formulation of active materials based on polymer composite systems where the peculiar characteristic of organics like easy handling and processing, flexibility, fast polymerization kinetics, and tunable consolidation mechanisms are combined with the high reactivity of inorganic materials. In particular, irreversible H_2O sorber materials are based on powders of one or more alkaline earth metal oxides chosen from oxides of calcium, strontium, barium, or magnesium, while reversible materials are generally based on aluminosilicates with particular reference to molecular sieves, zeolites, silica gel, or clay. Organometallic systems are also adopted but their thermodynamic instability and by-product evolution are limiting factors for their use in application.

Oxygen scavengers are based on substances or mixtures that chemically or enzymatically react with oxygen. They are generally composed of iron powder, ascorbic acid, unsaturated fatty acids like oleic and linoleic acids, enzymes like glucose oxidase, alcohol oxidase, hydrocarbons, unsaturated polymers, co-polyamides and transition metals, photosensitive dyes, and immobilized yeasts in solid materials. The formulation of homogenous polymer composite materials is strongly related to the achievement of host–guest interactions able to ensure a fine distribution of the active component. Frequently these interactions are limited at a physical level but dipole moments with the polymer chain or the organization of oriented structures can be particularly effective to promote getter dissolution at molecular level. When a similar interaction is not obtained, it is fundamental to across any filler–matrix incompatibility in order to avoid particles aggregation, segregation, or voids occurrence at interface level. In this regard, active particles can be modified by introducing a capping agent able to regulate the surface properties in tuning physical interactions to the polymer chains. In detail, a chemically tailored particle surface ensures high affinity with the matrix, no phase segregation, no leaking pathways within the substrate/scavenger interphase due to particles wetting, better transparency of films in the visible range due to reduced discontinuity of refraction index across the particle–matrix interface, and improved adhesive properties due to particle wetting. Moreover, when a reactive material is adopted in polymer composite active materials, the capturing mechanisms are mainly based on the chemical properties of the getter component. In this case the filler material reacts through a strong chemical bond with the undesired permeant species, and the transport process will be affected by a modification in composition and a resulting change in some factors that influence the diffusion and the solubility: filler particle size, surface and interfacial energies, matrix viscoelastic properties, and physical state of the permeant. The ability to compensate this evolution is one of the key characteristics of polymer composite materials that require a careful designing activity. Reversible getter materials can also be adopted where no modification in particle size or in other properties can be manages. In that case, it is fundamentally important to focus on reversible materials where the permeate–active material interaction is related to suitable energy sites. For instance, in nanostructured zeolites water is bound primarily to extra-framework cations, and for specific compositions, any desorption process at a temperature lower than 150°C can be excluded.

If inert particles are applied to produce an efficient permeation barrier, the sealant works like a passive homogeneous barrier and moisture permeability depends mainly on its diffusion into the sealant material. However, if suitable materials are introduced to trap the permeant through physical or chemical interactions, the permeate solubility in the barrier will be also modified. In this approach, the permeability becomes a more complex function of solubility and diffusion coefficients, while the transition time or lag time is strongly increased. Moreover, a new parameter is introduced, the **breakthrough time**, which describes the time that is needed for the permeate gas to pass throughout the barrier and start permeating into the protected device. Within the breakthrough time the WVTR is practically zero and no degradation mechanism occurs. In these transient conditions, the permeation evaluation is very complex and standard analysis techniques are not efficient, because they can measure only the WVTR in the quasi-stationary conditions with high uncertainty in the lag time measurements. Recently, a new spectroscopic method has been introduced where the investigation is based on the spatial mapping of the peaks in the FTIR spectrum that are correlated with the presence of moisture. A spatially resolved infrared transmittance signal of a hydroxyl (–OH groups) combination peak for typical stretching and bending frequencies can be used in defining a 2D or 3D map of the moisture diffusion front (see Fig. 9.14 for an example). So far SAES Getters S.p.A. is the only manufacturer of active polymer composites.

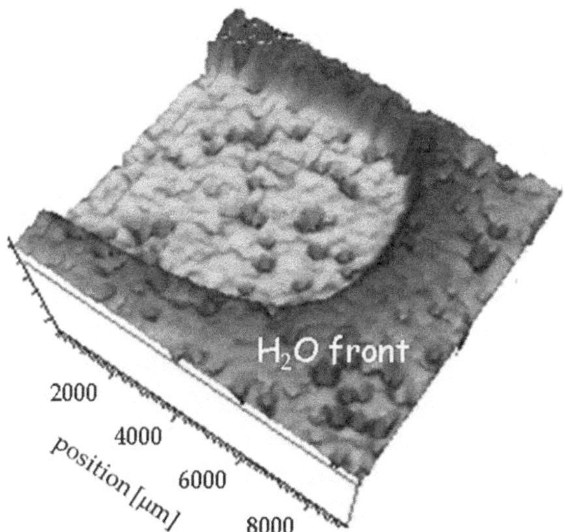

Figure 9.14 Three-dimensional image of water ingress into active barrier materials.

9.4.7 Types of Adhesives with Respect to Their Processing

To obtain high-performances barriers, every component of barrier adhesives has to be diligently tuned. Depending on the process requirements, different types of adhesive can be adopted: UV-curable adhesive, hot-melt adhesive, and pressure-sensitive adhesives (PSAs). Moreover, it is important to mention that the terms "adhesive" and "sealant" are frequently used as synonyms but they represent different material classes as their peculiar characteristics are different. In detail, a sealant is mainly adopted to seal any space that exists between surfaces where they are applied in order to present a watertight area, while an adhesive is used to bind one surface to another and not to seal the space in between. On the basis of different final applications, sealants and adhesives are formulated in different ways in their main structure and composition in order to obtain higher rigidity and strength for adhesives and higher malleability and lower strength for sealants, and higher viscosity for adhesive and easier processability for sealants. Despite this, the following classification of UV-curable, pressure-sensitive, and hot-melt systems can be adopted for both adhesive and sealant materials.

9.4.7.1 UV-curable adhesives

During the last few years the development of UV-curing technology has provided new emphasis to the formulation of performant adhesives and sealants due to some interesting benefits: low content of volatile organic compounds (VOCs), no time constraints, no waste, and a fast curing process. UV-curing adhesives can be classified into two major groups, cationic and free radical. In the case of free-radical adhesives, one of the major advantages is the very fast curing process related to the polymerization mechanism where the UV radiation–promoted bond scission in radical photoinitiator molecules induces free-radical production for a fast initiation process (curing start). A typical UV-curing process is performed by applying an irradiance of 100 mW/cm^2 for 15–30 seconds for a resulting UV dose of 1.5–3 J/cm^2. However, oxygen can work like a radical scavenger and an oxygen atmosphere can slow or even inhibit the curing process and requires a higher UV dose. For this reason, the employment of free-radial adhesives in air atmosphere has to be combined with optimized composition, high reactive photoinitiators, and suitable light sources.

The second class of UV adhesives is made of cationic formulations where no radical is produced during the polymerization process but, under UV light, a strong acid is activated, independently from the oxygen presence, to catalyze the polymerization process. For this reason, the curing process of cationic adhesives proceeds slower than free-radical systems. A thermal posttreatment is often recommended to complete the polymerization process but the cured formulations show improved thermal and mechanical stability. A typical UV-curing process is performed by using UV light sources with a minimum irradiance of 100 mW/cm^2 for 60–90 seconds for a resulting UV dose of 6–9 J/cm^2. Often, a UV light source with an emission maximum at 365 nm is used.

Epoxy resins are characterized by high chemical and corrosion resistance, good mechanical and thermal properties, outstanding adhesion to various substrates, low shrinkage upon cure, good electrical insulating properties, and the ability to be processed under a variety of conditions. They represent a versatile class of cationic sealing materials for barrier adhesive production. Commercial epoxy resins are generally composed of cycloaliphatic epoxy resins or epichlorohydrine-phenol formaldehyde resins contain aliphatic, cycloaliphatic, or aromatic backbones and they are formulated like solventless and single component products able to produce, after curing, a highly cross-linked network. According to the required physical and chemical properties, processing methods, and curing conditions, a wide variety of curing agents can be adopted, for example, Lewis bases such as tertiary amines or Lewis acids such as boron trifluoride monoethylamine. The photoinitiated cationic curing of epoxy resins has received great interest during the last few years for the application of coatings from solvent-free or high-solid systems. In this regard, salt photoinitiators like onium salts, ferrocenium salts or diazonium salts are frequently adopted for photocleaving activation for epoxy ring-opening polymerization, which represents a fundamental step for the curing and cross-linking reaction. Cationic UV-curable systems are preferred in comparison to free-radical systems because they do not show oxygen sensitivity and have excellent adhesion, moderate shrinkage, and a reasonable humidity resistance. No oxygen sensitivity means that a lower photoinitiator concentration can be adopted with resulting less VOC content, for example, benzene, propylene carbonate, diphenyl sulphide, propionaldehyde, and 2-propenal. The impact of VOCs can be very dangerous for final devices due to their chemical and physical reactivity, especially when in contact with organic semiconductors. For this reason, their content should be evaluated by headspace gas chromatography (HS-GC). Moreover, if some specific materials are placed in direct contact to cationic UV-curable resins, they will affect the polymerization kinetic up to totally inhibiting the curing process. Modified systems like epoxide-acrylate resins are preferred for application on plastic and metal coatings. In these systems, the combination of cationic and free-radical mechanisms can improve the curing process and the achievement of final properties. In general, UV cationic adhesives and sealants have to be selected carefully and their characteristics synchronized with process constrains and the application requirements.

Manufactures of passive UV-curable adhesives for organic electronics include the companies Henkel, Three Bond, Adhesive Research, DELO, and Nagase, while SAES Getters S.p.A. is a manufacturer of active UV-curable adhesives.

9.4.7.2 Pressure-sensitive adhesives

PSAs are nonstructural adhesives able to stick to a surface upon contact through the application of a moderate pressure. PSAs are viscoelastic solids where the hysteresis of the thermodynamic work of adhesion is established through a molecular contact to the surface and the only active interface forces are van der Waals forces. To ensure the right adhesion properties, adhesive composition has to be finely tuned and typical polymer matrices like rubber, acrylates, stryrene-based co-polymers, and silicones are adopted. Tackifying components are generally introduced to tune the adhesive–substrate interface properties, to modulate the glass transition temperature (T_g) and to optimize the dissipative properties. Cross-linking agents are used avoid flow and, once the adhesive has been coated, to prevent creep.

The main properties of PSAs are stickiness, strain hardness at a high level of strain, and good resistance to a continuous shear force. To be sticky upon contact to the surface, PSAs have to ensure tack properties at a given temperature and show high viscoelastic dissipation (typical broad glass transition phase with onset value 25°C–50°C lower than the usage temperature). Moreover, the stickiness to the touch is strongly dependent on the formation of bridging fibrils under pulling stress. In detail, during detachment, a process of cavity growth from the interface is induced and controlled by the elastic modulus of the adhesive, while the subsequent formation of fibrils arising from the wall cavities is strictly related to the elongation properties. In particular, the ability to strain-harden at a high level of strain is a fundamental property of PSAs to ensure a right level of elongation strain and can be obtained through the definition of the matrix composition in terms of the cross-linking degree and molecular weight distribution. Finally, the level of chemical or physical cross-linking is also fundamental in avoiding any Newtonian viscosity behavior at a fixed temperature and to determine the PSAs' ability to maintain a moderate level of stress for a long time.

A critical aspect in PSA technology is the protection of the adhesive protection before using it. By considering their sticky characteristics, dust particles could easily contaminate the adhesion surface and a protection layer is needed. Typically this issue is solved by combining the sticky side of the adhesive to a tape (high adhesion needed) and by covering the adhesive surface with a protective film (low adhesion) able to be peeled cleanly without damaging the adhesive properties. Selected polymers, including silicone matrices, polymers with long alkyl or fluoroalkyl side chains, fluoropolymers, and polyolefins, are used as protective films to provide a surface energy lower than the surface energy of PSAs in order to ensure the key feature of release materials. Generally poly(vinyl carbamates) are adopted for the back side of tapes.

9.4.7.3 Hot-melt adhesives

Hot-melt adhesives represent a class of thermoplastic materials that are completely solid at room temperature but melt and resolidify upon heating and cooling processing steps. When heated above their melting point, they become fluid and are able to wet the surface to which they are applied. A quantity of fluid hot melt is applied on the surface and it is brought in contact to a second surface. By cooling the adhesive below the solidification temperature, the substrates are bonded. Typically, a heating and lamination process is performed under vacuum to optimize the adhesion strength. In spite of other adhesive types, where solidification is obtained through thermal and UV-promoted polymerization mechanisms or through solvent evaporation, hot-melt adhesives convert from a hot fluid to a solid by a cooling process. Hot-melt materials may provide some advantages over the use of liquid adhesives in terms of higher lamination speeds and easier integration in roll-to-roll processing.

Hot-melt formulations contain polymers to provide strength and cohesion while being hot (providing resistance to separation while the adhesive is hot), tackifiers and low–molecular weight components to dilute the polymer entanglement network, modulate the glass-transition temperature, tune the viscosity, and improve wet-out properties that are related to the molecular contact of the adhesive with the substrate over the entire bonding area.

Typical organic matrices include polyamide, ethylene vinyl acetate (EVA), ethylene propylene diene rubber (EPDM), polyisobutylene (PIB), and other polyolefin varieties that are available in different forms and are selected according to the final application requirements. To define the hot-melt characteristics, the organic matrix is generally combined with inert fillers to set rheological properties in melt form and mechanical properties in the solid state. Hindered phenols and amines are added as aging resistors (antioxidant and UV stabilizer). Silane modifiers are used to allow filler–matrix interactions and to ensure suitable adhesion properties. Some hot-melt adhesives can remain permanently sticky, ensuring an excellent adhesive bond even when the glue is cold. In this case, the required contact pressure is the key to creating a sufficient contact area between the adhesive and the joined part. Because of that property, this group of adhesives is often referred to as pressure-sensitive hot-melt adhesives.

Main advantages of hot-melt adhesives are reduced amount of waste, little or no solvent component, rapid bonding action, high-volume processing, and low-cost production. Hot-melt adhesives are typically provided in steel drums and dispensed using a hot-melt applicator, but engineered formulations can be also supplied in pellet form suitable for direct application by using an extruder linked to a dispensing module. Manufactures of hot-melt adhesives include 3M, FPC Corp., Heartland Adhesives, Bostik Ink, Henkel Loctite, and SAES Getters S.p.A. for active hot-melt adhesives (with water/oxygen absorbing functionality).

9.5 Encapsulation Processing

This section gives an overview of the encapsulation process itself, meaning the process of combining an organic electronic device with an encapsulation product such as a glass cover or a barrier film. The section covers the preparation of the module components, methods for applying and coating the adhesives, and processing issues, especially for roll-to-roll processing of flexible devices.

9.5.1 Barrier Film and Device Preparation

9.5.1.1 Clean atmosphere

Depending on the process it is necessary to define clean-room class conditions (Table 9.4) during the coating as well as the encapsulation process. Clean-room conditions can be realized in the complete room where the processes are performed or only in the near surrounding of the process or the process plant itself. Laminar flow units are used to define the atmosphere and cleanliness inside a processing machine. With these units it is possible to realize zones with a low volume of dust and particle. In laminar zones a purified, vertical or horizontal air flow is applied. The size of these zones is defined with curtains or encapsulated areas. Laminar flow units allow minimizing the particle concentration in the air.

Table 9.4 Maximum allowed particle contamination (particles/m^3) for the different clean-room classes according to ISO 14644-1

	Size					
Class	**≥0.1 μm**	**≥0.1 μm**	**≥0.3 μm**	**≥0.5 μm**	**≥1 μm**	**≥5 μm**
ISO 1	10	2				
ISO 2	100	24	10	4		
ISO 3	1000	237	102	35	8	
ISO 4	10^4	2370	1020	352	83	
ISO 5	10^5	2.37×10^4	1.02×10^4	3520	832	29
ISO 6	10^6	2.37×10^5	1.02×10^5	3.52×10^4	8320	293
ISO 7				3.52×10^5	8.32×10^4	2930
ISO 8				3.52×10^6	8.32×10^5	2.93×10^4
ISO 9				3.52×10^7	8.32×10^6	2.93×10^5

9.5.1.2 Inert atmosphere

The realization of an inert atmosphere in the surrounding of the product is realized to protect the sensitive organic layer against oxidation and humidity-induced degradation. Mostly nitrogen is used as inert gas. To realize an inert concentration of approx. 2 ppm or less oxygen and water vapor, it is necessary to encapsulate the complete process plant with a hermetic casing (see Fig. 9.15). Very often, a so called glove-box design is chosen to give the operator of the plant the chance to make adjustments without opening the enclosure. Mostly the plant is divided into different sections to separate the wet-coating processes, the drying station, and the un- and rewinding unit from each other. Such machine design allows to change material rolls without contaminating the atmosphere in the coating or drying modules.

Figure 9.15 Inert casing of the winding, coating, and drying area in a wet coating line.

9.5.2 Application of Adhesives

9.5.2.1 Coating and printing techniques

In general, one distinguishes between coating and printing techniques depending on the layer design. The term "coating" refers to processes in which a closed, uniform layer has to be applied to the substrate. In the case of patterned layer design we speak about "printing" processes. Generally high performance and productivity of the coating or printing process are necessary. Solution coating processes for printed electronic need to fulfill the following requirements, in addition to applying virtually defect-free layers:

- Cross-web homogeneity of the layer thickness of less than ±1%
- Possibility for deposition of layers with a wet thickness less than 1 μm to a dry thickness less than 100 nm
- Down-web uniformity of the layer thickness over 24 hours of continuous processing
- Printing texture up to 2 pt. resolution

The coating technologies may be classified into self-metered and premetered techniques. Self-metered coating means that the applied coating weight depends on the process. Known are dip-coating processes where the speed of substrate defines the coating thickness. In the case of roller coating the speed of the roller and the gap between the rollers defines the coating thickness, and in the case of knife-edge coating the gap between the knife and the substrate defines the coating thickness. Premetered coating means that the applied coating weight does **not** depend on the process, for example, slot-die coating and spray coating, where the pumping speed of the dosing systems defines mainly the coating thickness.

Slot-die coating technologies are characterized by capillary forces, which are acting between the slot die and the substrate. The distance between the slot die and substrate is often less than 200 μm. A very low wet film thickness (~1 μm) is possible at substrate speeds of up to 50 m/min and wet film thicknesses can be obtained at substrate distances that are 300 times larger than the wet film thickness. Different slot-die operations are defined by the length of the gap between the slot die and the substrate. For many applications it is necessary to apply an intermitted design of the coating layer (see Fig. 9.16). KROENERT is offering two different intermitted slot-die coating processes. The coating material supply can be controlled and periodically switched on and off with either installing a rotating bar in the slot die or using a valve in front the die.

Roller technologies can be used for printing as well as coating processes. In the case of printing procedures the application roller is in contact with the substrate; during coating procedures the process can be contactless or with contact. The advantage of roller technology is its flexibility in use. Depending on the running direction of the application roller and the design of the roller it is possible to realize coating and printing operations with the same coating device. For the printing application an engravure of the roller needs to be defined for the highest resolution. Hexagonal cell and structure engravings are used and in the case of the coating application all kinds of line engraving may be applied.

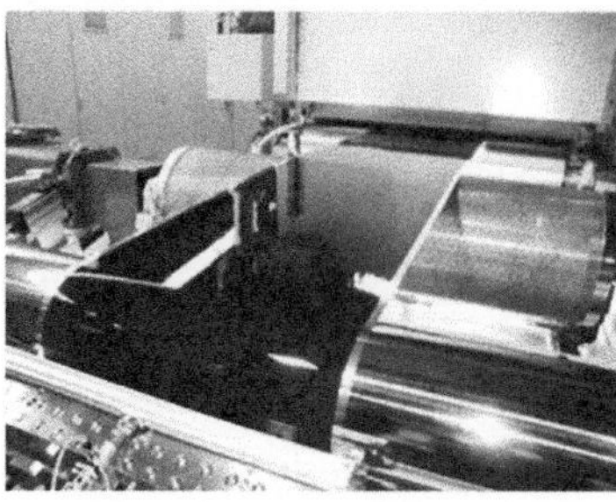

Figure 9.16 Switch from continuous to intermitted application with slot-die coating.

Alternatively, an indirect or flexo-printing technology is used for printing applications. An engraved roller transports the coating slurry out of a tank system and transfers it to the flexo roller in a high-speed printing process. Also rotary screen technology is used for printing processes. This technology is used mainly for thicker lines and designs for instance to interconnect solar cells with printed conducting lines.

9.5.2.2 Lamination processing

The coated device film and the barrier film are unwound and guided with guiding rollers to the lamination roller pair. Between the lamination roller pair the substrates are guided together. Generally three different process methods are defined for roll-to-roll barrier film lamination according to the distance of the lamination rollers. These are air lamination, gap lamination, and pressure lamination (Fig. 9.17).

1. Air lamination: For air lamination, the lamination rollers have a large distance. There is no pressure applied to the barrier film. The top substrate is pressed onto the lower substrate

only by its own weight. The adhesion is highly influenced by the substrate weight of the top film and is considerably low because of that. This technology is used for very sensitive substrates that are prone to damage when pressure is applied.

2. Gap lamination: In the case of gap lamination the distance between the lamination rollers is set to approximate the sum of the thicknesses of the two films that are to be laminated together. This technology is mainly used for a sensitive device structure because of the low pressure being applied to the two substrates.
3. Pressure lamination: In the case of pressure lamination the gap between the lamination rollers is set to be nearly zero. The substrates are pressed with a very high pressure toward each other. Because of the high pressure the adhesion will be very good in the case of good adhesive properties. However, pressure-sensitive systems and films will be eventually damaged in a pressure lamination process.

9.5.3 Substrate- and Device-Handling Requirements

Flexible substrates are required to produce bendable, conformable, and flexible organic devices. Mostly PET, PEN, and also polyimide (PI) films are used as basic substrates. These films are delivered as rolls to the machine. During the different process steps it is necessary to monitor the tension during all wet-coating as well as lamination processes.

9.5.3.1 Wet-coating processes

Often, the substrate is handled under different temperature conditions during wet-coating and drying processes. However, elastic properties of polymer materials are changing with increasing temperature. To realize a perfect even and constant coating layer thickness and coating film stability it is necessary to handle the substrate under different tension conditions. The coating process is usually done at room temperature without impairing the polymeric stability of polymer films. However, drying and curing processes can be critical as the substrate has to withstand temperatures up to 180°C, depending on the drying and curing process parameters.

The elastic modulus of the polymer is reduced at that temperature, which results in a much higher strain when applying the same web tension as at room temperature. For example, the elastic modulus of PET is reduced from 3700 MPa at room temperature to approx. 900 MPa at 120°, resulting in four times' higher elongation at the same tension. The larger strain potentially causes damage or even delamination of the coating. Reducing tension for high-temperature processing helps to reduce or eliminate damaging effects on the coating and the polymer.

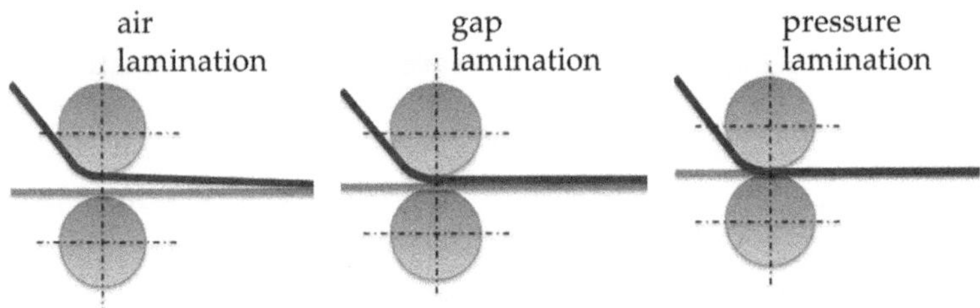

Figure 9.17 Overview of different lamination process types.

9.5.3.2 Encapsulation processes

The barrier or protective films are very often coated with thin ceramic or other inorganic layers to realize the requested high water and oxygen barrier properties. The use of these barrier films have to be realized in a very sensitive way because at too high tension the barrier layer will be destroyed very easily. Process experiences exist with a tension adjustment to nearly zero. Guiding the material through a (roll-to-roll) processing machine, however, requires at least a minimum web tension. Without tension, the substrate would move from left to right in an uncontrolled way. The substrate film itself has certain elasticity. A too high tension would result in the destruction of the barrier layer, which does not have the same elasticity as the substrate. A perfect tension control and monitoring is very important.

9.6 Conclusions

The term "encapsulation" refers to both materials and processes for packaging of organic electronic devices to protect them against extrinsic degradation factors such as reactive chemicals and gases (oxygen, water vapor), radiation (UV light), and mechanical impacts (scratching, punctual loads). While rigid devices are commonly packed between glass sheets, flexible—meaning bendable and conformable—devices rely on coated flexible polymer films as encapsulation material. As especially water vapor is a critical degradation factor for organic devices and as polymers are highly permeable for water vapor, much attention has been paid to reducing water vapor permeability of polymers by coating them with transparent metal oxide layers or layer stacks that consist of alternating oxide layers and polymer layers. These coated films are referred to as barrier films or encapsulation films. The figures of merit to compare different encapsulation films are the WVTR and the OTR. Understanding the measurement of these quantities and selecting the right measurement conditions and measurement time are essential to correctly judge and compare the performance of different barrier films. The same barrier film has different permeation rates under different conditions and when determining the WVTR with different methods and different evaluation algorithms.

It is important to understand that the WVTR of a barrier film does not translate directly into a lifetime of an encapsulated device. The relationship between barrier performance and device degradation is a hot topic in current research exploring the behavior of defects in the encapsulation, the intrinsic permeation through the system, and side diffusion through adhesives in the module with respect to the degradation behavior of the device. In any case, the device itself is the best measure for performance and applicability of an encapsulation system.

Water vapor and/or oxygen permeability is just one among several properties of an encapsulation system that determine the applicability of an encapsulation technology for a specific device type and application scenario. Optical properties such as transmission, reflection, and absorption behavior and haze play a major role in the performance of the encapsulated device. Mechanical properties such as bendability, crack formation under strain, and thermal expansion or robustness against scratches and punctual loads determine the stability of the system not only in processing but also in common application scenarios.

An organic electronic system or module consists at least of one or two barrier/encapsulation films, transparent and metal electrodes, adhesives and sealants, and the active device layers themselves. Looking at the interactions between these constituents rather than the properties of each part alone is one key to success for preparing long-living flexible organic devices that enable new design opportunities and application scenarios for the field of organic electronics.

Exercises

9.1 It is known that 20 mg water per square meter is sufficient to degrade an organic solar cell to 50% of its original efficiency. Calculate the required steady-state WVTR for a T^{50} lifetime of 5 years, assuming (a) a lag time of permeation of 0 hours and (b) a lag time of 5000 hours.

9.2 Calculate the water ingress into a glass-encapsulated system through the edge seal for a 5 mm long edge seal with a 50 μm thick adhesive having a 2 g/m^2day WVTR for a device of 20 × 40 cm^2, solely taking into account the diffusion through the adhesive, ignoring any potential interface diffusion.

9.3 Based on the previous exercise (Ex. 9.2), calculate the area (cm^2) of a 100 μm thick film of a getter paste to be introduced in the encapsulation package to sorb the permeating water during a period of 10 years (Getter paste characteristics: sorption capacity 12% wt; density: 1.02 g/cm^3).

9.4 Calculate the acceleration factors of the WVTR from the partial pressure difference between the ambient conditions: 23°C/50% relative humidity, 38°C/90% relative humidity, 60°C/90% relative humidity, and 85°C/85% relative humidity.

9.5 Select the right adhesive for an encapsulation process by considering the following requirements: the process has to be performed under a nitrogen flow in a single fast step by means of UV irradiation. The maximum curing time is 30 sec by adopting a UV light source with an irradiance of 100 mW/cm2. (Available products: (a) free-radical adhesive with a UV dose reactivity of 1.5 J/cm^2 and (b) cationic epoxy adhesive with a UV dose reactivity of 12.0 J/cm^2).

Suggested Readings

Burrows, P. (2001). Ultra barrier flexible substrates for flat panel displays. *Displays*, **22**, pp. 65–69.

Choi, K., and Jo, W. H. (1995). Effect of chain flexibility on selectivity in the gas separation process: molecular dynamics simulation. *Macromolecules*, **28**, pp. 8598–8603.

Fahlteich, J., et al. (2014). The role of defects in single- and multi-layer barriers for flexible electronics. *2014 Fall Bulletin of the Society of Vacuum Coaters*, pp. 36–43.

Glicksman, M. E., (1999). *Diffusion in Solids*, Wiley-Interscience, New York.

Grossiord, N. (2012). Degradation mechanisms in organic photovoltaic devices. *Org. Electron.*, **13**, pp. 432–456.

Hermenau, M., Schubert, S., Klumbies, H., Fahlteich, J., Müller-Meskamp, L., Leo, K., and Riede, M. (2012). The effect of barrier performance on the lifetime of small-molecule organic solar cells. *Sol. Energy Mater. Sol. Cells*, **97**, pp. 102–108.

Krebs, F. C. (2012). *Stability and Degradation of Organic and Polymer Solar Cells*, Chapter 10, pp. 269–329, eds., Müller-Meskamp, L., Fahlteich, J., and Krebs, F. C., John Wiley & Sons.

Kucukpinar, E., and Doruker, P. (2003). Molecular simulations of small gas diffusion and solubility in copolymers of styrene. *Polymer*, **44**, pp. 3607–3620.

Lim, S.-F. (2001). Correlation between dark spot growth and pinhole size in organic light-emitting dioses. *Appl. Phys. Lett.*, **78**, p. 15.

Massey, L. K. (2003). *Permeability Properties of Plastics and Elastomers, 2nd Ed.: A Guide to Packaging and Barrier Materials*, Plastics Design Library, Elsevier Science & Technology Books.

Nisato, G., et al. (2003). P-88: Thin film encapsulation for OLEDs: evaluation of multi-layers barriers using the Ca Test. *SID Symp. Dig. Tech. Pap.*, **34**, pp. 550–553.

Nisato, G., et al. (2014). Experimental comparison of high-performance water vapor permeation measurement methods. *Org. Electron.*, **15**, pp. 3746–3755.

Schaer, M., Nüesch, F., Berner, D., Leo, W., and Zuppiroli, L. (2001). Water vapor and oxygen degradation mechanisms in organic light-emitting diodes. *Adv. Funct. Mater.*, **11**(2), pp. 116–121.

Vacca, P. (2010). *Dispensable Polymeric Precursor for Transparent Composite Sorber Material*. US2013181163.

Vacca, P., Bonucci, A., and Scoponi, M. (2012). *Sealant Composition*. US2014166083 (A1).

Chapter 10

Printed Sensors and Sensing Systems

Giorgio Mattana[a] **and Danick Briand**[b]
[a]*ELORGA – IMS, Bordeaux, France*
[b]*EPFL-IMT SAMLAB, Neuchâtel, Switzerland*
danick.briand@epfl.ch

10.1 Introduction

In this chapter, we review the fundamentals and working principles of sensors, supported by some examples of printed sensors on foils. Integration of sensors into smart systems is also briefly addressed. This chapter is not an exhaustive review of the state of the art in printed sensors but an introduction to the field of sensors and their implementation using organic and printed electronics (OPE) technologies. After having introduced what a sensor is, its main characteristics, and its different working principles, we will focus on concrete developments of printed bio-, chemical, and physical sensors.

Since the early 1980s, we have seen a strong growth of microelectromechanical systems (MEMS) sensors based on silicon

Organic and Printed Electronics: Fundamentals and Applications
Edited by Giovanni Nisato, Donald Lupo, and Simone Ganz

ISBN 978-981-4669-74-0 (Hardcover), 978-981-4669-75-7 (eBook)
www.panstanford.com

semiconductor and micromachining technologies. Several of these devices, such as accelerometers, gyroscopes, microphones, and pressure, temperature, and humidity sensors, have reached the consumer electronics market. General speaking, sensing devices produced using OPE technologies cannot meet the level of performances obtained with silicon sensors. However, thanks to OPE, sensors and smart sensing systems with unique characteristics such as flexibility, conformability, transparency, biocompatibility, disposability, wearability, and possibility of fabrication over a large area can be envisioned. This is being done in complementarity to silicon technology and not in competition. OPE is characterized by the use of more eco-friendly materials and, also thanks to the employment of additive and low-temperature processes, is leading to the production of greener sensing and electronic systems. In the case of the production of a high number of sensor components or intelligent sensing surfaces, OPE can become very attractive with the use of cost-effective large-area manufacturing techniques.

The attractiveness of OPE to realize sensors has been reflected by a major growth in the number of scientific publications that has started in the mid-2000s. The organic and printed sensors market has been predicted to grow fast in the coming years: Several types of sensors—strain/pressure, temperature, light, humidity, gas, and biological—have been implemented on flexible substrates and using printing processes. Some of them, such as temperature, light, touch, and glucose sensors, are being commercialized.

Fully printed sensors are still at their infancy age and significant R&D and production efforts are still required. Most sensing systems are currently built using a hybrid integration effort, which involves the use of CMOS electronics components. In applications in which a high sensor production volume or a large sensing area will be required, one can expect that large-area manufacturing at low cost, going toward an all-printing approach, will be the chosen solution. However, components' encapsulation, which is very sensor-type specific, their integration into systems, and also their calibration are aspects that will not be addressed in this chapter, but deserve special attention for successful development of commercially exploitable sensor solutions.

10.2 Sensor Fundamentals

10.2.1 Definition and Classification

A sensor is a component part of a measurement system that employs physical or (bio)chemical properties in order to obtain quantitative data that can be processed in different ways. The input signal is called the measurand, which is the physical or (bio)chemical quantity to be measured (force, light intensity, ion concentration). A sensor is a transducer that allows to sense, to detect the measurand, by converting a physical or (bio)chemical quantity into an electrical signal. The output can be either digital or analog. In the case of digital sensors, the signal produced by the sensor is binary, while for analog sensors the signal produced is continuous, in time and amplitude, and proportional to the measurand. Sensors outputs are typically analog signals that require generally some processing. Sensors are either active or passive devices. Active sensors (or modulating sensors) require an external source of power (excitation voltage) that provides the majority of the output power of the signal (see Fig. 10.1a). For passive sensors (or self-generating/self-exciting sensors), the output power is almost entirely provided by the measured signal without an excitation voltage (see Fig. 10.1b). Examples of passive sensors are thermocouples, which produce an electromotive force (emf) from the difference in junction temperatures, and piezoelectric crystals, which, thanks to the piezoelectric effect, convert mechanical strain into electrical charges.

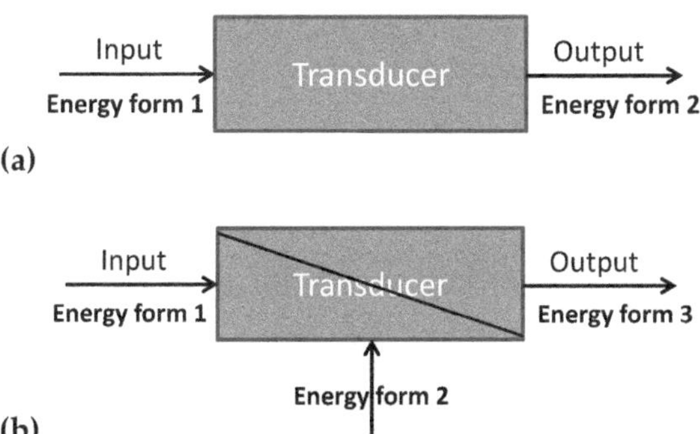

Figure 10.1 (a) Self-generating/active sensor and (b) modulating/passive sensor.

There are a number of different criteria that are used for classifying sensors; here is a nonexhaustive list:

- Physical or (bio)chemical effect/transduction principle
- Principal form of signals
- Measurand (primary input variable)
- Technology and material
- Application
- Size (nano, micro, milli, macro)
- Characteristics (sensitivity, accuracy, etc.)
- Cost

The selection of the criteria is strongly dependent on the type of user. For students, it might be adequate to use the classification by principle. For application engineers, it may be desired to class the sensors by measurand or application. For development engineers, technology, size, cost, and characteristics might be more of interest.

Classifying sensors according to the nature of the measured quantity provides the following categories:

- Physical sensors
- Chemical sensors
- Biological sensors

For all of them the measurement principles can be of a physical, chemical, or biological nature. For example, the measurement of the concentration of a given gas using the shift in frequency of a resonator (due to a change in mass) is a chemical sensor based on the physical principle.

If we look at the principal form of signals, we can distinguish seven classes:

- Mechanical
- Thermal
- Electrical
- Magnetic
- Radiant
- Chemical
- Biological

A number of examples for each class are given in Table 10.1.

Table 10.1 Classification of sensors according to the principal form of signals

Form of signal	Measurands
Mechanical	Length, area, volume, displacement, velocity, acceleration, amplitude, force, torque, pressure, mass, flow, acoustic wavelength, etc.
Thermal	Temperature, heat, heat flow, heat capacity, entropy, etc.
Electrical	Charge, current, voltage, resistance, conductance, inductance, capacitance, dielectric permittivity, polarization, electric field, frequency, dipole moment, etc.
Magnetic	Magnetic field, flux density, magnetic moment, magnetic permeability, magnetization, etc.
Radiant	Ultra-violet, visible, infra-red, micro-waves, radio waves, gamma rays, X-rays, etc.
Chemical	Humidity, concentration of odours, vapours and gases, pH levels, ions, heavy metals, pollutants, etc.
Biological	Sugars, antigens, proteins, DNA, hormones, neurotransmitters, etc.

Sensor behaviors and performances are described via a set of parameters that are derived from measurement sciences. These characteristics are reviewed in the next section.

10.2.2 Sensor Characteristics

We will review in this section the main parameters that can be used to characterize the behavior and performance of sensors. We will go through a set of definitions for the main static (sensitivity, accuracy, resolution, reproducibility, repeatability, linearity, saturation, hysteresis, drift, noise, range or span) and dynamic (transient response, response time, recovery time, frequency response, rise time) sensors' characteristics.

The response curve is a very valuable source of information about the features of a sensor. It provides information on the output response of the sensor as a function of the measurand (M) to which the sensor is exposed to. Figure 10.2 shows an example of a response curve with a linear behavior that gradually converges toward saturation. This type of behavior is called a Langmuir-like response.

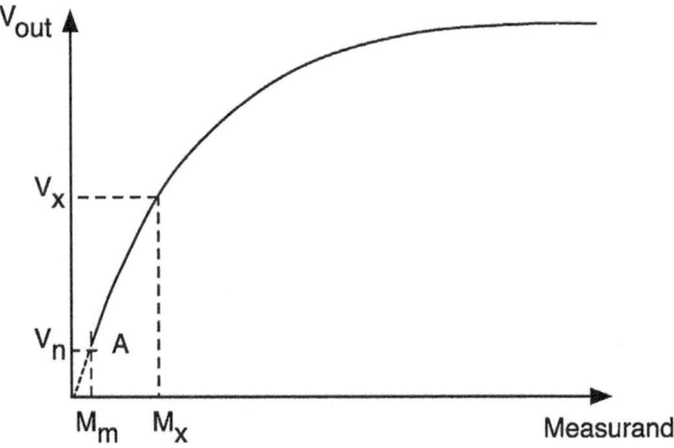

Figure 10.2 Example of a generic sensor response curve. M_m is the measurand value corresponding to a signal-to-noise ratio equal to 1. Reprinted from A. D'Amico et al. (see Suggested Readings), with permission from Elsevier, Copyright 2014.

The response of a sensor can be expressed considering either the absolute signal (V_{out}) of the sensor or the difference in the signal ($V_{out} - V_0$) due to the exposure to the measurand, V_0 being the output signal without the application of the measurand. The relative response $(V_{out} - V_0)/V_0$, which is unitless, is interesting to compare sensors having responses with different measurement units. The differential and relative responses can be used to counteract the drift of the signal baseline. The baseline is the electrical signal from a sensor when no measured variable is present, while drift is here the low-frequency change of the sensor baseline with time. The smallest measurand quantity measurable, the limit of detection for a chemical sensor, is determined by the noise level (V_N). Noise is a random deviation of the signal that varies in time.

We will now go through different definitions characterizing the response of a sensor.

- *Sensitivity*: Variation of the sensor signal as a function of the variation of the measurand. Sensitivity is the derivative of the response curve. For a nonlinear response curve, as in Fig. 10.2, the sensitivity is a function of the measurand value. When the response curve is linear the sensitivity is given by

$$S = \Delta V/\Delta M \quad (10.1)$$

- *Resolution:* The minimal amount of measurand change that produces a measurable signal. Resolution corresponds to the ability of a sensor to see small differences in readings. The resolution of a sensing system must be better than the final accuracy the measurement requires. The primary determining factor of resolution is electrical noise. Resolution is given in the unit of the measure of the measurand, and therefore at the opposite of the response curve and sensitivity, the resolution enables a comparison among sensors.
- *Accuracy:* The closeness of a measured value to a standard or known value.
- *Precision:* The closeness of two or more measurements to each other.
- *Reproducibility and repeatability:* Two extreme conditions of precision. Repeatability describes the minimum variability and reproducibility the maximum variability in sensor output.
- *Repeatability:* Variability between sets of measurements taken over a short time interval.
- *Reproducibility:* Variability between sets of measurements but performed over a long time interval or by different operators/laboratories.
- *Linearity:* Ability to reproduce the input characteristics symmetrically (Fig. 10.3). Linearity describes the maximum deviation of the output of a device from a best-fitting straight line through the calibration data.
- *Saturation:* Every sensor has its operating limits. Even if it is considered linear, at some levels of the input stimuli, its output signal no longer will be responsive. A further increase in stimulus does not produce a desirable output. It is said that the sensor exhibits a span-end nonlinearity or saturation (Fig. 10.3).

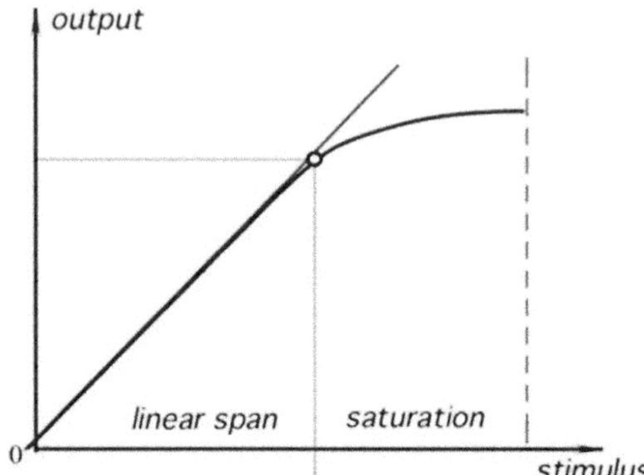

Figure 10.3 Response curve with two distinct response regions, linear span and saturation.

- *Hysteresis:* The characteristic that a transducer has in being unable to repeat faithfully, in the opposite direction of operation, the data that have been recorded in one direction.
- *Drift:* Slow variation in the sensor output over long periods. Methods to account for sensor drift include periodic calibration, predictive aging (or pre-aging the sensor), and historical data logging combined with the use of filters.
- *Noise:* In electronics, noise is a random fluctuation in an electrical signal.
- *Range or span:* The maximum and minimum value range over which a sensor works well. Every sensor is designed to work over a specified range. The design ranges are usually fixed and if exceeded result in permanent damage to or destruction of a sensor.
- *Dynamic range or span*: The range of input physical signals that may be converted to electrical signals by a sensor. Signals outside of this range are expected to cause unacceptably large inaccuracy (specified sensitivity is not guaranteed).
- *Transient response:* Response of a system to a change from equilibrium (Fig. 10.4).
- *Speed of response:* How fast a sensor can react to a change in input. The speed of response may be specified in terms of a

time constant τ, which is a measure of the sensor's inertia that can be determined experimentally from the transient response to step input. The sensor output reaches 63% of the final value after $t = \tau$ (see Fig. 10.4).

- *Response time*: The *time* for a *sensor* to respond from no load to a step change in load. Response time is usually specified as the *time* to rise to 90% of the final value, measured from the onset of the step input change in the measured variable.
- *Recovery time:* The time for a sensor to return to the baseline value after the step removal of the measured variable. It is usually specified as the time to fall to 10% of the final value after the step removal of the measured variable.

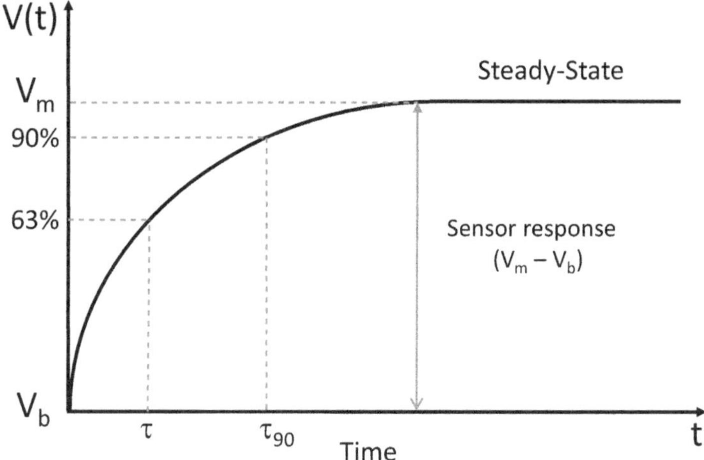

Figure 10.4 Transient response of an ideal sensor system, with a linear first-order behavior (V_m: maximum; V_b: baseline).

One essential point in sensor technology is therefore the calibration of the devices to assess the different parameters addressed above. Sensors must be calibrated against some standard in order that the results can be relied upon and compared to other observations. Another important aspect is the reliability and stability of the devices over time, which is of prime importance in most applications. Software and electronics can be used in several cases to compensate variations in behavior of a sensor over time.

10.3 Sensors' Working Principles and Examples of Printed Sensors

Sensors are devices able to provide a measurable output in response to the modification of a certain physical or chemical quantity. For this reason, they are ubiquitous in our life: any human activity that needs to identify and quantify a certain parameter relies on the utilization of some types of sensors. There exists an enormous number of sensors, so for practical purposes it is customary to group sensors into typologies: One of the most important and utilized criteria to classify sensors takes into account the *sensors' working principle*, that is, the chemophysical mechanism that is responsible for the detection of the stimulus and its transformation into a measurable electrical signal (*transduction*) that can be used to quantify the stimulus itself. In the following sections, the principal typologies of sensors, classified according to the working principle criterion, will be presented. Each section will be introduced by a concise description of the transduction mechanism and its mathematical formulation; then a brief overview of the application of printing techniques in the fabrication of that typology of sensors will follow. Finally, a meaningful example of printed sensor will be described more in detail for each section.

10.3.1 Resistive Sensors

The electrical resistance R (expressed in ohms, Ω) of an object is a physical parameter describing the opposition to the passage of an electrical current when a voltage is applied across it. Resistance depends both on the material composing the object and its geometrical shape. For an object with a uniform cross section area A and length l, its resistance can be expressed as follows:

$$R = \rho \frac{l}{A} \tag{10.2}$$

where ρ is an intrinsic property of the material called electrical resistivity (expressed in $\Omega \times$ m).

Sensors able to detect physical or chemical signals through a change in their electrical resistance are called *resistive sensors*.

Amongst the different types of resistive sensors, two examples will be described here.

10.3.1.1 Thermal resistive sensors (thermistors)

These sensors rely on the variation of electrical resistivity (and, as a consequence, of resistance) with temperature.

For many materials, for a given temperature range, electrical resistivity depends linearly on temperature and the relationship between these two quantities can be described as follows:

$$\rho(T) = \rho_0[1 + \alpha_0(T - T_0)] \tag{10.3}$$

where $\rho(T)$ is resistivity at temperature T, ρ_0 is resistivity at a fixed temperature T_0 (usually 25°C), and α_0 is the **temperature coefficient of resistance** (TCR) (measured in K^{-1} or in $C^{\circ-1}$).

Several different techniques and materials have been employed so far for the fabrication of printed thermistors; in most cases, flexible plastic sheets were used as substrates and metal nanoparticle–based inks were either screen- or inkjet-printed to fabricate the resistors. Molina-Lopez et al. realized inkjet-printed silver meander-shaped resistors on polyethylene terephthalate (PET) substrates; the effects of nickel electroplating was also evaluated. The nonelectroplated devices exhibit an extremely linear behavior in the considered range of temperatures (–10°C, +60°C), with a TCR up to $(6.52 \pm 0.05) \times 10^{-4}$ °C^{-1}; electroplated devices show a better TCR (up to $[1.82 \pm 0.06] \times 10^{-3}$ °C^{-1}), as can be appreciated in Fig. 10.5.

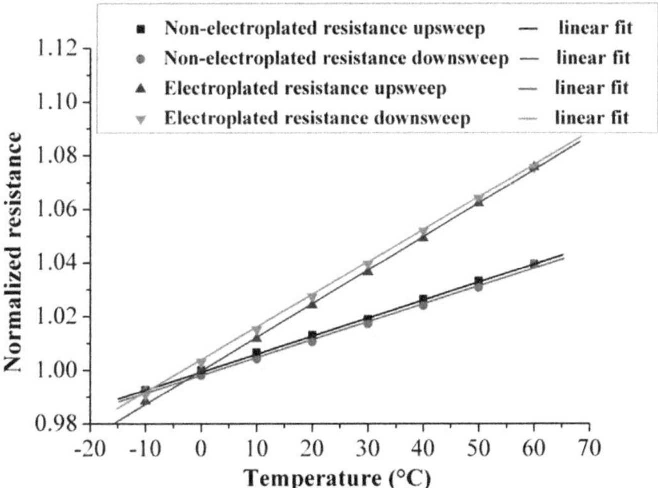

Figure 10.5 Normalized response of thermistors. Reprinted from Molina-Lopez et al. (see Suggested Readings), with permission from Elsevier, Copyright 2012.

Another example of printed resistive sensors is provided by Britton et al. The authors reported on the fabrication of a thick silicon film obtained by nanoparticulate silicon-based inks. These inks were fabricated by milling bulk silicon (either n-type or p-type doped) and reducing it into a nanoparticle powder (maximum nanoparticle diameter: 50 nm). This powder was then dissolved in ethanol and polymeric binders (such as cellulose acetate butyrate [CAB] or acrylic screen printing pastes) were subsequently added to the suspension in order to achieve the viscosity necessary for screen printing. The screen-printed layers were deposited on the top of low-temperature substrates such as paper. This technology was later patented for the fabrication of resistive silicon-based temperature sensors realized on paper substrates at room temperature. These devices are entirely screen-printed: the silicon ink is deposited on top of interdigitated silver electrodes (length of electrodes: 16 mm; gap between electrodes: 0.25 mm), giving a resistor with an approximate resistance value of 100 kΩ. Sensors were tested at a temperature range between 20°C and 60°C and exhibited a negative temperature coefficient (NTC) characterized by a beta value of 2000 ± 100 K. Such sensors began being commercialized by PTS Sensors Ltd. in 2010, and very recently (March 2014) Thin Film Electronics ASA and PST Sensors Ltd. have entered into a purchase and licence agreement, which will allow Thin Film Electronics to employ these printed temperature sensors for the fabrication of its temperature-sensing smart labels where PTS-printed thermistors will be integrated within a printed radio-frequency (RF) circuit with near-field communication functionality.

10.3.1.2 Chemiresistive semiconductor-based gas sensors

These sensors exhibit a change in resistance when exposed to specific gaseous species. The absorption of the target gas on the surface of the gas-sensitive material causes a reversible electron exchange, temporarily modifying the charge density within the bulk and/or surface of the gas-sensitive material and causing a resistivity variation.

At the moment, the majority of gas-sensitive materials employed in the field of printed electronics are represented by conductive polymers and metal oxides, deposited mostly by means of screen and inkjet printing.

Crowley et al. described the fabrication of a full-printed gas sensor for the detection of a toxic reducing gas, namely hydrogen sulphide (H_2S). The sensing semiconductive material was a polymer, polyaniline (PANI), doped with copper(II) chloride salts ($CuCl_2$). These sensors show a detection limit of 2.5 ppmv and exhibit a linear relationship between measured current and concentration over the 10–100 ppmv region (Fig. 10.6).

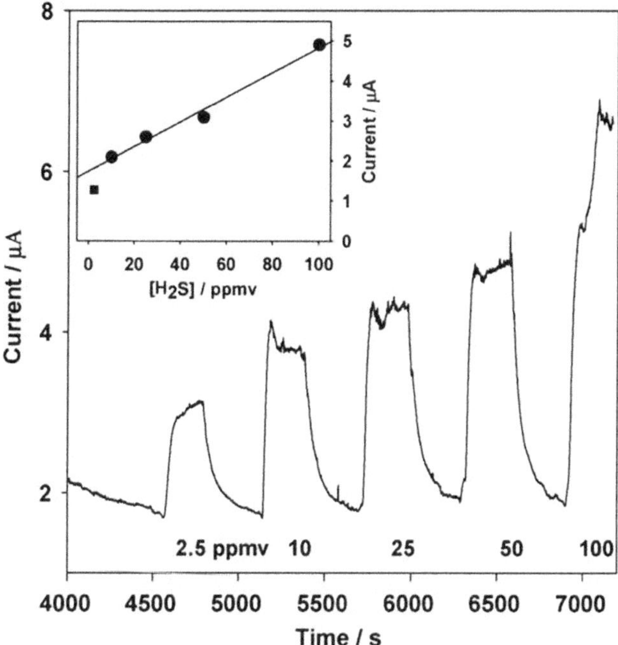

Figure 10.6 Current vs. time response of the H_2S sensors. Inset shows the calibration curve obtained by sampling the signal 180 s after each H_2S injection. Reprinted, with permission, from Crowley et al. (see Suggested Readings), © 2010 IEEE.

10.3.1.3 Piezoresistive sensors

Piezoresistivity is a well-known property characterizing metal, semiconductive, and CERMET (composite materials composed of ceramics and metals) films; it manifests itself as electrical resistivity variations when a mechanical stress and/or strain is applied to the film, according to the following law:

$$\frac{\Delta\rho}{\rho} = \Pi \cdot T \tag{10.4}$$

where the term on the right represents the relative resistivity variation, Π is the piezoresistive coefficient (expressed in Pa^{-1}), and T is the applied mechanical stress (measured in Pa). Resistivity variations are typically attributed to atom displacement within the material, which causes a modification of band transport of charges, increasing the material's band gap (positive resistivity variations) or decreasing it (negative resistivity variations). Printing fabrication techniques have been widely used for the realization of piezoresistive devices, since many piezoresistive materials (carbon-based ink, metallic inks, CERMETs, and, more recently, polymers and rubbers) are commonly available in liquid form and can be easily deposited by printing. A recent example was presented by Thompson and Yoon, who reported on the fabrication of a poly(3,4-ethylenedioxythiophene) polystyrene sulfonate (PEDOT:PSS) based piezoresistive pressure sensor realized by employing the aerosol printing technique. The sensors (meander-shaped resistors) were printed on a polyimide flexible substrate and resistance variations due to strain were recorded by connecting the sensors to a Wheatstone bridge and measuring the voltage drop between two midpoints of the circuit. Sensors were tested by applying cycles composed of 3000 sinusoidal strain (peak amplitude of 0.002 and frequency of 0.5, 1, and 2 Hz) and recording the resistance relative variation ($\Delta R/R_0$) at the same time. An example of resistance variation to strain is depicted in Fig. 10.7. Sensors proved to be able to follow the applied strain with a very small response time (less than 1 ms, regardless of the frequency) with a peak response between 0.1% and 0.15%. For a frequency of 2 Hz, the gauge factor had a maximum value of 0.53, although significantly lower than what is usually reported in the literature for nonprinted devices.

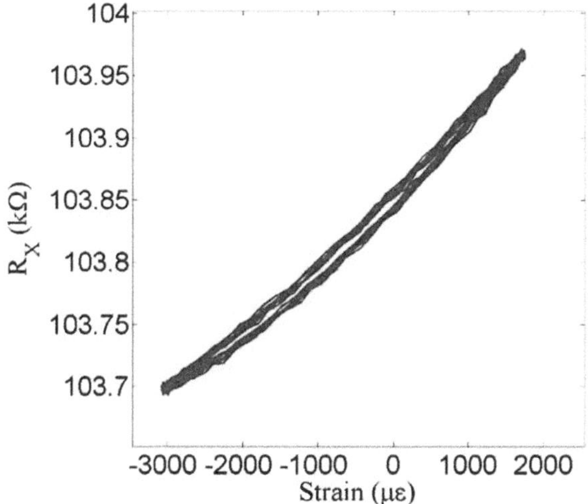

Figure 10.7 Resistance variations as a response to strain for 10 cycles at a frequency of 2 Hz. Reprinted, with permission, from Thompson and Yoon (see Suggested Readings), © 2013 IEEE.

10.3.2 Electrochemical Sensors

Electrochemical sensors (ECSs) are devices able to convert chemical information into an electrical signal. ECSs are mostly used to identify and quantify a certain analyte, typically within a liquid environment.

ECSs are electrochemical cells composed of three different electrodes: a reference electrode (RE) with a fixed redox potential, a working electrode (WE) where reaction with the analyte to be detected occurs, and an auxiliary electrode (AE), across which an electrical current flows.

Among the different types of ECSs, particular importance is attributed to the so-called *potentiometric sensors*, where the interaction between the analyte and the WE causes a variation of the electrochemical potential between the WE itself and the RE.

This potential variation may be expressed by the Nernst equation[1]:

---

[1]Equation 10.5 actually describes only *one* half-cell redox reaction. It should be noted that, in an electrochemical cell, *two* half-cell reactions take place at the same time. When the electrochemical cell is used as a sensor, however, only one of these two reactions should involve the analyte; the other reaction should be reversible and noninterfering.

$$E - E_0 = \frac{RT}{zF}\ln\left(\frac{a_{OX}}{a_{RED}}\right) \tag{10.5}$$

where $E - E_0$ is the potential variation, R is the universal gas constant (R = 8.314472 J K^{-1} mol^{-1}), T is the absolute temperature, F is the Faraday constant (F = 9.64853399 × 10^4 C mol^{-1}), z is the number of moles of electrons exchanged during the reaction, and a_{OX} and a_{RED} are the chemical activities of the oxidant and reductant species, respectively. The unknown concentration of the analyte can then be determined by its chemical activity.

Printed technologies have been successfully employed for the fabrication of ECSs; in particular, screen printing has been used for the fabrication of the three electrodes, with silver, gold, carbon, and copper sulphides being the most commonly used materials. Inkjet printing has also been used, as demonstrated by Crowley et al., who reported the fabrication of an amperometric sensor, for the detection of ammonium ions NH_4^+ in aqueous solutions. In this system, a silver wire (0.1 mm in diameter) was used as a pseudo-RE, a platinum wire (0.5 mm in diameter) as the AE, and the WE was fabricated by inkjet-printing a home-made solution of PANI nanoparticles doped with dodecylbenzene sulphonic acid (DBSA); all electrical connections were fabricated by screen-printing silver-based electrodes. These sensors were tested for the detection of ammonia in water solutions, showing excellent linearity (R^2 = 0.9999) within the range of 0–80 μM (corresponding to a range of 0–1.44 ppm), with a linear sensitivity of 2×10^{-8} A μM^{-1} and a detection limit of 2.58 μM (0.44 ppm) (see Fig. 10.8 for a typical sensor response).

10.3.3 Capacitive Sensors

Capacitive sensors are devices able to convert physical and chemical stimuli into a variation of an electrical parameter called *capacitance*.

The capacitance of an isolated conductive body is defined as the ratio between the electrical charge of that body Q and its electric potential V and is measured in Farad (F), even though submultiples (such as pF, nF, and μF) are most often used. Usually, capacitance is defined between two electrodes separated by a dielectric material; its analytical expression thus depends both on the geometry of the two conductors and on the electrical properties of the dielectric material placed between them:

$$C = \varepsilon \cdot G \tag{10.6}$$

where G is the so-called geometrical factor (which takes into account the conductors' geometry), while ε is the dielectric permittivity of the insulating material (measured in F/m).

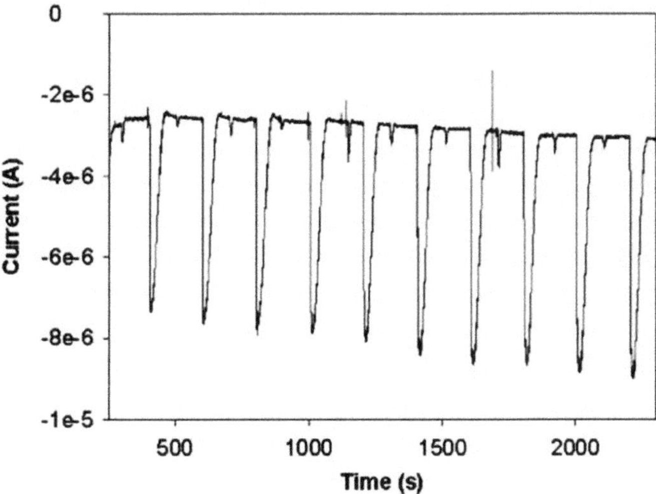

Figure 10.8 Current vs. time response to the injection of 200 μL of a 1 mM aqueous ammonium chloride solution of the amperometric sensor. Reprinted, with permission, from Crowley et al. (see Suggested Readings), © 2010 IEEE.

An important typology of capacitors is represented by the *parallel-plate capacitor*. In this case, the two conductors (called *plates*) are flat, parallel, and separated by a distance d; the overlap area between the two plates is indicated as A. In this system, G is simply

$$G = \frac{A}{d} \tag{10.7}$$

Several examples of printed capacitive sensors are reported in the literature, on a variety of substrates. In the case of capacitive sensors for the detection of gaseous analytes, inkjet printing has been successfully used both for the fabrication of silver interdigitated electrode (IDE) capacitors and for the deposition of the humidity-sensing layer (based on CAB) in the work of Molina-Lopez et al.

For pressure capacitive sensors, screen printing is the most frequently utilized technique, in particular for the fabrication of parallel-plate capacitors, which can be easily made with a multilayer approach. An example concerning this latter typology of devices is given by Narakathu et al., who reported on the fabrication of a full-printed pressure sensor array on a flexible plastic substrate (PET). Screen printing was used for the deposition of the bottom electrodes (fabricated using a commercial silver ink) and of the dielectric (polydimethylsiloxane, PDMS), while the top electrodes (also in silver) were gravure-printed. Each single sensor was characterized in terms of response to applied pressure. It was found that the lowest detectable pressure had a value of 800 kP, in correspondence of capacitance increases of 5%. The dynamic response of the sensors was also recorded (see Fig. 10.9) in a pressure range between 800 kPa and 18 MPa; capacitance increases with pressure, with a maximum increase of 40% observed for an applied pressure of 18 MPa.

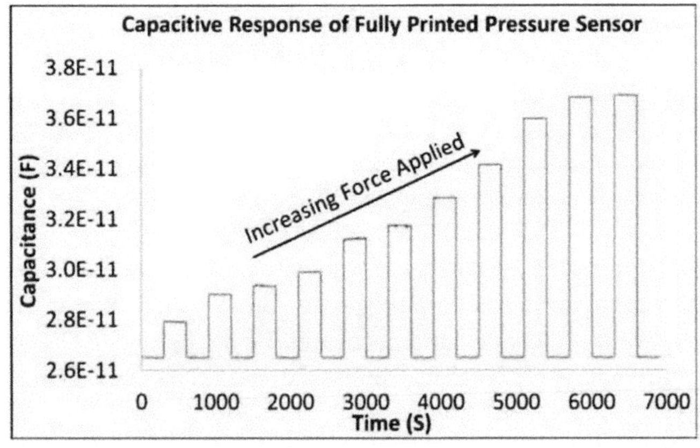

Figure 10.9 Capacitance vs. time response to the application of a perpendicular force in the printed pressure sensors. Reprinted, with permission, from Narakathu et al. (see Suggested Readings), © 2012 IEEE.

Capacitive sensors can also be employed to detect gaseous analytes, relying on the capacitance variations determined by the interaction of the capacitor with the analyte to be detected, usually caused by either modification of the structure's geometry (e.g., swelling) or modification of the dielectric electrical properties.

For this application, the most important device configuration is represented by the *interdigitated structure*, where the two electrodes, shaped as interdigitated fingers (IDFs), are coplanar and the dielectric is deposited on the top of them (and thus exposed directly to the analyte). In this case, it is not possible to obtain a closed, analytic expression to describe the capacitance, which becomes a function of several different parameters:

$$C = f(N, \varepsilon_{SUB}, \varepsilon_{DIEL}, W, G, L, t, h_{SUB}, h_{DIEL}) \quad (10.8)$$

where N is the number of fingers; ε_{SUB} and ε_{DIEL} are the dielectric relative constants of the substrate and of the gas-sensing material, respectively; W, G, L, and t are the electrodes width, gap, length, and thickness, respectively; and h_{SUB} and h_{DIEL} are the initial thicknesses of the substrate and the sensing material, respectively. As mentioned earlier in this section, interaction with the analyte causes both a swelling ($h'_{DIEL} = h_{DIEL} + \Delta h_{DIEL}$) and a modification of the electrical properties ($\varepsilon'_{DIEL} = \varepsilon_{DIEL} + \Delta\varepsilon_{DIEL}$) of the sensing dielectric deposited on the top of the electrodes so that capacitance variations may be expressed as follows:

$$\Delta C = f(\Delta h_{DIEL}, \Delta\varepsilon_{DIEL}) \quad (10.9)$$

and can in turn be used to measure the concentration of the target analyte. Printing techniques have been used for the fabrication of this typology of sensors, both for the deposition of the IDEs (mostly realized by using metallic nanoparticle–based inks) and for the deposition of the gas-sensitive top dielectric layer. Molina-Lopez et al. showed an all-additive process, based on inkjet printing, for the realization of humidity sensors where an Ag nanoparticle–based ink was used for the electrodes printing ($W = 95$ µm, $G = 105$ µm, and $t = 180$ nm) while a solution of CAB in hexyl acetate was inkjet-printed on top of the IDEs and used as relative humidity (RH)-sensing layer. The influence of a nickel layer (up to 5 µm thick) electroplated on the silver IDEs was also evaluated.

Nickel electroplated sensors showed better performance and better environmental stability than bare silver electrodes, with a sensitivity of 2.6 fF/% RH and a response time of 24 s. Figure 10.10a shows a picture of these inkjet-printed humidity sensors; the printed temperature sensors described in the paragraph of resistive sensors are shown as well. It is worth noting that devices were fully encapsulated in order to prevent contamination from the outside environment, as well as a form of protection against handling and scratches. In this case, a dry photoresist film (PerMX 3050TM 50 μm thick from DuPont®), in combination with the dry adhesive ARClear® 8932 (50 μm thick from Adhesives Research) deposited around the sensors, was used as a mechanical support on which to fix a gas-permeable porous acrylic copolymer (Versapor® 10000R from PALL) film, 135 μm thick (see Fig. 10.10b). It is important to notice that, unlike other devices (such as field-effect transistors [FETs] and light-emitting diodes [LEDs], for instance), which can be fully encapsulated (see Chapter 7 on encapsulation procedures), in the case of chemical sensors special care must be taken in order to allow exposition of the sensing layer/device to the analyte to be detected.

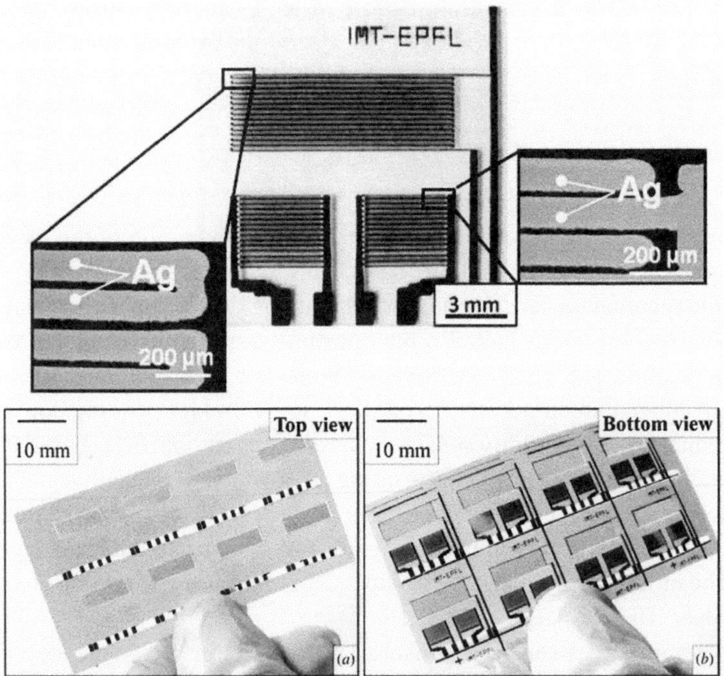

Figure 10.10 (Top images) Inkjet-printed resistive temperature sensors (detail on the left) and capacitive humidity sensors (detail on the right). (Bottom images) Encapsulated devices: (a) top view and (b) bottom view. Reprinted from Molina-Lopez et al. (see Suggested Readings), with permission from Elsevier, Copyright 2012.

10.3.4 Acoustic Wave Sensors

Acoustic waves are pressure perturbations that propagate along a certain medium with a finite velocity v and a frequency f. Pressure perturbations traveling along fluid media such as air or liquids (by far, the most common case) cause an oscillation of the particles composing the medium in the same direction as the wave propagation. For this reason, acoustic waves are a type of longitudinal wave.

Considering, for the sake of simplicity, only the one-dimensional case, a progressing acoustic wave traveling along the x direction through a certain medium can be mathematically described as follows:

$$p_{+}(x, t) = A_{+} \sin(2\pi ft - kx + \varphi_{0+}) \quad (10.10)$$

where A_+ is the wave maximum amplitude, f is the frequency (i.e., number of oscillations per unit time; $2\pi f$ is the angular frequency), k is the wave number (number of oscillations per unit length), and φ_{0+} is the phase (a parameter describing the amplitude of the pressure wave for t = 0 s and x = 0 m).

Acoustic wave sensors are devices able to detect and measure input stimuli that interact with the propagation of the acoustic waves along the tested medium, modifying the waves' characteristics (typically the frequency f).

So far, the contributions of printing fabrication techniques to the development of acoustic sensors may be roughly grouped into two main categories:

- Printing techniques have been used for the deposition of sensing materials on pre-existing acoustic sensors.
- Printing techniques have been used for the fabrication of the whole sensor.

An example belonging to the latter category is the one reported by Busch et al. The authors reported on the fabrication of inkjet-printing interdigitated conductive structures (length approx. 10 mm, line width of approx. 200 μm) on a pretensioned, poled polyvinylidene fluoride (PVDF) film (18 μm). A solution composed of 95% PEDOT:PSS, 5 wt% dimethyl sulfoxide (DMSO), and 0.1 wt% Triton X100 (surfactant) was used for the fabrication of the electrodes. Such sensors, with a resonance peak at 335 kHz, were used as mass sensors: they were loaded with a mass consisting of thin films of poly(vinyl alcohol) (PVA), obtained from 2 wt% solutions in distilled water and inkjet-printed on the underside of PVDF films (i.e., opposite to PEDOT:PSS electrodes); modification of the sensors' mass determined a consequent modification of the acoustic wave propagation, causing a shift in the speed versus frequency peak, as shown in Fig. 10.11.

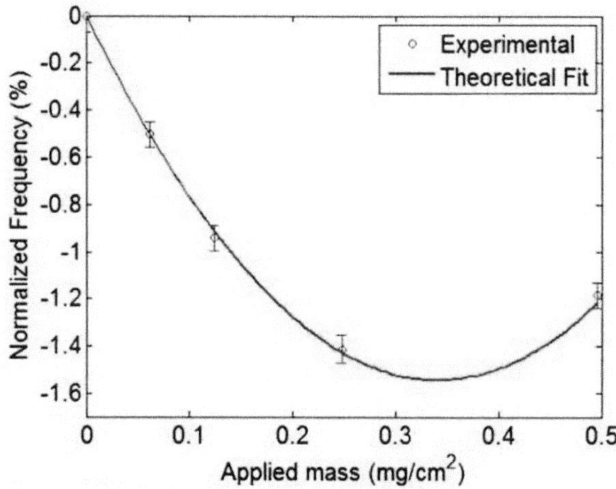

Figure 10.11 Normalized resonance frequency peak shift plotted as a function of the applied mass. Reprinted, with permission, from Busch et al. (see Suggested Readings), © 2012 IEEE.

10.3.5 Field-Effect Transistor Sensors

FETs are electronic devices where two electrodes called *source* and *drain* are separated by a semiconductive channel whose conductivity can be increased or decreased as a function of the potential applied to a third electrode called *gate*.

Very often, the gate electrode is placed on top of the channel and physically separated from it by a thin dielectric layer. This typology of FET is known by the acronym MOSFET (metal oxide–semiconductor field-effect transistor) since the dielectric placed between the gate and the semiconductor is usually an oxide.

In a MOSFET, two input voltages are used to control the device, the drain–source voltage V_{ds} and the gate–source voltage V_{gs}, and the output signal is the current flowing between the source and drain, I_{ds}.

In a MOSFET, two working regimes are distinguished, depending on the fact that the V_{gs} voltage is greater or smaller than the so-called *saturation voltage* V_{sat} (which is a function of the physical and geometrical characteristics of the device). When $V_{gs} < V_{sat}$, the transistor is in its *linear regime*, where it behaves as a resistor, I_{ds} being linearly proportional to V_{ds}. When $V_{gs} < V_{sat}$ the transistor is in the *saturation regime* and behaves as a current generator, the saturation current I_{dsat} being independent on V_{ds}. The relationship between voltages and current can be mathematically expressed as follows:

1. For the linear regime:

$$I_{ds} = \mu C_{ox} \frac{W}{L} (V_{gs} - V_{th}) \cdot V_{ds} \tag{10.11}$$

2. For the saturation regime:

$$I_{ds} = \frac{1}{2} \mu C_{ox} \frac{W}{L} (V_{gs} - V_{th})^2 \tag{10.12}$$

where μ is the charge mobility, C_{ox} is the capacitance per unit area of the MOS structure, W is the channel width, L is the channel length (distance between the source and drain electrodes), and V_{th} is the threshold voltage.

MOSFETs have been extensively used for the fabrication of either physical or chemical sensors for more than 30 years. Printing fabrication techniques have been successfully used on a large variety of substrates (plastic foils but also biodegradable materials) for the fabrication of electrodes, dielectric materials and semiconductors; this is true not only for inorganic semiconductors but also for organic-based materials, for the fabrication of the so-called organic field-effect transistors (OFETs). OFETs exhibit a structure that is similar to that of MOSFET, that is, an organic semiconductive channel separating the source and drain electrodes and capacitively coupled with the gate electrode. The main difference with respect to MOSFETs is that organic semiconductors are intrinsic, undoped semiconductors and therefore they work in *enhancement mode* (rather than *depletion mode* and population inversion, which is more typical of MOSFETs). Also, transport mechanisms are deeply different in the two FET categories, band-like transport in MOSFETs and hopping in OFETs, which leads to organic mobilities that are several orders of magnitude lower than the values commonly reported for silicon. Notwithstanding these differences, OFET output curves look very similar to those of an inorganic MOSFET (see the example in Fig. 10.12, taken from Ukah et al. 2011) and indeed they can be described conveniently by Eqs. 10.11 and 10. 12.

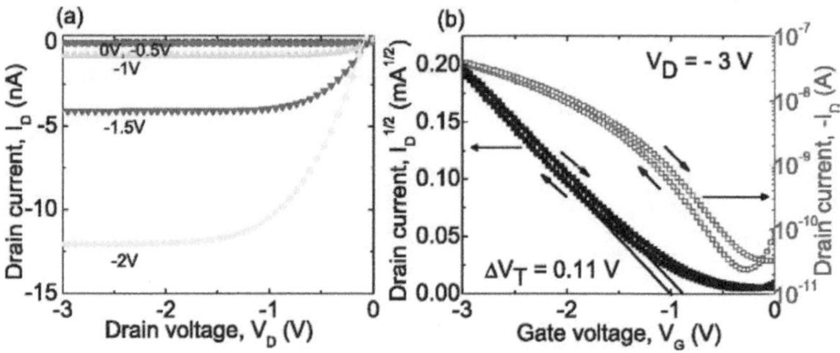

Figure 10.12 Output curves of a p-type OFET: I_{ds}–V_{ds} on the left and I_{ds}–V_{gs} on the right. Reprinted from Ukah et al. (see Suggested Readings), with permission from AIP Publishing LLC, Copyright 2011.

An interesting example of printed FETs utilized as pressure sensors was provided by Noguchi et al. in 2006. The authors reported on the fabrication of an array of transistors in which the electrodes (realized with silver nanoparticle-based ink) and the gate dielectric (polyimide) were deposited on the substrate by means of inkjet printing. The devices were then passivated with 5 μm of parylene-C on the top of which a suspended electrode constantly polarized at −100 V and covered with a polyethylene naphthalene (PEN) film is deposited. When pressure is applied, the suspended electrode touches the transistor's drain, thus activating the device and causing a measurable source–drain current increment, as shown in Fig. 10.13a; Fig. 10.13b shows an optical picture of the final devices.

10.3.6 Optical Sensors

Optical sensors are devices in which an optical signal (light beam) is changed in a measurable and reproducible way by an external physical or chemical stimulus.

The term "light beam" is used here in a very general way to indicate any electromagnetic wave. Under the hypotheses of a one-dimensional plane wave traveling along the positive direction of the

x axis of an inertial reference system, this electromagnetic wave can be mathematically described, as shown in Eq. 10.13:

$$E(x, t) = A(t) \cdot \cos(\omega t - kx + \vartheta(t)) \quad (10.13)$$

where $A(t)$ is the amplitude of the wave (measured in $V \cdot m^{-1}$ in the case of an electric wave and $A \cdot m^{-1}$ in the case of a magnetic wave), $\omega = 2\pi f$ is the angular frequency (f is the frequency measured in Hz), $k = 2\pi\lambda^{-1}$ is the wave number (λ is the wavelength measured in m or its submultiples), and $\vartheta(t)$ is the wave phase measured in radians or in degrees.

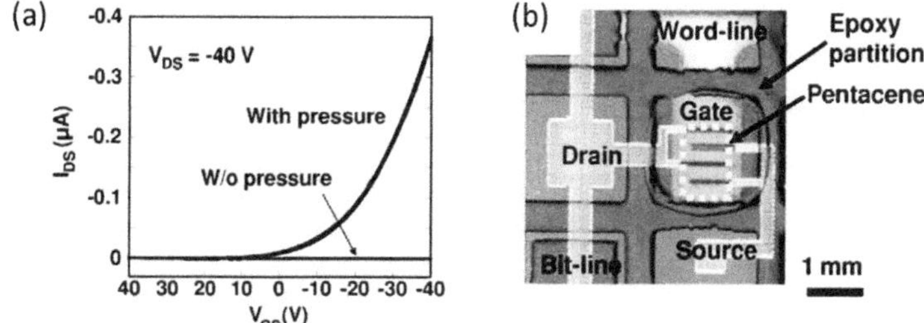

Figure 10.13 Drain current as a function of gate–source voltage with and without pressure applied on the sensor (a); optical picture of the pressure-sensitive OFETs (b). Reprinted from Noguchi et al. (see Suggested Readings), with permission from AIP Publishing LLC, Copyright 2006.

The application of printing technologies to the fabrication of optical sensors has been constantly expanding in these last years and printed photodetectors available on large-area flexible substrates are already commercially available (ISORG, France). Most of these devices consist of multilayered structures where the photosensitive material (either an organic compound/blend or, more rarely, an inorganic layer) is sandwiched between two transparent electrodes; applications for these sensors have a very wide range and include color, industrial, instrumental, medical, and large-area sensing.

Another important field where printing technologies have been successfully utilized is the fabrication of colorimetric gas or liquid sensors. Colorimetric sensors are essentially based on the change of the sensing material absorption spectrum under exposure to a certain analyte. When the sensor interacts with the analyte, the light beam reflected by the sensing layer exhibits a shift in the wave number Δk (causing a change in the color of the material). Printing techniques have been used especially for the deposition of the sensing materials, used for the detection of a large variety of analytes such as acetone and isopropanol, acetic acid, and volatile amines. An example of printed optical sensor is the one reported by Courbat et al. in 2011. The authors described a low-cost polymeric optical waveguide made on plastic (PET and PEN) foils for the detection of ammonia (NH_3), fabricated by means of additive, low-temperature processes. Transducers were based on micromirrors (realized using a UV-curable epoxy resin patterned with a PDMS mold), while the colorimetric gas-sensitive layer (the ammonia-sensing bromophenol blue, embedded in a matrix of PMMA and a hydrophobic plasticizer) was deposited on the substrate by means of inkjet printing. The sensing signal was represented by a differential current measured by one reference photodiode and another identical photodiode coupled with the colorimetric layer; interaction between the sensing layer and the target gas caused a shift in the layer's absorption peak, which was transformed in a measurable current signal by the photodiodes (and successively converted in a voltage signal by the amplifying read-out circuit). These sensors exhibited a sensitivity of 12.1 mV/ppm and a detection limit of 800 ppb. Recovery time was improved from 43 to 4 minutes by heating up the substrate at 60°C. Power consumption was also investigated, there being a trade-off between robustness to noise and detection limit; the best compromise, allowing a sub-ppm detection, was found to be 947 μW. A schematic of these sensors and an optical picture of the plastic waveguide are shown in Fig. 10.14.

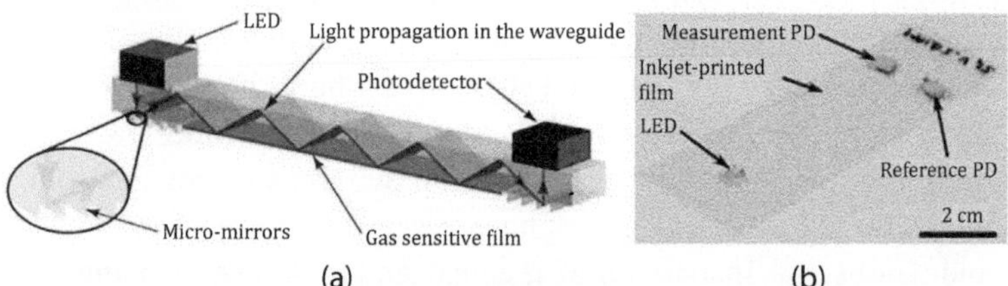

Figure 10.14 (a) Schematic of a colorimetric gas sensor using a plastic foil as an optical waveguide and (b) optical picture of the polymeric optical waveguide with an inkjet-printed ammonia-sensitive colorimetric film. Reprinted from Courbat et al. (see Suggested Readings), with permission from Elsevier, Copyright 2011.

10.3.7 Other Typologies of Sensors

The aim of this section is to provide some general information about other typologies of sensors that could not be described, because of space constraints, in the previous sections. It should be noted that the following categories of sensors are best classified according to the nature of the detected stimuli, rather than the working principle.

10.3.7.1 Accelerometers

Accelerometers are devices able to measure the acceleration experienced by a body. Typically, an accelerometer is equivalent to an oscillating, damped mass connected to a spring: by measuring the mass displacement it is possible to evaluate the mass acceleration.

Nowadays, most of accelerometers consist of cantilever beams equipped with a proof mass fabricated by using the microelectromechanical systems (MEMS) technology; these devices are typically fabricated on silicon wafers with conventional clean-room technologies. More recently, however, also printing techniques have been employed for the fabrication of accelerometers. Hense et al. fabricated a suspended inertial mass accelerometer in PEDOT:PSS, inkjet-printed on top of a cyclododecane sacrificial layer, previously deposited onto a metallic counterelectrode. The accelerometer was coupled with a printed capacitance-to-voltage converter, so oscillations of the mass along the z axis resulting in capacitance changes of the system could be converted into an output voltage signal. To improve the sensor's performance, an additional mass of 8.77×10^{-3} g was glued onto the PEDOT:PSS layer; this system showed an initial capacitance of 7.5 pF and a sensitivity of 0.235 pF g^{-1}. Another example is provided by Wei et al., who demonstrated the feasibility of fabricating a freestanding cantilever-based accelerometer on a textile substrate by means of screen printing. In their work, a common polyester-cotton fabric is used as the substrate, while a thermally curable, screen-printable silver paste is utilized for the fabrication of the electrodes and the cantilever. Such a paste can be cured at 80°C for just 5 minutes and presents therefore perfect thermal compatibility with the textile substrate on which it is printed. Before the fabrication of devices, the fabric is locally smoothened by screen-printing a UV-curable planarization paste, and then the bottom electrode is fabricated on the top of it. Between the bottom electrode and the cantilever, a sacrificial layer made of trimethyloethane (TME) dissolved in cyclohexanol and propylene glycol is also screen-printed; then the structural material (i.e., a UV-curable dielectric polymer, used as a cantilever) and the top electrode are printed. The sacrificial layer thermally decomposes at 160°C; therefore by annealing at such temperature the multilayered structure, it disappears, leaving a gap between the bottom electrode and the cantilever (which thus becomes suspended). Several different cantilever designs were tested: the highest sensitivity (0.0022 V m^{-1} s^2) is recorded at 30 Hz for a cantilever that is 12 mm long and 10 mm wide. The same sensors, fabricated on a textile wrapped around a person's forearm and connected to a capacitance-to-voltage converter able to transmit wirelessly recorded data to a PC, were successfully tested as sensors of human movement, showing performances comparable to those of a commercially available wireless accelerometer.

10.3.7.2 Biosensors

Biosensors are devices able to monitor biomolecular interactions in real time. In biosensors, one component (usually called *ligand* or *receptor*, which can be either a biological or a nonbiological molecule) is immobilized on a solid surface, while the component to be detected (the *analyte*, by definition a biological molecule) is dispersed into a solution put in touch with the surface. The interaction between the ligand and the analyte results in a change of the electrical properties of the solid surface, which can be measured in order to determine the amount and typology of analyte that has reacted with the ligand.

Typically, but not exclusively, biosensors are ECSs where the ligand is immobilized on the WE; reactions between the ligand and the analyte cause variations of the current measured between the AE and RE (*amperometric principle*) and such variations can be used to determine if and how much analyte is present into a certain test solution.

Printing fabrication techniques have been extensively used for the realization of biosensors, both for the fabrication of the three-electrode systems and for the deposition of the ligand. A well-known example of printed biosensors that have been already on the market for quite a few years is represented by the disposable glucose test strips used to measure the blood level of glucose.

10.3.7.3 Ionic and pH sensors

Ionic sensors may be defined as devices able to convert the activity of a specific ion dissolved into a certain solution into a measurable electric potential; they belong therefore to the class of potentiometric ECSs described in the previous sections. Ionic sensors are mostly used to detect and quantify the concentration of a certain ionic species into a given solution, with a very wide spectrum of potential applications, varying from the detection of certain toxic metallic cations in biological fluids to water quality control.

The majority of printed ionic sensors presented in the scientific literature are ECSs based on the ion-selective electrode (ISE) working principle, that is, the WE is equipped with an ion-selective membrane able to selectively allow the interaction of the WE with just one ionic species present in the analytic solution. When the membrane permits the interaction between the WE and the H_3O^+ ions dissolved in the solution, the system can be used to measure the solution's pH and is indeed called the *pH sensor*.

As with the other ECSs, printing techniques have been frequently used for the fabrication of three-electrode systems.

10.4 Integration of Printed Sensors into Systems

Sensor systems involve the integration of different components together and requirements and methods depend strongly on the type of sensors involved and the applications. Integration can involve the assembly of, among others, displays, memories, read-out and communication electronics, power management and supply (battery, solar cell), antennas, fluidic interfaces, etc. Different integration strategies can be applied: foil-to-foil versus the one-foil approach, all-printed or hybrid technologies, and so on. The choice is made according to systems and application requirements (cost, flexibility, and maturity of technologies involved, temperature limitations).

Following an all-printed approach, one of the most mature developments is the work performed by the company Thin Film Electronics and its partners. They demonstrated a smart temperature-sensing label made of printed memory, a sensor (from PST), organic read-out electronics (from PARC), and display components. A certain amount of work has involved the integration of sensing components on RFID labels. Temperature and humidity sensors have received significant attention with their integration on a high-frequency (HF) and ultrahigh-frequency (UHF) RFID polymeric sensors using notably inkjet printing and roll-to-roll gravure fabrication techniques. Tentzeris at Georgia Tech, USA, has reported on different types of wireless sensing labels made on low-cost paper substrates.

Heterogeneous integration, that is, the integration of active silicon technology with passive printed structures can provide the best from two technological worlds. It fulfills an important role in realizing added electronic functionalities to daily products where large-area integration and low cost are imperative. The technology has been investigated for applications such as food packaging, vaccines, and medical blisters.

For perishable food products and vaccines the purpose is to record the temperature and humidity history during transport and storing. This was realizing through a printed smart RFID label, as presented in Fig. 10.15, in the framework of the European project FlexSmell.

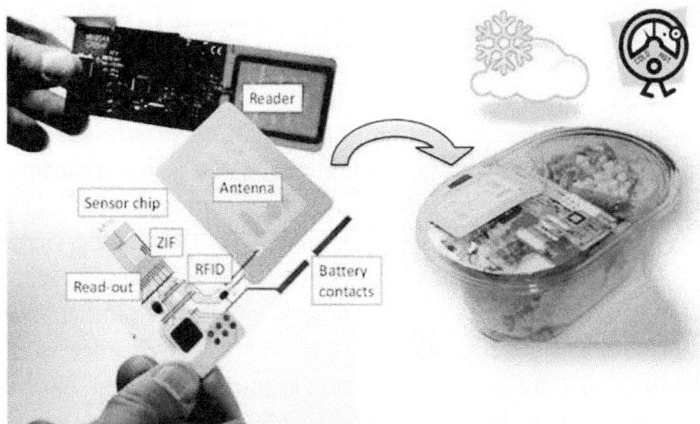

Figure 10.15 Inkjet-printed multisensor (temperature, humidity, gas) platform integrated on a screen-printed HF RFID label using the foil-to-foil approach realized within the frame of the European project FlexSmell.

With the integration of a thin-film battery, the passive smart sensing label becomes semi-active and allows data logging of the sensors parameters over time. The smart label data could be retrieved via a commercial reader.

Along with printed temperature and humidity sensors, printed gas sensors on polymeric foil for ammonia, and volatile organic compounds were integrated into the multisensor platform. These sensors were based on resistive and capacitive principles and were fabricated by inkjet printing on PET or PEN substrates. Silver nanoparticle ink is printed to form the electrodes of the sensors, the RTD resistor, and the heater followed by electroplating steps of nickel and gold. The devices were functionalized either by inkjet printing or vapor phase deposition polymerization (VDP) with different gas-sensitive layers: CAB, poly(acrylate acid) (PAA), and doped PANI films.

The RFID label was made by screen-printing silver on a PEN substrate to form an HF antenna and electrical tracks. A near-field communication (NFC)-compatible RFID, a μ-controller, and some electronic components were integrated onto the foil, enabling powering, communication, and sensor read-out. The multisensor platform was integrated on the smart RFID label using a foil-to-foil approach involving two possible techniques: through-foil vias (TFVs) filled with an isotropic conductive adhesive (ICA) and the use of an anisotropic conductive adhesive (ACA). The TFVs are formed by laser-etching an adhesive film laminated on the receiver substrate, which creates the electrically insulated vias, followed by their filling with the silver ICA using screen printing. The sensor foil was flip-chip, aligned, and pressed on the vias while curing. The ACA approach is simpler but less cost effective. The latter involves the stencil-printing of the more expensive ACA material, used for electrical connections as well as mechanical fixation. A controlled temperature ramp and applied pressure allow its activation.

In the case of medicine blisters, there is great interest in tracking the medicine usage of patients. At Holst Center in the Netherlands, as a proof-of-concept an intelligent blister was developed on the basis of heterogeneous integration. An image of such blister is given in Fig. 10.16.

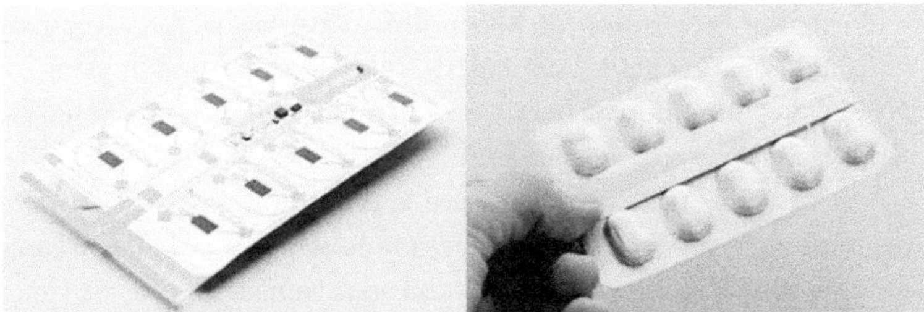

Figure 10.16 Printed smart blister used to track medicine compliance developed at Holst Center, the Netherlands. The system consists of an NFC-compliant radio chip, a microcontroller, and a printed resistor for each pill.

The blister requires a limited set of components to be integrated. The key components are logic, memory, timing, and radio connectivity, which are all addressed by conventional silicon-based components. Ideally all these components are integrated into a single-chip solution. In this case, however, due to availability, radio and logic functionality are divided over two chips and connected via a serial I^2C bus. In this case secure connectivity is provided through NFC to a reader or a suitable mobile phone. Power is provided by a small lithium battery embedded onto the footprint of the blister. Finally the electronic circuitry is realized by a combination of screen-printed conductive silver tracks and dielectric layers to provide bridging functionality. Essentially the blister relies on the mechanical breakage of conductive printed tracks during the process of pushing the pills through to the blister. The printed resistors connected in parallel, multiple pills are monitored continuously and simultaneously using an analog input, reducing the number of inputs and outputs required on the chip. Using PET foil base substrates in combination with screen-printed circuitry, it was found that substantial cost reduction could be realized, imperative for widespread usage of such technology in disposable products.

Polymeric and printed technologies are also of interest for the large-area manufacturing at low cost of microfluidics and lab-on-a-foil systems, for instance, using roll-to-roll processes, bringing new opportunities for the development of point-of-care and disposable diagnostic chips.

10.5 Conclusions and Future Perspectives

In this chapter, the fundamentals of sensors were first reviewed. This part was followed by the introduction of some of the main working principles with examples of implementation of sensing devices on foil and using printing. A sensor is a transducer that is used to sense/detect some characteristics of its environment. It converts physical or chemical measurands into a corresponding output, typically an electrical signal. Depending on the nature of the measured quantity, we can group sensors into three categories: chemical, physical, and biological sensors. The characteristics of a sensor depend on many different parameters such as the working principle used, materials and design involved, mode of operation, and read-out, which need to be optimized carefully.

OPE offers new opportunities in the field of sensing with unique characteristics such as flexibility, transparency, low form factor, low cost, biocompatibility, and disposability, among others. We have seen that a large set of working principles has been already applied to the development of printed and flexible sensors. The technology is particular suitable for the implementation of chemical and biological sensors. The additive and local character of their processing is beneficial for the multilayer functionalization of chemical and biological sensor systems. Significant progress, reaching the stage of commercialization, has been made on physical temperature, light, and pressure sensors. However, MEMS-based printed sensors are really at their infancy and major challenges remain in their successful deployment.

With the large volumes of produced sensors expected in the near future, we can expect that large-area manufacturing at low cost based on additive and printing processes will receive more and more attention. However, one should not neglect the different issues inherent to high-volume production. Sensor testing and calibration, and in some cases the specific encapsulation, which is strongly product and application dependent, are topics that need to be addressed with more effort. Stretchable and bioelectronics sensing technologies are raising a lot of interest for compliant design applications, but involving often an elastomeric substrate (e.g., PDMS), their realization based on large-area manufacturing and printing techniques remains problematic and would currently not really be advantageous according to the required system area and volume of production.

Finally, the sensor component alone is not a solution and its deployment will depend on how successful its integration into systems will be. For more computing-demanding applications, the first systems will be mainly based on a co-integration approach with silicon components. But if successful, all OPE systems, including electronics, memory, power supply, and communication, will enable a new generation of smart objects and large-area intelligent surfaces for applications we cannot even think about at the moment.

Exercises

10.1 List the ideal characteristics of a sensor in terms of:

- Response (V)
- Baseline (V_b)
- Response time (τ)
- Frequency bandwidth
- Working range ($V_{max} - V_{min}$)
- Sensitivity (S)
- Resolution

10.2 State the principal forms of signal energy and working principles that can be used to classify different types of sensors. For each type of signal energy, list the working principles that can be used and give two examples of sensors.

10.3 A thermistor is characterized by the following resistance versus temperature response:

The two points indicated as A and B in the graph have the following coordinates:

	A	*B*
Temperature (°C)	20	35
Resistance (kΩ)	49.07125	51.8575

Knowing that the resistor is 1 cm long and has a uniform cross-sectional area of 0.1 mm^2:

- Identify the metal the resistor is made of with the help of the following table:

Material	TCR @ 25°C (°C^{-1})
Nickel	0.005866
Gold	0.003715
Nichrome	0.00017

- Calculate the value of electrical resistivity measured at 50°C.

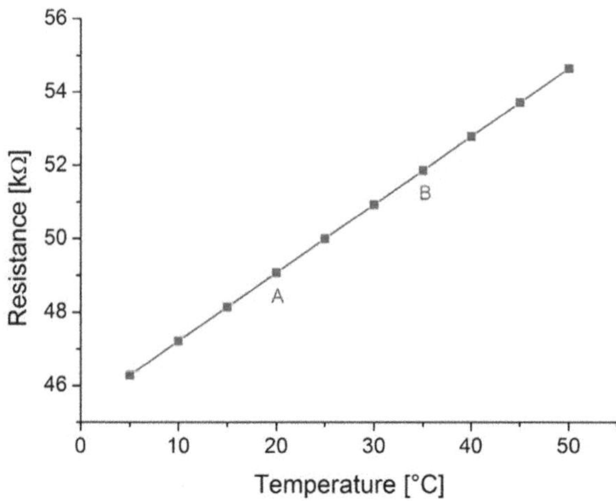

Figure E.10.1 Resistance vs. temperature response of the thermistor described in Ex. 10.3.

10.4 An aqueous solution of nickel at $T = 25°C$ is characterized by an activity coefficient of a_{Ni} of 10^{-5}. A potentiometer is placed inside the solution and is calibrated to produce an output of 0 V.

To this solution, an unknown quantity of solid silver is added. Silver dissolves and reacts with nickel ions according to the following reaction:

$$2Al(s) + 3Ni^{2+}(aq) \rightarrow 2Al^{3+}(aq) + 3Ni(s)$$

When equilibrium is reached, solid nickel precipitates on the bottom of the beaker and the potentiometer displays an output of +0.06 V.

- Calculate the activity of silver ions in the solution.
- Knowing that, for silver ions, $a_{Al} = [Al^{3+}]^2$, calculate the concentration of Al(III) ions dissolved in the solution.

10.5 An inkjet-printed n-type OFET ($\mu_n = 10$ cm^2/Vs, $C_{ox} = 3$ nF/cm^2, $W/L = 2000$) is used to detect a mechanical signal (pressure). When a force is applied to the transistor, its threshold voltage varies according to the following law:

$$V_{*th} = V_{th0} + 0.001 \times P$$

where V_{th0} is the threshold voltage when no pressure is applied and P is the applied pressure (expressed in kP).

The transistor is polarized in the linear regime by applying a V_{gs} = + 2 V and a first I_{ds}–V_{ds} curve is acquired; then, a positive pressure P_x is applied on the transistor and the same I_{ds}–V_{ds} curve is acquired (always keeping the V_{gs} at +2 V).

The two curves are shown in the following graph:

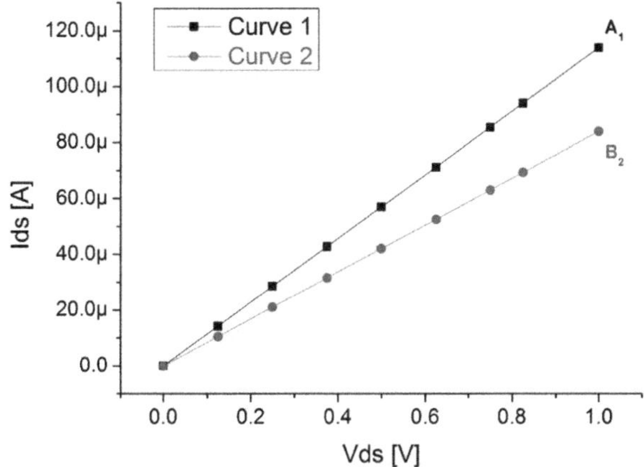

Figure E.10.2 I_{ds}–V_{ds} in the linear regime of the MOSFET described in Ex. 10.5. One curve is acquired when no pressure is applied on the device; the other is acquired when a pressure P_x is applied.

The two points $A1$ and $B2$ have the following coordinates:

	A1	*B2*
Voltage (V)	1	1
Current (μA)	114	84

- Calculate the threshold voltages corresponding to both curves.
- Which one of the two curves is acquired when a positive pressure P_x is applied? Justify your reply.
- Once you have identified which curve is acquired when pressure is applied, calculate the value of the applied pressure P_x.

Suggested Readings

Abe, K., et al. (2010). Inkjet-printed paperfluidic immuno-chemical sensing device. *Anal. Bioanal. Chem.*, **398**, pp. 885–893.

Amos, S. W. (2013). *Principles of Transistor Circuits: Introduction to the Design of Amplifiers, Receivers and Digital Circuits*, pp. 34–35, Elsevier, the Netherlands.

Bergveld, P. (1985). The impact of MOSFET-based sensors. *Sens. Actuators*, **8**, pp.109–127.

Busch, J. R., et al. (2012). Inkjet printed all-polymer flexural plate wave sensors. *IEEE 25th Int. Conf. Micro Electro Mech. Syst.*, pp. 571–574.

Carpenter, M. A., Mathur, S., and Kolmalov, A. (2012). *Metal Oxide Nanomaterials for Chemical Sensors*, p. xv, Springer Science & Business Media, Germany.

Ciferri, A. (2005). *Supramolecular Polymers*, 2nd Ed. (CRC Press, USA), pp. 527–529.

Courbat, J., et. al. (2011). Polymeric foil optical waveguide with inkjet printed gas sensitive film for colorimetric sensing. *Sens. Actuators B*, **160**, pp. 910–915.

Crowley, K., et al. (2008). An aqueous ammonia sensor based on an inkjet-printed polyaniline nanoparticle-modified electrode. *Analyst*, **133**, pp. 391–399.

Crowley, K., et al. (2010). Fabrication of polyaniline-based gas sensors using piezoelectric inkjet and screen printing for the detection of hydrogen sulphide. *IEEE Sens. J.*, **10**, pp. 1419–1426.

D'Amico, A., Di Natale, C., and Sarro, P. M. (2015). Ingredients for sensors science. *Sens. Actuators B*, **207**, pp. 1060–1068.

Facchetti, A., Yoon, M. H., and Marks, T. J. (2005). Gate dielectrics for organic field-effect transistors: new opportunities for organic electronics. *Adv. Mater.*, **17**, pp. 1705–1725.

Ferrari, V., et al. (2000). Multisensor array of mass microbalances for chemical detection based on resonant piezo-layers of screen-printed PZT. *Sens. Actuators B: Chem.*, **68**, pp. 81–87.

Focke, M., et al. (2010). Lab-on-a-foil: microfluidics on thin and flexible films. *Lab Chip*, **10**, pp. 1365–1386.

Glisson, T. H. (2011). *Introduction to Circuit Analysis and Design*, p. 32, Springer Science & Business Media, Germany.

Göpel, W., Hesse, J., and Zemel, J. N. (2008). *Sensors: A Comprehensive Survey*, 9 volumes, Wiley.

Harsanyi, G. (2010). *Sensors in Biomedical Applications: Fundamentals, Technology and Applications*, pp. 75–76, CRC Press, USA.

Hense, A., et al. (2010). Acceleration sensors based on polymer-electronic materials. *Procedia Eng.*, **5**, pp. 713–716.

Jung, W., et al. (2009). Acoustic wave-based NO_2 sensor: ink-jet printed active layer. *Sens. Actuators B: Chem.*, **141**, pp. 485–490.

Kanan, S. M., et al. (2009). Semiconducting metal oxide based sensors for selective gas pollutant detection. *Sensors*, **9**, pp. 8158–8196.

Klejwa, N., et al. (2011). A reel-to-reel compatible printed accelerometer. *Transducers*, pp. 699–702.

Laurberg Vig, A., et al. (2011). Roll-to-roll fabricated lab-on-a-chip devices. *J. Micromech. Microeng.*, **21**, p. 035006.

Lin, C., et al. (2011). Inkjet-printed organic field-effect transistor by using composite semiconductor material of carbon nanoparticles and poly(3-hexylthiophene). *J. Nanotech.*, **2011**, p. 142890.

Liu, C. C. (2000). *Electrochemical Sensors*, CRC Press, USA.

Ma, H., Jen, A. K. Y., and Dalton, L. R. (2002). Polymer-based optical waveguides: materials, processing, and devices. *Adv. Mater.*, **14**, pp. 1339–1365.

Manz, A., and Becker, H. (1999). *Microsystem Technology in Chemistry and Life Sciences*, p. 196, Springer Science & Business Media, Germany.

Mattana, G., et al. (2015). Polylactic acid as a biodegradable material for all-solution-processed organic electronic devices. *Org. Electron.*, **17**, pp. 77–86.

Mensing, J. P., et al. (2010). Inkjet-printed optical gas sensors based on metallo-phthalocyanine layers. *Int. Conf. Elect. Eng./Electron. Computer Telecommun. Inform. Technol.*, pp. 966–969.

Metzger, R. M. (2012). *The Physical Chemist's Toolbox*, pp. 541–548, Wiley & Sons, USA.

Molina-Lopez, F. (2014). *Inkjet-Printed Multisensor Platform on Flexible Substrates for Environmental Monitoring*. PhD thesis EPFL, no. 6191.

Molina-Lopez, F., Briand, D., and de Rooij, N. F. (2012). All additive inkjet printed humidity sensors on plastic substrate. *Sens. Actuators B: Chem.*, **166–167**, pp. 212–222.

Molina-Lopez, F., et al. (2013). Large-area compatible fabrication and encapsulation of inkjet-printed humidity sensors with integrated thermal compensation. *J. Micromech. Microeng.*, **23**, p. 025012.

Narakathu, B. B., et al. (2012). A novel fully printed and flexible capacitive pressure sensor. *IEEE Sens.*, pp. 1–4.

Noguchi, Y., Sekitani, T., and Someya, T. (2006). Organic-transistor-based flexible pressure sensors using ink-jet-printed electrodes and gate dielectric layers. *Appl. Phys. Lett.*, **89**, p. 253507.

Nomura, K., et al. (2004). Room-temperature fabrication of transparent flexible thin-film transistors using amorphous oxide semiconductors. *Nature,* **432**, pp. 488–492.

O'Toole, M., et al. (2009). Inkjet printed LED based pH chemical sensor for gas sensing. *Anal. Chim. Acta,* **652**, pp. 308–314.

Owens, R. M., and Malliaras, G. G. (2010). Organic electronics at the interface with biology. *MRS Bull.*, **35**(06), pp. 449–456.

Pal, B. P. (1992). *Fundamentals of Fibre Optics in Telecommunication and Sensor Systems*, p. 657, Bohem Press, Switzerland.

Regtien, P. P. L. (2012). *Sensors for Mechatronics*, p. 101, Elsevier, the Netherlands.

Ridley, B. A., Nivi, B., and Jacobson, J. M. (1999). All-inorganic field-effect transistor fabricated by printing. *Science,* **286**, pp. 746–749.

Rienstra, S. W., and Hirschberg, A. (2015). *An Introduction to Acoustics*, pp. 18–42, Eindhoven University of Technology.

Santos, J. L., and Farahi, F. (2015). *Handbook of Optical Sensors*, pp. 14–16, CRC Press, USA.

Sirringhaus, H. (2005). Device physics of solution-processed organic field-effect transistors. *Adv. Mater.,* **17**, pp. 2411–2425.

Sirringhaus, H., et al. (2000). High-resolution inkjet printing of all-polymer transistor circuit. *Science,* **290**, pp. 2123–2126.

Stetter, J. R., Penrose, W. R., and Yao, S. (2003). Sensors, chemical sensors, electrochemical sensors, and ECS. *J. Electrochem. Soc.*, **150**, p. S11.

Stumpel, J. E., et al. (2014). An optical sensor for volatile amines based on an inkjet-printed, hydrogen-bonded, cholesteric liquid crystalline film. *Adv. Opt. Mater.,* **2**, pp. 459–464.

Takeda, Y. (2012). *Ultrasonic Doppler Velocity Profiler for Fluid Flow*, pp. 23–24, Springer Science & Business Media, Germany.

Talapin, V. D., and Murray, C. B. (2005). PbSe nanocrystal solids for n- and p-channel thin film field-effect transistors. *Science,* **310**, pp. 86–89.

Thompson, B., and Yoon, H. (2013). Aerosol-printed strain sensor using PEDOT:PSS. *IEEE Sens. J.*, **13**, pp. 4256–4263.

Ukah, N. B., et al. (2011). Low-operating voltage and stable organic field-effect transistors with poly (methyl methacrylate) gate dielectric solution deposited from a high dipole moment solvent. *Appl. Phys. Lett.*, **99**, p. 243302.

Vásquez Quintero, A. (2015). *Low Temperature Integration Methods Using Conductive Adhesives and Dry Film Photoresist for Flexible Electronics.* PhD thesis EPFL, no. 6497.

Veres, J., et al. (2004). Gate insulators in organic field-effect transistors. *Chem. Mater.*, **16**, pp. 4543–4555.

Wei, Y., et al. (2013). Screen printing of a capacitive cantilever-based motion sensor on fabric using a novel sacrificial layer process for smart fabric applications. *Meas. Sci. Technol.*, **24**, p. 075104.

Whitaker, J. C. (1996). *The Electronics Handbook*, p. 2330, CRC Press, USA.

Wolfer, T., et al. (2014). Flexographic and inkjet printing of polymer optical waveguides for fully integrated sensor systems. *Procedia Techol.*, **14**, pp. 522–530.

Yang, L., Rida, A., and Tentzeris, M. M. (2009). *Design and Development of Radio Frequency Identification (RFID) and RFID-Enabled Sensors on Flexible Low Cost Substrates*, Morgan & Claypool, USA.

Chapter 11

Hybrid Printed Electronics

Marc Koetse, Edsger Smits, Erik Rubingh, Pit Teunissen, Roel Kusters, Robert Abbel, and Jeroen van den Brand
Integration Technologies for Flexible Systems, Holst Centre, High Tech Campus 31, 5656 AE, Eindhoven, The Netherlands
marc.koetse@tno.nl

Although many electronic functionalities can be realized by printed or organic electronics, short-term marketable products often require robust, reproducible, and nondisturbing technologies. In this chapter we show how hybrid electronics, a combination of printed circuitry, thin-film electronics, and classical silicon-based electronics, may give rise to new ways of realizing electronic products.

11.1 Introduction

Today, companies, both small and large, have become increasingly interested in printed electronics, often drawn by the promise of cheap manufacturing technologies and hence cheap products. They expect the technology to deliver at least the same quality at lower cost with improved functionality. Interviews conducted within

Organic and Printed Electronics: Fundamentals and Applications
Edited by Giovanni Nisato, Donald Lupo, and Simone Ganz

ISBN 978-981-4669-74-0 (Hardcover), 978-981-4669-75-7 (eBook)
www.panstanford.com

the framework of the COLAE project (FP7-ICT-2011-7, grant no. 288881, www.colae.eu) indicated that the typical application areas companies are interested in are lighting (organic light-emitting diode [OLED] lighting), packaging (sensors and lighting elements in labels for product and medical packaging), energy harvesting, and heating elements. This is very much in line with the type of projects that are run at research institutes and the customer requests we encounter. The main driver for companies to consider printed electronics in these application areas is foremost cost reduction, improved reliability, added functionality, and an improved form factor. In addition, they are looking for a technology that allows them to do something that could not be done before, thereby giving them a competitive edge.

The current research on flexible electronics focuses mainly on components (OLEDs, organic photovoltaics [OPVs], displays, for instance) rather than providing a complete solution. Such research is predominantly driven by combinations of academic groups such as universities, start-ups, and research institutes, as well as large-scale enterprises.

From an application point of view flexible electronics is a promising enabling technology to small and medium enterprises (SMEs) by simply improving upon existing products. A major challenge, however, for SME companies is their requirement of short-term integration of new technologies, typically within a year. This means that to succeed, sets of technologies are required that are relatively mature and reliable and can be introduced without disturbing the overall manufacturing methods and supply chain. At the same time they should bare the general advantages of printed electronics, namely low cost, simple manufacturing, thinness, light weight, etc.

Hybrid electronics, a mixture of state-of-the-art printed and thin-film electronics, combined with classic electronics, may offer this. The manufacturing can mostly be done on existing infrastructure and simultaneously it offers exciting possibilities for new and improved products. This chapter will explore the most common technologies used today to achieve functional products and prototypes based on hybrid printed electronics (HPE). This includes the printing of circuitry, the assembly of components, and some typical finishing techniques, where meaningful technologies that are less common, or more disruptive, in the manufacturing world will be discussed. This will be followed by a case study and a discussion of the design rules and the challenges that industry may face adopting this technology.

11.2 Basic Technologies

11.2.1 Basic Process Flow

At a high level, the basic process flow for HPE consists of three important groups of processes, as depicted in Fig. 11.1. After preparation of the substrates, the manufacturing typically starts with printing electronic circuitry on a flexible substrate. This is followed by the assembly of passive and active components, and finally the product is finished, for instance, encapsulated or packaged.

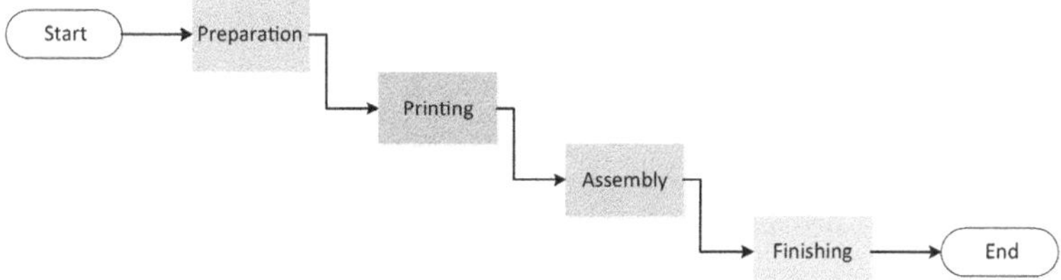

Figure 11.1 Schematic representation of the sets of processes typical for hybrid printed electronics.

This process flow is very generic. It is based on a number of manufacturing technologies that have been available for several years and are mature. What is new is the combination of printing circuitry and classical components in combination with the fact that all processes have to be performed at relatively low temperatures. For instance, conventional soldering is typically not compatible with low temperatures. Other challenges lie in the reliability of the systems and in some cases the application of the processes in a roll-to-roll process flow. The remainder of this section will describe the most important processes in more detail.

Figure 11.2 shows the generic process flow in more detail. As can be seen, the number of process steps is significant, and a number of them may have to be repeated in order to obtain a product with the required functionality. The blocks under the processes show the materials and components that are being used per step. In the following sections we will explain these process steps in more detail.

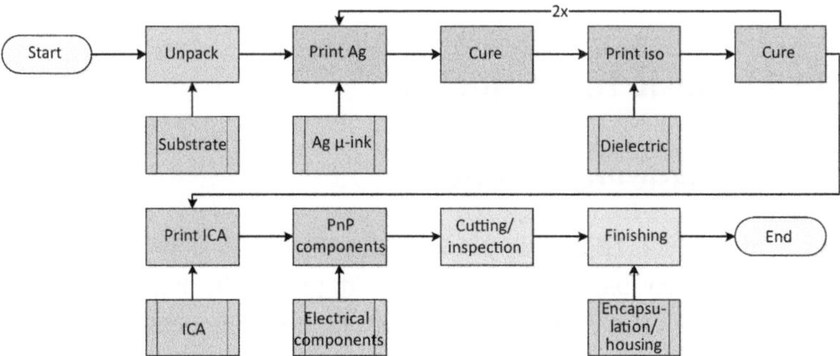

Figure 11.2 Schematic representation of a generic process flow for a hybrid printed electronics product.

11.3 Printing Circuitry

The manufacturing of membrane switches and touch panels for graphical user interfaces is probably the most mature industry using printing electronic circuitry as its base technology. Circuitry and basic sensors such as capacitive touch sensors are routinely printed. The most important difference, with respect to HPE, is the typical line width (>250 μm in industry vs. <100 μm for HPE). This is mainly due to the fact that the (semi-)products made by printing today do not, or hardly, require any component integration. From a cost point of view, printed circuitry is in fact most interesting for exactly those products that require large areas and a limited number of components (for a more detailed discussion on this topic see Section 11.6.2). Screen printing is probably the most important and widely used technique to manufacture printed circuitry. Alternatively, inkjet printing is a very versatile technology, especially if small single-layered batches are required, for instance, for prototyping. Gravure and flexographic printing are also used, especially for very-high-throughput single-layered applications that do not require high conductivity, for instance, ultrahigh-frequency radio-frequency identification (UHF RFID) antennas.

11.3.1 Screen Printing

As stated above, screen printing is probably the most widely used technique to manufacture printed circuitry in industry. It is a versatile and the foremost cheap technique that allows for the deposition of relatively thick layers, depending on the type of ink between 1 μm and 15 μm per print step. The types of inks that can be applied are rather broad as long as the viscosity is not too low (100–80,000 mPa·s).

Multilayered printing is very common. In the case of printed circuitry, this typically involves the printing of conductive and isolating materials. A stack containing five or more conductive layers can be realized relatively easily, provided the overlay accuracy is high. A 10 μm overlay accuracy can be achieved in a research environment using optimal screens and substrates. In a manufacturing environment an overlay of 30 μm is more realistic. Features as small as 40 μm have been obtained by using a carefully tweaked process with specific surface treatments, inks, and screens [3].

Screen printing can be applied in two ways. Flatbed (sheet-to-sheet) printing is typically used for relatively small batches in a semi-automatic way. Up to 3000 sheets per hour can be printed this way (Fig. 11.3). Alternatively, screen printing may be applied in a roll-to-roll fashion (Fig. 11.4). Instead of a flat screen, a rotary screen is used. Such a setup runs at a speed of at least 4 m/min, the maximum speed being dependent on the capacity of the drying and sintering equipment. Speeds up to 100 m/min have been reported in industrial environments.

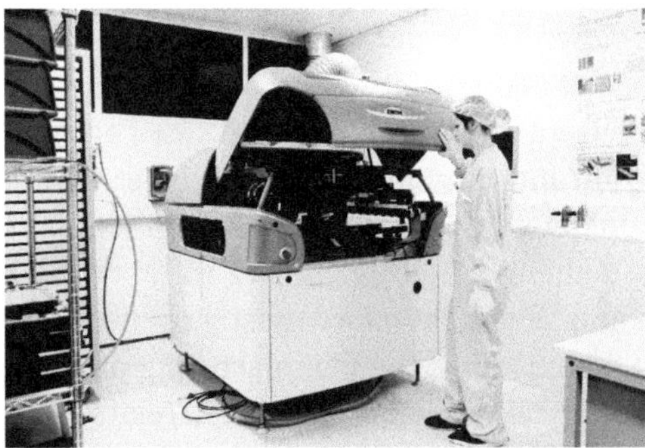

Figure 11.3 Flatbed screen printer with an optical alignment system at the Holst Centre.

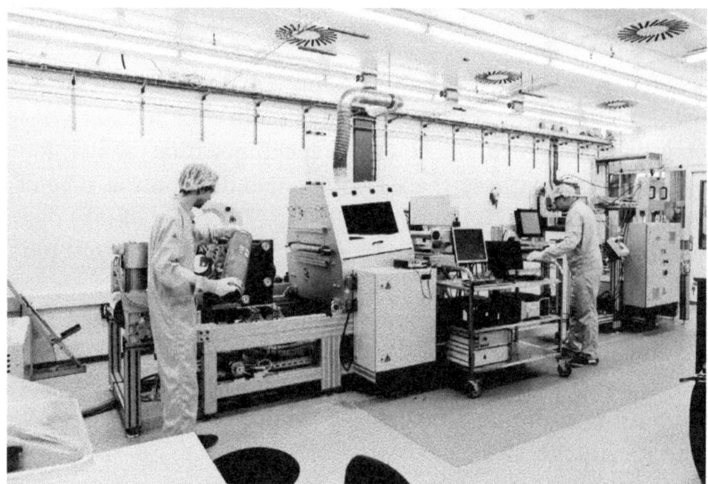

Figure 11.4 Roll-to-roll printing line for the printing of printed electronic circuits at the Holst Centre. The line consists of a rotary screen-printing unit on the left and inkjet-printing units are located on the right. The photonic curing unit is located in the middle.

11.3.2 Inkjet Printing and Flexogravure Printing

Inkjet printing is another important technology for manufacturing electronic circuitry on cheap foil substrates. The results of this method are very similar to circuitry made with flexogravure printing. In both cases inks with nanoparticles are used and the resulting metal lines have similar conductivity and thickness. Depending on the ink and the substrate, a line thickness up to 1 μm can be reached. Line widths down to 20 μm and even smaller have been shown. The advantage of inkjet printing is that it is a digital method, which makes it very useful for prototyping and small batches. Applications that require thin and very smooth conductive structures may benefit from either inkjet or flexogravure printing. The latter technology does require some tooling (the gravure roll) but can be run at very high speeds and seems therefore mostly useful for very large area applications, for example, the manufacturing of bus bars in OLEDs and photovoltaic devices and also the circuitry and antenna for RFID tags. Multilayered printing with these techniques may be problematic because the structures that are deposited are typically pure silver and good adhesion of these structures on substrates and adhesion of subsequent layers on the metal surface are not evident.

In the past, inkjet printing has been investigated intensively for use in the manufacturing of integrated circuits (ICs), thin-film transistors, etc. The focus has, however, recently shifted more toward thin-film technologies in combination with large-area coating (slot-die coating) for the manufacturing of display backplanes and simple IC applications such as RFID chips, the main motivation being that printing organic semiconductors with sufficient reliability, yield, and stability to realize complex circuits were found to be challenging. Instead, state-of-the-art display technologies such as photolithography and oxide semiconductors using a mix of conventional fabrication methods are used [5]. Ultimately such technology, while promising lower costs, will be challenging to upscale on a short term and will require, contrary to HPE, large investments costs.

11.3.3 Substrates

The most important advantage of using print technology, apart from it being an additive technology, is the fact that it is very compatible with cheap substrates. The most important requirement for the substrate is that it is dimensionally stable under thermal cycling, mechanical stress, and changes in environmental conditions. The success of multilayered printing and the various assembly steps rely strongly on the overlay precision of the layers. Therefore mechanical relaxations (e.g., irreversible shrinkage) after the substrate's exposure to elevated temperatures should be minimized. It was found that polyesters are particularly suited as substrates because they combine a relatively high glass transition temperature (T_g) and can be heat-stabilized, minimizing mechanical relaxations. In addition the material is optically transparent, which is useful for numerous applications. Polyesters are medium priced: much cheaper than the workhorse in flexible circuit boards (flexible printed circuit [FPC]), polyimide, although they are more expensive than polyolefin foils. The latter materials are too mechanically unstable to serve as useful substrates for precise printing. In the case of HPE an overlay accuracy of about 50 μm is sufficient for most applications. Note that for thin-film electronics, and especially thin-film transistors (TFTs), manufacturing with overlay accuracies on the order of a few microns is required. Still heat-stabilized polyethylene naphthalene (PEN) foils have been demonstrated to be well-suited substrates for such applications if process temperatures are kept below 150°C.

Figure 11.5 shows a diagram with the temperature behavior of heat-stabilized PEN. As expected for a semicrystalline material, the deformation is strongly dependent on the direction in which the foil was stretched during manufacturing. It also shows that even in a heat-stabilized product, the shrinkage can be significant.

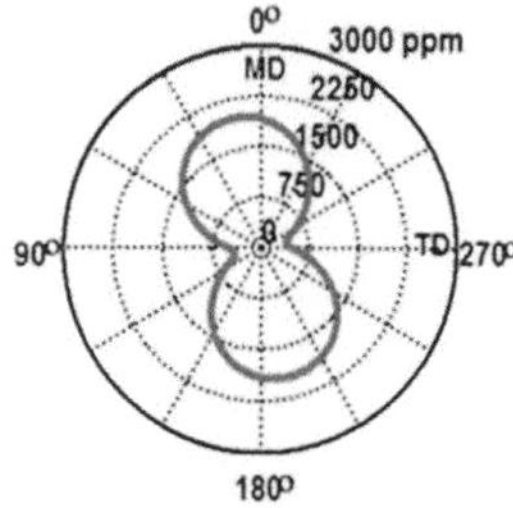

Figure 11.5 Polar plot of shrinkage behavior of PEN taken after annealing for 10 min at 200°C.

Polyethylene terephthalate (PET), a cheaper alternative to PEN, has a similar behavior and is more commonly used as a low-cost substrate. Other substrates, such as polyimide or polyether etherketone (PEEK) are occasionally used as well, especially for high-end applications where high temperatures (>350°C) are needed and a low thermal coefficient of expansion, for instance, in combinations with traditional thin-film technologies. The use of paper as a substrate for HPE has also been quite successful since it is one of the most common substrates used for (graphic) printing, facilitating the adaptation by the industry, especially print houses (Fig. 11.6).

11.3.4 Inks

In most cases silver inks are used to manufacture printed circuitry. It is the industry standard since the early onset of printing conductive materials. Silver is relatively noble and therefore stable inks can be obtained using a dispersion of small metal particles. Because of the increasing price for bulk silver, quite some research effort is ongoing for developing cheaper alternatives, some of which are emerging in industry. Examples of such inks include dispersions with silver-coated copper particles, Cu particles, Ni particles, and metal complexes.

Figure 11.6 Paper substrate with rotary screen-printed Ag ink, cured with photonic curing (http://www.ropas-project.eu/).

Silver nanoparticle inks are available for a wide range of printing techniques, including screen printing, inkjet printing, and flexographic printing. The inks have, after curing, a high specific conductivity (up to 20% of bulk silver). After curing the printed structures consists of almost pure metal, which may give issues with reliability. For instance, adhesion to the substrate or overlaying layers may be problematic. Also the remaining layers are relatively thin, typically on the order of 1 μm.

Micron size–particle inks on the other hand still contain a substantial amount of binder. The absolute conductivity is therefore lower, up to 10% of bulk silver, but the adhesion to the substrate and overlaying layers is much better. The layer thickness is typically 5 to 10 μm and is strongly dependent on the ink and process settings.

One of the strong points of printing a circuitry is the ability to incorporate functional components in the circuit in the same process. Examples of such components include resistors (using carbon inks) (Fig. 11.7), capacitors (combination of metal and dielectrics), capacitive sliders, and resistive sensors (using chemically sensitive conductive inks).

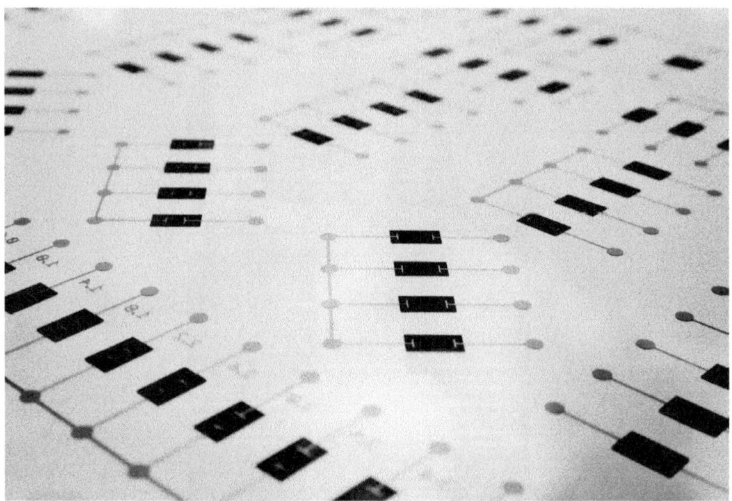

Figure 11.7 Printed Ag ink circuit in combination with carbon-based resistors.

11.3.5 Curing and Sintering

To render metal inks conductive, thermal treatment is required. Although the method and especially the equipment are uniform, two distinct processes are applied for this. Nanoparticle inks are sintered, fusing the particles to form a continuous metal phase. For bulk silver a temperature above the melting temperature is needed, which is not compatible with the foil substrates used for HPE. For nanoparticles, however, the high surface-to-volume ratio allows for a considerable lower temperature to be used. Typically the sintering temperature is around 130°C for about 15 min.

Micron-size particles are already too big for this short cut. Inks that contain these larger particles are therefore cured, and the solvent and the binder are partly removed under elevated temperature, yielding a matrix of cured binder where the metal particles are in contact with each other but not sintered. This explains the improved adhesion to the substrate as well as the lower conductivity.

Although curing and sintering with elevated temperatures are still the most common way, photonic curing and sintering are becoming more widely available. Because this process is extremely fast (seconds rather than minutes), it is especially useful for roll-to-roll processes. Photonic sintering makes use of the fact that (visible) light is absorbed by the noncured ink but not by the substrate. This allows to selectively heat up the ink to temperatures that exceed the processing temperature ceiling of the substrate. Figure 11.8 shows the temperature and resistance profile of a conductive ink during a flash sequence. Also it shows how the cross section of the line changes during the process. During the first second the solvent is evaporated; after 1.5 s, the silver particles are aggregated or fused together, dramatically increasing the electrical conductivity of the printed structure. The sintering is complete after 2.5 s. In the case of nanoparticle inks, the particles would be fused together (sintering), while micron-size-particle inks, rather, aggregate.

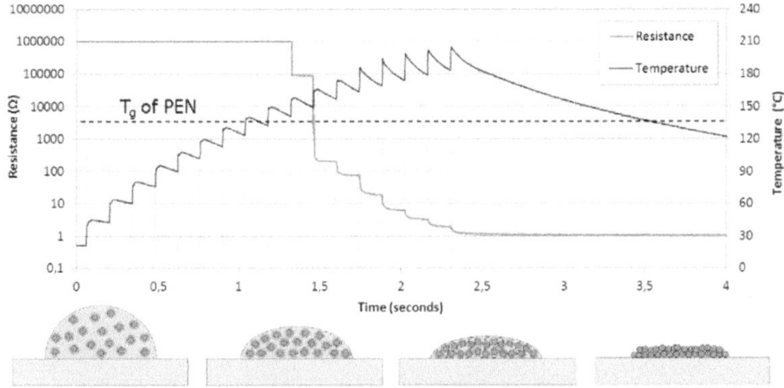

Figure 11.8 Temperature and resistance profile of a printed silver ink during a photonic flash sequence. The bottom shows a schematic representation of the line cross section.

11.3.6 Alternatives to Direct Printing

Although the printing circuitry using silver inks conductive is very attractive and certainly useful for products that require large-area substrates with little functional components, functionalities that require fine features (<50 μ) or alternative metals (Cu, Ni, Au) are probably better made with alternative deposition techniques. For instance, fine interdigitated electrodes for resistive sensors or transistors can be obtained from classical thin-film deposition techniques such as sputtering in combination with photolithography and etching. Alternatively processes based on printing a seed layer followed by electrodeless plating have also been employed.

11.4 Assembly

Assembly is a combination of processes that lead to the interconnection of electronic components on a printed circuit. In the overall process flow for HPE, it is the next set of technologies needed to achieve a functional hybrid printed system. Assembly aims to bond passive and active components to the printed circuit in a reliable and robust way (Fig. 11.9).

Figure 11.9 Assembled 0402 size resistors using an isotropic conductive adhesive.

Classically, components are soldered to printed circuit boards (PCBs) and FPCs. The typical process steps are stencil printing of the solder paste, picking and placing of the components, and reflowing of the solder, providing a metallic joint. The processes have been optimized such that today high reliability and speed can be combined with very high yields.

The low-cost substrates that are normally used for HPE will not maintain their mechanical stability during a solder reflow step as a result of the high temperatures required to melt a typical solder eutectic such as SnAgCu. Therefore, electrically conductive adhesives are used to adhere the components to the circuit, providing both electrical and mechanical interconnection. These adhesives can be cured at relatively low temperatures, typically <130°C, making them particularly suited for HPE. Today many interesting developments are going on in the field of adhesives, but also progress is made in developing low-temperature-reflowing solders such as SnBi. This chapter, however, will focus on conductive adhesive systems. For conductive interconnection two types of adhesives are available, isotropic conductive adhesives (ICAs) and anisotropic conductive adhesives (ACAs). Additional mechanical stability is often realized using a nonconductive adhesive (NCA), also known as underfill or glob top, which is much cheaper and better suited for its purpose. The adhesives can be applied in different ways, depending on the type of adhesive but also the substrate and the size of the required dots.

11.4.1 Isotropic Conductive Adhesive

As the name indicates, an ICA conducts, after curing, in all directions. Compositionally it is very similar to screen-printing pastes containing a high percentage of relatively big silver particles in a polymer matrix. Because of this, it must be applied with high precision to avoid shorting.

The material is usually applied by means of stencil- or large-mesh screen printing, similar to how solder paste is generally applied. This can be done with a screen printer that has sufficient overlay accuracy. It allows to deposit spots of the ICA down to at least 0.1 × 0.2 mm, the single-contact-pad footprint of the smallest commercially available component to date, a 01005 passive component. Recently we demonstrated that this technique could be used to interconnect bare light-emitting diode (LED) chips with contact pads of 80 μm. Another method to achieve small spots is pin transfer, where a small dot of adhesive is applied to the substrate by picking it up with a pin, as the name of the technique suggests. This method allows for a minimum drop size of 80 μm. It should be noted that the latter technology is still in the development phase.

Alternatively, the adhesive may be applied using a dispenser or by jetting. Both methods are useful for drops that exceed 250–300 μm. Jetting is much faster than dispensing but requires much more expensive equipment. Furthermore, adhesives need to be optimized for these processes to avoid clogging of the jetting nozzles.

For fine pitch jetting, laser-induced forward transfer (LIFT) is an interesting option. Using this technique, spot sizes down to 75 μm could be realized. With this latter technology fine pitch flip-chip applications with ICAs come into the range of possibilities next to the assembly of suface mount technology (SMT) discrete components.

In general ICA materials cure at temperatures ranging from 70°C to well above 160°C in regular ovens for a few minutes to hours. This implies that bonding and curing are separate processes and have a certain impact on the applications, for example, special handling of preassembled products or very accurate component topology needed.

11.4.2 Anisotropic Conductive Adhesives

Contrary to ICAs, ACAs have the ability to conduct in only the vertical direction. An ACA allows for electrical interconnections between components and the substrates where the contact pads are very close together, beyond the patterning ability. Anisotropic conducting materials are widely used in the bonding of flex cables (e.g., displays). The materials are also applied for components that require small pitches, for instance, bare dies, especially more complex ones such as microcontrollers or application-specific integrated circuits (ASICs) that require landing pads with a pitch of <40 μm. This process, where the chip is bonded upside down to the substrate, is often referred to as flip-chip die bonding, instead of face up as is the case in a classic wire-bonded chip package.

With current state-of-the-art industrial screen printing, stencil printing, and dispensing technologies of ICAs, the spot sizes and gaps that are required are not easily achieved. ACAs, however, are perfectly suited since the application of the adhesive does not have to be that precise and the overlay of the pick-and-place machine determines the alignment accuracy, not the adhesive.

ACAs are adhesives with a very low particle fill factor (~10%). The particles that give conductivity are mostly metal (Au)-coated polymer spheres or micron-size Ag particles. When trapped between the bump on the chip and the landing pad, one or a few of these particles will provide the electrical interconnect (*z* direction). There are, however, too little particles in the adhesive to also form a conductive path along the substrate (*xy* direction), as shown in Fig. 11.10.

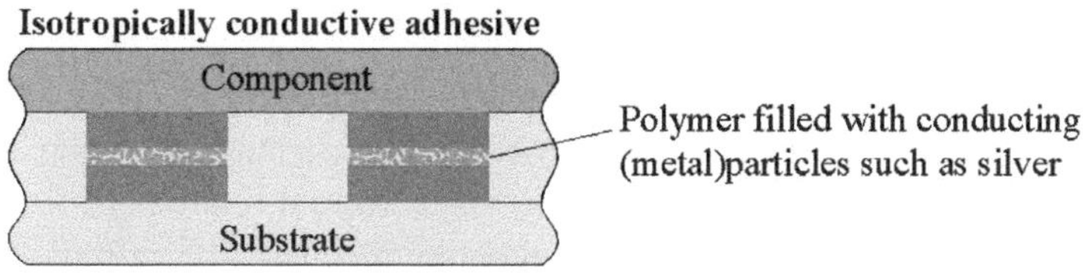

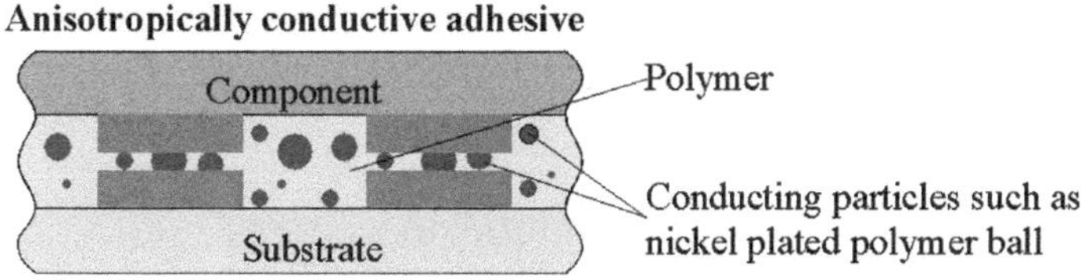

Figure 11.10 Schematic representation of how an ICA (top) and an ACA work.

The material can be obtained in two forms, as a liquid or as a tape-like film (anisotropic conductive film [ACF]). In liquid form it can be applied using the techniques discussed for ICAs; the film is typically transferred but requires prebonding at low temperature.

Different from ICAs, ACAs and ACFs need to be cured under pressure to keep the conductive particles under tension to form the conductive path. This has an impact on the processing equipment, making it more complex and expensive. Curing temperature is commonly above 150°C; however, bonding time ranges from a few milliseconds to minutes.

11.4.3 Pick and Place

After application of the adhesive, the components are placed on the substrate. This has to occur with good accuracy in order to ensure proper functionality. Standard pick-and-place machines can be used for this process. These machines are mostly equipped for nonflexible substrates, so foil-based substrates are normally laminated to temporary carriers. This is a common technique that is applied widely in the field of flexible electronics. The state-of-the-art equipment allows to assemble the smallest SMT or bare die (e.g., LEDs, RFID chips) components at very high throughputs, up to 15,000 units per hour (uph).

For flip chip of microprocessors generally throughput can go up to 1500 uph. Recently interest is growing in pick-and-place equipment and processes for roll-to-roll production. Some equipment is available, especially for RFID inlay and LED strip manufacturing. The process is characterized by the use of a limited number of simple components with a limited amount of bonding pads (typically 2 or 3) and the lack of high precision. In general, most machines have a web width limited to 10 cm, although machines capable of handling wider webs have been demonstrated. It should be noted that current web solutions are predominantly stop and go rather than continuous.

11.4.4 Assembly of Components in Systems

The use of a combination of the above-mentioned techniques allows for the manufacturing of very thin and very flexible but highly complex systems. Figure 11.11 shows a picture of a hybrid printed system. It consists of a multilayered print on PEN: two conductive layers with bare thinned dies, a microcontroller, and a near-field communication (NFC) radio, assembled using an ACA. The components (01005) are interconnected using an ICA. Because the ICA does not give enough mechanical stability a so-called underfill material is used to provide the necessary bonding power. This material is typically dispensed or jetted. Figure 11.12 below shows a transparent glob-top material that is used to protect the chips from shear forces.

11.4.5 Foil-to-Foil Lamination

A process typical for HPE is foil-to-foil lamination. It mostly is concerned with the integration of two functional foils that have been manufactured using different technologies. An example of such a system is shown in Fig. 11.13. Here a small foil with lithographically made sensors (ultrasmall lines and gap) is connected to a circuit. Despite the large line width and pitch of the connectors an ACF is used to create the interconnection. The main reason for this is that the ACF can also provide the mechanical interconnection. If an ICA was used, both the ICA and an NCA would have to be applied in a patterned way. This is rather difficult and not an easy strategy and therefore an ACF is the method of choice.

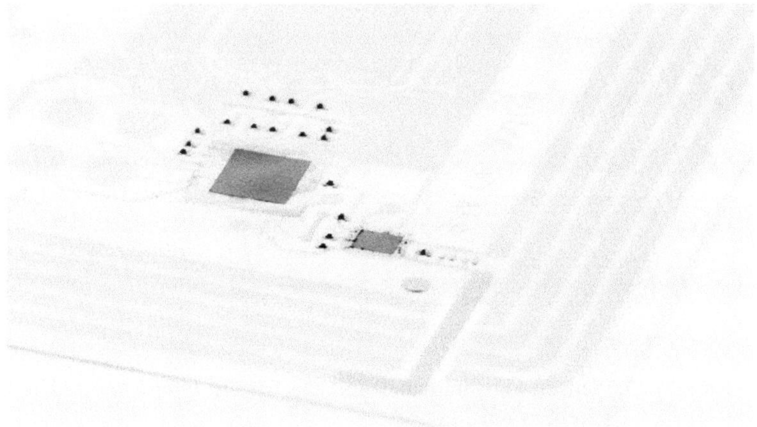

Figure 11.11 Close-up photo of a hybrid printed electronics system.

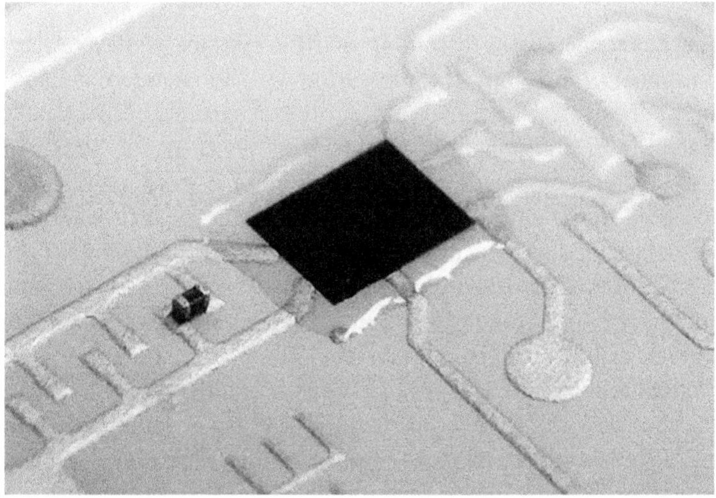

Figure 11.12 Close-up picture of a glob-top-protected bare die.

With the large pitches, as exemplified by Fig. 11.13, overlay alignment is not the main issue. There are examples, however, where much higher accuracies are required and an alternative approach has been recently shown. By combining fluidic self-assembly with laser-defined trenches, capillary forces could be used to accurately place foils within a range of a few micrometers.

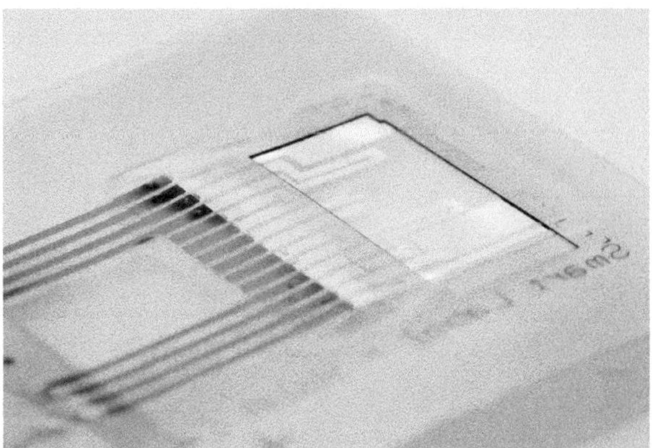

Figure 11.13 Sensor foil laminated and electrically interconnected to a circuitry foil.

11.4.6 Reliability

The challenge in the assembly of HPE systems is the variety in components and chips in combination with nonstandard substrates and adhesives. The reliability of these systems, especially the assembly of components, should be as good, if not better than existing technologies, mainly solder joints if products are to be accepted in the market. For many applications, where reliability is less an issue, the requirements can be met most of the time. Typical methods to validate the reliability are tests for the electrical behavior under elevated temperature/humidity, temperature shock, flexing, and shear forces. These tests are very comparable to those used for classical electronics, and specifications are similar as well.

The most common failure modes that are encountered include the delamination of the inks, both from the substrate as from the interfaces in multilayered systems. Another important failure mode is the release of components from the substrate and circuitry.

Figure 11.14 shows the results of a bending test for 01005 (top) and 0402 (bottom) components on a printed Ag circuit. The substrates were bent 1000 times over a rod ranging from 25 to 5 mm in diameter. The substrate, ink, and ICA were the same in all cases.

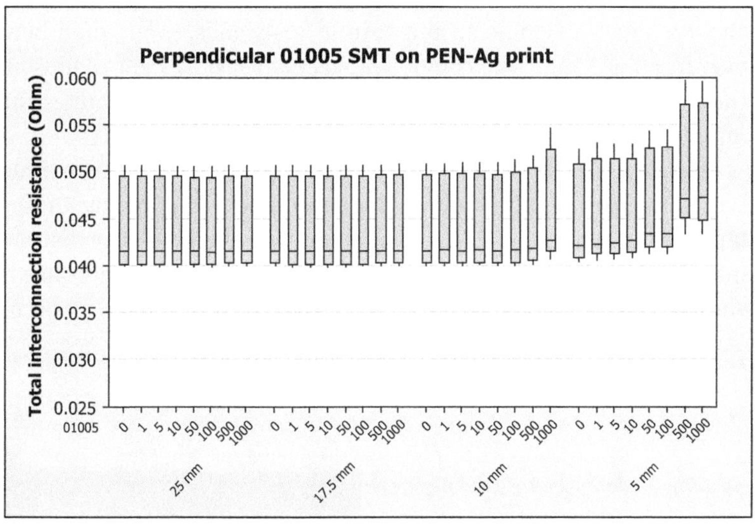

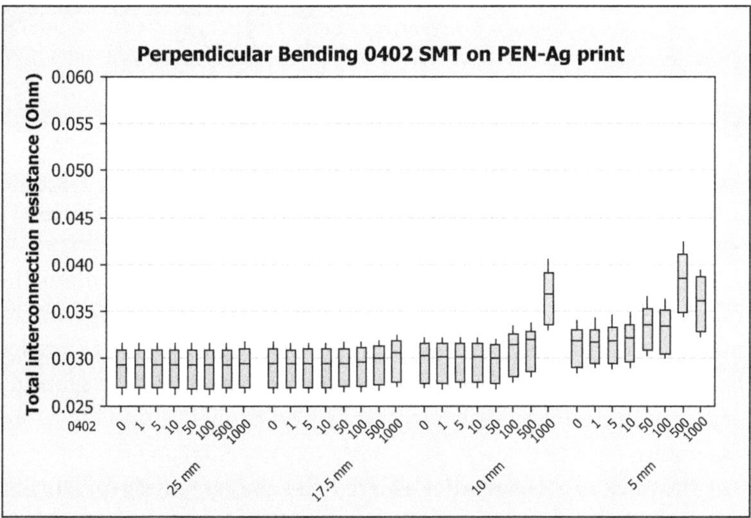

Figure 11.14 Resistance development of an adhesive interconnect between 01005 components (top) or 0402 components (bottom) and printed Ag circuitry.

As expected, the reliability is not only dependent on the inks and other materials that are used but also on the size of the components used. The initial resistance of the 01005 components is slightly higher because the pad size of these components is significantly smaller than for 0402 components. By bending over 10 mm the 0402 components start to fail, whereas the 01005 components only show a small insignificant increase in resistance, even when bent over a rod of 5 mm. The components actually act as rigid islands on the flexible substrate and the size of these islands determines the minimum bending radius.

Figure 11.15 shows pictures of substrates after a shear test. It can be clearly seen that the failure mode can differ significantly. On the left side, Ag printed on PEN, the assembly failed on the ink substrate interface. On the right side, failure through the adhesive interface is shown. Depending on the inks and adhesives used, failure can occur at any interface.

Figure 11.15 Results of a shear test showing two possible failure modes. Left: Delamination of the silver track from the PEN substrate. Right: Delamination of the component from the adhesive.

Optimizing the materials and processes to improve upon the reliability of HPE systems is key in getting the technology to the market. As discussed previously, important improvements can also be obtained by making use of well-known solutions such as glob-top coatings, underfill adhesive materials, and proper packaging of the assembly.

The thickness of bare dies also reduces the sensitivity for shear. If chips are thinned and treated for stress relief, they may become very flexible and bending multiple times over a radius of 10 mm is very feasible. Mechanical and electrical properties of ultrathin chips and flexible electronics assemblies during bending have been previously demonstrated.

11.5 Finishing

The finishing for HPE encompasses basically all steps that are taken after assembly of electronic components. This includes packaging, making of the desired shapes, and the assembly into the final product.

For the packaging and protection of the electronic system, a number of techniques have been shown. Lamination with thermoplastic materials such as thermoplastic polyurethanes (TPUs) (Fig. 11.16) gives thin flexible systems that can easily be further processed in final products. The encapsulation provides both mechanical stability and adequate protection against moisture and handling. Using vacuum lamination, lamination of wearable electronics, silicones are common encapsulating materials. They are hydrophobic and very flexible, making them suited for wearables. Mechanically silicones are somewhat less stable as they tend to rupture upon repetitive strain. Silicones are typically casted or molded, which is compatible with HPE as long as the temperature is kept below approximately 130°C.

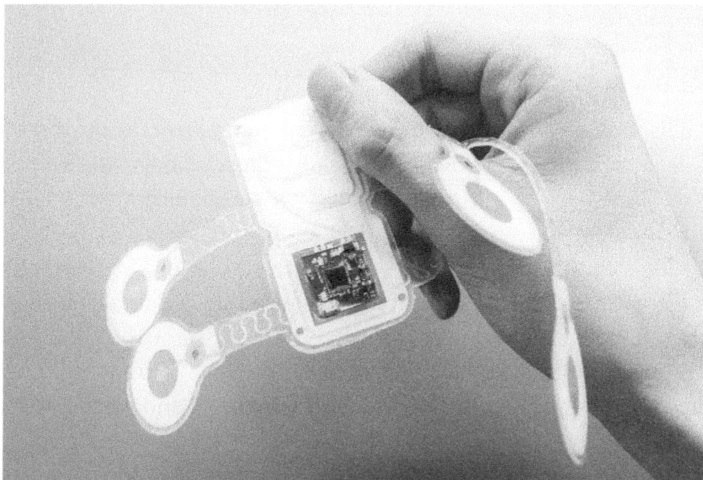

Figure 11.16 Hybrid printed electronics health patch encapsulated with thermoplastic polyurethane.

Finishing also includes the processes that shape the product. In the industry, die cutting is the method of choice for small products at a large scale. Laser cutting is often used for large products because of the scalability and thus price of the tooling required for die cutting. Laser cutting is also more frequently used for small batch manufacturing as it relies on a digital method, making it easy to adjust the patterns. In most cases a CO_2 laser is used for cutting as this is the cheapest type of laser. It works well for many substrates; however, it is not compatible with all materials, for instance, polyimides, leaving large heat-affected zones. A laser can be mounted on an *xy* stage, determining the accuracy of the system. For high-end applications, solid-state lasers have become more prevalent. While solid-state lasers, such as Nd:YAG are generally more expensive, they allow for better control of the energy dosage and line width and they are therefore very suitable for sensitive materials and fine features. When such a laser is combined with a galvano-scanning system, the speed and accuracy of the system can be very high. Recently fiber lasers have begun to replace both CO_2 and Nd:YAG lasers due to their improved performance, lower cost of ownership, and reliability.

11.6 Applying the Technology: A Case Study

11.6.1 Flexsmell Label

Labels with electronic functionalities present one of the most sought after applications for HPE. Typically, these labels are required have some sort of RFID functionality in combination with sensing functionality.

Classically, the antennas on the most simple and cheap RFID labels are not printed. They are made either by die cut or by etching metal foils. For more advanced labels, often using UHF frequencies, printing, especially R2R inkjet or flexo-gravure is frequently used. The chips that have to be placed are very simple and small and have typically two bond pads. The required pick-and-place accuracy is therefore relatively small, and since only one conductive layer is used, overlay accuracy is not an issue. These labels are made high speed.

More elaborated labels, having electronic functionalities, such as sensors, monitoring capabilities, and lighting elements, are much more complex to fabricate. Within the framework of the European project Flexsmell (FP7 Marie-Curie Initial Training Network [ITN], grant no. 238454)) strategies were set out to develop a smart olfactic label that combines most the technologies that have been discussed in this chapter.

Figure 11.17 shows the finished label in closed and open form. The label is required to communicate with the outside world over an NFC protocol. The tag is designed to contain the necessary hardware (chips) enabling the read-out of capacitive and resistive sensors possible. Finally, the sensor label should be able to work in monitoring mode, that is, it should have sufficient memory and battery power for autonomous functioning.

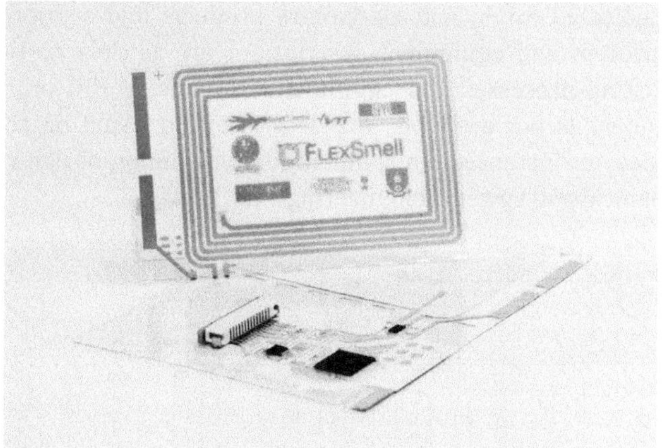

Figure 11.17 Flexsmell label with assembled components.

Screen printing was used as the basic printing technology because of the required resistance of the antenna (suited for communication at 13.56 MHz, so primarily inductive coupling). The required functionality made that a number of chips had to be integrated, and with these chips, numerous passive components were necessary as well. The most important active components are an EEprom/NFC chip, a capacitive front end, and a microcontroller. A detailed description of the manufacturing of the foil-based sensor chip can be found in Ref. [8].

11.6.2 Total Cost of Ownership

Because the above-described label is a good example of a typical HPE product, the total cost of ownership of the product was calculated as part of a feasibility study within the framework of the COLAE project. This calculation not only gives a good impression of the total cost of manufacturing but also provides valuable information on the cost buildup and shows where important improvements can be made, for example, what the expensive process steps are or where it may be helpful to look at the used equipment or components.

The model that is used for the calculations is based on the process flow. For every process step, we calculate the fixed cost, the operational cost, and the cost of the materials and components used (Fig. 11.18). The database currently comprises over 150 different kinds of equipment and more than 250 processes, most of which are dedicated for hybrid electronics products and semiproducts. The process and equipment descriptions are as close as possible to existing processes and equipment in industry. If a process or equipment is not available, the description is based on research facilities, for instance, in a combination of common design rules in mechanical and systems engineering.

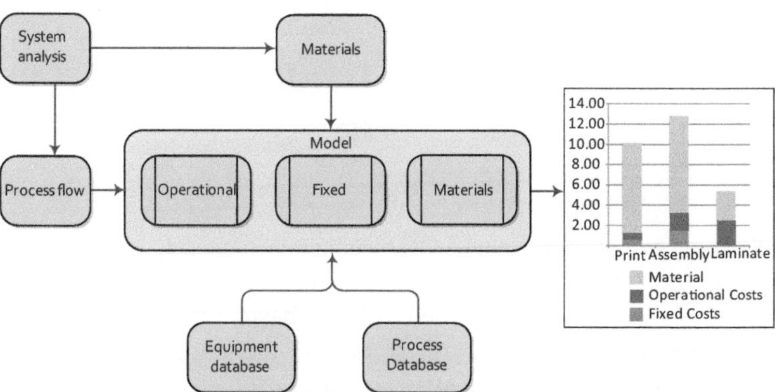

Figure 11.18 Schematic representation of the model used to calculate the total cost of ownership for hybrid printed electronic (semi-) products.

For the Flexsmell label we used the system design as described above. The bill of materials was determined on the basis of the used components and materials using prices for a batch size of 100,000 good produced units (gpu). The size of the label is 70 × 70 mm in open form. It will be folded in the final stage of the product (see Fig. 11.17).

The process flow used for this case is very similar to the generic process flow, depicted in Fig. 11.2. Because of the amount of products, the total foil area is only 490 m^2, and sheet-to-sheet processing is chosen. Although sheet-to-sheet processing is much slower than roll to roll, the equipment and tooling cost are much higher for the latter and the current batch size is not large enough. Also, pick-and-place equipment suited for such complex systems is not readily available. This makes that sheet to sheet is a much more likely strategy to be used for this application.

The process flow starts with four printing steps, followed by assembly. The sensor is placed by foil-to-foil lamination and assembly. For the sensor foil we used the flow described in Molina-Lopez et al. (2013).

The total cost of ownership for the label, including the sensor foil, is about €6.50 per gpu. Figure 11.19 shows the results of the calculations in euros per cost item. As typical for HPE, the price is mainly determined by the cost of the consumables, in this case the battery (approx. €2.00) and components (approx. €3.50). That leaves only one euro for the sensor and the foil. The cost of labor is relatively high because of some manual steps that are assumed in the process flow.

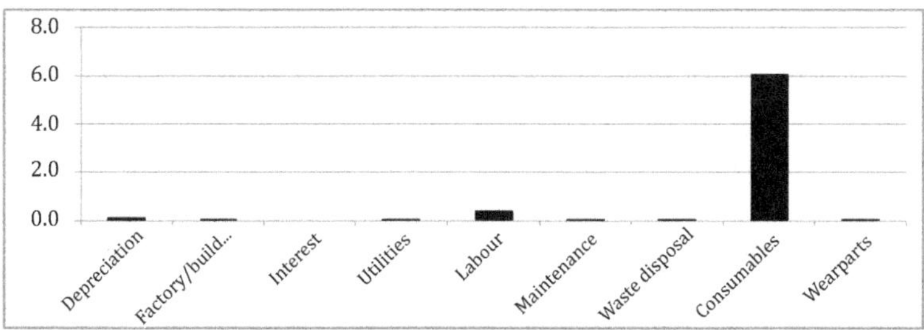

Figure 11.19 Cost of ownership of the Flexsmell label in euros per cost item.

Figure 11.20 shows the cost of ownership per process step. In presenting the cost breakdown in this way, the large impact of the components and battery become evident yet again. Also it clearly shows that the printing of the substrate only has a minor contribution to the overall system.

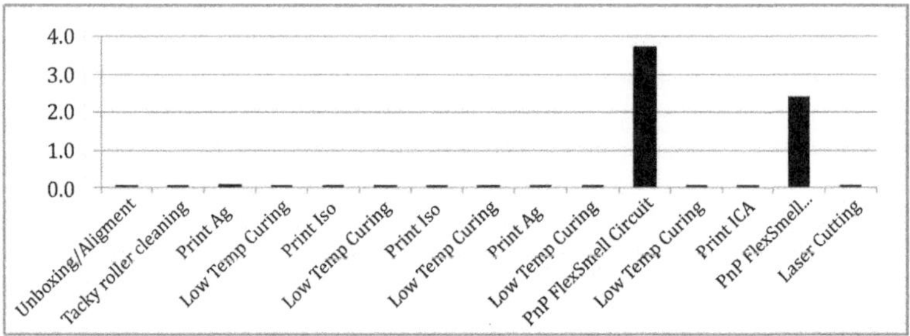

Figure 11.20 Cost of ownership of the Flexsmell label in euros per process step.

The findings of this study are representative of many HPE products. First of all, roll-to-roll manufacturing of printed circuitry only makes sense if the application demands large-area substrates; in all other cases sheet-to-sheet production, also found most frequently in industry, is preferable. It should be noted that there may be applications that require to be roll-manufactured for ease or cost further in the manufacturing chain. The use of printed substrates as such makes sense if the area is relatively large and the number of components is low. In the above example a double-layered FPC would add approximately one euro to estimated costs (based on FPC prices given in the iNEMI road map 2013, www.inemi.org), which is significant. Therefore printing does make sense for this label. In most cases the components, especially the active components, represent the bulk of the total cost. Making the product cheaper means reducing the functionality of the product.

11.7 Examples and Challenges of Realized Product Prototypes

This chapter started with stating that the areas of interest for printed and large-area electronics were in lighting (OLEDs), energy (solar, harvesting, and heating), packaging, and health care. HPE has been applied in most of these application areas. For instance, LEDs assembled on foils have been shown that may be used to produce freeform lighting objects. This can even be done with bare die LEDs, LEDs without packaging, to obtain truly flat, flexible lighting surfaces.

There are ample examples of the applicability in packaging-related products of which the discussed Flexsmell label is a typical example. Electronic solutions are often too expensive for fast-moving consumer goods where a label is allowed to cost cents instead of euros. Currently the trend is going into active packaging for high-end products and marketing samples using heating and lighting components.

With the ongoing wearable electronics boom, a lot of interest is directed toward the application of HPE in wearable devices. At the Holst Centre, particular attention is given to disposable health patches such as those shown in Fig. 11.16. The challenge here is to obtain thin and flexible devices that are cheap to manufacture and unobtrusive when worn by the customer or patient. The unobtrusiveness requires stretchable and conformable electronics in combination with low cost and the use of environmentally friendly materials. An alternative approach is to embed the electronics in garments. In this use case, aspects such as wearability, reliability under use, and especially reliability in the washing machine are of prime interest.

11.8 Conclusions

In this chapter we have described the state-of-the-art technologies that are available for the manufacturing of HPE. Discussions with potential customers and companies indicated that most are very interested in printed electronics as long as it solves their immediate problem and the technology is not disruptive in existing manufacturing strategies. With HPE, these needs are served in an important way because it delivers the advantage of printed electronics combined with the reliability of classic electronics.

After a description of the typical process flow for HPE, including printing, assembly, and finishing, we describe the techniques used for the printing of metal, mostly silver tracks, for printed circuits. For reasons such as conductivity, process speed and throughput, reliability, and the availability in industry, screen printing is often the method of choice. For specific applications that require very thin and flat lines and where high conductivity is of lesser importance, inkjet printing may be a good alternative. Flexographic and gravure printing creates similar features at much higher throughput. Examples include gridlines for OPV and UHF RFID antennas. For application in backplanes and integrated circuits, (inkjet) printed structures cannot provide the required performance yet.

The second step involves a set of processes that aims to interconnect classic electronic components, both passive and active, electrically to the printed circuit. We discuss the most important types of adhesives that can be used for this process and show how they can be used to assemble passive components down to 01005 size and bare die chips that are thinned to 20 μm. We discuss strategies to apply the adhesives in the small footprints needed for these components. Also small functional foils that have been fabricated with nonprinting thin-film technologies can be elegantly integrated using conductive adhesives. The reliability of these processes and the resulting semiproducts are currently the biggest hurdle to get the technology widely accepted in industry, although important improvements have been made in recent years.

The last steps in the process flow deal with the finishing of the semiproducts. Laminating, encapsulating, and also giving the product is final form factor belong to this group. Most of these technologies are readily available in industry.

The technologies described in this chapter can be used in a wide range of applications. The example given here deals with an olfactic label that was developed in the framework of a European project (Flexsmell). The label is a perfect example of HPE. It uses a printed circuit that is indeed more cost effective than an FPC of similar size and functionality. To obtain the required functionality many expensive components had to be placed. By means of a total-cost-of-ownership study we have shown the needed components actually make up the bulk of the cost. From experience we know that this is true for most of the products and demonstrators, which is being worked on today.

The technology needed for the successful introduction of HPE in the market is at a sufficient level. The reliability of, especially, the assembly still needs attention and the technology is mature enough to serve the first applications. Some companies are taking important steps today and the first products are expected in the near future. At the same time, HPE opens up application areas that have been difficult to address with classical electronic technologies. Unique properties that yield very thin, very flexible, and even stretchable assemblies make this possible. The application areas include, but are not limited to, the domain of wearable, unobtrusive, high-quality electronic devices.

Exercises

As an exercise and to get acquainted with HPE, it might help to make a conceptual design of the product in mind and try to analyze the technologies needed to meet the specifications. What is possible with the current state of the art, and where does one have to resort to classical components? Try to make choices not only on technological matureness but also on the cost involved.

11.1 The Flexsmell label (50 × 70 mm) discussed in the text contains a printed NFC antenna, which is probably the most challenging electronic feature that has to be printed. Especially the requirement of high conductivity and high inductance (high number of coils) is challenging. Determine what print technology would be preferred to manufacture the label.

   a. Calculate the resistance per centimeter of a line printed with a width of 1 mm using inkjet printing with a typical thickness of 0.5 μm. (Assume a specific conductivity of 20% of bulk silver.)
   b. Calculate the same for a screen-printed line (thickness 7μm, 10% of bulk silver).
   c. One of the requirements for the antenna is resistivity < 25 ohms. How many coils can one make with an inkjet-printed and a screen-printed line?
   d. What print technology would you recommend to print an NFC antenna for this label?

11.2 The label contains important chips that are of importance for the function, an NFC chip (ST M24LR64-e) and a microcontroller (TI MSP430F1611).

   a. On the basis of the information on the sizes you can find in the datasheets of these chips, what would be the best, most practical technology for printing of the electronic circuit for these chips and the deposition of the adhesives? Assume the most compact package.
   b. What would be the preferred adhesive to mount these devices and why?
   c. To obtain an ultrathin label, both chips can be mounted as bare dies (see Fig. 11.11). What implications would this

have on the design, manufacturing steps, and materials used?

11.3 For the mass manufacturing of labels, roll-to-roll printing could be of interest. That also implies a curing or sintering procedure that is compatible with high speed.

   a. Explain the difference between curing and sintering.
   b. Rotary screen printing has a minimum speed of 5 m/min with a typical curing or sintering time of 15 min. This would require a 75 m oven. What would be the required length of a curing station if photonic sintering were used? (Use the information from Fig. 11.8 as input.)
   c. Imagine a machine with a web width of 300 mm would run for one working year (1 shift, 1800 h). What would be the total number of tags that are printed? What would be output if the machine would run 24 × 7, 360 days a year?

Suggested Readings

Buffat, P., and Borel, J.-P. (1976). Size effect on the melting temperature of gold particles. *Phys. Rev. A,* **13**, p. 2287.

Fjelstadt, J. (2007). *Flexible Circuit Technology*, BR.

Ju, M., et al. (2012). Double screen printed metallization of crystalline silicon solar cells as low as 30 μm metal line width for mass production. *Sol. Energ. Mater. Sol. Cells*, **100**, pp. 204–208.

Kamyshny, A., and Magdassi, S. (2014). Conductive nanomaterials for printed electronics. *Small*, **10**, pp. 315–335.

Kerry, J., and Butler, P. (2008). *Smart Packaging Technologies for Fast Moving Consumer Goods*, John Wiley & Sons.

Koski, K. (2012). Inkjet-printed passive UHF RFID tags: review and performance evaluation. *Int J. Adv. Manuf. Technol.*, **62**(1–4), pp. 167–182.

Li, F., et al. (2014). Integration of flexible AMOLED displays using oxide semiconductor TFT backplanes. *SID Symp. Dig. Tech. Paper*, **45**, p. 431–434.

Molina-Lopez, F., et al. (2013). Large-area compatible fabrication and encapsulation of inkjet-printed humidity sensors on flexible foils with integrated thermal compensation. *J. Micromech. Microeng.*, **23**, p. 025012.

Myny, K. (2014). A thin-film microprocessor with inkjet print-programmable memory. *Scientific Reports*, **4**, Article number 7398.

Chapter 12

Environmental Aspects of Printable and Organic Electronics

Dirk Hengevoss, Yannick-Serge Zimmermann, Nadja Brun, Christoph Hugi, Markus Lenz, Philippe F.-X. Corvini, and Karl Fent
University of Applied Sciences Northwestern Switzerland, School for Life Sciences, Institute for Ecopreneurship, Gründenstrasse 40, 4132 Muttenz, Switzerland
dirk.hengevoss@fhnw.ch

12.1 Introduction

In this chapter an introduction to selected environmental issues of electronic technologies will be given.

The learning targets are twofold—first of all, to understand the main environmental issues in the life cycle of established electronic technologies, and second, to be able to explain the potential main environmental issues in the emerging field of organic and printed electronics (OPE).

In the emerging field of OPE a variety of organic but also inorganic materials and combinations thereof are applied. For instance, novel printing inks containing nanoparticles are developed and used in organic field-effect transistors (OFETs), organic light-emitting

Organic and Printed Electronics: Fundamentals and Applications
Edited by Giovanni Nisato, Donald Lupo, and Simone Ganz

ISBN 978-981-4669-74-0 (Hardcover), 978-981-4669-75-7 (eBook)
www.panstanford.com

diodes (OLEDs), organic photovoltaics (OPV), and organic sensors. OPE promises to be an environmentally more efficient technology, contributing to saving fossil fuels and rare metals. However, for instance, nanosilver formulations developed for roll-to-roll (R2R) printing demonstrate that the importance of some metals will actually increase. While the mere availability of silicon dioxide and silicates as raw material for established technologies will not be a challenge in the future, energy necessary to refine these materials will be an issue. Crude oil—the common feedstock for organic chemistry—is becoming scarcer, posing concerns for both energy and raw materials for OPE. Therefore, progress in green chemistry, which focusses on renewable resources as well as resource and energy efficiency, will help make OPE environmentally sound.

Often little is known regarding potential impacts of novel developed substances for OPE. These may have negative impacts on ecosystem quality and human health if released to environmental compartments (e.g., surface waters) during production, the use phase, or at the end-of-life disposal. Societies are increasingly concerned about potential environmental risks of new technologies. This is taken into consideration by politics underlined by the requirement for environmental appraisals in research calls. Environmental risks of upcoming OPE materials need to be investigated in fate and ecotoxicity studies as well as environmental benefits and losses of OPE devices in life cycle assessments (LCAs). This allows for efficient corrective measures at an early OPE development stage, if necessary.

For a comprehensive appraisal of environmental impacts, all environmental burdens in a technology's life cycle need to be taken into account, including the influence of gray energy, fossil and renewable resource demand, rare metal usage, greenhouse gas and material emissions, land use, etc. To do so, LCA is the method of choice. Yet LCA requires substantial information on effects of the emissions, which are often not available for new substances, and laboratory studies on fate and ecotoxicity are an essential complement for a comprehensive assessment.

To understand the environmental benefits and losses of OPE, in the following Section 12.1.1 some key environmental issues of established electronic technologies are briefly discussed. In Section 12.2 the rationale for an environmental assessment of OPE will be discussed. In Section 12.3 some basic methodologies used for LCA and fate and ecotoxicity studies will be explained in theory and in a case study for OPV.

12.1.1 Environmental Issues of Established Electronic Technologies

Many established electronic technologies (including photovoltaic [PV] cells) make use of highly pure silicon and can contain scarce (rare earth elements [REEs]) or toxic (cadmium, lead) metals, as well as problematic organic substances such as halogenated flame retardants, which can affect the hormonal system of living organisms. Each step in their life cycle generates environmental burdens affecting human health and ecosystem quality, that is, from the exploitation of raw materials and production of components and products until the use phase and final disposal. In the following sections selected issues in the life cycle of established electronics are presented: In Section 12.1.1.1 the issues of raw materials with an example of indium are dealt with, in Section 12.1.1.2 the potential risks of nanomaterials, in Section 12.1.1.3 the resource-intensive production of electronic-grade silicon and the energy-intensive vacuum evaporation process, in Section 12.1.1.4 the environmental burdens of energy consumption and short lifetimes of communication and multimedia products, and in Section 12.1.1.5 the environmental issues arising at the end of life and the disposal of electronic products.

12.1.1.1 Raw material sourcing: Indium

In contrast to silicon, indium is rare in the earth's crust and only appears in concentrations up to about 100 parts per million in zinc ores. Zinc mining is the main source of globally produced primary indium. Since the mid-1980s the importance of indium is increasing. Due to its particular characteristics, that is, semiconducting and simultaneously applicable as transparent layers, it is used as indium tin oxide (ITO) in liquid crystal displays (LCDs) and touch screens. Moreover, indium is used in PV such as copper indium gallium selenide (CIGS) cells. According to the US Geological Survey 795 tons of indium were produced in 2012, 55%–85% of which were processed to ITO and further to thin ITO films by vacuum evaporation (see also Section 12.1.1.4).

Firstly, mining of indium-containing ores is leading to environmental concerns since mining might take place in countries with low enforcement of environmental regulations, so auxiliary materials such as strong acids often end up in the environment. Secondly, negative health effects of indium residuals on miners are still unknown. Furthermore, indium also poses some risks to countries' economies: Since indium is scarce and expensive, the reserves are not distributed homogeneously among countries and no comparable alternative has been found yet; the major global economies (USA, EU, and Japan) classified indium to be critical.

12.1.1.2 Potential risks of new forms of raw materials: Nanomaterials

In the past 50 years, nanotechnology has gained increasing importance and is found in countless commercially available products. Nanomaterials are defined as materials containing particles (unbound state, aggregates, or agglomerates), which for ≥50% (size distribution) appear in one or more external dimensions in the range of 1–100 nm. Nanomaterials are not only man-made but are emitted as well naturally such as from combustion processes (e.g., forest fires), simple erosions, or volcanic dust.

The small particle size brings various unique and desirable properties. Nanoparticles have a very large surface area compared to bulk material, offering a much larger area for chemical reactions and adsorption capacity. Even though from a physicochemical viewpoint nanomaterials are interesting in different fields of application, a general concern has risen that nanomaterials might be taken up more easily by organisms through ingestion, inhalation, skin lesions, or gills in fish. In addition, the unique nanoparticle properties may result in enhanced dissolution, redox reactions, and generation of reactive oxygen species, which ultimately can lead to adverse effects in exposed organisms.

Among the many different nanoparticles, metallic nanoparticles pose a high concern. Organic nanoparticles such as fullerenes are widely used in different applications (e.g., as semiconductors) and are manufactured in large quantities. Although fullerenes may have adverse effects at high concentrations, they are of less concern for potential (eco)toxicological effects. The asbestos-like effect of carbon nanotubes (CNTs) is a cause of concern, particularly for human pulmonary exposure.

In general, nanoparticles are taken up by organisms mainly by epithelia in the lung and by gills in aquatic organisms, as well as in the intestine. These organs may also be sinks for nanoparticles but other tissues can be affected as well, including the brain. Ecotoxicological effects of nanoparticles have been described in aquatic and terrestrial organisms, but such studies mainly focused on acute toxicity to various species, whereas long-term effects are rarely investigated.

The uptake of nanoparticles is dependent on the concentration and the size of the nanoparticles. Passive diffusion seems most important, although in the gut and other tissues, vesicular uptake (endocytosis) may be equally important. Toxicological modes of action involved in adverse outcome of nanoparticles include cellular oxidative stress, cell membrane disruption, alteration of transcriptional gene expression, and induction of apoptosis and necrosis. All of these molecular and cellular effects translate to adverse effects in cell physiology and, ultimately, to cytotoxicity. Recently, it has been shown that silver and silica nanoparticles lead to induction of endoplasmic reticulum (ER) stress.

In addition to potential toxicity, nanoparticles can be accumulated in organisms, leading to food chain magnification. Bioaccumulation has been demonstrated for different nanoparticles (e.g., gold) and in different experimental food chains, such as algae-animal or bacteria-protozoa.

12.1.1.3 Production issues energy and material efficiency: Electronic-grade silicon and high-vacuum-evaporation process

Electronic-grade silicon [53], the basic material in semiconductor production, is produced in sophisticated chemical plants. As silicon is the second-most abundant element in the earth's crust the availability of silicon dioxide and silicates is nearly unlimited. The production of electronic-grade silicon, however, starts with energy-intensive reduction processes at high temperatures. The produced metallurgical silicon has a purity of about 99%. In the following purification steps electronic-grade silicon (only 1 ppb contaminants) is produced using hydrochloric acid and heat.

The main environmental impacts of these processes stem from the immense energy demand. Electricity and heating energy generated from fossil fuels (oil, coal, peat, or natural gas) are a major source of greenhouse gases and pollutants. Additionally, these fossil fuels are limited and becoming scarcer.

Many of the production steps for electronic components are based on vacuum evaporation process technology such as sputtering of ITO on screens, layering of compact discs, functionalization of semiconductors, and production of thin-film PV (CIGS, cadmium-telluride [CdTe] cells). The evacuation of vacuum chambers down to 10^{-3} to 10^{-7} mbar requires substantial pumping energy and in the case of ITO sputtering additional heating and cooling energy. Furthermore, the yield of the target material that will be sputtered is very low and high losses occur, for example, by adsorption to the vacuum chamber walls.

12.1.1.4 Use-phase energy and lifetime issue: Communication and multimedia products

Environmental burdens from the power generation (e.g., radioactive waste, CO_2 emissions from fossil fuels, etc.) for operating communication and multimedia consumer goods, such as smartphones, computers, and flat-screen TVs, and short lifetimes are the main environmental issues in the use phase. The energy efficiency of the devices is steadily is increasing, for instance, power consumption of today's flat-screen TVs is 50% to 70% lower at a display size of 80 cm compared to old television tubes. But the overall environmental benefit is reduced or even outweighed; since screen diameters are increasing, lifetimes of the devices get shorter and broken devices are not or cannot be repaired. As a consequence, the amount of electronic waste is increasing globally. Moreover, the increasing demand for, for example, REEs, is a secondary effect of short lifetimes. Therefore, increasing supply of cheap and short-dated products is not only increasing electronic waste piles but also puts additional pressure on raw materials reserves and consequently environmental compartments. Products' lifetime, upgradability, and reparability are therefore becoming key elements to reduce environmental burdens.

12.1.1.5 End-of-life issues: Disposal of electronic products

Electronic waste is the fastest-growing waste fraction in the world. These increasing amounts of electronic waste are a major source of environmental burdens. Even though toxic substances such polychlorinated biphenyls (PCBs) or cadmium are banned/limited to be incorporated into electronic devices nowadays (except for PV), the stock of harmful substances in wastes is still enormous. Treatment of the waste and recycling of useful substances, so far mainly metals, are very energy and labor intensive and require costly pollution control measures. As one example, antimony is the most often used flame retardant in consumer products. Even though many of its compounds are toxic, it is not recovered at the end of the products' lifetime and therefore ends up in incineration plants or landfills. Manual recycling is a common practice in many developing and emerging countries. Without any pollution control and human-protecting measures toxic emissions might lead to serious health problems. On a global scale, a major environmental objective is to close material cycles efficiently with a minimum of pollution generation and a minimum of valuable raw material loss. Urban mining, that is, mining of raw materials from secondary sources like waste, is therefore becoming a major focus for material sourcing.

12.2 Rationale for an Environmental Assessment of OPE

After presenting environmental issues of established electronics, OPE shall be discussed. OPE may contribute to mitigate some environmental burdens of established electronic products. For instance, the dependency on rare metals might be minimized by substitution with organic components. Energy demand for vacuum evaporation processes might be lessened using R2R printing. However, new technologies may entail new risks and unexpected environmental impacts. Therefore, upcoming OPE materials need to be investigated in fate and ecotoxicity studies as well as new OPE devices should be assessed in LCAs and compared with established electronic products at an early development stage. Like this, environmental performance of OPE can be further improved and costly mitigation measures avoided at a later stage. In the following Section 12.2.1, potential problems arising from OPE along with their life cycle are presented.

12.2.1 Potentially Arising Environmental Issues of OPE

New environmental problems may potentially arise in all life cycle phases of OPE products. Their relevance will increase with rising production volumes of OPE substances in future. Poly(3,4-ethyl enedioxythiophene):polystyrenesulfonate (PEDOT:PSS) applied in touch screens and OLEDs, as well as poly(thiophenes) such as poly-(3-hexyl-thiophene) (P3HT), applied in, for example, OPV, are examples of organic OPE components. Furthermore, inorganic nanosilver formulations allow printing of transparent and very fine electrode structures and shall substitute ITO. Despite this, the amount of substances in OPE functional layers is very low compared to substrates and encapsulation layers on which they are printed. In the case of OPV, over 90% of the material share consists of polyethylene terephthalate (PET), also widely known as packaging material for mineral water.

In the following sections, selected environmental issues specific for OPE in the production and in the use- and end-of-life phases are presented. Section 12.2.1.1 deals with environmental impacts of purchasing bulk chemicals in emerging and development countries, Section 12.2.1.2 with the use of solvents and resource efficiency in chemical production, Section 12.2.1.3 with potential emissions of UV blockers in the use phase, and Section 12.2.1.4 with risks arising from unsound disposal practices.

12.2.1.1 Raw material sourcing: Purchasing of bulk chemicals from developing countries

European standards for chemical safety are high, but basic bulk chemicals—relevant inputs for OPE material synthesis—are purchased on the global market. In 2013 the EU-28 countries imported around 3.2 million tons of organic and inorganic chemicals from India and China. In these countries environmental standards are often incomplete or are not enforced sufficiently. Therefore, for a sustainable production of OPE it is essential to assess the whole supply chain of chemicals with the associated emissions and control measures in place. The common practice to trust in environmental standard certificates is in general not sufficient to guarantee acceptable environmental impacts.

12.2.1.2 Production issues: Use of solvents and resource efficiency

Solvents represent the largest share of chemicals that are used in both OPE material synthesis and in R2R printing. Typical solvents used for reactions, dilutions, purifications, separations, and recrystallizations are methanol, ethanol, tetrahydrofuran (THF), hexane, acetic acid, chloroform, chlorobenzene, and difluoromethane. In the production of, for example, active polymers, the ratio between reactants and solvent can be up to 1:90 or even higher just for one of several synthesis steps. In R2R printing solvents used typically are isopropanol, ethanol, butanol, and water. The solvent content of ready-to-print inks can be up to 98%.

Many solvents are liquid at room temperature but evaporate quickly due to their volatility (volatile organic compounds [VOCs]). Emissions of organic compounds to air, water, and soil might be harmful. For instance, under sunlight and the presence of NO_x emitted from traffic, heating, and industrial furnaces, summer smog is generated. This so-called photochemical ozone formation depends on the type of emitted solvent (see also the section on LCA). Besides this, chlorinated solvents negatively affect air, water, soil, and biota, if released. According to GlaxoSmithKline's (GSK) solvent selection guide some of them (e.g., chloroform, dichloromethane) must be substituted because an environmental, health, and safety regulatory ban applies.

The level of environmental protection standards for each solvent depends on the classification according to REACH[1]. For transportation, handling, storage, and use in production, legal requirements on operational health and safety, incident management, emission control, and disposal must be followed—at least in the European chemical industry. VOC emissions to air and solvent emissions to wastewater must be treated until required limit values are fulfilled.

[1]Regulatory framework for the management of chemicals (REACH), European Chemicals Agency http://echa.europa.eu/home.

Resource efficiency in chemical production is influenced by the number of synthesis and purification steps, achieved yields, production quantities, and process parameters applied. The production volume for OPE substances is small compared to other chemicals. Even for large-scale mass production, the annual demand of OPE substances would only be in the range of several hundred tons globally. The low costs of basic chemicals in relation to high labor and installation costs as well as the high margins in product sales are the decisive factors, making a resource efficiency optimization unattractive from a business point of view. As a consequence, production yields are lower and material inputs into processes higher than for bulk chemicals. The generated wastes will mostly be incinerated since low volumes are not worth being recycled. Large companies often operate their own incineration plants and depending on the heat value of the waste thermic recycling is worthwhile.

12.2.1.3 Use phase: UV blocker

During the use phase of OPE it is of utmost importance that the product be safe for handling by consumers. Some potentially harmful substances cannot be avoided in OPE, since they are particularly crucial for their functionality. Therefore, it has to be ensured that their encapsulation is tight and strong enough so that proper handling does not lead to the release of any of its components. Furthermore, substances that form hazardous decomposition products during fire should be avoided, for example, as in some PV cell types applied on buildings.

Aging processes due to UV irradiation, exposure to pollutants, high surface temperatures, hot-cold and dry-wet cycles, and mechanical stress and damage might still lead to leaching of certain OPE compounds. To prevent UV degradation of OPE active layers, UV-absorbing organic chemicals (UV stabilizers) are incorporated into encapsulating plastics. Therefore, they may potentially be the first compounds being prone to leaching during the use phase. For instance, widely used benzotriazole UV stabilizers can lead to bioaccumulation in aquatic organisms due to high lipophilicity and persistence. The hormonal activity of a series of UV stabilizers was tested, and as one example, 2-(2-hydroxy-5-methylphenyl) benzotriazole (UV-P) did not show estrogenicity and androgenicity but significant anti-androgenic activity. Furthermore, the effect of UV-P was assessed in hatched zebrafish embryos. This UV stabilizer led to the induction of the aryl hydrocarbon receptor (AHR) pathway with dose-related induction of genes, which may result in developmental and reproductive toxicity and carcinogenicity.

It should be noted that such effects have been observed at concentrations much higher than those occurring in the environment. Therefore, the risks of negative effects associated with UV stabilizers leaching from OPE seem low. Nevertheless, since long-term chronic effects are largely unknown, leaching of UV stabilizers in the OPE use phase needs to be avoided.

12.2.1.4 End-of-life disposal: PET and nanotoxicology

After use, OPE materials may be disposed on landfills, incinerated, or dumped uncontrolled in cases of a weak waste management system of country. This might result in a release of potentially hazardous substances into the environment. Organic molecules applied in OPE can be divided in polymers and associated chemicals such as plasticizers and UV stabilizers. PET, for example, used as the encapsulation in OPV, is considered to be nonbiodegradable and is eventually disintegrating into microparticles. As such, they may be taken up by organisms resulting in food chain bioaccumulation or starvation. Phthalates used as plasticizers may leach from PET. They were shown to have anti-androgenic activity but at concentrations that are above levels expected to occur from leaching and weathering of OPE materials.

Furthermore, well-established PET-recycling systems are already in place all over the world due to the predominant use for drinking bottles. Therefore, such a recycling process might be adapted for the PET share of OPE as well.

Besides polymers and their additives, nanoparticles might be released when OPE materials are landfilled, incinerated, or uncontrollably disposed. Current (eco)toxicological data indicate that nanoparticles exhibit moderate to low acute toxicity, depending on the physicochemical nature, metal composition, and size of the particles. So far, ecotoxicological knowledge exists mainly on acute toxicity and primarily of fullerenes, CNTs, and metallic nanoparticles. Acute toxic effects to aquatic organisms are found in the millgram per liter concentration range, with only some nanoparticles (i.e., silver nanoparticles) exhibiting adverse effects at lower concentrations. Besides acute toxicity, also indirect effects via sorption and physical blocking of uptake epithelia were detected. For instance, in fish high concentrations of single-walled CNTs act as a respiratory toxicant irritating the gills. In general, gills are a major target organ in fish for many nanoparticles, since there the uptake takes place. Nanoparticles may also negatively affect the embryonic development of organisms such as fish. The toxicity of metallic nanoparticles often originates from the metals released by the nanoparticles and not from the nanostructure itself. This was, for example, clearly shown for zinc oxide nanoparticles used in OPV.

In conclusion, the present knowledge is not sufficient for a comprehensive hazard and risk assessment of nanoparticles used in OPE. Current data indicate that potential environmental risks mainly stem from silver nanoparticles, which are extensively used in inks for printing of OPE. However, concentrations of released nanoparticles under current disposal scenarios are expected to be much lower than effect concentrations for terrestrial and aquatic organisms.

12.3 Case Study of OPE: Organic Photovoltaics

In this case study, methodologies and results of an environmental assessment of flexible OPV are presented. With LCA the overall environmental impact was compared with other PV technologies. Complementary fate and ecotoxicity aspects of new substances developed for OPV and OPE applications were studied by means of lab experiments.

Learning targets:

- To know the methodologies of LCA and fate and ecotoxicity studies for OPV
- To be able to interpret the results from the environmental assessment of OPV

12.3.1 Description of Flexible OPV P3HT:PCBM (2.8% Efficiency)

The OPV material described in this case study is a semitransparent, flexible, single-junction polymer cell with an efficiency to convert sunlight into electricity of 2.8% on an active surface of 60%. For the following LCA, the module performance of 30 $W\cdot m^{-2}$ for a lifetime of 20 years (behind glass) was assumed. The OPV material is produced in Europe. This is relevant as certain emissions related to support processes like electricity production might depend on the production location. OPV active layers are encapsulated in a transparent PET substrate acting as protection against UV light, moisture, and oxygen. The adjacent ITO layer (90% In_2O_3, 10% SnO_2) acts as the anode. The active layer is consisting of a nano-ZnO (nZNO) hole-blocking layer, a light-absorbing composition of P3HT (electron donor) and fullerenes (6,6)-phenyl-C_{61}-butyric acid methyl ester ($PC_{60}BM$; electron acceptor), and a PEDOT:PSS electron-blocking layer. The structure is finished by a silver grid cathode. Altogether, the cell thickness lies below 0.5 mm.

12.3.2 Life Cycle Assessment

LCA is an established methodology to compare the environmental impacts of processes and products for decision making in research, industry, and politics. Section 12.3.2.1 introduces the methodological framework of LCA and Section 12.3.2.2 presents the application in a case study for PV technologies.

12.3.2.1 Methodological framework of a life cycle assessment

In an LCA, all environmental burdens of a product's or process's life cycle are taken into account, including the influence of gray energy; fossil and renewable resource demand; rare metals; greenhouse gas emissions, pollution emissions to air, soil, and water; land use; etc. LCA methodology is standardized in ISO 14040 and consists of four main phases, which are briefly described in the following paragraphs:

Phase 1: In the **goal and scope definition phase**, the aim of an LCA must be defined and central assumptions and system choices have to be taken. The goal and scope can be addressed by answering a number of questions:

- What kind of product shall be assessed?
- Shall it be compared to similar products or is it a single product assessment?
- What is the function of the product? Does it have the same function as the similar product but perform in a completely different way?
- What environmental issues are of concern for society?

The goal of the LCA presented in Section 12.3.2.2 is to compare the environmental impact of OPV with established PV technologies from "cradle to gate," that is, taking into account the environmental impacts from raw material exploitation to the point when the module is leaving the production facility. Unlike the cradle-to-gate LCA in a cradle-to-grave LCA all phases of the life cycle are considered, that is, also the required infrastructure for a PV power plant installation, maintenance, and emissions in the phase of electricity generation, infrastructure, and emissions of different end-of-life scenarios (Fig. 12.1). A cradle-to-grave LCA will be chosen when power generation technologies with different sources of primary energy shall be compared with PV technology, such as thermal power plants, wind power plants, etc. A useful functional unit to compare different modules is the environmental impact derived from the production of the required module area for 1 kilowatt peak (kWp) performance. A lifetime of an OPV module of 2/3 compared with established PV technologies requires a normalization step. One option is to take account of 3/2 OPV area to produce the same amount of electricity during the lifetime.

Theoretically, a full LCA would include all upstream and downstream processes associated with the product. In reality, the study is focussed on identifying the processes that are relevant to answer the key questions of the analysis, that is, the ones having significant environmental impacts. An appropriate cutoff point has to be defined to keep the system manageable: typically, processes that contribute to less than 5% of the environmental impact are neglected. To not lose information this must be done very carefully and cannot be done on a mass share only. For instance, many OPV substances have very low mass shares within a flexible module but cannot be neglected, because the production is very complex and resource intensive. Additionally, in a cradle-to-grave LCA the whole infrastructure of the power plants must be included. For PV technologies this is called the balance of system (BOS), which may heavily influence the LCA results. The BOS consists of inverters, mounting systems, frames, and electric installations and contributes, roughly estimated, about 50% to the environmental impact of PV power plants.

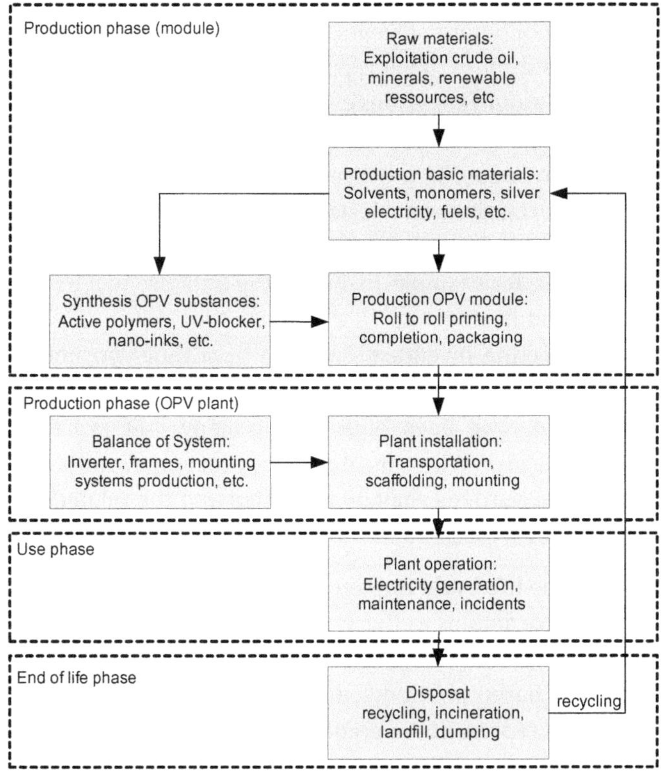

Figure 12.1 The life cycle of a power plant with flexible OPV includes the production phase with sourcing of the materials and energy, the electricity generation, and maintenance in the use-phase and end-of-life scenarios. In a cradle-to-gate LCA on the module level only the production phase is considered.

Phase 2: The **life cycle inventory** (LCI) involves gathering data on the emissions into the environment and burdens associated with the products and processes. As mentioned before, only the processes within the system boundary are considered.

A process consists of input materials, energy flows, the product and nonproduct output composed of waste, and emissions to water, air, and soil. The smallest unit is the so-called unit process. For simple macrolevel studies, there may be only a few unit processes, for example, a factory transportation and disposal of a product; for complex studies, there may be several unit processes such as a production step cascade. In the case of OPV, one unit process can be the synthesis of a polymer with the following typical input materials (resources): monomers, solvents, catalysts, heat, and cooling energy. The product output of the unit process is the polymer itself and the nonproduct output (emissions): spent solvents and catalysts, other coproducts (e.g., H_2O, BrH), VOCs, waste heat, etc. Unit processes are usually presented in a process flow diagram to visualize their linkages. Importantly, to enable data interpretation and comparisons, an LCA relates each of these environmental burdens to the same functional unit, for example, to 1 kg of the final product or to 1 kWp of the OPV module.

Since collecting inventory data can be a time-consuming task, it is common practice to use established LCA software tools and databases that have been built up for many unit processes and materials already. With an iterative procedure most relevant environmental burdens shall be identified and the related flows on an adequate level assessed.

Phase 3: In the **life cycle impact assessment** (LCIA), materials and energy resource data, as well as waste and emission data, are linked to impact category indicators, also known as midpoint indicators. This calculation is based on modeled factors, so-called characterization factors, which represent the predicted contribution to an impact per unit emission or resource consumption. In Fig. 12.2, the translation steps from emissions to category indicators are explained by means of selected airborne emissions.

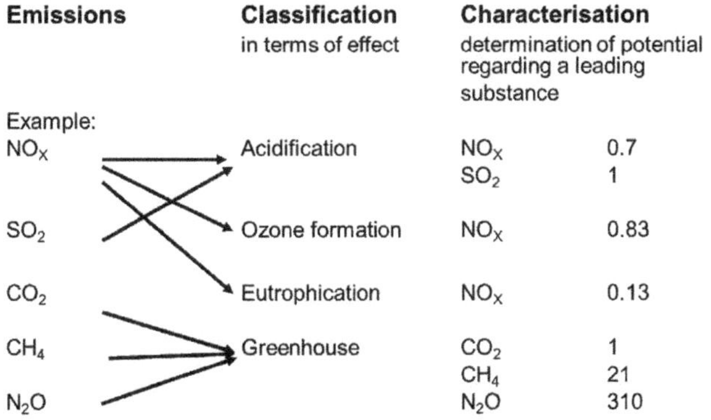

Emissions	**Classification** in terms of effect	**Characterisation** determination of potential regarding a leading substance	
Example: NO_X	Acidification	NO_X	0.7
		SO_2	1
SO_2	Ozone formation	NO_X	0.83
CO_2	Eutrophication	NO_X	0.13
CH_4	Greenhouse	CO_2	1
		CH_4	21
N_2O		N_2O	310

Figure 12.2 In LCIA emissions such as methane are assigned to the greenhouse effect classification. Nitrogen oxides (NO_x) have different environmental effects and must be assigned to the respective classes using fate models. The potential emission effect is characterized in relation to a leading substance. For instance, the same amount of methane (CH_4) has a 21 times higher greenhouse gas potential than the leading substance carbon dioxide.

The determination of the category indicators for toxic substances such as heavy metals, dust, and organic chemicals, as well as resource scarcity, follows similar rules. The final LCIA results are calculated by multiplication of each characterization factor with the emission amount. Category indicators are science based, have relatively small uncertainties, and allow for the comparison of benefits and losses of two products (e.g., different PV technologies) in different impact categories. The LCIA handbook recommends a set of midpoint indicators, which should be applied for LCIA in the European context. Additionally, guidelines from the international energy agency give recommendations for which indicators shall be applied for LCIA of PV technologies. The following are the indicators used for the LCA in Section 12.3.2.2 are introduced:

- *Intergovernmental Panel on Climate Change (IPCC) greenhouse gas* [1, 26]: The characterization of different gaseous emissions according to their global warming potential as well as the aggregation of different emissions in the impact category "climate change" is one of the most widely used methods in LCIA. Three emission categories are distinguished with different characterization factors:

- Direct emissions of greenhouse gases
- Emissions due to deforestation
- Biogenic CO_2 emissions. The characterization factor of biogenic CO_2 and CO emissions is zero. Biogenic methane emissions have the same factor as fossil methane emissions.

- *Cumulative energy demand (CED)* aims to assess the energy consumption throughout the life cycle of a product or process. This includes the direct applications as well as the indirect or gray energy consumption due to the use of, for example, construction materials or raw materials. CED values can be used to compare the results of a detailed LCA study to others in which only primary energy demand is reported. Finally, CED results can be used for reliability checks because it is easy to judge on the basis of CED comparisons whether major errors were done.
- *Energy payback time (EPBT)* is the time span in which a PV module or power plant produces as much power as it consumed to produce it. It can be calculated from the CED and a standard solar irradiation in cradle-to-gate LCA or the CED and effective solar irradiation of the dispatch region in a cradle-to-grave LCA.
- *ReCiPe (terrestrial acidification)*: Acidifying substances such as ammonia, nitrogen oxide, and sulfur dioxide are typically airborne and potentially damage ecosystems (e.g., forest decline) if deposited by atmospheric processes to the ground.
- *USEtox (human toxicity, carcinogeneicity)*: The model behind the human toxicity indicator reflects the damage to human health in life lost and health effects caused by the intake of toxic substances such as heavy metals, halogenated hydrocarbons, and others by air inhalation and ingestion.
- *USEtox (ecotoxicity)*: The model behind the ecotoxicity indicator reflects the damage to ecosystem by change in the potentially affected fraction of species due to change in concentration of toxic substances.

- *ReCiPe (photochemical oxidant formation)*: This indicator is used for tropospheric ozone formation also known as summer smog. It is formed in the presence of VOCs, carbon monoxide, and nitrogen oxides under solar radiation.
- *CML (ozone depletion)*: This indicator reflects the depletion of the ozone cap caused from halogenated hydrocarbon and chlorofluorocarbon concentrations in the stratosphere. The resulting increased level of UV-B radiation has effects on ecosystem and productivity (e.g., wood, fish, and crop production) and human health (e.g., skin cancer). The World Meteorological Organisation (WMO) reports the ozone depletion potential of the relevant substances.
- *ReCiPe (metal depletion)*: This cost-based indicator reflects the consequences of the depletion of nonrenewable resources such as metal-containing minerals.

With these indicators PV technology–specific insights can be found, such as the production of 1 kWp PV technology (A) has a, for example, 20% lower impact on acidification than 1 kWp PV technology (B) with the same efficiency.

In a multi environmental indicator comparison, it is difficult to reach a conclusion on which PV technology has the lowest environmental impact if no technology excels in all the relevant indicators. For instance, one question arising is whether climate change, acidification, human toxicity, etc., are the worst environmental problems and have priority to be mitigated.

With the help of endpoint indicators such questions can be addressed. The impact categories from the midpoint level are aggregated and weighted on the endpoint level. The ReCiPe methodology summarizes 14 scientific-based midpoint indicators to the 3 endpoint categories—(1) damage to human health, (2) damage to ecosystem diversity, and (3) damage to resource availability—which are weighted and normalized with values depending on subjective cultural perspectives. They are finally aggregated to a single endpoint indicator such as the "ReCiPe world, total."

A caveat: The ReCiPe methodology does not consider uranium as an energetic resource and radioactive waste. Therefore, in an LCA with the goal to compare electricity from nuclear energy with PV electricity the sensitivity of the results must be checked with other endpoint indicators such as the "Swiss methodology for environmental scarcity." The latter weighting system is oriented on the Swiss environmental legislation. Therefore all applied methods must be screened to properly account for the relevant emissions of all systems investigated.

Phase 4: Throughout an LCA, it is necessary to review and, if appropriate, revise the scope of the study by considering the results emerging from the inventory analysis and impact assessment. This process, **interpretation**, is iterative and might, for example, identify data quality issues such as data gaps, which need to be closed.

Generally, this activity would already form part of the initial goal and scope definition phase. The procedure typically involves the examination of the sensitivity of results, performance of a scenario analysis, a review of data quality, and a comparison of the results with the original goals of the study.

12.3.2.2 Life cycle assessment of OPV and established PV

The objective of the LCA is to compare established CdTe and multisilicon (multi-Si) PV cells with printed flexible OPV on the module level. The question whether the environmental impacts of OPV modules are higher or lower than those of CdTe and multi-Si modules should be answered.

The flexible OPV [6,6]-phenyl C61-butyric acid methylester (PCBM):P3HT (2.8% eff.) is described in Section 12.3.1. The **CdTe module** has en efficiency of 10.9%, performs with 110 $W{\cdot}m^{-2}$, has a lifetime of 30 years, and is produced in Europe. The **multi-Si module** performs with 210 $W{\cdot}m^{-2}$, has an efficiency of 21%, has a lifetime of 30 years, and also is produced in Europe. A module consists of 60 solar cells based on silicium wafers.

The **functional unit** is expressed in relation to 1 kWp of the respective system under standard radiation assumptions adjusted for different life spans. Any difference in performances at diffuse light or high temperature is not considered.

The **system boundary** includes all steps from cradle to gate (see Section 12.3.2.1). The data for the LCI were provided by Belectric OPV GmbH (Nuremberg, Germany) within the Sunflower project (http://sunflower-fp.eu), taken from the literature, the ecoinvent database 2.1, or are own estimations. The indicators chosen for LCIA are discussed in Section 12.3.2.1, Phase 3.

The **results of the LCIA** on the midpoint level are presented in Fig. 12.3. It shows the environmental impact of eight midpoint indicators and error bars with a confidential level of 68%. Each indicator is normalized to the PV module with the highest impact, that is, 100%, and discussed in the following next:

- *CED*: CdTe is around 50% lower than multi-Si (100%). The main impact stems from electricity consumption during silicium production. OPV has a slightly higher CED than CdTe (+15%).
- *IPCC greenhouse gas emission*: OPV has the lowest impact. It is around 40% lower than multi-Si (100%) but only 5% lower than CdTe. The main impact for multi-Si stems from electricity generation with fossil fuels.
- *CML ozone depletion*: CdTe has the lowest impact. It is about 85% lower than multi-Si (100%) and 15% lower than OPV. The main impact for multi-Si stems from vinyl fluoride, which is required for cell production.
- *ReCiPe (photochemical oxidant formation)*: CdTe is the best and has a 35% lower impact than multi-Si (100%). The OPV impact is within the statistical variation of multi-Si and around 25% higher than CdTe. The main impact for OPV stems from silver refinery and for multi-Si from the production tetrafluoroethylene, which is required for cell production.
- *ReCiPe (metal depletion)*: Multi-Si is the best and has about a 55% lower impact than CdTe (100%). The main CdTe impact stems from the copper demand. The main OPV impact, which is 20% lower than for CdTe, stems from the silver for the electrodes.
- *ReCiPe (terrestrial acidification)*: All technologies show similar results.
- *USEtox (human toxicity, carcinogeneicity)*: No relevant differences were observed due to large statistical variations. The main OPV impact is from the silver refinery for the electrode.
- *USEtox (ecotoxicity)*: CdTe shows a 30% lower impact than multi-Si (100%). The large statistical variations do not show a relevant difference between the technologies. The main OPV impact is from the silver refinement.

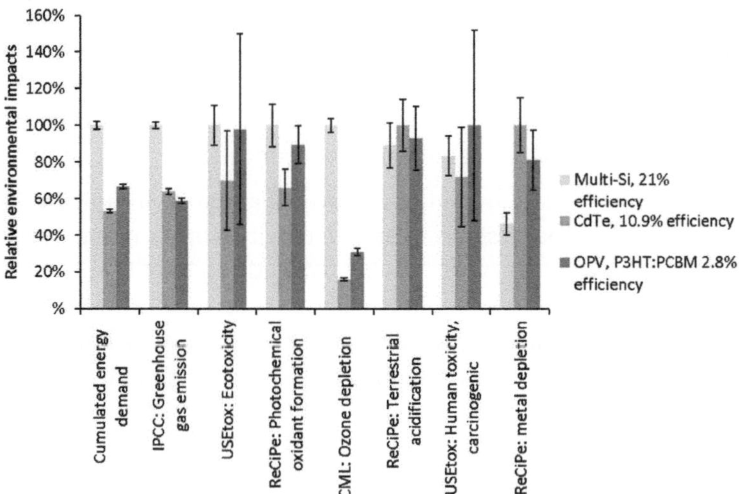

Figure 12.3 The relative environmental impacts (normalized to the highest impact technology in the respective category) of the compared PV technologies show the lowest impact for OPV in the category "IPCC: Greenhouse gas emission."

In summary, the eight midpoint indicators show the tendency that multi-Si (21% eff.) has the highest impact in most of the assessed categories and CdTe (10.9% eff.) shows the lowest impact. The energy demand and emissions related with the production of silicium are the main cause for the relative high environmental impact of the multi-Si (21% eff.), even if only half of the area is needed to produce the same amount of electricity than with CdTe (10.9% eff.). The efficiency of the OPV (2.8% eff.) is one-fourth of the efficiency of the CdTe (10.9% eff.) and the lifetime one-third of the CdTe. Therefore the OPV needs a factor 5.5 of the area of CdTe to produce the same amount of electricity, which results in the most impact categories and a higher environmental impact in production. For a final decision an additional indicator on the endpoint level is required. As discussed in Section 12.3.2.1, endpoint indicators are aggregated from different impact categories. It is important to remember that the aggregation is weighted by targets of an environmental policy or from cultural perspectives.

In Fig. 12.4 the LCIA is presented with the two endpoint indicators "ReCipe (H/A) world, total" and "Ecological scarcity 2006, total".

With both indicators, CdTe (10.9% eff.) shows the lowest impact. For multi-Si (21% eff.) and OPV (2.8% eff.) the "ReCipe (H/A) world" indicates about the same environmental impact. However, "Ecological scarcity 2006: total" shows that OPV (2.8% eff.) has compared with CdTe (10.9% eff.) a 45% higher and compared with multi-Si (21% eff.) a 30% higher environmental impact. The main environmental impact stems from refinement of the silver for the electrodes.

The following conclusions for the low-performing flexible OPV (2.8% eff.) can be made: Taking into account all results from mid- and endpoint indicators, the overall environmental impact of OPV is similar to the multi-Si (21% eff.). The CdTe (10.9% eff.) shows overall the lowest environmental impact. The main share of the environmental impact of OPV stems from the silver refinement. An expected impact of the scarcity of indium is not reflected in the applied indicators. The amount of required indium is very low compared to the copper demand for coating and contacts in CdTe and for the silver demand of electrodes in the flexible OPV.

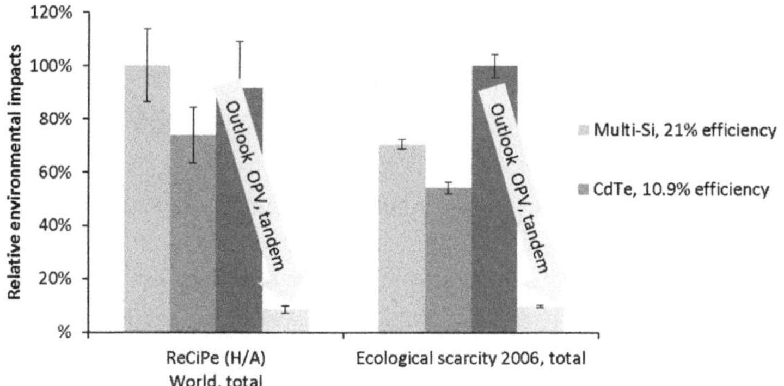

Figure 12.4 Both endpoint indicators "ReCiPe (H/A) world, total" and "Ecologiocal scarcity 2006, total" show the lowest environmental impact for CdTe (10.9% eff.) without considering a future tandem OPV (10% efficiency). The next generation of OPV will thus be less resource intensive, in which case both endpoint indicators point toward a potential environmental impact reduction of up to a factor of 10.

In current research the performance of OPV will be improved. The efficiency of OPV is predicted to increase up to 10% (efficiency on the cell level). ITO electrodes will be replaced by silver, and fine silver grids with a thickness of several nanometres might considerably reduce the amount required for the electrodes. As a result, resource savings, less pollutants, and less greenhouse gas emissions will further reduce the environmental impact of OPV. The predicted reduced environmental impact of a printed flexible tandem OPV (10% eff.) in both endpoint indicators (Fig. 12.4) is up to a factor 10 compared with a flexible OPV (2.8% efficiency) and 5 to 10 compared with the multi-Si (21% eff.) and CdTe (10.9% eff.) It can be expected, even if production of established PV technologies will be more optimized and their efficiencies increase, that printed flexible OPV will be one of the greenest PV technology choices in the future.

12.3.3 Fate Studies

For investigating the environmental impact of a product such as OPV it is of utmost importance to carry out so-called **fate studies**. These cover the potential pollutants that might be emitted from OPV and under which scenario and to what extent they are emitted. Furthermore, it is of interest to know how such pollutants behave in the environment, which chemical/biological transformation reactions they undergo, and to which environmental compartment they partition. For instance, one may study to which extent silver is released from the electrode during use or at the end of life. In the environment, silver might undergo further reactions due to exposure to chemicals. The chemical state of silver, the so-called **speciation**, eventually determines the effect silver has in the environment. For instance, silver might be transported in surface waters, adsorbed to soils and sediments, or taken up by organisms. This elucidation of predicted environmental concentrations (PECs), speciation, and partitioning pathways is the basis to reveal ecotoxic effects.

12.3.3.1 Fate of OPV modules during and after the use phase

At the end of an OPV material's lifetime, disused cells might face different scenarios. OPV materials were shown to be satisfyingly stable in accelerated aging experiments as well as under outdoor conditions for more than one year. When they reach their end of life, recycling, incineration (with flue gas treatment), and controlled landfilling are considered to be proper disposal strategies. Recycling should be preferred, since the vast majority of an OPV material's weight is PET encapsulation, for which recycling processes are already well established. On the other hand, inadequate disposal via uncontrolled landfilling or uncontrolled burning might occur.

Until now, no legislation as well as no processes exist in order to recycle PV cells. At least, the EU has included PV into its Waste Electrical and Electronic Equipment Directive (WEEE). From 2014 on, PV cells have to be returned—as other electronic equipment—instead of being disposed of as normal waste. Furthermore, since precious metals are contained in PV cells, the development of appropriate recycling methods will be driven by a financial incentive, especially when produced in on a large scale. However, recycling of OPV materials may require organic solvents to separate single components, which might themselves be of environmental concern. Since so far such a recycling strategy does not exist, nowadays controlled incineration of OPV materials in state-of-the-art incineration plants is assumed to be the best solution. Like this, not only metals might be recovered, but the energy bound in plastic chemical bonds can be thermally recovered generating electricity and heat. Nevertheless, in places where disposal infrastructure is not yet fully established, OPV-based products might encounter improper disposal. In uncontrolled landfilling, for instance, OPV materials would be exposed to harsh environmental conditions such as UV radiation, water, high temperatures, and chemicals. When PET is incinerated, a variety of toxic substances can be formed in the airborne soot as well as residue ash, for example, polycyclic aromatic hydrocarbons and heavy metals.

One major aspect is UV radiation, which leads to aging of OPV materials by causing photo-oxidation and yellowing of organic polymers. Aging of the encapsulation can result in loss of flexibility, crack formation, and delamination, which finally leads to emission of OPV components. Moreover, since plastic encapsulation is not entirely impermeable for oxygen and water, ambient humidity and oxygen might penetrate OPV materials in low concentrations. Mechanical stress during use, while landfilling, or when distributed into water systems can result in deterioration of OPV materials. Once mechanically damaged or even decomposed to small fragments with a large interface exposed to ambient conditions, functional materials can be prone to washout.

In conclusion, there is a wide variety of factors that might have a detrimental effect on OPV material integrity, potentially leading to cell weathering, aging, degradation, and possible release of compounds into the environment.

12.3.3.2 Leaching and fate of inorganic OPV compounds to the environment

At their end of life, mechanically damaged OPV materials might be exposed to rain when still installed on a roof or on a landfill. Further, illegally disposed OPV may as well end up via surface waters in lakes and the sea. Therefore, leaching studies were accomplished in order to calculate the PECs of emitted pollutants so that their ecotoxic potential can be assessed. In particular, nanosized components were of interest due to their still not well-known environmental fate and ecotoxic potential (see before). To give an example of another field: Outdoor façade paint does often contain TiO_x nanoparticles as whitening pigment, which can be washed out by rain, ending up in wastewater treatment plants or, even worse, in the aquatic environment. During wastewater treatment, TiO_x was of little concern, since about 98% of Ti was removed in the biological treatment step through sorption to natural organic matter and subsequent sedimentation. However, up to 20 $\mu g \cdot L^{-1}$ TiO_x nanoparticles (4–30 nm) were still found in the effluent and finally reach aquatic systems like rivers and lakes. Therefore, it is important firstly to know a pollutant's PEC (to compare with toxicity data) and secondly to understand the behavior of such particles in the environment (to make a prediction on partitioning). The form of such metallic leachates has a tremendous effect not only on its toxic potential but also on its environmental behavior. Under environmentally relevant conditions, the form and aggregation of nanoparticles depend on a variety of factors. For instance, acidic pH conditions would stabilize ZnO nanoparticles released from OPV materials in monodisperse form and natural organic matter can adsorb to ZnO, resulting in negative surface charges preventing aggregation. On the other hand, divalent cations (e.g., Ca^{2+}) are leading to aggregation of particles due to charge neutralization and larger aggregates are not taken up and accumulated by organisms to the same extent. In conclusion, different factors are influencing the accumulation and bioavailability of nanoparticles and defining their fate and toxicity. Sedimentation of larger particles or accumulations leads to less availability for pelagic organisms compared to benthic organisms. Nanosilver, for instance, showed high toxic potential in countless studies (see later). But recently it has been shown that in sewage systems, nanosilver is immediately sulfidized to Ag_2S particles (practically insoluble in water); thus the ecotoxic potential under environmentally relevant conditions is probably lessened. Examples like these show that not solely concentrations (PECs) are important but also the pollutant's speciation.

For OPV, leaching of metals and metalloids under different scenarios were carried out, amongst other for rooftop runoff (Fig. 12.5). Measured metal concentrations in solution were alarming at first sight. However, these concentrations strongly depended on the experimental setup, that is, the ratio of rain to OPV and experimental duration. A more appropriate mean to estimate ecotoxic potential is by calculating the PEC. For this, the following assumptions were made: The amounts of metal(loid)s leaching from different PV cells in laboratory experiments were projected to a roof fully covered with PV cells and then divided by the amount of rain precipitating on this area in the same time span as the experiment for arid and humid climate zones. Conclusively, the calculated PEC for OPV did not exceed the most restrictive limits, that is, drinking water limits, as regulated by the World Health Organisation (WHO). On the other hand, identical leaching from CIGS cells showed high concentrations of Se, Mo, and Cd being released. In the case of Cd, which is banned from incorporation into electronic equipment except for PV, PEC exceeded WHO limits by a factor of 58 as well as acutely toxic concentrations for freshwater organisms such as *Daphnia magna* by a factor of 7. Also PECs for Se and Mo were problematic in the context of bioaccumulation or toxicity.

To distinguish leachates between dissolved and nanoparticulate metals, a so-called time-resolved single-particle inductively coupled plasma mass spectrometry (TRSP-ICP-MS) method can be applied (Fig. 12.6). Briefly, when the metals are in a truly homogeneously dissolved state, ionization of the continuously arriving metals in the ICP leads to a continuous flow of charged metal ions recorded as a constant signal (flat baseline) by the MS detector. On the other hand, when nanoparticulate metals are present, single nanoparticles will discontinuously arrive in the plasma where ion accumulations (clouds) are produced. These are being detected as peaks. Like this, in the case of OPV, it was shown that except for Zn the leachate metallic compounds are mainly in dissolved form, so potential nanotoxic effects can already be excluded from the start.

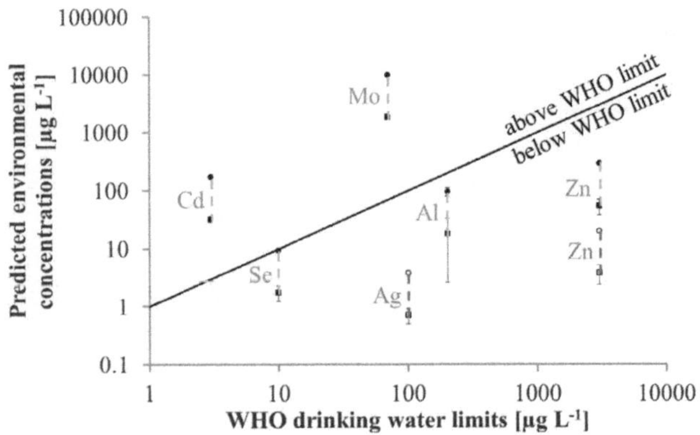

Figure 12.5 In a rooftop acidic rain runoff scenario, the PEC calculated from Cd and Mo leaching from CIGS (green) exceeded WHO drinking water limits by far for arid (circles) as well as humid (squares) climate zones, whereas Ag and Zn from OPV (red) did not pose any risk to the environment. Reprinted with permission from Zimmermann et al. (2013). *Environ. Sci. Technol.*, **47**, 13151–13159. Copyright 2013 American Chemical Society.

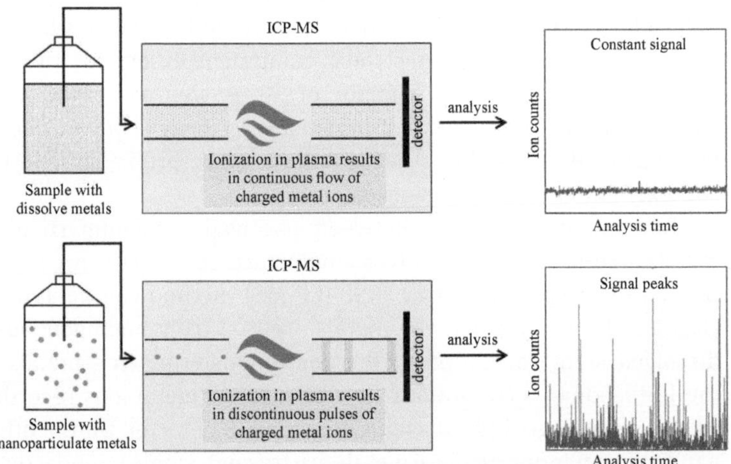

Figure 12.6 Time-resolved single-particle ICP-MS method to distinguish between dissolved metal ions (top) and nanoparticulate metals (bottom). For details refer to text.

12.3.3.3 Leaching and fate of organic OPV compounds to the environment

In comparison to metallic leachates, organic compounds might also be released from OPV materials. Especially PCBM is of interest, since such fullerene derivatives are increasingly used in industrial applications and their behavior and toxic potential are not well understood. Recently, pure C_{60} and C_{70} buckyballs have been detected in surface water and wastewater. Since fullerenes are emitted not only from anthropogenic sources but also by burning (e.g., forest fires) and volcano eruptions, their origin is often difficult to determine. Numerous studies exist about the behavior and toxicity of C_{60}, showing that there is a considerable toxic potential but strongly depending on its surface conjugation. The pristine C_{60} cage itself is highly hydrophobic and stable but becomes surface hydroxylated over time under toxic/aqueous conditions. Like this, it becomes more polar and easier soluble in water, which can result in transportation by surface waters instead of sedimentation/sorption. In consequence, toxicity for pelagic organisms may be more important than for benthic species. PCBM investigations are thus particularly interesting, since the change induced by environmental conditions (water, oxygen) might completely change its environmental fate and toxicity.

In leaching experiments described earlier for metals, no PCBM was detected. The main reason might be its persisting high hydrophobicity for which it does simply not dissolve in water and is not released from OPV materials. To estimate PCBM's hydrophobicity, the so-called log K_{ow} value was determined. It is defined as

$$\log_{10} K_{\text{OW}} = \log_{10} \frac{c_{\text{octanol}}}{c_{\text{water}}}$$

The log K_{ow} of a compound is determined by mixing it with octanol (a hydrophobic, non-water-miscible solvent) and water before equilibrium concentrations are determined. Compounds' environmental partitioning is strongly related to their hydrophobicity, wherefore log K_{ow} is routinely used to predict the latter.

With highly hydrophobic compounds such as PCBM, however, this method is not appropriate since the minute amounts dissolving in water are difficult to determine properly. The log K_{ow} of such highly hydrophobic compounds can be determined by liquid chromatography instead (Fig. 12.7). Here, the retention time of standard chemicals (with different hydrophobicity and therefore different log K_{ow}) is determined on a chromatographic column and a linear correlation between retention time and known log K_{ow} is calculated. By using PCBM's retention time in this correlation a log K_{ow} value of about 10.5 was determined, meaning that the amount of PCBM dissolving in hydrophobic octanol is $10^{10.5}$ (31 billion) times larger than in water. From this, it can be concluded that even if released from OPV materials, resulting concentrations would be extremely low. These minute amounts then will mainly be sorbed to organic material. In the long term, PCBM might be hydroxylated on the cage surface, which tends to result in a decreased log K_{ow}. Carefully concluding from bare C_{60}, which still have a log K_{ow} of >6.5 after 13 days of exposition to water, one would still expect aqueous concentrations of fully hydroxylated PCBM to be very low.

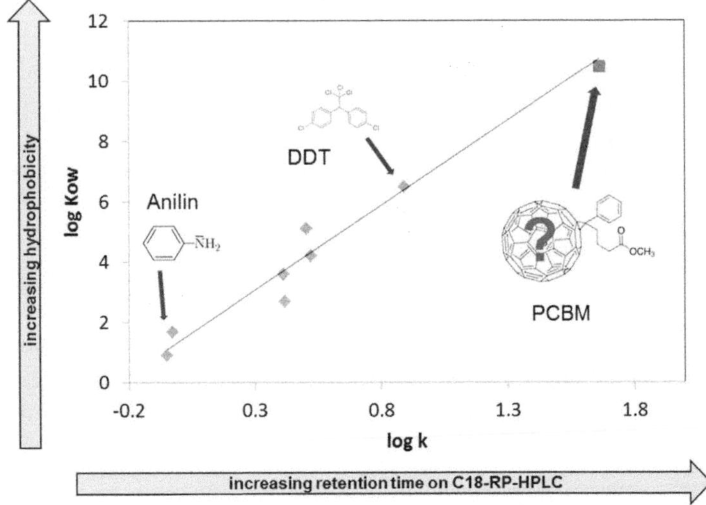

Figure 12.7 Investigation of the log K_{ow} value of PCBM contained in OPV materials. From standard chemicals such as anilin or DDT with a known log K_{ow}, the log K_{ow} for PCBM is extrapolated.

12.3.4 Ecotoxicity Studies

Ecotoxicology is a research field in which the effect of anthropogenic compounds on the living environment is investigated. The research sets its focus on a compound's possible adverse effects in relation to its environmental abundance and concentration. The aim is to determine its effects on molecular, cellular, organism, and population levels to foresee possible risks and prevent them. Most of the products used in everyday life are lacking an ecotoxicological assessment. Especially data on chronic effects are missing. Once products or their leaching compounds enter the environment (aquatic, terrestrial, or air), there their chemical behavior is crucial in determining adverse effects. Persistence in the environment, toxic activity, and potential of accumulation through the food web are some alarming characteristics of a compound. Regarding OPE materials, some of the components are fairly new and need to be investigated for their ecotoxic potential in order to ensure that there is no risk for the environment upon leaching. Potential toxicity of OPV compounds is discussed next.

12.3.4.1 Toxic potential of single OPV components

As mentioned earlier, **PET** as an embedding material has the highest weight share of OPV materials. Due to its resistance to environmental degradation (including biodegradation), large volumes of persistent PET waste are created. Usually, PET is undertaking fragmentation (by heat, oxidation, light, ionic radiation, hydrolysis, and mechanical shear) into smaller pieces down to <5 mm, called microplastics. For even smaller fragments (polymers), the transport across biological membranes is limited due to their size. Therefore, polymers as well as microplastics are not considered as directly toxic. Nevertheless, effects from ingestion of microplastics, including suffocation or blocking of the digestive tract, causing death, were documented for fish, birds, turtles, and marine mammals. Ingested microplastics are translocated from the gut to the circulatory system in mussels and finally transported to hemocytes. Despite its widespread use, only a few studies were conducted combining PET leaching with toxicity tests. These indicated contamination of mineral water with phthalates, antimony, and other endocrine-disrupting chemicals originating from plastic packaging materials. The acute toxicity of plastic product leachates to *D. magna* is very low (EC_{50}[2] at 5–80 g plastic material per liter or above). Therefore, although PET has the

[2]Concentration where 50% of the tested population shows an effect.

highest mass share in OPV, microplastics or leaching of phthalates and antimony from PET is regarded as minor toxicological relevance.

ITO is a sintered mixture composed of In_2O_3 and SnO_2 (9:1 wt:wt). Only a few studies were carried out on the ecotoxicity of ITO, although it is commonly used for electrodes. ITO particles have a pulmonary toxic effect such as pulmonary inflammation, as observed when ITO particles suspended in water were repetitively instilled intratracheal into hamsters. The toxicity of ITO particles and that of single components (In_2O_3 and SnO_2) were comparatively assessed in vivo (rats) and in vitro (lung epithelial cells, macrophages). The sintered ITO was significantly more toxic than its single components or the unsintered blend, triggering a strong inflammatory reaction, genotoxicity in vivo, a strong cytotoxic response, and production of reactive oxygen species (ROS) in vitro. Blaise et al. exposed ITO nanoparticles to several taxonomic groups of aquatic organisms and found acute toxicity at concentrations lower than 5 mg/L with the most sensitive response of *Hydra attenuate.* Here, the EC_{50} is 0.1–1 mg/L (96 h exposure). In fish, the soluble indium is suggested to be the toxic moiety and causing oxidative stress and inflammatory effects as well as disturbing protein folding. Currently, the toxicity of ITO to fish and its molecular or cellular mode of action are unknown. ITO may be replaced in novel OPV by a printed silver (nanoparticle) current collecting grid.

Silver nanoparticles (AgNPs) are contained in manifold consumer products, including antimicrobial textiles, microelectronics, and healing creams. The present annual worldwide production of AgNPs is estimated to be at 500 tons. Silver might enter the aquatic environment by several routes such as during synthesis, manufacturing, and incorporation of the nanoparticles into goods; through washing of impregnated textiles; and disposal. Modeled and experimental data suggest that concentrations present in the surface waters are in the nanogram per liter range and exponential increases are predicted due to increase in usage. The toxicity of particulate silver and nanosilver has already been investigated. Biological mechanisms were investigated using the microbe *Escherichia coli*, showing expression alterations in envelope and heat shock proteins, affecting membrane potential and culminating in the loss of cell viability. Bacterial communities are crucial in wastewater treatment and suppression of such species might impede treatment plant efficiency. The highest concentration of a toxic compound that still does not cause an adverse effect is referred to as the **no observed effect concentration (NOEC)**. The NOEC for a model aquatic invertebrate, *Daphnia* spp, was 0.001 μg/L, whereas for freshwater and seawater algae NOEC values between 2 mg/L and 0.002 mg/L were reported with the suggestion of silver ion as most toxic form. In fish embryos, AgNPs affect spinal cord deformities, cardiac arrhythmia, hatching success, and survival. Christen et al. showed the induction of ROS by AgNPs. Moreover, they showed for the first time a strong ER stress response in zebrafish liver cells and embryos. The activation of the ER stress response can lead to activation of apoptotic (programmed cell death) and inflammatory pathways. Overall, AgNPs showed a more pronounced effect compared to equal mass concentrations of $AgNO_3$, suggesting a nanoparticulate effect. Additionally, in zebrafish and rainbow trout AgNPs are incorporated into blood vessels, skin, brain, heart, and yolk, while Ag ions were concentrated in organelles, the nucleus, and the yolk only.

The organic polymer **PEDOT:PSS** has recently been suggested as an alternative to ITO in OPV. PEDOT as a single compound and PEDOT doped with PSS showed no cytotoxic nor inflammatory response in mouse or human cells as well as in rats. PEDOT and PSS are not considered to be toxic. However, when comparing varying lengths of PEDOT nanomaterials, the cytotoxic effect and ROS production increased with decreasing length. No information exists on toxicity of PSS itself and therefore further investigation is needed.

Currently, **P3HT** is the best-fitting organic polymer with the highest commercial relevance for the light-adsorbing (photoactive) layer of OPV. No toxicological analysis has been performed so far on this polymer. In general, polymers are too big to be taken up into cells; therefore toxic effects are not expected. However, UV irradiation and degradation mechanisms could lead to smaller-sized molecules (e.g., sulfoxides, sulfones, and sulfinate esters), which may have a potential for ecotoxicological risks.

PCBM (C_{60} or C_{70}) is the most commonly used electron acceptor in OPV. Due to the significantly higher absorption in the visible spectrum down to 700 nm, $PC_{70}BM$ is considered to be a new alternative for the more commonly used $PC_{60}BM$. The initial size of a PCBM cluster used as P3HT:PCBM active material in OPV is 20 nm, but after thermal annealing of the P3HT:PCBM nanocomposites, needle-like PCBM crystals are formed with a size of up to 20 μm. Fullerenes have been studied for their ecotoxicity. They may occur at 0.018–0.19 ng/L in surface waters and 3.69–25.1 ng/L in sewage treatment effluents. As described in the previous chapter, fullerenes are highly insoluble in water but hydroxylation render them hydrophilic over time. The presence of natural organic matter (NOM) has an effect on the physicochemical properties of nC_{60} and therefore environmental behavior; aggregation decreases with increasing NOM concentration. Despite the fact that NOM is ubiquitously present in aqueous environments, most ecotoxicity experiments on fullerenes were conducted without its presence. No measurable toxicity of nonhydroxylated C_{60} was noted to *E. coli* and *Bacillus subtilis* even at very high concentrations of 342 mg/L, but an altered membrane phospholipid composition was noted in *Pseudomonas putida* and *B. subtilis*. Overall, the antibacterial mechanism of C_{60} is still under debate. Significant sublethal but no lethal effects were found in the waterflea *D. magna*. C_{60} nanoparticles taken up into the digestive tract by *D. magna* are excreted as large agglomerates. In zebrafish embryos (*D. rerio*), C_{60} aggregates did not alter the mortality, and global gene expression changes were relatively minimal. In fathead minnows (*P. promelas*) no significant acute toxicity was found; however, PMP70 (transport protein involved in lipid metabolism) expression was suppressed. Thus far, the toxicity of the derivative PCBM as well as C_{70} has not been assessed. Conclusively, the effect of fullerenes is dependent on its surface derivatization and therefore direct transfer of toxicity data of C_{60} to the derivative PCBM present in OPV is not possible.

In OPV manufacturing, nano-ZnO (nZnO) is sputtered in the form of nanoparticles in a thin layer. Upon entering the environment, accumulation and bioavailability of nanoparticles are influenced by the amount of NOM, ionic strength, pH, surface area, and shape. Furthermore, their small size enhances uptake and interaction with biological tissues. Presently, the fate and behavior of nanomaterials in the environment and its potential adverse effects are limited. The acute toxicity of nZnO is known for a wide variety of organisms. The LD_{50} (lethal dose, 50% of the tested population dies) varies considerably depending on the test organism and exposition time.

The LD_{50} was found 21.1 mg/L in the bacteria *E. coli* after 2 h of exposure, 68 μg/L in the algae *P. subcapitata* after 72 h of exposure, and 1.793 mg/L to 3.969 mg/L in zebrafish *D. rerio* embryos after 96 h of exposure. Cellular uptake occurs by diffusion, caveolae, or endocytosis. Nanoparticles of 60 nm are not taken up by zebrafish embryos, whereas nanoparticles of 12 nm can be taken up via egg shell (chorion) pore canals. When metal-based nanoparticles enter the cell, they deliver free metal ions, resulting in metal-related toxicity. In fish, nanoparticles enter via gills and the gut epithelium. Then they may be distributed to other organs, including the brain. At the cellular level, surface-reactive nanoparticles lead to induction of ROS, as shown, for instance, for nZnO in zebrafish. ROS generation can trigger deoxyribonucleic acid (DNA) and membrane injury and oxidative stress that lead to phase II enzyme induction, inflammation, and mitochondrial perturbation with apoptosis. Particle dissolution and Zn^{2+} shedding of nZnO can lead to pro-inflammatory and cytotoxic effects, which leads to lysosomal damage, triggering of intracellular $[Ca^{2+}]$ flux, mitochondrial perturbation, and cytotoxicity. This mechanism was proven to take place in organisms, as intracellular ROS were measured in the embryonic cell suspension from zebrafish embryos exposed to nZnO. In summary, much is known on nZnO toxicity, but certain questions, including the molecular action and target toxicity, are still open. The potential nanotoxicity effects of nZnO used in novel OPV were assessed by Brun et al. and detailed methods as well as results are described next.

12.3.4.2 Methodology: General description of gene expression analysis and laser ablation

Environmental pollution has eventual adverse effects on cells, tissues, organisms, populations, and, ultimately, entire ecosystems. The biological response related to exposure of environmental chemicals is measured by one or several sensitive indicators of toxic effects, so called **biomarkers**. Most biomarkers are based on alterations in biochemical parameters (e.g., enzyme activity, hormones in blood, specific protein levels) or cytological or histological abnormalities (e.g., chromosome aberrations, micronuclei, imposex). Recent advances in molecular biology promoted biomarkers based on the analysis of transcription of stress-related genes, the so-called gene expression analysis.

On the molecular level, the presence of a toxicant triggers a response, which is starting with a receptor detecting the presence of the toxicant (e.g., hormone, metal). The receptor becomes activated and binds to specific DNA sequences in the cell nucleus. The DNA sequence upstream is then transcribed to so-called messenger ribonucleic acid (mRNA), which is then exported to the cell cytoplasm. There, its information is translated to synthesize proteins, helping to readjust the cells metabolism. Gene expression biomarkers assess the amount of transcribed mRNA of certain target genes.

Ribonucleic acid (RNA) is extracted from small individuals, tissues, or cell lines and quantified by polymerase chain reaction (PCR). In the PCR approach, target DNA is detected at the end of the analysis, whereas by using devices able to monitor the process of PCR in real time (quantitative real-time PCR, or qRT-PCR) the target DNA is detected and quantified simultaneously. qRT-PCR is more precise and reproducible and allows processing of a large number of samples. For relative quantification, the target gene is compared to an internal reference gene, which is equally expressed in both toxicant-affected and nonaffected cells. The expressions of target genes from toxicant affected cells are then presented as fold change of control samples. As gene quantification data from qRT-PCR are derived from rations of exponential functions, the data are not normally distributed. Therefore the data need to be normalized (e.g., logarithmic transformation) before statistical analysis (e.g., ANOVA, Student's *T*-test) can be conducted.

Bioimaging of metals in tissue sections or whole organisms by laser ablation inductively coupled plasma mass spectrometry (LA-ICP-MS) is an elegant technique but still in its infancy. It gains increasing importance not only to visualize uptake and spatial distribution in metal exposed organisms but also to study metals and metalloproteins playing a major role in organ function. In this technique, a focused laser beam ablates 5–100 μm wide sections of a biological sample while ICP mass spectra of the ablated material are recorded continuously. It can be used for single-point measurements to investigate concentration profiles in depth or line scan measurements for 2D imaging. Samples are mostly prepared as cryo-sections or paraffin sections mounted on a glass slide.

12.3.4.3 Example for toxicity studies: Assessment of nZnO

Zinc is an essential trace element for organisms but induces toxicity at elevated concentrations. In organisms, cellular Zn ion fluctuations are mainly regulated by Zn binding proteins called metallothioneins (MTs). In the case of an excess supply of Zn, the MT gene is increasingly transcribed and MT proteins translated. If MTs are not able to bind the excess of Zn, ROS formation may be provoked. Induction of ROS formation was observed for both Zn and nZnO. Excessive production of ROS can induce pro-inflammatory and cytotoxic effects. nZnO and associated released Zn may ultimately lead to apoptosis and acute toxicity at high concentrations. However, induction of an inflammatory response has not been investigated before.

For the OPV case study, effects of nZnO and equal concentrations of released dissolved Zn were compared in experimental exposures of zebrafish eggs. The zebrafish embryos were exposed from 3 h postfertilization (hpf) to 96 hpf. An environmentally relevant exposure media was prepared by adding salts and naturally occurring alginate to nanopure water. Alginate is a polysaccharide-based organic material previously shown to stabilize nanoparticles and reduce their aggregation rate. The Brunauer–Emmet–Teller (BET) analysis showed a mean size of 9.4 nm under dry conditions and between 196 and 336 nm in exposure media (Nanosight), indicating a considerable particle aggregation. ICP-MS analysis was performed to quantify Zn ions released from nZnO. An appropriate amount of $ZnCl_2$ was used to obtain a medium without nanoparticles but the equivalent concentration of soluble Zn. Zebrafish embryos were exposed to 0.2, 1, and 5 mg/L nZnO, as well as to the corresponding concentrations of 0.27, 1.30, and 5.74 mg/L $ZnCl_2$. Exposure to nZnO or Zn affected the hatching rate, depending on the exposure concentration. At 72 hpf, hatching was significantly delayed in exposed embryos at all exposure concentrations of nZnO and Zn^{2+} compared to controls. At 96 hpf, almost all individuals of the lowest-dose groups hatched, whereas only a few embryos hatched in the mid- and high-dose groups. No significant differences occurred between nZnO- and corresponding Zn-dissolved concentrations (Fig. 12.8). Therefore the dissolution of Zn from nZnO, rather than the nanoparticles, is suggested to be the key determinant for delayed and inhibited hatching. An inhibitory effect of nZnO on hatching of embryos was previously described.

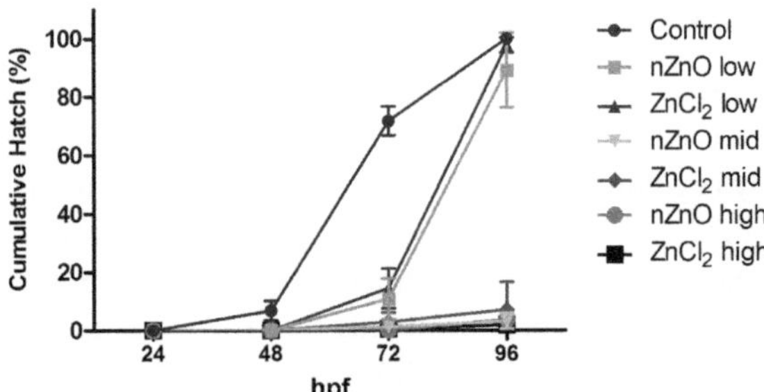

Figure 12.8 Zebrafish embryo hatching success at 24, 48, 72, and 96 hours postfertilization (hpf) when exposed to nZnO (0.2, 1, and 5 mg/L) and $ZnCl_2$ (0.27, 1.30, and 5.74 mg/L). The low concentration delays hatching, whereas the middle and high concentrations almost completely inhibit hatching. Figure from Brun et al. (2013). *Sci. Total Environ.*, 476–477, 657–666.

LA-ICP-MS analysis demonstrated that Zn is accumulated in exposed zebrafish embryos at 120 and 168 hpf. The prominent target organs for Zn uptake were the retina and pigment layer of the eyes, brain structures, spinal cord, and yolk sac (Fig. 12.9), possibly due to the higher activity and presence of uptake transporters in these tissues.

Gene expression analysis by qRT-PCR was performed using 48 h and 96 h embryos. For the two time points, mRNA levels of genes related to MT (*mt2*) and oxidative stress (*Cu/Zn-sod*) were determined. Transcripts of *mt2* were significantly up-regulated at both time points and almost all concentrations, displaying an enhanced need for metal binding. Expression of *Cu/Zn-sod* showed a similar temporal pattern for both nZnO and Zn, with an up-regulation at 48 hpf and a down-regulation at 96 hpf at the highest concentrations (Fig. 12.10). Superoxide dismutase (SOD) and catalase are enzymes protecting the cells against superoxide free radicals. While SOD catalyzes the breakdown of superoxide into oxygen and hydrogen peroxide, catalase (Cat) then catalyzes the decomposition of hydrogen peroxide to water and oxygen. The results indicate a reaction to elevated ROS levels and thus an induction of oxidative stress at 48 hpf. In contrast, the down-regulation at 96 hpf may be related to a negative feedback mechanism. Similar alterations in gene expression were found for nZnO and $ZnCl_2$.

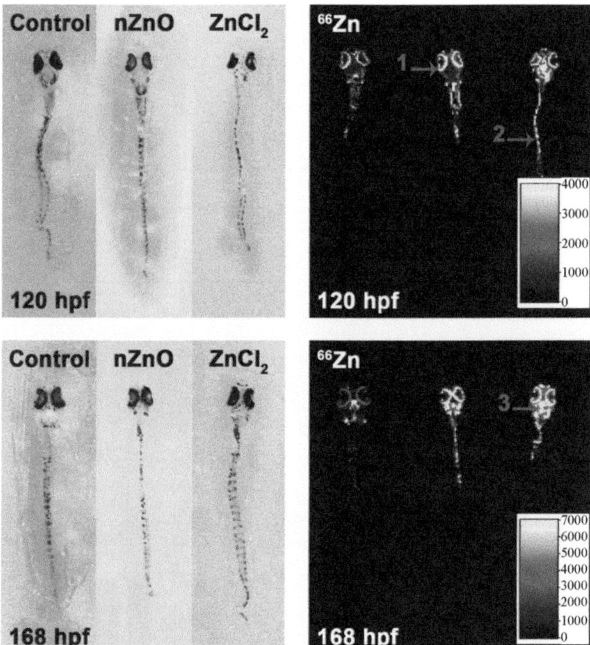

Figure 12.9 Pictures showing embedded embryos before LA-ICP-MS analysis (left) and Zn organ distribution measured by LA-ICP-MS (right). In each picture coronal zebrafish embryo sections of controls (embryo on the left), nZnO 5 mg/L (centre) and $ZnCl_2$ 5.74 mg/L (right) exposure groups at 120 and 168 hpf, respectively, are shown. Spots of accumulation are labelled as (1) retina and pigment layers of the eyes, (2) spinal cord, and (3) brain. Color bars show the intensity in counts per seconds. Figure from Brun et al. (2013). *Sci. Total Environ.*, 476–477, 657–666.

The results demonstrated dose-dependent effects of nZnO and Zn^{2+}. A fast release of Zn^{2+} from nZnO in water is very likely to make the free metal ions the primary source of the observed effects. LA-ICP-MS analysis has demonstrated that Zn released from both Zn sources became equally accumulated in zebrafish embryos. Both nZnO and Zn showed a similar hatching interference, and they induced similar effects on gene expression. This leads to the conclusion that the nanoparticle nature itself contributed very little to the effects and the biological activity on nZnO is mainly based on the release of free Zn.

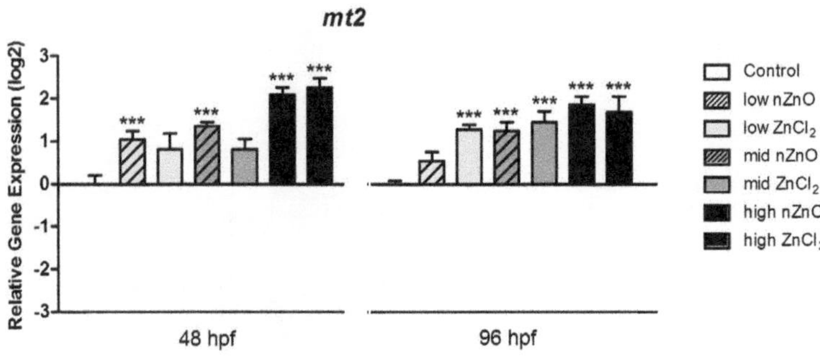

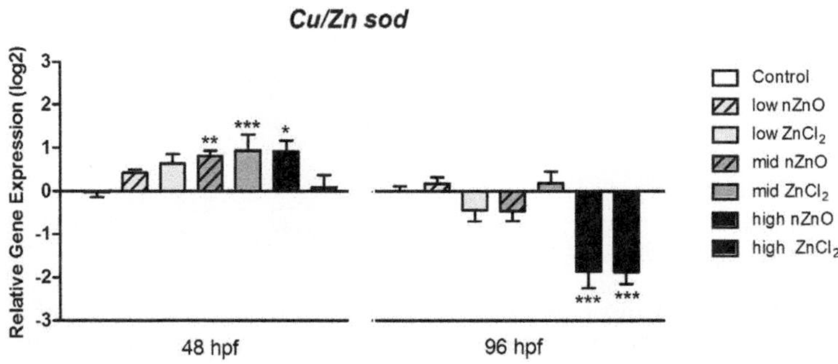

Figure 12.10 Gene expression of metallothionein (*mt2*) and oxidative stress related gene Cu/Zn superoxide dismutase (*Cu/Zn sod*) in zebrafish embryos (at 48 and 96 hpf) exposed 0.2, 1, and 5 mg/L nZnO and corresponding concentrations of free zinc by adding 0.27, 1.30, and 5.74 mg/L $ZnCl_2$. The asterisk represents a significant difference between treatment and control with $p < 0.05$. Figure from Brun et al. (2013). *Sci. Total Environ.*, 476–477, 657–666.

In sewage treatment plant effluents, concentrations of nZnO are expected to be in the range of 0.3–0.4 μg/L. Despite relatively high nZnO concentrations used in this study, the data contribute to the understanding of potential hazards and effects of nZnO in the aquatic environment. At the lowest tested concentration, still about 500 times higher than expected concentration in sewage treatment plant effluents, effects on the gene expression level diminished or were not observed at all.

12.3.5 Concluding Assessment of OPV Environmental Impact

OPV is a promising innovative electricity-generating technology offering novel application possibilities. In comparison to classic PV modules based on silicon, a wide variety of chemical substances are incorporated into OPV layers, for example, organic polymers or nanoparticulate metals. Such novel substances, also applied in OPE, are so far not well studied concerning their life cycle as well as their environmental fate and ecotoxicity.

LCA studies showed that total environmental impacts of OPV are at the current stage of development similar to multi-Si cells but higher than for CdTe cells. Since the main environmental impact of OPV stems from the electrodes, ITO replacement by silver and improved efficiency will considerably decrease their environmental impact below multi-Si and CdTe cells.

OPV leaching studies showed that metals are released from OPV materials, yet at such minute amounts that the PEC calculated would meet even stringent drinking water criteria. PCBM, which may have an adverse effect, was shown to be extremely hydrophobic, so even if released, it would most probably bind strongly to soil/sediments, rendering its bioavailability.

In ecotoxicity experiments it was shown that nanoparticulate ZnO leads to adverse effects. However, such effects were not elevated in comparison to ionic Zn(II), which is occurring naturally. Furthermore, effects were only problematic in concentrations far above concentrations that would result from OPV leaching.

In summary, from the current knowledge point of view, OPV bears the potential to become a truly green technology.

Exercises

12.1 The energy payback time (EPBT) indicator is very often used to compare PV technologies (see also Section 12.3.2.1). It focuses only on energy aspects and does not consider environmental impacts from emissions. In this exercise, EPBTs shall be calculated for multi-Si (21% eff.), CdTe (10.9% eff.), flexible OPV (2.8% eff.), and future tandem OPV (10% eff.). Normalization is required because of an expected lifetime of 30 years for multi-Si and CdTe and 20 years for OPV. The assumed harvested electricity is 1000 kWh per year and kWp. Table E.12.1 shows the CED for the production of PV technologies.

Table E.12.1 Cumulated energy demand for the production of 1 m^2 of a PV module

	Multi-Si (21% eff.)	**CdTe (10.9% eff.)**	**OPV (2.8% eff.)**	**OPV tandem (10% eff.)**
CED (MJ/m^2)	2900	840	180	73

12.2 Silver leaching: In a laboratory experiment over 4 months you determine that 100 mg of silver is leaching from 1 kg of OPV cells (=2 m^2). Calculate the PEC of silver in a scenario where OPV cells are installed on a roof (80% coverage; all rain water collected in one tank) in Madrid, Spain (average annual precipitation 300 mm; assumed to precipitate regularly over the year). Compare the PEC to the WHO drinking water limit.

12.3 Toxic components of the analyzed OPV: List the potentially toxic components of the analyzed OPV and their toxic effects. Metals have been elucidated as potentially toxic materials in OPV. Where in the environment do such hazardous materials end up? And what are the uptake routes in organisms?

Suggested Readings

Brun, N. R., Christen, V., Furrer, G., and Fent, K. (2014). Indium and indium tin oxide induce endoplasmic reticulum stress and oxidative stress in zebrafish (Danio rerio). *Environ. Sci. Technol.*, **48**, pp. 11679–11687.

Brun, N. R., Lenz, M., Wehrli, B., and Fent, K. (2014). Comparative effects of zinc oxide nanoparticles and dissolved zinc on zebrafish embryos and eleuthero-embryos: importance of zinc ions. *Sci. Total Environ.*, **476–477**, pp. 657–666.

Emmott, C. J. M., Urbina, A., and Nelson, J. (2012). Environmental and economic assessment of ITO-free electrodes for organic solar cells. *Sol. Energy Mater. Sol. Cells*, **97**, 14–21.

Espinosa, N., García-Valverde, R., Urbina, A., and Krebs, F. C. (2011). A life cycle analysis of polymer solar cell modules prepared using roll-to-roll methods under ambient conditions. *Sol. Energy Mater. Sol. Cells*, **95**(5), 1293–1302.

European Commission, Joint Research Centre, Institute for Environment and Sustainability. (2011). *International Reference Life Cycle Data System (ILCD) Handbook: Recommendations for Life Cycle Impact Assessment in the European Context*, 1st Ed., EUR 24571 EN. Luxemburg. Publications Office of the European Union.

Fabrega, J., Luoma, S. N., Tyler, C. R., Galloway, T. S., and Lead, J. R. (2011). Silver nanoparticles: behaviour and effects in the aquatic environment. *Environ. Int.*, **37**, pp. 517–531.

Fent, K., Chew, G., Li, J., and Gomez, E. (2014). Benzotriazole UV-stabilizers and benzotriazole: antiandrogenic activity in vitro and activation of aryl hydrocarbon receptor pathway in zebra fish eleuthero-embryos. *Sci. Total Environ.*, **482–483**, pp. 125–136.

Fent, K., Weisbrod, C. J., Wirth-Heller, A., and Pieles, U. (2010). Assessment of uptake and toxicity of fluorescent silica nanoparticles in zebrafish (Danio rerio) early life stages. *Aquat. Toxicol.*, **100**, pp. 218–228.

Grossiord, N., Kroon, J. M., Andriessen, R., and Blom, P. W. M. (2012). Degradation mechanisms in organic photovoltaic devices. *Org. Electron.*, **13**(3), pp. 432–456.

Handy, R. D., Owen, R., and Valsami-Jones, E. (2008). The ecotoxicology of nanoparticles and nanomaterials: current status, knowledge gaps, challenges, and future needs. *Ecotoxicology*, **17**, pp. 315–325.

Huijbregts, M., Hauschild, M., Jolliet, O., Margni, M., McKone, T., Rosenbaum, R. K., and Van de Meent, D. (2010). USEtoxTM User manual, www.usetox.org.

Institute for Applied Ecology (2012). Energieverbrauch von Fernsehgeräten, www.oeko.de/uploads/oeko/forschung_beratung/themen/nachhaltiger_konsum/infoblatt_fernseher.pdf (accessed October 30, 2014).

Isaacson, C. W., Kleber, M., and Field, J. A. (2009). Quantitative analysis of fullerene nanomaterials in environmental systems: a critical review. *Environ. Sci. Technol.*, **43**, pp. 6463–6474.

Jafvert, C. T., and Kulkarni, P. P. (2008). Buckminsterfullerene's (C60) octanol-water partition coefficient (Kow) and aqueous solubility. *Environ. Sci. Technol.*, **42**, pp. 5945–5950.

Kaegi, R., Voegelin, A., Sinnet, B., Zuleeg, S., Hagendorfer, H., Burkhardt, M., and Siegrist, H. (2011). Behavior of metallic silver nanoparticles in a pilot wastewater treatment plant. *Environ. Sci. Technol.*, **45**, pp. 3902–3908.

Laborda, F., Jiménez-Lamana, J., Bolea, E., and Castillo, J. R. (2011). Selective identification, characterization and determination of dissolved silver(I) and silver nanoparticles based on single particle detection by inductively coupled plasma mass spectrometry. *J. Anal. At. Spectrom.*, **26**(7), pp. 1362–1371.

Nel, A., Xia, T., Mädler, L., and, Li, N. (2006). Toxic potential of materials at the nanolevel. *Science*, **311**, pp. 622–627.

Nowack, B., and Bucheli, T. D. (2007). Occurrence, behavior and effects of nanoparticles in the environment. *Environ. Pollut.*, **150**, pp. 5–22.

Schwarzenbach, R. P., Gschwend, P. M., and Imboden, D. M. (2003) *Environmental Organic Chemistry*, 2nd Ed., John Wiley & Sons, Hoboken, New Jersey.

Zimmermann, Y. S., Schäffer, A., Corvini, P. F.-X., and Lenz, M. (2013). Thin-film photovoltaic cells: long-term metal(loid) leaching at their end-of-life. *Environ. Sci. Technol.*, **47**, pp. 13151–13159.

Zimmermann, Y. S., Schäffer, A., Hugi, C., Fent, K., Corvini, P. F.-X., and Lenz, M. (2012). Organic photovoltaics: potential fate and effects in the environment. *Environ. Int.*, **49**, pp. 128–140.

Chapter 13

Innovation Management

Ambarin Khan,[a] Silvia Massini,[a] and Chris Rider[b]
[a]*Alliance Manchester Business School and Manchester Institute of Innovation Research, The University of Manchester, Booth Street West, Manchester M15 6PB, United Kingdom*
[b]*Electrical Division, Engineering Department, University of Cambridge, 9 JJ Thomson Avenue, Cambridge, CB3 0FA, United Kingdom*
ambarinasad.khan@manchester.ac.uk,
silvia.massini@manchester.ac.uk, cbr24@cam.ac.uk

13.1 Organic and Printed Electronics: An Emerging Technology

Organic electronics refers to a new materials set as a technological differentiator; *printed* electronics, on the other hand, refers to a new process technology as a technological differentiator. Finding one simple name for the technology is complicated because inorganic functional materials may be printed and some organic materials are deposited by evaporation through a shadow mask. In essence though, organic and printed electronics (OPE) is a new way of making electronics enabled by a new materials set. It is a new manufacturing paradigm. Printing offers some generic benefits,

Organic and Printed Electronics: Fundamentals and Applications
Edited by Giovanni Nisato, Donald Lupo, and Simone Ganz

ISBN 978-981-4669-74-0 (Hardcover), 978-981-4669-75-7 (eBook)
www.panstanford.com

such as the potential for additive manufacture with a reduced number of process steps and less waste than subtractive processes. It also turns out that the new materials are largely low-temperature-processed, which means that low-temperature substrates, such as plastics and paper (which are flexible) can be used. The result is that electronic devices made under the new paradigm may exhibit form factor benefits such as flexibility, light weight, thinness, and robustness when compared to conventional silicon or circuit-board-based electronics. The actual electronic functionality that is being produced is not new: photovoltaics, lighting, displays, and sensors can all be implemented in other conventional ways. In the case of lighting, for example, organic light-emitting diode (OLED) offers a fourth-generation approach after incandescent bulbs, fluorescent tubes, and light-emitting diodes (LEDs). Innovation using OPE is almost always characterized, therefore, by considerations relating to the incumbent technology. There can be significant benefits, however, in using organic material architectures or printing to produce electronics for a particular application, and it is part of the skill of the business developer and the technologist to identify those applications where there is real value in using the new approach. This could relate to a performance or form factor benefit or it might relate to a manufacturing benefit of which the end user will be unaware.

OPE is an emerging technology, that is, science-based innovations with the potential to create a new industry or transform an existing one. Science-based businesses emerge at the intersection of multiple bodies of science. Innovation studies discussing the emergence of new technological and knowledge-based fields like biotechnology and nanotechnology have identified common dynamics that characterize different stages of development: uncertainty, complexity, heterogeneity of actors, distributed nature of knowledge, particularly relevant in the first stage of the industry life cycle, emergence of a dominant design, and convergence of a broad range of technological and scientific fields in a later stage.

Technological progress within emerging technological fields is mostly attributed to a large number of small entrepreneurial technology-based firms. However, as the technology develops, the domain of innovation shifts from small de novo firms to large established players and the nature of innovation evolves from advancement in scientific and technological fundamentals to commercial applications. Another important aspect of new technologies is their potential to be disruptive to the existing market and its value network. Products based on disruptive technologies tend to be simpler and cheaper initially, often inferior in terms of performance metrics as valued by mainstream customers, and offer different attributes (new value proposition) that become apparent and valued only once they develop from emerging markets into noticeable market penetration, from niche to mainstream.

OPE, though considered to have high potential for numerous applications, for example, in displays, lighting, photovoltaics, smart systems, sensors, batteries, radio-frequency identification (RFID), and smart textiles, is still developing and faces challenges as the supply chain is considered to be unbalanced with many players involved at producing the components rather than final product. At present there are approximately 3000 organizations active in the field located in Europe, USA, Japan, South Korea, Taiwan, and China, and they include universities, research institutes, large organizations, and start-ups. The advancement of material and other related technologies indicates greater technology push than market pull and no killer application has been identified yet to drive the whole market for OPE, although OLED display for mobile devices has now broken into the mainstream. It is estimated that 97% of the companies are materials, equipment, or component providers; only 3% make products or do integration.

The lack of articulated demand (by end users, and also by business customers) combined with a lack of articulated directions for product development creates a situation where actors are reluctant to invest and results in waiting games, which occurs when everybody waits for somebody else to commit to the new technology. Analysts argue that the lack of a one-size-fits-all solution in the printed electronics market and the variety of technologies each sitting in its own niche based on its own attributes delays the realization of high-volume markets that would enable true low cost.

The progress and development within printed electronics offers close comparison to the other science-based businesses like biotechnology and nanotechnology, which are characterized by a long period of risky investments and uncertainty and where entrepreneurial ventures and start-ups are involved in scientific discoveries owing to stronger technical challenges; reviewing them through innovation lenses tends to be all the more important.

In the next section we first introduce and explain the evolution of the OPE business sector and the complexity of the value chain (Section 13.2). Section 13.3 presents fundamental concepts from the innovation management literature that are relevant for understanding and managing the development of OPE: the systemic model of innovation, types of innovation (product, process, radical, incremental), and the sailing ship effect. Section 13.4 focuses more on the dynamics of developing technological innovations through collaboration within an ecosystem, referring to specific examples in the OPE industry, while Section 13.5 discusses several issues affecting commercialization of a new technology: technological and market uncertainty, testing a new market, developing dynamic capabilities, complementary technologies and products in the system, business model selection, intellectual property (IP), and the role of public policy in accelerating and fostering an emerging industry.

13.2 Introduction to the Organic and Printed Electronics Sector

Organic electronics is widely considered to have been born in 1977 with the discovery of conducting polymers and developed over the next two decades in a wave of optimism regarding the potential for the technology to reduce costs, to replace conventional electronics, and to create completely new markets. Plastic Logic, in a presentation given in 2002, stated, "Plastic Logic's technology [based on inkjet printing of active electronic circuits using organic semiconductors] has the potential to change radically the economics of key segments of the semiconductor industry. Capital costs will be significantly lower than for silicon by eliminating conventional photolithography, vacuum processing and high temperatures." Writing in *Nature* in 2004, Professor Steven Forrest, at the University of Michigan, one of the pioneers of OLED technology, said, "Organic electronics are beginning to make significant inroads into the commercial world, and if the field continues to progress at its current, rapid pace, electronics based on organic thin-film materials will soon become a mainstay of our technological existence. Already products based on active thin-film organic devices are in the market place, most notably the displays of several mobile electronic appliances. Yet the future holds even greater promise for this technology, with an entirely new generation of ultralow-cost, lightweight and even flexible electronic devices in the offing, which will perform functions traditionally accomplished using much more expensive components based on conventional semiconductor materials such as silicon." Moving forward 10 years to 2016 and, apart from OLED, which has become a market worth over $10 billion, largely driven by high-resolution color mobile displays on glass, there has been little to show for the much-heralded organic electronics revolution. Why is this?

Several factors characterize the development of markets for OPE and it is against the light of these realities that our discussion of general principles for innovation management for OPE will be set.

13.2.1 Allow Time for Big Materials Performance Gains

In the case of OLED displays, it took 32 years from the first publication of the bilayer OLED device by Professor Ching Tang, then of Kodak, in 1981 to the passing of the $10 billion market size milestone in 2013. Early OLEDs suffered from many issues, including very short lifetime (of the order of minutes), poor efficiency, and poor color reproduction. Process tools for vacuum deposition and patterning of OLED layers uniformly and rapidly over large areas of glass did not exist, and in addition, the thin-film transistor backplane (see, for instance, Chapter 5) needed to drive OLED displays required significantly higher performance and uniformity than the available amorphous silicon technology was then able to provide. Over the next 20 years, enormous resources were channelled into understanding and overcoming the causes of degradation and into the development of new generations of organic materials that provided higher efficiency, increased lifetime, and improved color. Process tools for mass manufacture were developed and one by one the technical challenges of providing a suitable backplane were overcome. All this has taken time and it is a common theme of organic electronics that lifetime and performance issues are often addressed by materials development over a period of years. Nevertheless, we see from the pioneering organic electronics technology of OLED that enormous improvements are possible and that consumer electronics products based on the technology can be advantaged in the marketplace.

Other organic electronics technologies should be able to leverage the learning from OLED and shorten the time to market.

13.2.2 The Incumbent Technology Is Still Evolving

Given the very long time it took for OLED to move from laboratory to market, it was possible for significant improvements to occur in the incumbent technology. For instance, in the case of active-matrix OLED (AMOLED) display, the incumbent technology was active-matrix liquid-crystal display (AMLCD). In the early 2000s, pioneers such as Kodak first started to commercialize OLED displays with AMOLED, promising major advantages over AMLCD: much lower power consumption, better off-angle performance, thinner construction, and more vivid color. This point was not lost on the tier 1 manufacturers of AMLCD and dramatic improvements in performance ensued over the next 10 years, particularly in color and off-angle viewability, so the performance gap between OLED and LCD has narrowed considerably and the business case for scaling up the technology became much more of a strategic matter for the leading display companies. For any new OPE technology, we must consider carefully the trajectory of improvement for the incumbent technology or (in the case of lighting and photovoltaics) technologies over the likely timescale for materials development and manufacturing scale-up.

13.2.3 The Challenges of Scale-Up Add Cost

In the early days of organic electronics, there was a perception that solution-based processing and, in particular, printing would give a cost advantage in manufacturing. For this reason, OPE has often been presented as a low-cost technology. The reality is that OPE presents other technical challenges and that when these are addressed in a high-yielding manufacturing facility, significant cost is added such that, again, the cost differential gap between the incumbent and the OPE technology is reduced. For example, although it may be highly cost effective to additively pattern organic materials by printing, some devices require feature sizes that are far below printing resolution so that other more expensive patterning approaches must be used. In the long term, solution processing of organic materials may offer a cost advantage in a high-volume facility, but for early applications of a given OPE technology, when volumes are low, this may not be the case. Another example relates to yield; although it is highly attractive to use a printing press to print electronic devices onto packaging materials, dust particles in the air may be of the same size as the features being printed such that yield can be dramatically reduced unless the whole press is placed in a clean room or an enclosed environment. This, of course, adds significant cost.

13.2.4 All System Components Must Be Ready before Product Sales

In the case of AMOLED, the scale-up of a suitable transistor backplane of sufficient quality turned out to be more challenging than the scale-up of the OLED frontplane. Yet, without the backplane, AMOLED displays are not possible. For OLED, as well as organic photovoltaics (OPV), a suitable moisture and oxygen barrier is needed. AMOLED uses glass, which has excellent performance as a substrate for organic electronics for a whole variety of reasons, including flatness, transparency, cost, and barrier properties. Roll-to-roll production has been considered as a route to reduce manufacturing cost for OLED lighting and for OPV, but to enable this, a flexible barrier is needed. Although thin flexible glass is now becoming available, it is not a suitable substrate for reasons of cost and robustness. Instead, a transparent plastic film with an integral moisture and oxygen barrier is being developed by several organizations worldwide, but as yet, there is no low-cost solution. Thus the low-cost flexible barrier is part of a roll-to-roll-produced OPV product, and even if the OPV device itself has suitable performance for its intended applications and can be manufactured at high yield on a roll-to-roll line, the product cannot reach its cost targets until the low-cost plastic flexible barrier film has been commercialized.

13.2.5 Multifunctional Systems Are Needed for Some Applications

In applications for OPE, such as intelligent packaging, anticounterfeiting and smart objects, systems are needed and these might need a sensor, some signal processing, a display, and on-board power in the form of an energy-harvesting module or a thin primary battery. Large end-users may want a company to put the whole system onto a label, which can then be applied to products at high speed using conventional converting equipment. Frequently, however, individual items in the system are produced by small companies and there are few companies that will integrate the whole system from multiple suppliers onto the delivery vehicle, such as a label. The need for system integrators is in some cases contributing to the slow growth of the market.

Returning now to the question posed earlier as to why the current market for OPE has not developed in a more sizeable one, we can see that there are two broad reasons:

- First, timescales are longer than expected because of the need for improved materials, new process tools, and the integration of complete systems. Many companies active in OPE are now close to or are actually selling products into first applications. In time, volumes will grow, yield will improve, costs will reduce, and greater opportunities will develop. We can expect market numbers to increase significantly over the next five years.
- Second, some companies did not survive or exited because they were not able to develop products offering sufficient value for early markets compared to the incumbent technologies.

13.2.6 Complexity in the Value Chain

The value chain of OPE is still evolving (Fig. 13.1). Apart from traditional players like the chemical industry, other players are from industries as diverse as mechanical engineering, printing, packaging, and consumer goods. Some players are dedicated to OPE technology development, while others are attracted by the huge potential of this emerging technology in their markets. The evolution of OPE is not limited only to knowledge development and accumulation within companies, but is also about formation of clusters and, ultimately, formation of a global community.

With this as a background, we now introduce basic innovation management concepts that can be applied to the commercial development of OPE.

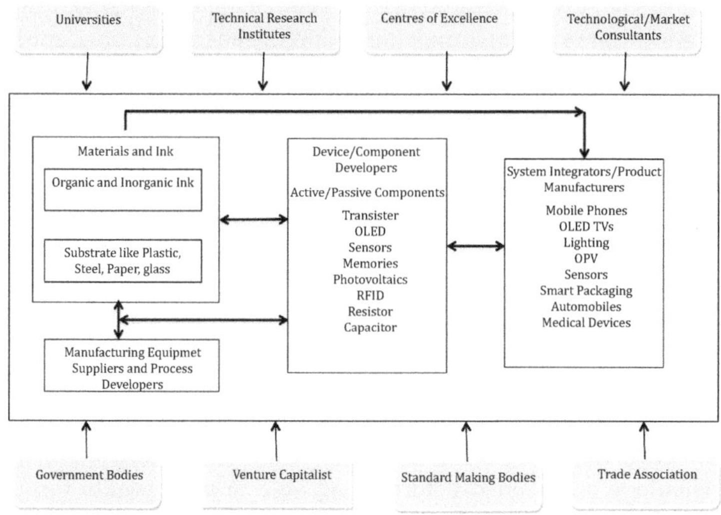

Figure 13.1 Value chain.

13.3 Innovation Processes and Management

To understand the main issues and challenges in developing successful new products and technologies it is important to identify the main characteristics, patterns, and dynamics of innovation. There are many definitions of innovation, but the most widely accepted in the innovation management community is that "innovation is the successful exploitation of new ideas." This implies conceptualizing innovation not only as an object or outcome of a research and development process, but also as the process for its development, commercial exploitation, and eventually wider diffusion in markets and economies, resulting in wealth creation and profitability.

13.3.1 Innovation Processes: From Linear to Systemic Models

Over the last 40 years innovation studies, as a discipline at the interface between economics, business, management, and strategy, have made substantial progress in developing understanding of innovation and its sources, challenges, and management. This progress has given origin to five conceptual models, or generations, of innovation, starting from simple and linear models of science or technology push and demand pull, which reflect the post–World War II environment, to integrated, systemic models closer to the more complex reality of developing technological innovation today.

The first, and simplest, model of innovation is the science push, which is based on the capacity of relatively small elites, often based in universities and public research institutes, to make scientific discoveries that underlie new scientific industries, as observed in the 1960s. This model reflected the success of big science and can be exemplified by the Manhattan Project and the development of the atomic bomb. Today, this basic model is often used to justify public investment in new sciences, such as biotechnology and nanotechnology, where substantial investments in developing new technology are needed before being able to reap profits through commercialization, and therefore provide economic incentives to private firms to invest in emerging technologies.

By the 1970s the pace of scientific discoveries was beginning to slow, and the optimism of the 1960s was in decline. The organization of technological innovation by the private sector began to change, with the emergence of the demand-led second generation of R&D management, which lasted from the 1970s to the mid-1980s. In this model, R&D became much more accountable and R&D centers were often downsized, with a greater emphasis on problem-solving development, which was increasingly decentralized to firms' production sites. Innovation as a whole became much more responsive to market trends rather than creating new markets. As a consequence of this change in the mind-set of corporations, R&D became increasingly short term, project based, and solutions based, with a stronger focus on marketing as a strategic organizational capability for successful innovation.

The third-generation model combined both the science push and market pull models, which preceded it, as the marketing–R&D coupling model of innovation, but whilst recognizing the importance of both supply and demand in innovation, this model was still highly linear or sequential, emphasizing the development of new products, rather than implementing their production.

A more fundamental re-conceptualization of innovation as a process and how to manage it came from the extraordinary success of Japan. Japanese firms developed an approach to innovation that integrates developments in various *functional areas* within a firm, for example, marketing, R&D, production engineering, manufacturing, etc., all undertaken at the same time, in parallel, rather than in a linear and sequential manner. This approach sees innovation as a distributed system and much more inclusive and incremental, rather than imposed and radical. This fourth generation of R&D, which dates from the late 1980s, was based around attempting to balance short- and long-term financial needs by maintaining a portfolio of short- and long-term projects, undertaken in partnership with the operating divisions of the company and sometimes with other strategic partners, such as universities and public research units.

The success of the Japanese highlighted that innovation is increasingly distributed between firms and other organizations, rather than dependent on a single large firm acting alone. This understanding gave rise to the current strategic integration, fifth-generation model of innovation, which observes that innovation is increasingly dependent on cross-functional integration within the firm, in combination with close interrelations with external agents, such as suppliers, customers (or users), and possibly, in addition, commercial partners and other sources of technology, such as universities or public research institutes. The development of OLED seems to fit very well with this systemic and distributed view of innovation within an ecosystem made of a variety of partners, due to the complex nature of this technology and the multiplicity of potential applications, whereby the earlier simple linear models of innovation would not allow to capture the full dynamics and interactions of developing and managing OPE.

13.3.2 Types of Innovation over the Industrial Life Cycle

In recent decades, it has been observed that technologies and industries are increasingly blending together, with innovations increasingly crossing industry boundaries: a new fiber developed by the textile industry has potential for building materials and medical equipment. This technological fusion blends incremental technical improvements from several previously separate fields of technology, generating products that revolutionize markets. Optoelectronics and mechatronics are examples of areas of technological fusion.

It has been recognized that in several markets for OPE, it will be necessary to combine both printed and silicon-based electronics to form working systems. For example, if it is required to store more than a few tens of bits of digital data, it is currently necessary to use silicon memory. Similarly for more than the simplest processing of information, silicon microprocessors are required. An example of this would be for printed sensor systems, some elements of which may be printed with other elements implemented in conventional silicon. The challenge for these hybrid systems is to maintain the form factor benefits of thinness and flexibility that are enabled by printing electronics on flexible substrates.

Individual companies, however large, are unlikely to be able to master and remain abreast of developments in several technological fields at once. Consequently, firms need to build research consortia and long-term ties, both with other companies and with public sector organizations such as universities, and this requires the development of collaboration skills, in what we now call an *ecosystem*, which, crucially, requires reciprocity, mutual trust, and responsibility, to enjoy shared rewards and reduce opportunistic behavior.

The need to engage in collaborative arrangements for innovation seems to be increasing in recent years for several reasons. First, the relative cost of R&D is rising faster than general business costs because R&D is labor intensive and wages tend to rise faster than other costs. Second, markets have become less certain and are quicker to change. This, coupled with the decline in the mass production of highly standardized products, means firms seek to reduce the risks inherent in innovation by engaging in collaborations and strategic alliances. Moreover, the emergence of new competitors in the global business landscape, mainly high-tech companies in emerging economies such as South Korea, Singapore, and Taiwan, has put additional pressure on long-established companies in Western economies.

Collaborations between industry and the science base of universities and government research organizations have attracted considerable attention in recent years among academic researchers and policy makers in most advanced economies, including continental Europe, USA, and Japan. Encouraging greater collaboration is not just a question of fixing the institutional arrangements; it is also a matter of developing trust upon which collaborative relationships are based. It may also involve universities changing their research priorities to make them more relevant to industry, although this does not necessarily mean making research directly responsive to the requirements of industry.

An important distinction is between radical and incremental innovation. Incremental innovations build on existing products and the knowledge required to design and manufacture them; they tend to enhance existing competences, allow existing products to stay competitive, and are characterized by low uncertainty. Most innovations are incremental. Radical innovations require technological knowledge that is very different from existing knowledge, which in turn can become obsolete along with the products on which they are based. Radical innovations are therefore competence destroying; they are characterized by higher uncertainty and risk and occur less frequently. However, when they do appear in the market, they can dramatically change the competitive industrial landscape. Incumbents may outperform new entrants in some radical innovations because, although their technological capabilities may become obsolete, their stronger market capabilities may remain intact.

Managing innovation in OPE is particularly challenging because it is a pervasive technology that can be applied in a variety of industries, in multiple markets that are at different stages of development. For example, OLED display is already being produced at scale, but OLED lighting is still in its infancy and may not succeed from a business point of view. Overall, the dynamics of competition are very different in different markets. OPE is by and large a new way of doing something that is already done. Therefore competition is not only among companies developing and producing OLED but also against producers of the technology it could replace.

One interesting phenomenon is the observed increase in the rate of (often incremental) innovation of an incumbent technology as a response to the introduction of a new technology to a market. This phenomenon, known as the *sailing ship effect,* refers to advances made in sailing ships in the second half of the 1800s in response to the introduction of steamships. In fact, in the 50 years after the introduction of the steam ship, sailing ships made more improvements than they had in the previous 300 years. More generally, the sailing ship effect applies to situations in which an old technology is revitalized, experiencing a "last gasp" when faced with the risk of being replaced by a newer technology. Three possible explanations have been suggested as the cause of the sailing ship effect: (1) Old technologies improve in an attempt to avoid being replaced by the new ones, (2) components of new technology spill over and improve incumbent technologies, and (3) old technologies enjoy new notoriety generated by the new technologies. The effect is more likely to occur if the new technology experiences delays in the emergence of a dominant design or in the development of economies of scale. An example of this would be the significant improvements in LCD display technology over the last 15 years to close the gap on the perceived advantages of the emerging OLED display technology.

13.4 Developing Innovation through Collaboration: Role of the Ecosystem

One of the reasons discussed earlier for slow market growth is the absence of system integrators and the silo approach of component developers, that is, the lack of an integrated manufacturer (in most segments) who can assemble all the different components and develop a device for the end user. The industry is addressing this challenge and barrier to commercialization by adopting an ecosystem strategy and collaborating with diverse partners, as commented by an industry expert in an interview conducted in year 2012.

> *We've had many, many different companies trying to live out of individual components. We had the battery companies. We had display companies and so on and so on. And they started to talk to each other. And then no one knows who is taking the lead, I mean who is doing the integration? Is it we, or the other one?*
>
> *But I think that that has to mature and you have to find your roles. But that's the only way to get this out on the market, is the collaboration. But I'm not sure we have found the right form of collaborations yet. I mean, it's very often bilateral, that two companies talk to each other and then they try to do something.*

The concept of an ecosystem, that is, the collaborative arrangements through which firms combine their individual offering into a coherent customer-facing solution, suggests a third form of organizing economic activity, in addition to the market itself and the hierarchy in a highly integrated firm, and has its roots in complexity and chaos theory. A business ecosystem consists of interdependent heterogeneous actors linked by either cooperative or competitive relationships that are dependent on each other for their survival and success. The importance of building an ecosystem (or collaborative networks or value networks, as they are also called) varies across industries and tends to be critical for emerging technologies, owing to higher market and technological uncertainty, dynamism, complexity, and distributed nature of knowledge and increased interdependence with complementors.

An important characteristic of ecosystems is that they are not limited to particular industries and their boundaries are thus difficult to identify. Organizations that are part of a particular ecosystem may also belong to other networks and ecosystems at the same time. Loose coupling tends to be the important feature of successful business ecosystems. Participation in an ecosystem offers advantages such as reciprocal commitments from other members, shared vision, distributed cost, and a common fate. It enables the players to direct their efforts toward performing continuous innovation and establishing markets for their offerings rather than working on independent paths uncertainty.

The ecosystem approach seems to be promising for rapid technology development and commercialization within OPE. It has been adopted especially by small and medium-size players who specialize in particular knowledge areas but, at the same time, suffer from liability of newness and liability of adolescence and are constrained by limited financial and human resources. Prominent examples of the ecosystem approach are those by Pragmatic Printing (U.K.) and Thin Film Electronics (TFE) (Norway), which are presented and discussed next.

Pragmatic Printing specializes in the field of imprinted logic circuits that enable functionality and interactivity to be applied across a range of applications such as greeting cards, smart packaging, and security products. The company was established in 2010 after the acquisition of Nano ePrint is headquartered in Cambridge (U.K.), and employs a licensing model. Since its inception the company has been able to leverage the networks developed earlier around Nano ePrint and is working toward the development of an ecosystem with diverse and heterogeneous partners to demonstrate the possibilities of the technology. The Pragmatic Printing ecosystem exists at three different levels (as shown in Fig. 13.2) around core technology (i.e., imprinted logic), around systems consisting of other complementary technologies, and at the product level with brand owners. The networks around core technology mainly consist of equipment providers, material suppliers, designers, and integrators, while at the system level, networks are more fluid and include suppliers of other technologies such as printed batteries, displays, antennae, or photovoltaics that may be required to enable particular system functions. The requirements of the product level are driven mostly by brand owners and their needs for product features and performance.

> *We have been talking about ecosystems for 3 years. There is a growing acknowledgement that companies with their own technologies need to work with partners who have complementary technologies to deliver the solution.*
>
> *[. . .] with the state of adoption of PE, we inevitably have to stimulate activities across the whole value chain not just focusing on developing our technology but working with end users and brand owners, who won't necessarily be a licensee but who would be the one who is actually placing the demands on licensees and their existing supply chain.*[1]

The company initiated its "Pragmatic Printing Pilot Program" initiative (P4) with a vision to transition from small-scale fabrication of product demonstrators to pilot production and collaborated with a range of ecosystem partners, in particular the Centre for Process Innovation (CPI), a UK-based technology center that now houses the P4 line and acts as an intermediary. Intermediary organizations, including brokers, third parties, and agencies, are involved in supporting the innovation process. As innovation activities tend to be increasingly distributed, the role of intermediaries has evolved

[1]Interview conducted by one of the authors with Pragmatic Printing management during the period 2011–2013.

from being passive to more active, such as the co-development of innovation, network facilitation, and network governance.

Commenting on the need for moving toward pilot production and field trials that otherwise may not be considered a necessary step for organizations planning to license their technology to mature industries, a respondent at Pragmatic Printing elaborated:

> *The turnover rate [of field trials is] not fast enough - projects need to be allowed to fail quickly. And then from the experience you will learn something, rebuild and do things slightly differently. There's often too much focus on the process technology rather than getting field trials started and using these to identify problems for volume commercial deployment.*

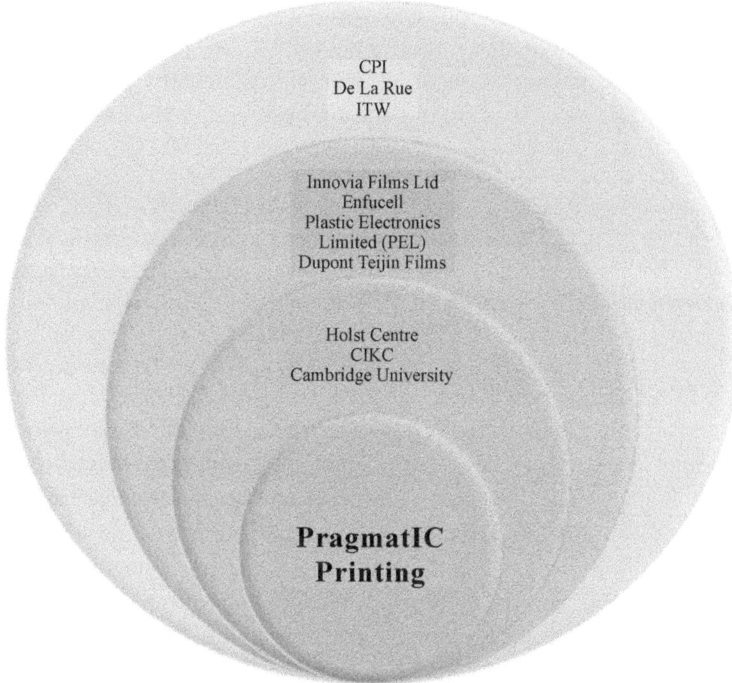

Figure 13.2 Pragmatic printing ecosystem (2013).

Innovation in its early stages often requires long periods of experimentation and multiple iterations, including probing and learning processes for technologies that have multiple applications in diverse markets. Probing with earlier versions of immature technology involves a variety of ways and may take the form of demonstration projects, field trials, and prototypes. Demonstration projects and trials (DTs) reduce the technical and market uncertainties associated with new technologies, aid in the technology development and validation, and enable scale-up and commercialization.

TFE, having its origins in 1990, headquartered in Oslo, Norway, is another example of the development of an ecosystem. TFE's founding expertise is in ferroelectric memories but owing to its entrepreneurial orientation does not want to limit itself simply to licensing or supplying components to integrators. Since 2010, under the leadership of Davor Sutija, the company has been keen to establish itself as a product company and is working toward integrated printed systems. TFE's vision is to accelerate its participation in the "Internet of Things" megatrend and it believes that building an ecosystem is a key enabler.

> *So, the entire idea of creating an ecosystem is that it allows you to become world class . . . before you make commitments of capital to make a particular product. So, our idea about creating an ecosystem is on the one side to figure out what is technologically possible in printed electronics; we'll learn from our partners and create a win-win business model, and then get business partners (Source: Interview[2]).*

TFE's vision is to apply electronics to disposable items. After thorough market analysis, TFE selected three product systems where printed electronics offers unique advantages: temperature sensor tags, display tags, and open systems enabling communication with devices by radio frequency (RF). To achieve these goals, strategic collaborations were established with Acreo (electrochromic displays), PST Sensors, Imprint Energy (batteries), and Polyera (n-type semiconductors) and existing relationships with PARC, InkTec, and Solvay were further strengthened. In January 2014, TFE completed the acquisition of Silicon Valley industry leader, Kovio, with its near-field communication (NFC) technology. The strategic move enables TFE's expansion of IP assets and ensures rapid progress toward its vision of "Internet of Things" and adding functionality to

---

[2]Interview conducted by one of the authors with TFE during 2011–2013.

everyday objects. In addition to that, the NFC Innovation Center has also been established in Kovio facilities at San Jose employing more than 20 Kovio team members.

TFE has pursued not only technological partnership, as is the case with most of the technology and component developers in the printed electronics space, but is also working toward the development of a broad range of commercial opportunities. TFE has established strategic relationships with fortune 500 companies such as Bemis, Brady, and Hasbro to enable commercialization and reduce the speed to market. Figure 13.3 demonstrates TFE ecosystem approach:

> *The fundamental risk is that the technology either doesn't succeed or that another technology replaces it, or that intended application goes away because of something . . . So a delay in getting a product to market, increases risk" (interview DT).*

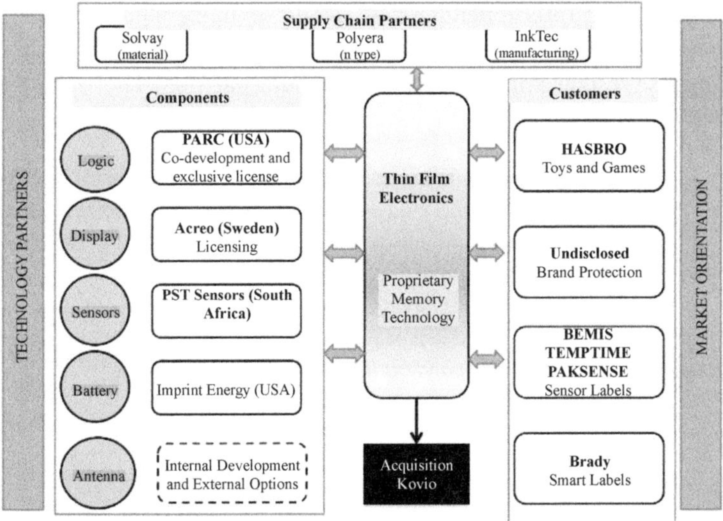

Figure 13.3 Thin Film Electronics ecosystem (2013).

13.4.1 Challenges of Innovating within an Ecosystem

Whereas the ecosystem approach enables continuous innovation due to increased cooperation, the formation of an innovation ecosystem is not without risks, and even if alliances can mitigate risk, organizations struggle to get value out of their alliance portfolios. Due to increased interconnectedness that characterizes the ecosystem, the challenges do not reside only within the integrator firm but are also related to technological barriers offered by other component providers as well.

As increased complexity and uncertainty surround the final product, the technology provided by partners and component providers in the ecosystem might be developing at a different rate than desired. A change in one part of the system may trigger changes throughout and hamper the success of the final innovation, slowing the overall pace of technology development and finally resulting in lost opportunity. Constraints in the ecosystem, however, are not only technological but there are other relational challenges such as opportunistic behavior by the partners; lack of motivation, transparency, reciprocity, and trust; misalignment of goals and objectives; and changes in priorities during the collaboration. These constraining interdependencies of one or more alliance partners have a negative influence on the performance of entire ecosystems and also restrict knowledge transfer among partners. These concerns were also raised in interviews conducted by one of the authors:

> *[. . .] The challenges around logic technology ecosystem . . . are concerned with partners and suppliers buying into the same vision, so there are always risks that this is not going to happen or times will slip. They put a lot of effort into something that they don't necessarily get immediate benefit from. That's about having open, honest and realistic discussions and finding partners that are willing to embrace this and work with you from an early-stage (respondent RP).*[3]

13.5 Considerations for Commercialization

It is not our intention in this section to provide detailed analysis and recommendations concerning all the issues that should be considered before and during the commercialization of OPE technologies but rather to provide an overview of those that are particularly relevant.

---

[3]Interview conducted by one of the authors during period 2011–2013

13.5.1 Uncertainty and Market Dynamics

New products based on new technologies are in general characterized by a high degree of uncertainty, both technological and commercial. Innovating companies face uncertainty about developing a new product that will actually work and perform better than the incumbent product; they also cannot always tell whether consumers will be willing to buy the novel product. This double uncertainty has a strong impact on the investments that companies, financial institutions, and venture capitalists are willing to commit in new technology-based activities. On the other hand, even once the necessary financial resources have been made available, early adopters of the new technologies are crucial to further progress as they will allow increasing production and therefore economies of scale that will make the new product more affordable and attractive to other adopters and allow incremental improvements in the new technology, which again may make it more attractive to other buyers.

These technological and market uncertainties are evident in the case of OPE. Despite the initial hype and promises, participants in the industry have not been able to find niche markets or a killer application and meet the initial expectations of the actors, and it has been argued that, despite having developed fairly sophisticated knowledge of organic electronic materials, the industry appears more like silicon in the 1960s than silicon today, and a shorter development time should not be expected.

Uncertainty about technological developments in the field of OPE increases the risks for actors to commit resources. This translates into a situation where actors are waiting for the other parts of the system to be developed. Lack of articulated demand, and the differences across sectors, exacerbates the situation. Key actors like material suppliers and printing equipment manufacturers anticipate changes to their current business models but are reluctant to move.

13.5.2 Testing Business Assumptions

OPE is a new way of making electronic devices. For the most part, OPE devices offer the same broad functionality as conventional electronics: displays, lighting, sensors, logic, photovoltaics, etc.

Neglecting for the moment the question of manufacturing cost, which usually has a strong dependence on production scale, OPE offers a subtly different value proposition from conventional electronics. The form factor may be thin, lightweight, and, in some cases, flexible, although glass is often a substrate of choice, owing to its excellent flatness, temperature resistance, and barrier properties. Tunability of performance may be achieved through optimizing the molecular design of organic materials. Product customizability may be achieved if digital (maskless) manufacturing techniques are used, such as inkjet printing or direct-write patterning techniques. Other benefits may include low energy production, low capital cost for a manufacturing process, environmentally benign materials, and large-area deployment where electronic functionality may be printed on surfaces separated by very large distances compared with a silicon wafer or a printed circuit board or a sheet of display glass. When assessing a new business opportunity that is proposed for an OPE implementation, it is essential to articulate clearly why OPE should be used over conventional electronics: the "why OPE" question. If there is no compelling reason why OPE can offer better value than conventional electronics, then it is almost certain that it will be cheaper and faster to implement a solution in conventional electronics.

13.5.3 Developing Dynamic Capabilities

Having strong or superior technology does not guarantee commercial success. It is important to maintain full awareness of the products being developed by competitors and be able to respond quickly to unexpected changes in the industry. Sometimes this is not enough and a fast second mover may launch a more successful (cheaper and/or better) product than the original one launched by the first mover.

Konarka, a highly prominent start-up with a focus on solution-processed OPV, was founded in 2001, raised over $170 million during its life, built a factory able to produce 1 GW of photovoltaics, and filed for bankruptcy in 2012. Its strategy was to build volume from early niche products and using its low-cost manufacturing processes to coat the active materials on cheap plastic such that it would be able to undercut the cost of electricity production of silicon photovoltaics (the incumbent). Much has been written about the failure of Konarka, but two clear points emerge:

- It was early in the field of OPV and therefore had to build its business on early active materials, which had relatively low efficiency and a short lifetime.
- There was an unprecedented drop of around 3× in the price of silicon modules during the period from 2007 to 2011 such that Konarka's cost target advantage was all but eliminated, making further investment extremely challenging.

Konarka, together with many other photovoltaic manufacturing companies of both conventional and thin-film technologies, experienced the fallout from the unexpected and dramatic impact of photovoltaic market dynamics.

Plastic Logic, like Konarka, was also an early pioneer of organic electronics and developed a flexible transistor backplane technology to drive bistable black-and-white displays for the new eReader market. Its first product, the Que, boasted a large screen size with touch capability, physical robustness, good outdoor readability, and long battery life. It was announced at the Consumer Electronics Show (CES) in January 2010. Some three months later, Apple launched the iPad, effectively creating the tablet category. Although the iPad did not have such a large screen, such a long battery life, or such good outdoor readability, it performed well as an eReader and additionally was a high-performance computing tool with a very high-quality, touch-sensitive, video-rate color display and, significantly, was cheaper than the Que. In August 2010, Plastic Logic withdrew the Que, without having shipped a single device. The introduction of the iPad completely changed the market landscape. Pure eReader device sales peaked in 2011 and have been declining since then. Plastic Logic has now changed its strategy and is focusing on new opportunities.

Both these examples illustrate the need for market vigilance and the ability to change direction quickly should the unexpected occur.

13.5.4 Supplying a System

We have already discussed the importance of developing an ecosystem with other suppliers so that a complete product system may be brought together comprising multiple functional parts each supplied by a leader in the provision of those parts. An important question for every new commercial venture is how to define its participation strategy in the markets it chooses to serve. Will it supply a single component to a system integrator? Will it supply a subsystem comprising some of its own technology and some technology components brought in from other suppliers? Will it supply complete systems to a product company? Will it supply complete products to an end user? Will it provide some element of a service offering with or without an accompanying product?

In OPE, the technology always resides in a larger system. Even large-area products like photovoltaics or lighting, which would at first appear to be single-function products, are almost always supplied with other system components. For a photovoltaic product, as well as the photovoltaic element itself, there may well be an electronic protection circuit to stop damage to the device, there may be a circuit to enable the device to operate at its maximum power point, there may be a voltage converter to increase the output voltage, there may be a charge store (battery or supercapacitor), and there may also be some circuitry to enable safe and efficient charging of the charge store. Then there is the mechanical mounting or deployment of the device and its electrical connection system that must be designed. With all these system components, the prospective supplier must consider whether it will supply the single component, a subsystem, a fully working but unpackaged system, or a full product. If it chooses to sell only a part of the full system, then it must work with its partners to ensure that all the system parts work well together and that the integrator knows how to get the best out of the system. With more complex systems, such as wireless sensing labels, the end user may simply require a roll of labels that can be attached to product packaging using conventional converting equipment, and may not be interested in manufacturing the labels. A system on a label must then be supplied, and if there is no obvious integration company in the supply chain to assemble all the system elements onto the label, this becomes a key barrier to commercialization. This particular issue has been widely recognized in the OPE industry, where there are currently relatively few integrators. An increasing number of interventions by government through R&D support are enabling this gap to be closed.

Once a company has decided where it will participate in the value chain and when it has understood the boundaries of the system that the end customer requires and the price the end customer is prepared to pay for the system, it is possible to assess where the system value is really created and whether the company can secure suitable returns on its investment. It can be at this point that it becomes apparent that more value could be captured if the company is willing to enlarge its offering to include more value-generating parts of the system. Which strategy to adopt will depend on the firm's capabilities, as well as the opportunity itself. A company may decide to change its capabilities by acquisition of staff or even a business if it feels it cannot participate in the best opportunity due to mismatch of capability, or it could consider outsourcing part of its operation to a partner with the right capability. Continuing, then, with the photovoltaic example, it may become apparent that rather than supplying just the photovoltaic element to a system integrator, it may be much more commercially attractive to supply the photovoltaic element and a thin rechargeable battery attached to it. The company may then decide to source a suitable battery from a partner and develop the attachment and connection processes to produce the new system component.

13.5.5 Business Model Selection

The selection of its business model is a key step for a new company. For OPE businesses, where there is often a significant physical technology content, the three most common choices are:

1. Manufacture and sell goods based on the new technology.
2. Develop the new technology to the point that capability to manufacture at production scale has been demonstrated and licence the manufacturing process to companies that wish to manufacture themselves.
3. Develop the technology to the point that the technology has been demonstrated (at a small scale) to be able to provide the anticipated performance and/or cost benefits and then licence the technology (patents or know-how) to a manufacturing company that has the capability to both scale-up and manufacture the new technology.

The pros and cons of the three options relate significantly to the following dimensions:

- Investment and resource required (with investment for 1 > 2 > 3)
- Time required (with time to money of 1 longer than 2 longer than 3)
- Return on investment, dependent on the specifics of the opportunity being considered
- The number of potential licensees and manufacturing partners with the capabilities and motivation to manufacture

If a company happens to be in an established business that already owns a suitable or nearly suitable manufacturing plant, option 1 could be a very good choice. If, however, it is a new business, in the current business climate, it can be difficult to raise venture capital for building a manufacturing line (option 1) and so it may be pragmatic to consider either to outsource manufacturing to someone else (buy versus make) or to use someone else's manufacturing line for development activities to the point that the technical risk is reduced sufficiently that venture capital may be raised or a licensing deal may be done (option 2). Option 3 requires much less investment, but because of the relatively higher level of risk required by licensees versus option 2, the returns are likely to be significantly lower. Options 2 and 3 depend, of course, on the availability of potential licensees or manufacturing partners with the capability to manufacture. For new-to-the-world technology, this capability may not exist.

In some cases there is a fourth option that may apply, particularly to new businesses that are still working out what their business will do and in which markets they will participate. This involves supplying high-value custom products employing the technology. Each product is customized to the specific needs of the individual customer (usually a large company), and because volumes are therefore very small, products can be made by hand or at least using very small-scale manufacturing equipment. This can be an attractive way to start a new business, because it generates early revenue, provides early experience of real customers and their market needs, and enables a much more accurate determination of where the value of the new technology or product concept really lies. Some businesses that start life in this project consultancy mode change to a manufacturing mode once they understand where the profitable opportunities lie in particular markets.

Many books have been written on the topic of business model selection and the reader is referred to them for more detail. (See, for example, *The Innovator's Solution*, by Christensen and Raynor.)

13.5.6 Intellectual Property

For OPE businesses, where there is often a significant new technology element, IP strategy becomes an important consideration. Registered and legally protected IP rights include patents, which cover technological novelty; registered designs, which cover the external appearance of a technology; and trademarks, which cover the commercial aspects of a technology (e.g., its brand name, logo, color). Elements of the IP strategy will include:

1. Protecting a business by stopping others copying a proprietary technology. A patent is a right that prevents use of inventions developed by others and thus creates a strong economic incentive for innovation. Broad patent coverage of key enabling inventions adds significant value to a business because it provides a competitive edge. Patents are also business assets in their own right and may be bought and sold or licensed. Filing, securing, and maintaining patents is expensive, however, so it is best to focus on those inventions that uniquely enable significant commercial value, for example, through large cost reduction or substantially improved performance. Inventions that can be easily circumvented provide little value, unless all routes to circumvention add significant cost to imitators.
2. "Freedom to operate." This term refers to a situation where a business may sell products and services without infringing third-party patents or design rights or trademarks. Simply by owning patents on an invention that is being used in a product does not mean that a company is not inadvertently using an invention in its product that has already been patented by someone else. To ensure that a business has freedom to operate, it is usual practice to employ a patent specialist to conduct a freedom-to-operate search once the key elements

of the product and process are known. If a piece of prior art is identified in the search with claims that fall directly onto a feature of the product to be sold or the processes used to make it or its function in use, several options arise, including the following:

- Investigate whether the patent in question is still in force and that it is in force in the jurisdiction of planned sales or manufacturing.
- If the patent is in force, investigate whether it can be avoided, for example, by redesigning the product or changing the manufacturing process or by using it in a different way.
- Investigate whether the patent is valid. If it can be shown that the relevant claim is invalid, it can be ignored. A common way to do this is by finding an example in the public domain of the same invention being practised anywhere in the world before the patent's priority date.
- Take a licence. If there are no other options, it may be necessary to take a licence from the patent holder. This often involves payment of a royalty on goods sold.

It is outside the scope of this text to go into this topic in detail; however, it is strongly recommended that professional advice be taken regarding IP matters.

13.5.7 Maximizing Resources

13.5.7.1 Networks

We have already discussed the commercial loss that can occur if a product is too late to market. A new business venture can save time if it does not attempt to do everything itself. By focusing only on the most important areas to keep in-house and working with partners for other things, it is possible to move more quickly and with less resource required. It can therefore be valuable to make the most of networks such as the COLAE network, funded by the European Union to support the commercialization of organic and large-area electronics technologies by linking the major clusters across Europe. Other similar networks and trade organisations exist around the world and provide a great opportunity to meet potential industrial partners and intermediate technology organizations that may provide access to pilot-scale manufacturing lines.

13.5.7.2 Finance

In the early stages of a developing a new business venture it is particularly important to maximize the impact of the limited financial resources available. We have already discussed the value of working with partners in an ecosystem and it is important to emphasize the value of contracts, especially in the period of technology development prior to product sales, from other companies in the value chain and from potential end users, not only as a source of income, but particularly because they provide significant information about end-user and value chain partner requirements and this, in turn, may be used to refine business plans and technology development targets. It is often possible to involve the end customer in a specification or validation role. The financial support for such projects significantly de-risks the venture for all partners and provides a vehicle to assess how well the consortium can work together prior to procurement and joint venture contracts being signed.

Other areas where some national governments provide support include:

- Capital grants or tax arrangements in support of manufacturing investment
- Research and development tax credits
- Funding to enable companies to work with universities or to transfer staff between companies and universities for dissemination of knowledge
- Public procurement, for example, for health service or the military

Finally, there are a number of charitable foundations, such as the Gates Foundation, which provide funds for particular areas of global need or health research charities that have particular aims and objectives.

13.5.8 Public Policy Initiatives

Since OPE is considered as an enabling technology with significant potential to create commercial value and jobs, there has been increasing financial support from government, venture capital, and other organizations like the National Science Foundation (USA), the Department of Energy (USA), the Defense Advanced Research Projects Agency (DARPA, USA), and the New Energy and Industrial Technology Development Organization (NEDO) in Japan. According to the Lux Research report *Printing for Profits: Investments and Opportunities in Printed, Flexible, and Organic Electronics*, around €5.6 billion were invested by venture capitalists within flexible OPE during the period from 2006 to 2012. Display and smart packaging received the largest share of the funding, accounting to 37% and 23%, respectively. A large number of R&D projects amounting to more than €100 million have been financed under the European Union sixth Framework Programme (FP6, 2002–2006) and 51 projects amounting to more than €220 million have been funded under the seventh research Framework Programme (FP7, 2007–2013) with an intiative to build European leadership in the industry and prevent the drain of resources towards Asia. Additionally, many national governments provide financial support for collaborative research and development through the provision of grants, often in a competitive process. These schemes can be ideal for progressing the development of a system in which several companies provide technology elements.

Frequent events and conferences are organized at national and international level by consultants like IDTechEx, Intertech Pira, and Plastic Electronics Foundation with an aim to bring the players together onto a common platform and promote collaboration, build a supply chain, create awareness, and generate market pull. Technological road mapping and standards development initiatives are pursued by associations, prominent ones being the the Organic Electronics Association (OE-A), Photonics 21, the IPC Printed Electronics Management Council steering committee, the Printed Electronics Arena (PEA), FlexTech Alliance, the Korean Printed Electronics Association (KoPEA), the Japan Advanced Printed Electronics Technology Research Association (JAPERA), and iNEMI.

As the field of OPE is gaining momentum, the stakeholders have stressed the importance of developing standards. In 2011 the IEC TC119 initiative was launched to address standardization within the space of printed electronics. Initial activity in this domain started in 2004 and was established within the Institute of Electrical and

Electronics Engineering (IEEE), which resulted in the development of IEEE 1620™ and IEEE 1620.1™ standards.

> Historically, the adoption of standards has shown that it facilitates the growth of an emerging field and reduces the burden placed on individual companies to invest significant resources in the development of company specific compliance documentation[4].

13.6 Conclusions

- Emerging technologies and radical innovation are crucial for the development of the economy and growth of an organization but require a long period of development and large investment in terms of financial capital.
- Electronics technology breakthroughs enabled primarily by new active materials may exhibit lifetime and performance issues that require sustained materials development over a period of many years. In the case of OLED displays, it took 32 years from the first publication of the bilayer OLED device to the passing of the $10 billion market size milestone in 2013.
- Challenges faced by emerging technologies such as OPE are:
    - There is rapid improvement of the incumbent technology during the course of the time when technology is developing, also referred to as the sailing ship effect.
    - Although OPE promises a significant cost advantage there are many factors involved in scale-up, which may reduce this in relation to incumbent technologies.
    - The OPE industry is still emerging and several key enablers are not widely available yet. This makes business development and commercialization more challenging.
    - The need for system integrators is in some cases contributing to the slow growth of the market.
- Innovation is increasingly dependent not only on cross-functional integration within a company, but also requires

[4]http://www.ipcoutlook.org/pdf/printed_electronics_innovation_ipc.pdf

close collaborations with external agents, such as suppliers, customers (or users), and possibly, in addition, commercial partners and other sources of technology, such as universities or public research institutes.

- The development of OPE seems to fit very well with this systemic and distributed view of innovation within an ecosystem made of a variety of partners due to the complex nature of this technology and the multiplicity of potential applications.
- Whereas the ecosystem approach enables continuous innovation due to increased cooperation, the formation of an innovation ecosystem is not without risks.
- Constraints in the ecosystem are both technological as well as relational such as opportunistic behavior by the partners; lack of motivation, transparency, reciprocity, and trust; misalignment of goals and objectives; and changes in priorities during the collaboration.
- Devising processes for coordination and governance are crucial.
- Developing dynamic capabilities enhances chances of survival. Having strong or superior technology does not guarantee commercial success. It is important to maintain full awareness of products being developed by competitors and be able to respond quickly to unexpected changes in the industry.

Exercises

13.1 Why is it important to build an ecosystem in the case of emerging technologies?

13.2 (a) What are the reasons why it may be important to secure patent protection for inventions in OPE?

(b) Why might a decision be taken after consideration not to file a patent on a particular invention?

Suggested Readings

Adner, R. (2006). Match your innovation strategy to your innovation ecosystem. *Harvard Bus. Rev.*, **84**(4), pp. 98–107.

Adner, R., and Kapoor, R. (2010). Value creation in innovation ecosystems: how the structure of technological interdependence affects firm performance in new technology generations. *Strategic Manage. J.*, **31**, pp. 306–333.

Bozeman, B., Laredo, P., and Mangematin, V. (2007). Understanding the emergence and deployment of "nano" S&T. *Res. Policy*, **36**, pp. 807–812.

Brüderl, J., and Schüssler, R. (1990). Organizational morality: liabilities of newness and adolescence. *Admin. Sci. Quart.*, **35**, pp. 530–547.

Christensen, C. M. (1997). *Innovator's Dilemma: When New Technologies Cause Great Firms to Fail*, Harvard Business School Press, Boston, MA.

Christensen, C. M., and Raynor, M. E. (2003). *Innovator's Solution: Creating and Sustaining Successful Growth*, Harvard Business School Press, Boston, MA.

Connell, D., and Probert, J. (2010). *Exploding the Myths of UK Innovation Policy: How 'Soft Companies' and R&D Contracts for Customers Drive the Growth of the Hi-Tech Economy*, Centre for Business Research, University of Cambridge.

Darby, M. R., and Zucker, L. G. (2005). Grilichesian breakthroughs: inventions of methods of inventing and firm entry in nanotechnology. *Annales d'Economie et de Statistique*, pp. 143–164.

Das, R., and Harrop, P. (2011). *Printed, Organic & Flexible Electronics Forecasts, Players & Opportunities 2011–2022*, www.IDTechEx.com.

Day, G. S., and Schoemaker, P. J. (2000). A different game. In *Wharton on Managing Emerging Technologies*, pp. 1–23.

Frost and Sullivan (2010). World Printed Electronics Market.

Frost and Sullivan (2011). Opportunities in Printed Electronics. Technical Insights.

Govindarajan, V., and Kopalle, P. K. (2006). Disruptiveness of innovations: measurement and an assessment of reliability and validity. *Strategic Manage. J.*, **27**(2), pp. 189–199.

Gueguen, G., Pellegrin-Boucher, E., and Torres, O. (2006). Between cooperation and competition: the benefits of collective strategies within business ecosystems. The example of the software industry. *EIASM 2nd Workshop on Competition Strategy*, Milano, Italy.

Hellman, H. (2007). *Probing Applications: How Firms Manage the Commercialisation of Fuel Cell Technology*, Doctoral dissertation, Delft University of Technology.

Henderson, R. M., and Clark, K. B. (1990). Architectural innovation: the reconfiguration of existing product technologies and the failure of established firms. *Admin. Sci. Quart.*, pp. 9–30.

Hendry, C., Harborne, P., and Brown, J. (2010). So what do innovating companies really get from publicly funded demonstration projects and trials? Innovation lessons from solar photovoltaics and wind. *Energy Policy*, **38**(8), pp. 4507–4519.

Howells, J. (2006). Intermediation and the role of intermediaries in innovation. *Res. Policy*, **35**(5), pp. 715–728.

Iansiti, M., and Levien, R. (2002). *Keystones and Dominators: Framing the Operational Dynamics of Business Ecosystem*, Working paper.

Kettunen, J., Kaisto, I., van den Kieboom, E., Rikkola, R., and Korhonen, R. (2011). *Promoting Entrepreneurship in Organic and Large Area Electronics in Europe*, http://www.vtt.fi/inf/pdf/tiedotteet/2011/T2579.pdf.

King, Z. (2009). *Plastic Electronics: Putting the UK at the Forefront of a New Technological Revolution*, Aim Research.

Kodama, F. (1995). *Emerging Patterns of Innovation: Sources of Japan's Technological Edge*, Harvard Business Press.

Lengnick-Hall, C. A., and Wolff, J. A. (1999). Similarities and contradictions in the core logic of three strategy research streams. *Strategic Manage. J.*, **20**(12), pp. 1109–1132.

Mokyr, J. (1990). *The Lever of Riches*, Oxford University Press, New York.

Moore, J. F. (1993). Predators and prey: a new ecology of competition. *Harvard Bus. Rev.*, **71**(3), pp. 75–86.

Moore, J. F. (1996). *The Death of Competition: Leadership and Strategy in the Age of Business Ecosystem*, John Wiley & Sons.

Moore, J. F. (2006). Business ecosystems and the view from the firm. *Antitrust Bull.*, **51**(1), p. 31.

Parandian, A., and Rip, A., and Kulvete, H. (2012). Dual dynamics of promises and waiting games around emerging nano-technologies. *Technol. Anal. Strategic Manage.*, **24**(6), pp. 265–582.

Parise, S., and Casher, A. (2003). Alliance portfolios: designing and managing your network of business-partner relationships. *Acad. Manage. Exec.*, **17**(4), pp. 25–39.

Peltoniemi, M. (2006). Preliminary theoretical framework for the study of business ecosystems. *Emergen. Complex. Organ.*, **8**(1), pp. 10–19.

Raven, R. P. J. M., and Geels, F. W. (2010). Socio-cognitive evolution in niche development: Comparative analysis of biogas development in Denmark and the Netherlands (1973–2004). *Technovation*, **30**(2), pp. 87–99.

Rosenkopf, L., and Schilling, M. A. (2007). Comparing alliance network structure across industries: observations and explanations. *SEJ*, **1**(3–4), pp. 191–209.

Rothwell, R. (1992). Developments towards the fifth generation model of innovation. *Technol. Anal. Strategic Manage.*, **4**(1), pp. 73–75.

Roussel, P. A., Saad, K. N., and Erickson, T. J. (1991). *Third Generation R&D: Managing the Link to Corporate Strategy*, Harvard Business School Press, Boston, MA.

Stinchcombe, A. L. (1965). Social structures and organizations. In *Handbook of Organizations*, March, J. G., ed., Rand McNally, Chicago, pp. 142–193.

Teece, D. J. (1986). Profiting from technological innovation: implications for integration, collaboration, licensing and public policy. *Res. Policy*, **15**(6), pp. 285–305.

Tidd, J., and Bessant, J. (2013). *Managing Innovation: Integrating Technological, Market and Organizational Change*, John Wiley & Sons.

Van der Valk, T., Moors, E. H. M., and Meeus, M. T. H. (2009). Conceptualizing patterns in the dynamics of emerging technologies: the case of biotechnology developments in the Netherlands. *Technovation*, **29**(4), pp. 247–264.

Ward, W. H. (1967). The sailing ship effect. *Bull. Inst. Phys. Phys. Soc.*, **18**, p. 169.

Yildiz, H. B., Henchion, M., and Hewitt-Dundas, N. *Exploring Innovation Intermediaries*, http://www.berenschot.nl/...enchion_hewitt-dundas.doc.

Chapter 14

Market Perspectives and Road Map for Organic Electronics

Donald Lupo,[a] Wolfgang Clemens,[b] Sven Breitung,[c] and Klaus Hecker[c]

[a]*Department of Electronics and Communications Engineering, Tampere University of Technology, Korkeakoulunkatu 3, PO Box 692, 33100 Tampere, Finland*

[b]*PolyIC GmbH und Co. KG, Tucherstr. 2, 90763 Fürth, Germany*

[c]*OE-A (Organic and Printed Electronics Association), VDMA, Lyoner Str. 18, 60528 Frankfurt am Main, Germany*

donald.lupo@tut.fi

In the first 11 chapters you have learned the basics of organic electronics technology and should now have some familiarity with the materials, processes, and devices that are most important. You should also now be familiar with some of what are currently seen as the key applications for organic electronics and the technical requirements for them. In Chapter 12 you learned something about the innovation process and how it relates to developing products with organic electronics. In this chapter we want to look at what the organic electronics community sees as perspectives for the

Organic and Printed Electronics: Fundamentals and Applications
Edited by Giovanni Nisato, Donald Lupo, and Simone Ganz

ISBN 978-981-4669-74-0 (Hardcover), 978-981-4669-75-7 (eBook)
www.panstanford.com

field and the challenges to be encountered to get there. This kind of exercise is called road mapping and is a key activity in many fields, not just organic electronics. Therefore we will try to explain the procedures as well as the results. For this we will use primarily the results of the most recently completed road map developed by the Organic and Printed Electronics Association (OE-A, www.oe-a.org), an international industrial association with over 200 industrial and academic organizations as members. This chapter is an adaptation of the summary article and white paper for the fifth edition of the road map, published in the OE-A brochure in 2013, and is used with permission of the OE-A.

14.1 Introduction: What Is Road Mapping?

You have seen in the earlier chapters that organic electronics is based on the combination of new materials and cost-effective, large-area production processes that open up new fields of application. One of the reasons that organic (and printed) electronics has attracted much interest, in addition to the unique properties of such devices as organic light-emitting diodes (OLEDs), is the possibility to manufacture thin, lightweight, flexible, and environmentally friendly electronics. Due to the possibility of low-cost, high-volume manufacturing processes such as printing and coating, including reel-to-reel production, there is potential for high-volume, low-cost products with a unique form factor and low environmental impact.

Organic electronics is still, however, a nascent industry, which has enormous potential but for which many applications are only beginning to enter the market or be prepared for the market and which has undergone many changes—and will continue to do so as the field develops.

Especially in fields where technology development and applications are changing, it is important to gain an understanding of the past and present situation and to apply this to make some predictions of future developments and challenges in the field. Since the result of this process is a kind of map for where the field is going and critical work that needs to be done, this process is referred to as *road mapping*.

Road mapping is an established process in technology industries, and one of the best known one is the one done regularly for the semiconductor industry, the *International Technology Roadmap for Semiconductors. The International Energy Agency Photovoltaic Power Systems Program* (http://www.iea-pvps.org/; *Photovoltaic and Solar Forecasting Report*, for example). The road mapping is done by stakeholders in the industry; some key tasks are to (1) show the current state of the technology and how it has developed since the last edition, (2) predict the future development of the field, and (3) identify the developments that cannot be achieved with a simple linear extrapolation of the development to date.

The developments that cannot be achieved without major breakthroughs or new approaches are referred to as "red brick walls." Identification of these red brick walls is particularly important because it points out the areas in which effort in new research and technology development is urgently needed. In the rest of this chapter we will discuss how organic electronics has been road-mapped and what the current road map (published in 2013) looks like.

14.2 Road Mapping for Organic Electronics

A key aspect of the SEMI road map is the existence of clear scaling parameters based on Moore's law. This law was formulated by Gordon Moore in 1965 and predicted that the density of transistors would double every 24 months and the overall performance per cost ca. every 18 months. Surprisingly, this postulation has been shown to be valid even up to today and to apply not only to transistor density but also to other factors such as density of memory, processor speed, and cost of computing power or memory. Identifying the points at which continuing application of Moore's law is expected to reach a dead end (due to process limitations, quantum effects, etc.) has been the driving force in identifying the red brick walls in the SEMI road map.

Such a clear scaling law has not yet been identified in organic electronics. Furthermore, the range of applications is huge, as has been shown in the earlier chapters of this book, and the requirements can vary vastly from one kind of application to another. This presents significant challenges for road mapping across the industry. One possible approach to deal with this, which was chosen by the Organic and Printed Electronics Association (OE-A), is to define clusters of applications and to first do dedicated road maps for these clusters. These sub–road maps are useful in themselves for the application clusters, but the following step is to look for conclusions that are valid across applications clusters and thus relevant for the whole field. Applications were grouped into the following clusters: OLED lighting, organic photovoltaics (OPV), flexible displays, electronics and components (printed memories, batteries, active devices and logic, and passive devices), and integrated smart systems (ISSs) (sensors, smart objects, and smart textiles).

In the absence of clear scaling laws like Moore's law, an alternative approach has to be taken. The route chosen in this OE-A road map was to assemble experts and stakeholders in each application cluster and to predict first the development of future applications for the short, the medium, and the long term and the associated performance requirements visible to the user (so-called application parameters). These were then related to more fundamental properties of devices and systems that determine these performance levels, called technology parameters. Additional groups of experts looked at the status of technology and the predicted improvements based on normal incremental technological progress. This was done for functional materials, substrates, and printing/patterning processes. Cases in which predicted technology development did not correspond to the expected performance requirements were identified as the red brick walls to be addressed. The results of the work are summarized in the rest of this chapter. In addition to summarizing the results, we have tried to explain how the process worked, so the reader can consider how this process might be applied to other branches of organic electronics or other technology fields entirely.

14.3 Application Road Maps

Individual road maps with detailed analysis of application and technology parameters and red brick walls were made for each of the application clusters. For reasons of space, it will only be possible to give a brief summary for each application and go into some more detail for the overall road map, but the interested reader can obtain the full white paper from the OE-A.

14.3.1 OLED Lighting

While a complete OLED lighting system requires additional components such as electronic drivers and control software, the focus in the road map was on the OLED device hardware itself, especially on white OLEDs for general lighting, rather than some of the other applications for colored light, for example, decoration, automotive, etc. This was done to keep the scope of the road map manageable. Furthermore, this is currently the main target in OLED lighting, and the first commercial products are also based on white OLEDs.

As we saw in Chapter 5, OLEDs are based on very thin single or multiple layers of organic semiconductors and light emitters. This makes them unique among bright light sources, since other sources such as incandescent lamps, fluorescent tubes, and light-emitting diodes (LEDs) are almost always point or small-area sources and require fixtures and/or optics to create light over an area. OLED lighting products promise novel features in the longer term, such as large area, flexibility, diffuse light emission, thinness, high efficiency, and tunable color. New lighting applications can be expected to take advantage of OLED properties, for example, embedded lighting or homogeneous area lighting (Fig. 14.1).

OLEDs for lighting have attracted interest for some time, but until late in the first decade of the 21st century much work was research and development, frequently in precompetitive projects, such as the EU projects OLLA and OLED100.eu and national initiatives in the U.S. and Japan. Companies driving the development during this time include Novaled, Philips, Osram, Sumitomo Merck, BASF, and General Electric. Products in the form of lighting tiles and designer lamps began to be introduced in 2009; over the last few years brightness and whiteness have improved and cost has come down while a number of designers have started to use OLEDs in their products.

Figure 14.1 Interactive mirror with integrated OLED lighting (Source: Philips Lighting).

The road map group considered both small-molecule and polymer OLEDs and concluded that small-molecule OLEDs will be dominant for a while due to higher efficiency and that there will be a move with time from rigid glass to flexible polymer substrates, opening up new product and design options. On the basis of the analysis of the market status and expected development, a prediction for future generations of OLED lighting products was made, as shown in Fig. 14.2.

On the basis of the analysis of the product requirements in the future, the following key application parameters were identified for OLED lighting: lifetime (time to 80% of the initial luminance), efficiency (lumens/watt), size/scalability to large area), luminance, and cost (both per lumen and per watt). These led to identification of the key underlying technology parameters such as electrical efficiency (uniformity and driving voltage), quantum efficiency (photon generation/current), extraction efficiency (photons emerging from the OLED), lifetime, and color stability (change in color coordinates with luminance, temperature, or time).

A comparison of the requirements with technology developments led to the identification of the following red brick walls for OLED lighting: encapsulation (need for lower-cost methods for complete exclusion of water and air), outcoupling (need for cost-effective ways to enhance light outcoupling), standardization (reporting standards allowing comparison of products in a reliable way), manufacturing costs (orders-of-magnitude reduction in price needed to reach the general lighting market), and investment (high cost of investment in manufacturing equipment for large-scale production).

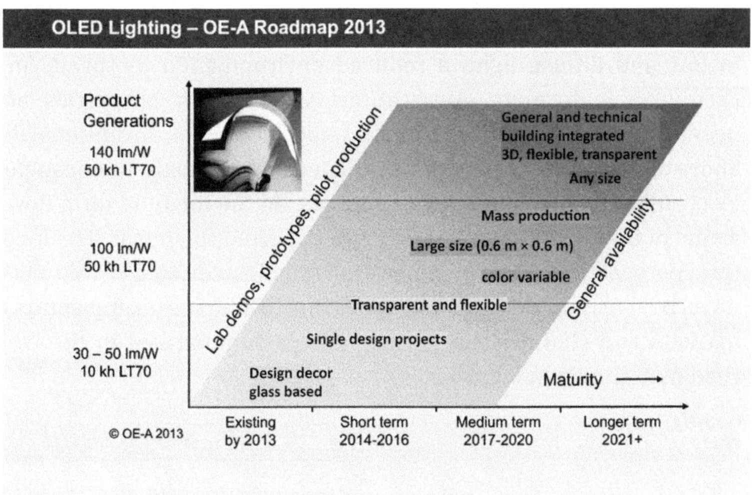

Figure 14.2 Road map for OLED Lighting, 5th edition (Source: OE-A).

14.3.2 Organic Photovoltaics

The OPV working group focused on two types of systems comprising organic materials: those based on dye-sensitized titania and those based on organic semiconductors. While some inorganic technologies have become printable as well, for example, CIGS, it was decided to restrict the analysis to systems including organic functional units for clarity.

Flexible OPV (Chapter 6) modules have been available for some years now, and products integrating flexible OPV modules have been commercially available since 2010. These applications target low-power consumer applications, for example, modules for battery chargers for mobile electronics such as cell phones, photovoltaics (PV)-powered computer keyboards, or rechargeable flashlights. Despite the difficult market for PV in general in the last couple of years and the significant setback of the bankruptcy of key OPV pioneer Konarka, both technical and commercial development has continued. However, competition with conventional Si and thin -film PV continues to be strong.

The road map group concluded that the main product opportunities in the short to the medium term would leverage some of the specific advantages of OPV, such as light weight, flexible design with options for color and semitransparency, good performance in low and diffuse light, a reduced environmental footprint, and customizable formats. Applications where such properties are needed do not place OPV in direct competition with conventional PV. Short-term applications were predicted to be mostly in consumer electronics and portable power sources. In the medium term novel forms of building integrated OPV are expected to appear. The long-term perspective of energy generation remains a driving vision, while novel business models are also appearing to address unconventional markets and channels to market. This is summarized in the OPV road map shown in Fig. 14.3.

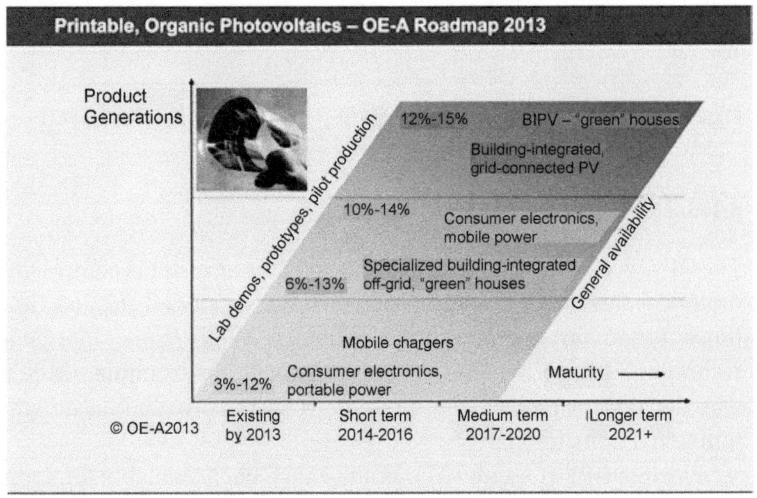

Figure 14.3 Road map for OPV, 5th edition (Source: OE-A).

Following the same procedure as the OLED lighting group, the key application parameters for further market development were identified as efficiency (especially for modules in production), power output per year (dependent on location but also technology, where OPV can be good for diffused light), lifetime (especially lifetime in realistic environments), optical transmission (including semitransparent modules), process for production (maturity, i.e. readiness for real production), and product production costs (€/W_p, €/m^2). The related technology parameters were identified as active p- and n-type layers (absorption, charge transport, valence and conduction band levels), electrode materials (especially transparent conductive electrodes), and barrier/packaging layers and materials. In addition, though it could not be quantified as a parameter, the issue of suitable production processes (coating, printing, low-temperature evaporation) was identified as important. On the basis of the analysis above, the key red brick walls were identified as improved photoactive layers, increased stability/lifetime, and overall costs, dominated by active materials and encapsulation. While it has not yet been realized on a large scale, the processing options for OPV were seen as a path to lower overall costs in the future.

14.3.3 Flexible Displays

Flexible displays are an extension of flat-panel displays, which have had tremendous success in replacing conventional displays such as cathode ray tubes (CRTs) for use in computers and televisions and in enabling new products such as laptop and tablet computers, e-readers, and smart mobile phones. Flexible displays can dispense with some key issues of current flat displays, such as the presence of (breakable and relatively heavy) glass and the inability to be bent, rolled, or used with other than flat form factors. The requirements for a flexible display depend strongly on the intended type of use, so some simplifications were required to deal with such a range of applications. The road map team focused on the following key types of use and looked for the most universal set of requirements possible: information and signage (conformal and light weight is more important than being bendable or rollable), reading (ruggedness, light weight, and optional bendability or rollability are desired), or entertainment/multimedia (where video rate, color, and resolution are critical, in addition to the above factors).

While there was tremendous hype about flexible displays in the late 1990s and early 2000s, the market has thus far not developed as rapidly as expected, and some companies have either faced bankruptcy or been forced to look for new business models. Nonetheless, there have been some successes, such as commercial installation of flexible price labels based on E-Ink's electrophoretic materials in a Finnish electronics supermarket chain store by Marisense, flexible watch displays by E-Ink and Sonostar, and numerous impressive e-reader and OLED prototypes, as well as first somewhat bendable OLED displays in TVs. On the basis of development to date and expected future markets, a road map covering the overall field of flexible displays was prepared, covering technical development of prototypes as well as market penetration. It is expected that products will grow in complexity and size as well as in degree of flexibility. The road map is shown in Fig. 14.4, including products that have already appeared commercially (price labels, smart cards) or in pilot scale (rugged e-readers), as well as more demanding products expected in the future.

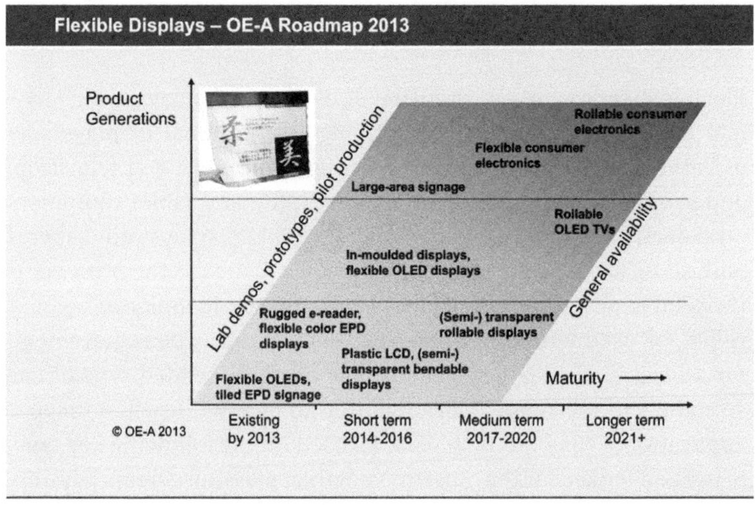

Figure 14.4 Road map for flexible displays (Source: OE-A).

The group focused on application parameters that are specific to flexible displays in order to keep the number of parameters at a manageable level. These were bend radius (radius to which the display can be bent), lifetime (under typical duty cycle conditions for the application), resolution (pixels per inch), update speed (time to a new image), and transparency (for see-through displays, for example, in automotive applications). The associated technology parameters were semiconductor mobility (for displays using organic thin-film transistor [OTFT] backplanes), maximum process temperature, barrier property requirements, registration (alignment of successive layers), driving voltage, on/off ratio (for OTFTs in the backplane), and switching speed per stage of transistors (for both backplanes and future row drivers). The key challenges/brick walls were identified as flexible encapsulation materials (at an acceptable price) and lowering of process cost (premium cost over rigid display), with operational temperature range a possible challenge for some applications as well.

14.3.4 Electronics and Components

As mentioned above, the electronics and components cluster includes printed memories, batteries, active devices and logic, and passive devices, which are the building blocks or toolkit that will be needed to make future organic electronics–based products. Sub–road maps were developed for each of these areas by teams within the OE-A. For brevity only a short summary of the sub–road maps can be given.

Printed memory is needed for any application where the user needs to store and process information, for example, stored value tickets, sensors, or measuring devices. If the user wants to change the information stored in the memory after production a rewritable memory, either write-once read-many (WORM) memory or rewritable random access memory (RAM) is necessary. Furthermore, for many applications without constant power, the memory needs to be nonvolatile (NV). Identification (ID) devices and promotional cards using read-only memory (ROM), WORM memory, or NV-RAM were already hitting the market at the time of the last road map and have continued to make headway. Reference designs of toys using NV-RAM memory have been launched, and applications using printed memory for brand protection (i.e., antifraud, anticounterfeit uses) have also emerged recently. Printed memory will be an important component for the ISS application cluster, and technology development is proceeding this way. For example, in late 2012 Thin

Film Electronics ASA demonstrated, together with PARC, successful integration of a sensor, a display, memory, and complementary OTFT-based logic.

The road map for printed memory cannot be shown here for space reasons, but the trend is expected to point toward more complex systems, moving from simple gaming and antifraud applications to ticketing, display memory, and electronic products. The memory alone will not be the product, but the memory will be an enabler in systems that comprise memory along with other functions such as sensors, power supplies, circuitry, and displays, that is, ISSs. Key application parameters for enabling this are manufacturing cost, production capacity, disposability (lack or strategically or environmentally problematic materials), retention time (in years), environmental stability, and bit error rate. Key technology parameters were film uniformity, drive voltage, switching endurance (number of read/write cycles), compatibility with drive transistors, printing methods, and electrode conductivity (related to access time). The red brick walls were identified as decoder complexity (capability of printing sufficiently complex decoder circuitry, rather than memory itself) and overcoming limitations in printing methods, for example, to lower memory voltage.

Most organic electronics applications target mobile devices rather than products that are connected to a mains power supply. Although passive radio-frequency identification (RFID) tags require power only when being addressed, most applications will require a more or less continuous source of power. While there are other options, such as energy scavenging and supercapacitors, the main focus in the fifth edition of the road map is on *flexible batteries* as a key technology. A large variety of thin and even printed batteries are commercially available. They are available for discontinuous use today and will be constantly improved in capacity, enabling continuous use. Currently, nonrechargeable zinc-carbon batteries are predominant for printed batteries but there is significant development in rechargeable batteries, for example, based on lithium, as well as research activity on printed miniature supercapacitors, which are a kind of cross between batteries and conventional capacitors.

The road map group predicts that there will be a progression from batteries that use printed parts, through batteries that are fully printed in separate processes, to batteries that are printed as part of an integrated process for printing electronic systems, as well as a progression from single charge through rechargeable batteries and from single cells to multicell integration. In the longer term, batteries will also be integrated directly in textiles and packages. The key application parameters important for this development were found to be energy density (energy stored per volume), power density (maximum power output per volume), voltage, peak current, lifetime, cycle life (for rechargeable batteries), temperature range, and bending radius. The associated technology parameters are layer thickness (affects energy density and lifetime), ionic conductivity (affects maximum power density), thermal stability, encapsulation, and flexibility of battery and housing. The key red brick walls that require breakthroughs to enable future market development were found to be the need for an inert atmosphere in production (especially for Li ion batteries, technical and economic challenges) and sufficient encapsulation materials.

Active devices include those key components based on a semiconductor that create some kind of active response to electrical input, but for the purposes of the road map specifically excluding those devices that are already treated as applications such as OPV or OLEDs. In particular, the group looked at diodes, transistors, logic circuits, and display elements (not treated here), with a focus on integration demonstrating functionality rather than single devices. Ring oscillators were an early test circuits for OTFTs and can also be used as clocking circuits. Poly–integrated circuits (ICs) demonstrated RFID transponder chips up to 8 bits based on printed polymers. Microprocessors based on up to 4000 small-molecule OTFTs have also been demonstrated. A number of complementary simple circuits based on n- and p-type transistors have also been demonstrated, either using p- and n- type organic semiconductors or using p-type organics and metal oxides for the n-type materials. However, no fully commercial products based on organic transistors had appeared at the time of the road map.

The road map took account of the current status but predicts that there will be products entering the market and that the trend will be to more complex circuits and increased use of complementary circuit designs. Key application parameters will be switching speed, driving current, supply voltage, circuit complexity, and lifetime, while the technology parameters were identified as materials properties (semiconductor mobility, conductor conductivity, and insulator dielectric constant and stability), lateral resolution (especially source/drain electrodes), and active layer thickness. The biggest challenges, not including the always present competition with well-developed Si CMOS electronics, were seen to be improvement in lateral resolution; achievement of thinner, defect-free active and insulating layers; step changes in materials' properties (together with reproducibility); and improved availability of both p- and n-type materials.

Printed passive components were surveyed briefly, covering printable materials for resistors, capacitors, and inductors/antennas, as well as pigment-based electroluminescent films. Some more detailed study was made of the nascent application field of transparent conductive films, especially for replacement of indium-doped tin oxide (ITO) in displays, OLEDs, and OPV, as well as touch-screen sensors, which have become extremely widespread with the huge success of touch-screen smart phones and tablets. A number of approaches have yielded promising results, including finely patterned metal meshes, nanocarbon materials, metal nanowires, and conductive polymers, with no clear frontrunner yet. Application parameters covering such a wide range of fields were difficult to define, but properties such as conductivity and resolution were seen as generally important, with the key challenges going forward related to increased resolution, improved conductivity/transparency trade-off, and integration of components into a system.

14.3.5 Integrated Smart Systems

ISSs (sensors, smart objects, and smart textiles) refer to objects that combine diverse functions such as power, sensing, logic, memory, and output circuits in order to perform a desired function in a printed, flexible system. In a sense, ISSs are the real goal of organic electronics, so the importance of looking at integrated systems will only grow in the future. Applications are expected in fields as diverse as automotive, aeronautics, clothing, environmental monitoring, and health care. As with the components cluster, we will only briefly summarize the results here, and refer to the full white paper listed in the references for more detailed information.

As has been discussed earlier in this book, *sensors* can operate on a number of principles. The market development has started, as printed electrodes and test strips are already available, and the expectation is that the future will see printed integrated sensor systems where functions like power, memory, or display read-out are on board. The key challenges to be faced are related to integration of different components and especially interfacing to printed electronic circuitry.

The more generic field of smart objects includes such future sensors but also such applications as printed RFID, printed dynamic price labels, and a host of applications, including brand protection, games, and gimmicks. Although fully printed RFID as an alternative to Si-based RFID has not fared well due to the strong market and standards position of Si and the decreases in the cost of Si chips, RF-enabled applications such as product cards and ticketing have begun to appear, as have simple printed memory cards. The team reached the conclusion that a simple direct road map was difficult for such a broad field, but that, in a similar way to other applications, the trend in the future will be to incorporate more different functions as well as more complex system elements.

Wearable electronics, for example, in the form of applications such as heart rate monitors for smart watches, has already entered the market but the flexibility, and hopefully in the future also stretchability, of organic electronics makes it appear to be an enabler for a wide range of future *smart textile* applications. Such textiles can alter their characteristics to respond to external stimuli (mechanical, electrical, thermal, and chemical). In addition, functionalities such as communication, displays, sensors, or thermal management can be integrated into fabric to enable wearable electronics. It is expected that a combination of printed/organic components and conventional technology will be the way forward for some years, but the working group predicted that advances in performance and robustness in organic electronics will enable a higher level of integration in the future. A key challenge for direct integration into textiles can be ruggedness to washing as well as the need to be not only flexible but also stretchable.

14.4 Technology Road Map

As the OE-A is a primarily industry-oriented organization, the emphasis in the road map was on development of applications, products, and markets. However, the success of this depends also on successful technology development, so technological progress needs to be mapped as well. This helps to identify the red brick walls where development "as usual" will not solve the problems and breakthroughs are needed. The OE-A road map team looked especially at materials, including substrates, and at printing and patterning processes.

As *functional materials*, semiconducting, conducting, and dielectric materials were all surveyed. Overall it was found that it cannot yet be decided whether small-molecule/vacuum-processed materials or polymeric or other solution-processable materials will win out or whether both types of materials and processes will exist in parallel. Regarding semiconductors, both p-type and n-type materials have been developed, with strong research and progress especially in n-type materials in recent years, due to their importance in the production of CMOS circuits in combination with p-type materials and matching dielectrics. The increase in mobility has been dramatic, with transistor mobility for both p- and n-type materials reaching or even exceeding 1 $cm^2 V^{-1} s^{-1}$, at least on a laboratory scale. This means that at least in principle organic materials can now be compared with amorphous Si, though there are still challenges to match or exceed polycrystalline Si. At the same time there has been progress in both vapor-deposited and printable metal-oxide semiconductors. There has been some speculation as to whether charge carrier mobility may turn out to be a scaling law for organic electronics in a similar way to Moore's law for conventional electronics. The rapid progress in materials properties offers some evidence that this could prove to be the case, but it is still too early to be sure.

In any case, another key finding, especially among companies trying to commercialize organic electronics–based products, is that mobility is not always the most important parameter for a material; stability, ability to be processed, and reproducibility of the materials themselves and devices made from them are at least as important for realization of viable products. This has been one of the things holding back commercialization so far, and progress in this area makes it likely that real products will reach the market soon.

Materials for OPV are a special class of organic semiconductors, and there has been strong progress in this area as well. Efficiencies in lab-scale devices have reached 10% and above for both solution- and vacuum-deposited devices, as well as for dye-sensitized hybrid cells, that is, they have caught up with amorphous Si. However, translation into large-area, multicell modules with similar efficiency and sufficient stability has proven to be challenging, and work is still ongoing. OLED materials are also a subset of organic semiconductors and dyes but were not covered in this edition of the road map.

As a new class of materials creating much interest, nanostructured carbon was also looked at. In particular graphene has attracted much interest due to reports of extremely high mobility in transistors as well as high conductivity combined with high transparency. Interest has arisen both for transparent conductive films and electrodes in energy storage devices. Numerous methods to prepare graphene and make films have been reported, including solution processing using graphene oxide. At the moment commercial applications are still limited and the long-term cost development is unclear, but graphene was seen by the team as an important material for future development.

The team surveyed metal-based conductive inks only briefly, looking at advances in flexibility and 3D printability and new developments in nanoparticle- and precursor-based silver and especially copper inks. Photonic sintering was seen as a promising tool for future integration into high-volume, high-speed processes. The number of commercial suppliers of these novel inks has increased, though it cannot yet be said to be a truly high-volume market. Transparent conductors were viewed in somewhat more detail because of the importance for touch screens, OLEDs, and OPV. Continued improvement of conductivity in the polymer poly(3,4-ethylenedioxythiophene):poly(styrenesulfonate) (PEDOT:PSS) has enabled these materials to be qualified for applications such as electrochromic displays and touch screens. At the same time, inorganic materials based on carbon nanotubes, graphene, or silver nanowires have gained significant traction as well and are being seen as possible alternatives to ITO in many applications. However, ITO still remains the main material used in large-scale commercial devices, though increasing penetration of these novel materials is expected in the future.

The functional materials group looked briefly at other materials as well, such as dielectrics, where tuning properties such as dielectric constant and interface to the semiconductor are critical, for example, for OTFTs. However, no road mapping as such was done for these materials. The need for improved encapsulation materials was also discussed and some of the classes of materials mentioned, but here, too, no real road mapping of future development was possible.

Since a key advantage of organic electronics is expected to be the low cost and light weight possible due to the use of low-cost, flexible *substrates*, this was also treated in the technology road map. The requirements vary depending on the application as well as the materials and processes used in manufacturing. For example, metal-oxide inks require higher sintering temperature and therefore have higher requirements for temperature stability. In all cases, the surface energy of the substrate needs to be compatible with the materials to be deposited, and frequently surface modification layers or processes are needed. For many applications, planarized substrates, or substrates that are temperature-stabilized for lower mechanical distortion, are needed. There has been significant activity in the field on both novel substrate materials and improved substrate manufacturing, which is expected to continue. The group summarized the major substrate materials used so far in organic electronics and their suitability for various applications, as shown in Fig. 14.5.

Printing and patterning processes are critical for organic electronics, as the key selling point is expected to be the possibility to manufacture using low-cost, high-volume processes and integrate with other processes such as graphic printing. The road map group looked in some detail at conventional mass-printing processes such as gravure, offset, screen, and flexo printing, and the range of conditions under which each works, but as this has been covered in the chapter on printing methods we will not repeat the analysis here. The group identified that there has been progress in improving resolution with these processes, with features down to 10 μm being realized now. New materials for screens and for flexo forms and novel engraving methods for gravure cylinders have been important in enabling this progress, though improved inks have also played a role. Progress in inkjet printing was also documented. This has been both in the form of scale-up to larger printers with more heads and nozzles for higher throughput, as well as the development of increasingly higher-resolution printing heads. This includes also new developments, still not ready for large-scale manufacturing, that promise feature sizes down to 1 μm and even below.

	Glass	Metal	Paper	PET	PEN	PC	PI	PES	PEEK	Textiles
Lighting (OLED)	++	+	(+)	+	+	(+)	x	x	X	(+)
Flexible Displays	+	(+)	(+)	++	+	x	x	x	x	x
Flexible Batteries	+	+	(+)	+	+	x	+	+	x	X
Passive Devices	-	-	(+)	++	(+)	(+)	-x	x	(0)	x
Active Devices & Logics	-	-	(+)	++	(+)	(+)	-	x	(0)	x
Printed Memory	-	-	x	++	+	(+)	x	x	x	X
Smart Objects	-	-	++	++	+	x	x	x	x	+
Sensors	-	-	(+)	++	(+)	(+)	-	x	(0)	(+)
Smart Textiles	-	-	x	+	(+)	(+)	-	x	x	++

++ Standard + Suitable 0 Option, not widely used
- Not suitable × Not known ( ) Only for selected applications

Figure 14.5 Suitability of different substrate materials (Source: OE-A).

Alternative approaches were also investigated. There has been significant progress in large-area vacuum deposition, so this process is looking potentially more cost effective than previously thought. Large-area photolithography has also made progress. Novel approaches such as laser ablation and forward transfer, soft lithography, and nano-imprint lithography (NIL) have also made progress. In general, it was found that there is a trade-off between resolution and throughput for processes used in organic electronics, though this is not an absolute, and some method, such as NIL, show some suitability for roll-to-roll processing. However the status for typical production processes as of the fifth edition of the road map is summarized in Fig. 14.6.

Related to both materials and processing is what the road map defined as *technology levels*. This refers to the kinds of processes that are used, on a continuum from processes very similar to conventional

Si to processes very similar to conventional printing and coating. The *wafer level* refers to batch processes similar to Si, such as vacuum deposition, spin coating, photolithography, and subtractive etching processes. "Hybrid" was used to refer to combining some wafer-type processes, such as large-area photolithography or printed circuit board technologies, with flexible substrates and some later printing-based processes as well as laser-based processes or inkjet printing. (Note: This is not the same as *hybrid integration*, in which printed components and Si-based components are co-integrated and which was not discussed in detail in this edition of the road map.) Fully printed refers to continuous (roll-to-roll or sheet-to-sheet) automated mass production–type processes based on printing and coating. Most "champion" results so far have been obtained using modifications of wafer-level processes, and this could be said to apply to most commercial OLED displays as well. However, impressive demonstrators have been made using both hybrid and fully printed processes, and some commercial OPV products have been made roll to roll. The expected, or at least desired, development for the future is for products to move more and more into fully printed processes as materials and process technology improves and increased markets make really large-area manufacturing highly attractive.

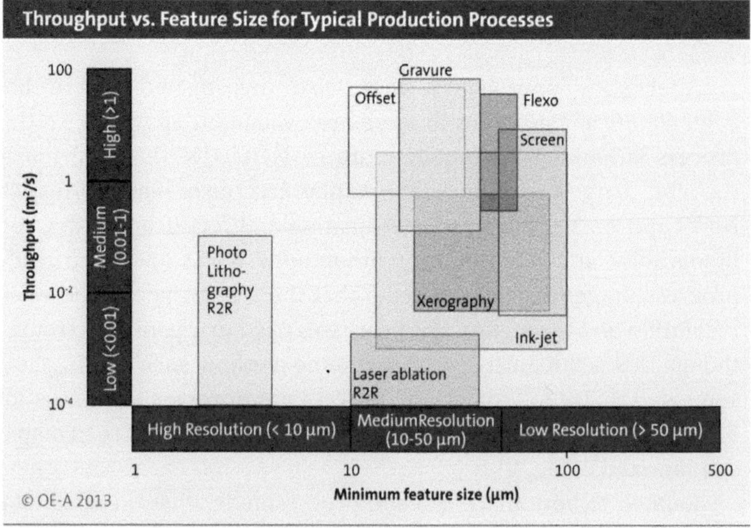

Figure 14.6 Throughput and feature size for some key patterning technologies (Source: OE-A).

14.5 Overall Road Map

At the time of the previous road map, products had just started to enter the market. One overall result of this edition of the road-mapping process was to note that despite some setbacks and the lack of widespread introduction of organic electronics in "killer applications," the progress into the market had indeed continued. Commercial products are available in all of the key technology clusters, with further market introduction expected within a few years. First organic electronic products reached the market a number of years ago, for example, passive ID cards that are mass-printed on paper and are used for ticketing or toys. Flexible lithium polymer batteries—produced in a reel-to-reel process—have been available for several years and can be used for smart cards and other mobile consumer products. Printed electrodes for glucose test strips or for electrocardiograms are common. OPV modules integrated into bags to charge mobile electrical devices are commercially available. Printed antennae are commonly used in (still Si-based) RFID tags. Large-area organic pressure sensors for applications such as retail logistics have recently been introduced. First OLED lighting-based products became available just before the fourth edition of the road map and but grew since then, with the number of both commercial OLED lighting products and large-scale installations at lighting trade fairs much larger than at the time of the last edition of the road map. Smart cards with built-in displays for one-time-password applications have started to be commercialized. New products such as flexible, roll-to-roll-produced e-paper price labels have been commercially installed into stores, printed RF-driven smart objects have become commercially available, and printed nonvolatile (NV) memories are being sold to product developers. Recently, printed systems incorporating organic electronics have also become commercial; for example, a rechargeable battery-powered flashlight containing OPV to recharge the battery, first shown as an OE-A demonstrator in 2011, can now be purchased from Mekoprint.

One of the challenges of road mapping such a broad, diverse field is trying to find the common themes in terms of market development and also the core parameters that are crucial for success over the whole field. However, on the basis of the work of the individual teams and the core road map team (the authors of this chapter), some overall trends were identified. As the road map is an important guide, not only for active researchers in the field but also for industries considering entering the field and public funding organizations identifying future trends and needs for research, it was found important to distill the multiple application road maps into an overall view of the future of organic electronics products. This overall road map is shown in Fig. 14.7.

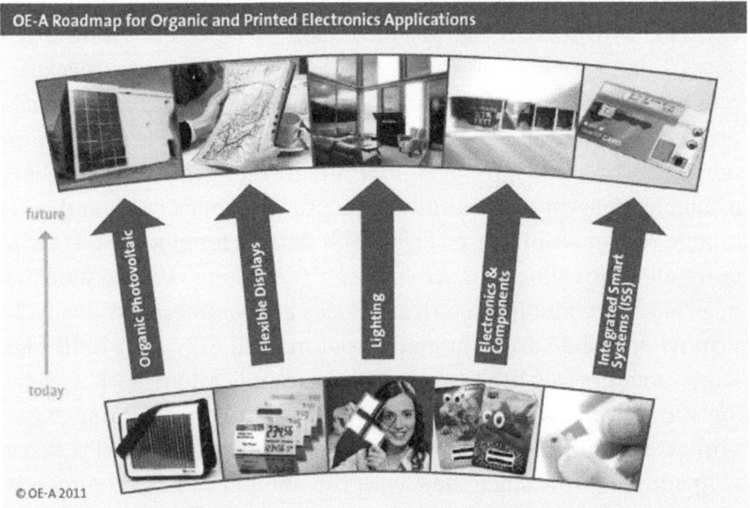

Figure 14.7 Overall summary road map for organic electronics, 5th edition (Source: OE-A).

As we saw above, there was quite a range of application and technology parameters identified for the individual clusters. The final task was to distill these into the most important parameters across the clusters. Some common threads could indeed be found. The most generally relevant application parameters and the reasons were found to be:

- *Device complexity*: Both the complexity of the circuit (e.g., number of transistors) as well as the number of different devices (e.g., circuit, power supply, switch, sensor, display) integrated into one product. This can strongly affect the reliability and production yield.
- *Operating frequency*: More complex applications require higher switching speeds.

- *Lifetime/stability/homogeneity/reliability*: Both shelf life and operational lifetime are critical for any application, and reliable, reproducible results are equally critical and can depend on both materials and processes.
- *Operating voltage:* This is critical, especially for devices powered by batteries or energy harvesting.
- *Efficiency*: This refers both to conversion efficiency of light to electricity or electricity to light (important for OPV and OLEDs) and to power efficiency of circuitry, which is critical, especially for lightweight, mobile applications or those based on energy harvesting.
- *Cost*: Although most applications target new applications and markets rather than replacements, costs have to be low. For some applications, such as rollable displays, a cost premium over conventional rigid displays may be accepted, while for other applications, for example, in packaging, low cost will be a major driving factor.

Similarly, by examining the technology parameters for the clusters, it is possible to extract the most generally relevant ones. In the fifth edition of the road map, these were found to be:

- *Mobility/electrical performance* (threshold voltage, on/off current): This depends on the individual material properties, for example, charge carrier mobility and dielectric properties, as well as the device design and processing.
- *Resolution/registration*: The performance of devices (voltage, current, speed) depends not only on the materials but also critically on the feature sizes of the devices and the ability to accurately register layers on top of each other. Improved resolution and registration will also be important for reducing space requirements in more complex systems.
- *Barrier properties/environmental stability*: Proper encapsulation at a reasonable price will be critical for almost all organic electronic products, even though the strictness of the requirements may vary by orders of magnitude.
- *Flexibility/bending radius*: Thin form factors and flexibility of the devices are key advantages of organic electronics.
- *Fit of process parameters* (speed, temperature, solvents, ambient conditions, vacuum, inert gas atmosphere):

Especially in complex systems, the different materials and process parameters for each component must be compatible in order to achieve reasonable yield and cost.

- *Yield*: Low-cost electronics in high volumes are only possible when the processes allow production at high yields. This includes safe processes, adjusted materials, and circuit designs, as well as in-line quality control.

As mentioned at the beginning of the chapter, one key goal of road mapping is to identify the developments that cannot be achieved with a simple extrapolation of the development to date, that is, red brick walls. This was also one of the key goals of the OE-A road map. The way to identify these for the overall field of organic electronics is to look both at the red brick walls identified within the cluster road maps and at the relationship between overall application and technology parameters and the requirements for increasingly complex products. In the future more components with higher and higher performance will have to be fit into smaller and smaller areas, while for other applications high-performance components will have to be placed precisely over large areas, up to a few square meters. In the long run, a wafer-based approach is not likely to deliver the cost points needed for large-scale market penetration, so fully printed or at least hybrid process technology will be necessary.

On the basis of such considerations and the results of the individual application cluster and technology road maps, the core team concluded that major breakthroughs in the following areas are needed:

- *Materials*: Improved electrical performance (mobility of 5–10 $cm^2\ V^{-1}\ s^{-1}$ for both n- and p-type materials), processability (uniform devices, nontoxic solvents), and stability.
- *Processes*: Improved resolution, registration and uniformity (especially for large-area applications), and high-throughput inline characterization.
- *Encapsulation*: High-performance, defect-free transparent barrier films at low cost.
- *Standards and regulations*: Current CMOS-based standards are not suitable; defining new standards is challenging (though international activity is ongoing in this area).

It is clear that there is interaction between these challenges and that advances in one may contribute to solving another one, for example, processablity and uniformity are closely related. However, these identified challenges point out some of the key work that needs to be done, and the intention of publishing such road maps is partly to encourage work to address the key challenges.

Another goal of the road-mapping process is to assess the state of the industry. This may involve looking also at the past as well as the present and future. In the case of organic electronics, the massive hype of the early 2000s led to some disillusionment when organic electronics did not deliver massive markets quickly. However, careful analysis of the situation showed that the field is indeed growing, if not always in the ways initially expected. The technology has achieved enough market penetration to start to qualify as an industry, albeit a nascent one. It was found that the number of products and the range of applications where organic electronics is reaching the market have increased. Mass markets are not yet there, with the possible exception of OLED displays for mobile phones, but could be reached in the near future. While some fields have not developed as quickly as expected, like printed RFID, others have emerged that were less expected, such as novel touch screens. Some key trends were observed in the field:

- Rapid improvement in materials' properties (charge carrier mobility, OPV conversion efficiency)
- Scaling of patterning processes to smaller dimensions
- Hybrid processing and heterogenous integration of printed and silicon-based components
- Entry of flexible, lightweight, mobile electronic products based on organic electronics in the market
- Appearance of applications in intelligent packaging and smart textiles

However, some questions remain open; for example, so far it is not possible to define a simple Moore's law for organic electronics or identify a "killer application" in the longer term, though there are indications that progress in mobility and in resolution of printing and patterning could possibly lead to scaling laws in the future. Although there are many fascinating applications it is still too early to tell which of these—or new ones we have not yet thought of—will turn into a killer application or if gradual market penetration in a variety of areas within the diverse and complex field of organic electronics is more likely. It was concluded that optimism in the area of organic electronics was still appropriate. Further market development will be straightforward in some areas, while breakthroughs will be needed for others. The road map is a dynamic entity and is regularly updated. The sixth edition will be published during year 2015.

14.6 Conclusions

- Road mapping is an important process in evolving technologies and aims to predict the future development of a field and the challenges to achieve this development.
- The OE-A has been developing and biennially updating an organic electronics road map since 2005.
- Organic electronics is a very broad field without a clear scaling law analogous to Moore's law and required a different approach from classic semiconductor road maps.
- Key steps are:
    - Looking at future development
    - Clarifying the application and technology requirements (parameters) affecting this
    - Identifying the red brick walls where breakthroughs beyond normal progressive improvements are needed
- Strong technical progress has been achieved but some red brick walls related to materials performance, processes, encapsulation, and standardization were identified.
- Organic electronics has begun to enter the market in a variety of application areas, with further growth expected.

Acknowledgments

The OE-A road map is a group effort of all the member organizations and their representatives. The process described here and the results summarized in this chapter are the result of the contributions of many people, especially the speakers of the application and technology groups. We especially thank Giovanni Nisato for helpful discussions and suggestions for improvements to this chapter.

Exercises

14.1 What are the key goals of the road-mapping process? Why do you think they are important for evolving technologies?

14.2 The approach taken by the OE-A was somewhat different from road mapping, for example, in the semiconductor industry. Do you think this approach was the most suitable? Why or why not?

14.3 Do you think that a killer application will be found for organic electronics?

Suggested Readings

International Energy Agency Photovoltaic Power Systems Program, *Photovoltaic and Solar Forecasting Report* (http://www.iea-pvps.org).

International Technology Roadmap for Semiconductors, ITRS, 2013 (www.itrs.net).

Moore, G. E. (1965). Cramming more components onto integrated circuits. *Electronics*, **38**, pp. 114–117, reprinted in *Proc. IEEE*, **86**, 82 (1998).

White Paper: *OE-A Roadmap for Organic and Printed Electronics*, 5th Ed. OE-A, Frankfurt am Main, Germany, 2013.

Chapter 15

Experiments

Peter H. J. M. Ketelaars, Jan P. C. Bernards, and Martijn H. A. van Dongen
Thin Films and Functional Materials, Fontys University of Applied Sciences, Rachelsmolen 1, 5612 MA Eindhoven, The Netherlands
j.bernards@fontys.nl

In this chapter four experiments are described. Two experiments can be performed in an applied science laboratory that is equiped for bachelor education. For two experiments specific equipment is necessary.

15.1 Organometallic Light-Emitting Diode

An organic light-emitting diode (OLED) is a thin light-emitting device (LED) that consists of at least three layers on top of each other. Light is emitted by passing an electric current. The current in this case consists of hopping electrons. The light-emitting polymer layer is sandwiched between two electrodes. One of the electrodes is light transparent. This electrode is mostly made of indium tin oxide (ITO)

Organic and Printed Electronics: Fundamentals and Applications
Edited by Giovanni Nisato, Donald Lupo, and Simone Ganz

ISBN 978-981-4669-74-0 (Hardcover), 978-981-4669-75-7 (eBook)
www.panstanford.com

and is used as the anode. The other electrode is placed on top of the device and is used as the cathode. Aluminum is often used.

The variant discussed here is an organometallic LED. The light-emmiting layer consists of ruthenium complex ions arranged in a polymer matrix. Müller et al. (2004) describe the synthesis and working principle in their article. The working mechanism is visualized in Fig. 15.1.

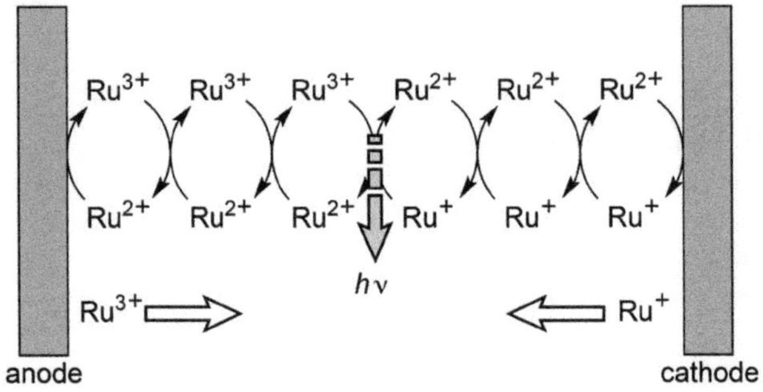

Figure 15.1 Electron transfer through a ruthenium film. At the anode oxidation ($Ru^{2+} \Rightarrow Ru^{3+}$) takes place and at the cathode reduction ($Ru^{2+} \Rightarrow Ru^{+}$). Reprinted (adapted) with permission from Müller, S., Rudmann, H., Rubner, M. F., and Sevian, H. Using organic light-emitting electrochemical thin-film devices to teach materials science. *J. Chem. Educ.*, 81(11), pp. 1620–1623, Copyright (2004) American Chemical Society.

At the anode there is oxidation of the Ru^{2+} complex into Ru^{3+}; at the cathode there is reduction of the Ru^{2+} complex into Ru^{1+}. Now light is emitted when Ru^{1+} and Ru^{3+} combine to form two Ru^{2+}. One of these is in an excited state.

In this context several experiments on light-emitting thin organic-metallic diodes can be carried out by second-year students of bachelor's degrees, certainly students of applied science, chemistry, or physics:

- Synthesis of tris(2,2′-bipyridine) ruthenium (II) tetrafluoroborate
- Characterization of the synthesized material by FTIR, ultraviolet-visible (UV-Vis), and fluorescence
- Constructing a simple three layer organometallic LED with ITO and aluminum as the electrodes and with ruthenium complex ions in a polymeric matrix layer as the light-emitting layer

- Characterization of the device by *I,V; V,L* and *I,V,L* experiments

15.1.1 Synthesis of tris(2,2′- bipyridine) Ruthenium (II) Tetrafluoroborate

The website http://education.mrsec.wisc.edu/nanolab/index.html discusses the synthesis ot the solid-state organic light-emitting diode based on tris(2,2′-bipyridine) ruthenium (II) complexes; this is explained with video support. The first step in the synthesis is the sequential reaction of complexation of Ru^{3+} by bipyridine. Next the Ru^{3+} is reduced to Ru^{2+} with an acid. After strirring and refluxing, this complex is precipitated with sodium tetrafluoroborate. The overall reactions of the tris-ruthenium complex are given below:

1. $2H_2O + NaH_2PO_2 + 12Bpy + 4RuCl_3$
   $\rightarrow 4[Ru(Bpy)_3]Cl_2 + 3HCl + H_3PO_4 + NaCl$
2. $[Ru(Bpy)_3]Cl_2 + 2NaBF_4 \rightarrow [Ru(Bpy)_3](BF_4)_2 + 2NaCl$

The formation of the tris-ruthenium complex is visualized in Fig. 15.2.

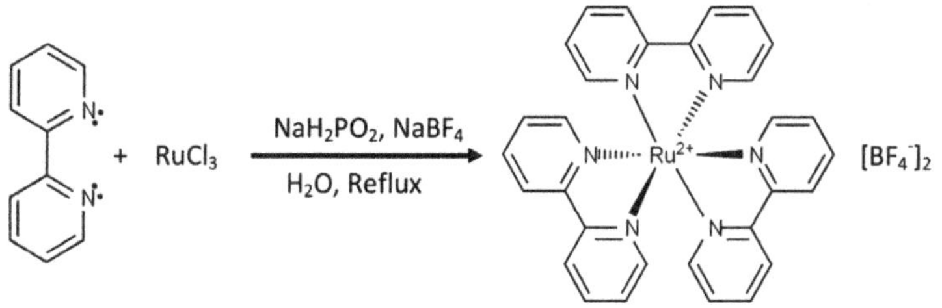

Figure 15.2 Formation of tris(2,2′bipyridine) ruthenium (II). Reprinted (adapted) with permission from Müller, S., Rudmann, H., Rubner, M. F., and Sevian, H. Using organic light-emitting electrochemical thin-film devices to teach materials science. *J. Chem. Educ.*, 81(11), pp. 1620–1623, Copyright (2004) American Chemical Society.

15.1.2 Characterization of the Synthesis Material

With standard lab equipment the characterization by FTIR, UV-Vis, and fluorescence analyses can be performed. Characterizing of the Ru complex is done by analysis of the absorption, emission, and excitation spectrum. In this specific situation three analysis devices are used:

- An FTIR from Nicolet (iS5) is used to analyze the specific structures of the complex. The chemical structure of the ruthenium complex can be mapped by evaluation of the different vibration bonds. An example of the spectrum is shown in Fig. 15.3.

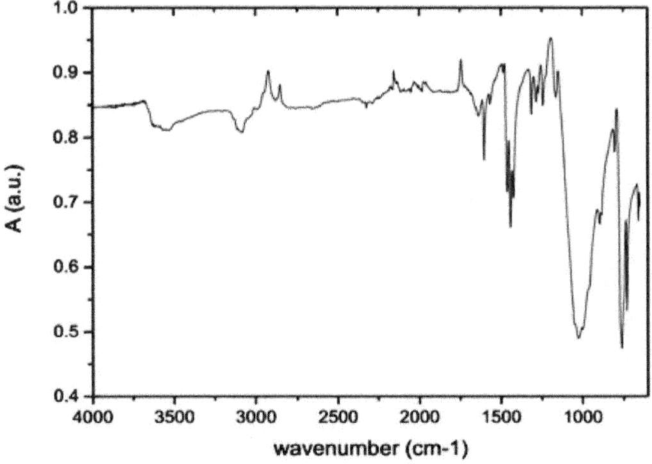

Figure 15.3 FTIR-ruthenium complex in ethanol.

Table 15.1 Overview of FTIR vibration bonds

Wave numbers (cm^{-1})	Chemical vibration bond
3628.84	~~O-H~~, N–H stretch
3081.25	~~O-H~~, N–H stretch, C–H stretch
1603.83	C=C stretch, C=N stretch
1464.49	C–H vibration, ~~O-H vibration~~
1444.72	C–H vibration, ~~O-H vibration~~
1423.96	C–H vibration, ~~O-H vibration~~
1027.55	~~C-O stretch~~, C–N stretch , C–C stretch
766.69	C–H vibration, N–H vibration ~~C-Cl , C-S stretch~~

In Table 15.1 the wave numbers and all known corresponding chemical structures are shown. The structures that are not to be expected are crossed out.

- The excitation and emission are measured with a spectrofluorometer of Jasco (FP-6200). The fluorescence measurements are shown in Fig. 15.4. As can be seen from the spectra, the excitation is intense at a wavelength of 290 nm and the light emitted by this ruthenium complex has a peak at 590 nm. This OLED is therefore red colored.

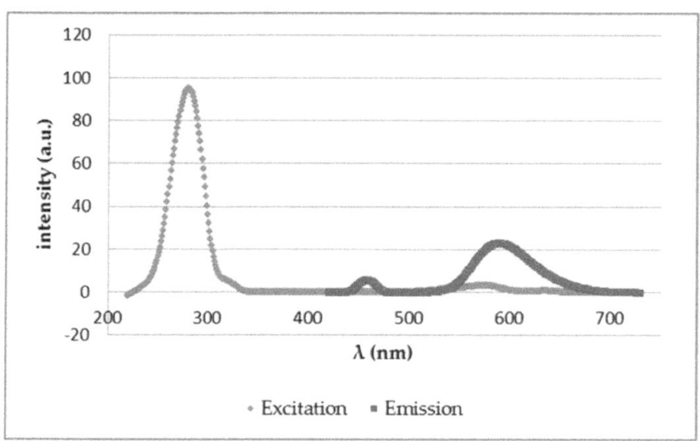

Figure 15.4 Excitation (blue) and emission (red) spectrum of ruthenium complex.

- A Jenway, model 6715, is used for the UV-Vis analysis. In the UV-Vis measurement UV light and visible light is radiated on a sample and the absorbance is measured (see Fig. 15.5).

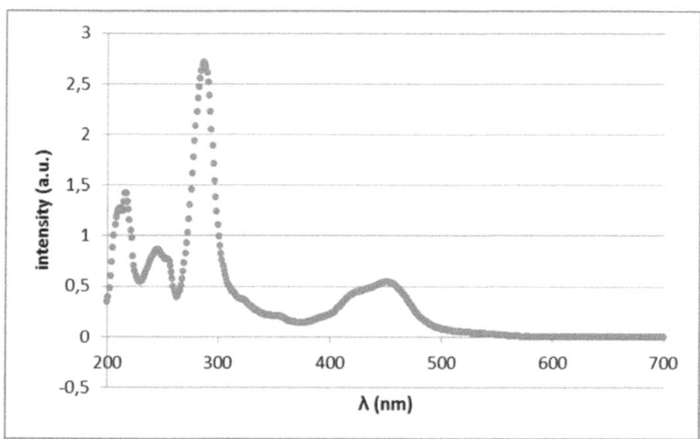

Figure 15.5 UV-Vis measurement of ruthenium complex.

15.1.3 Constructing a Simple Three-Layer OLED

ITO-covered glass slides, 5.0 × 5.0 cm; conductance <= 10 Ω/sq.; and thickness 150 nm can be used. The ITO is used as the anode in the OLED. To prevent a short circuit later, ITO is removed at one edge of the glass slides by etching with a concentrated hydrochloric acid solution. Sputtering of ITO on a clean glass slide is also possible if an ITO target and a vacuum sputtering system are available. Masks for patterning can be applied. In this case the ITO had a thickness of 150 nm as well, but the conductance was 25 Ω/sq.

A simple cleaning procedure is followed to make the ITO-covered glass slides fat and dust free. A small strip of adhesive tape is placed on a small part of the cleaned ITO. The ITO under this strip serves later as the direct contact for the positive pole of the power supply.

Next 1.0 mL of the ruthenium complex, dissolved in poly(vinyl alcohol), is applied by spin-coating onto the ITO. Spin-coating of the complex is repeated two more times. As suggested on the websites, the spin coaters are made of cooling fans from old PCs and surrounded by a splash shield. Afterwards, the spin-coated ruthenium complex is air-dried. An aluminum cathode layer is deposited by physical vapor deposition (PVD) onto the ruthenium complex. The evaporation is carried out in vacuum (10^{-5} mbar). A mask containing four gaps is placed on top of the slide with the complex before the PVD starts. Finally the adhesive strip is removed. In a simple electric circuit the four ruthenium OLEDs are connected to a power supply. The power, the electrical current, and the light emitted are measured. If the electric power is high enough the ruthenium complexes are reduced/oxidized, as explained before, and the emission of red light starts. Two working examples are shown in Fig. 15.6 and Fig. 15.7.

15.1.4 Characterization of the OLED by Electrical Current, Electrical Potential, and Light Emission

The physical characterization of the four OLEDs consists of *I,V* and *V,L* measurements. With data acquisition of National Instruments and Labview the characteristics are measured simultaneously. The illuminance is measured in $lm.m^{-2}$ = lx. The sensor used for this purpose is a TSL250R, a photodiode from Farnell, with a built-in amplifier. A conversion factor must be used to convert $V/(\mu W.cm^{-2})$

into lm.m^{-2}. At a wavelength of 550 nm is the number of lumens per watt, the largest, namely 683 lumens per watt. The light that is emitted by the OLEDs, see the fluorescent diagram in Fig. 15.4, is mainly around 590 nm.

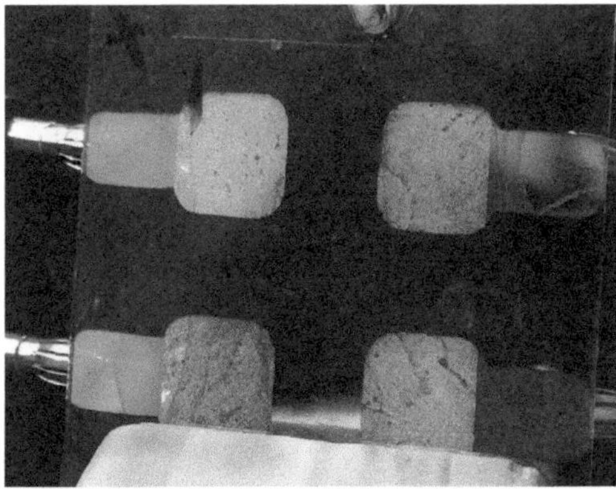

Figure 15.6 Working OLEDs with commercial ITO.

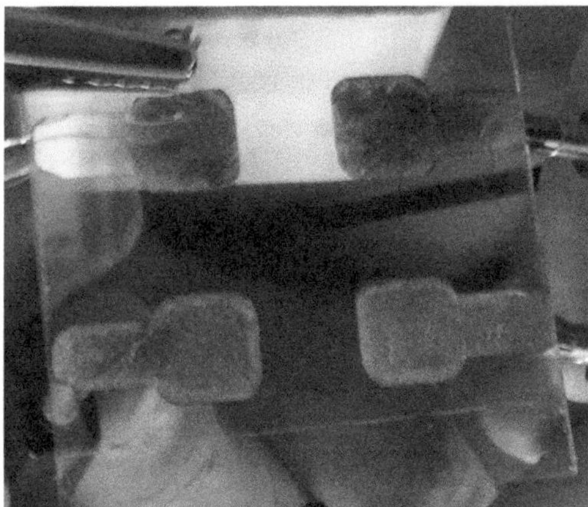

Figure 15.7 Working OLEDs with sputtered ITO.

The resistance of the OLEDs is of course not linear to the applied voltage, because it is a diode. So the electrical current is also not linearly proportional to the electrical potential. It also showed that at a constant applied voltage the electrical current crawls to an end value (see Fig. 15.8). The measurements are therefore always done after a certain interval of time.

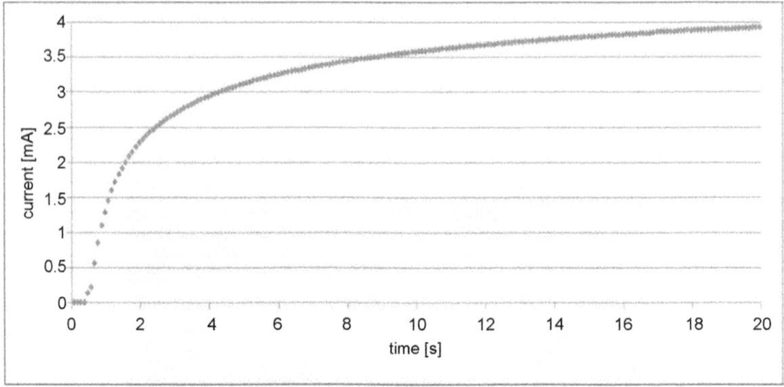

Figure 15.8 OLED with ruthenium complex: electric current in time at an applied voltage of 9.0 V.

The *I,V,L* characteristic is obtained by raising the voltage in steps of 0.2 V. The control signal is provided by a PC. To ensure that the OLED is fed with sufficient electric power, an amplifying circuit is placed between the PC+National Instruments board and the OLED. This circuit contains an OpAmp (μA741) with transformer 2 kΩ and 1 kΩ resistors for a strengthening of the voltage by a factor of 2. After this, there is a power buffer (LT1010) to ensure that the circuit feed is with a high-enough power to the OLEDs. The circuit is shown in Fig. 15.9.

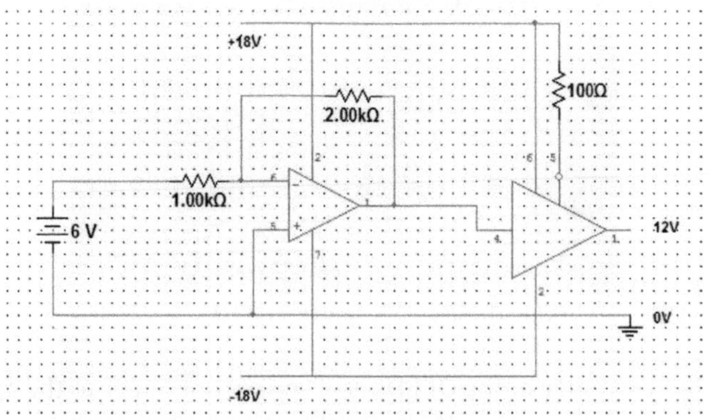

Figure 15.9 Electrical circuit to feed the OLEDs.

The direct current (DC) voltage of 6 V in Fig. 15.9 is only an example. The feeding voltage is with steps of 0.2 V raised up to 20 V.

The OLEDs were placed at a few millimeters above the measuring photodiodes. Some students results are given in Fig. 15.10 and Fig. 15.11.

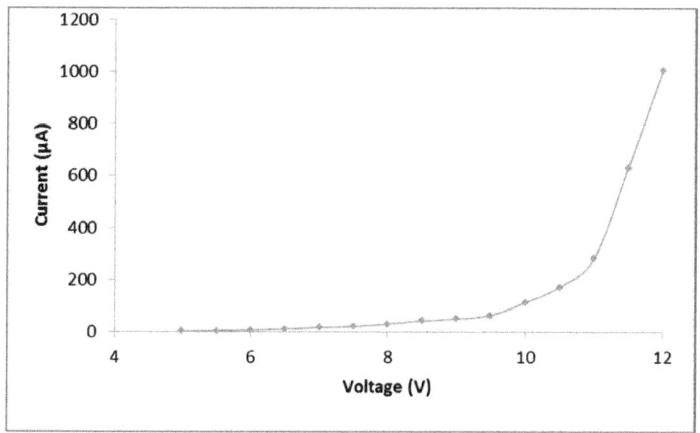

Figure 15.10 *I,V* characteristic of an OLED with ruthenium complex.

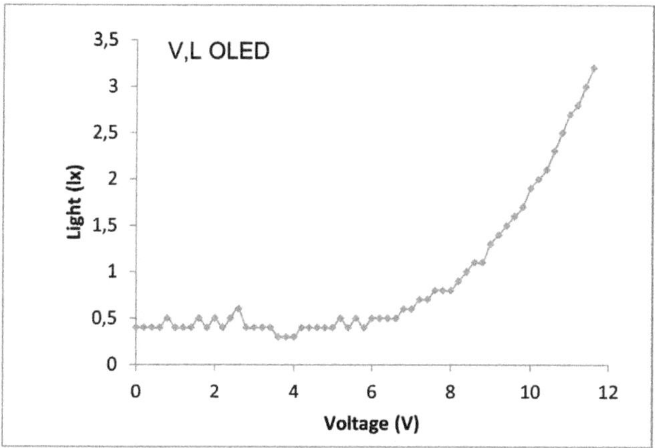

Figure 15.11 *V,L* characteristic of an OLED with ruthenium complex.

15.2 RFID Tags

If a package is moved from one place to another, it is good that it is recognized in a limited time frame at a few meters' distance. Radio-frequency identification (RFID) is a method of using radio waves for the purpose of transferring data, identifying and tracking so-called tags. RFID tags that do provide specific information are constructed differently. These tags contain digital information stored inside an integrated circuit (IC) present in the tags and connected to an antenna. RFID tags can be passive or active. An active RFID tag is connected to a battery, so it is activated continuously. A passive RFID tag only works when activated by an antenna, which is irradiated by an electromagnetic wave. Here, we only describe the making of a passive RFID tag. A typical identification system consists of an RFID tag and an RFID reader. The construction of the antenna in the RFID tag is explained in this part of the chapter. The RFID reader and the RFID IC are commercially available. An example of an RFID tag consisting of a dipole wire antenna with a stub and an IC is given in Fig. 15.12.

Figure 15.12 Dipole wire antenna with an IC placed in a stub.

The operation of the identification system is described here briefly. The reader emits an RF wave, which is received by the RFID tag antenna. The antenna converts this electromagnetic wave into an alternating current (AC), which powers the IC. The activated IC modulates the AC of the antenna to transmit its stored data to the reader. RFID tags can operate in different frequency bands. For the purpose of recognizing packages, ultrahigh-frequency (UHF) tags are applied. For Europe this UHF band is between 865 MHz and 868 MHz. For North America RFID tags operate between 900 MHz and 930 MHz. Therefore, an iDTRONIC UHF USB RFID reader operating at a range from 830 MHz to 950 MHZ is used. A commercially available IC from NXP, TSSOP8, is used in this experiment. This chip is EPC class 1 Generation 2. Its impedance according to the datasheets is:

$$Z_{IC} = 16 - j148\ \Omega \text{ at } 915\text{ MHz and } Z_{IC} = 26 - j160\ \Omega \text{ at } 866\text{ MHz}$$

The total power transfer P_{IC} received by the IC is dependent on the potential of the antenna, V_A, and the impedance of the antenna and IC, respectively, Z_A and Z_{IC}:

$$P_{\mathrm{IC}} = V_{\mathrm{IC}} \cdot i = \frac{V_{\mathrm{A}} \cdot Z_{\mathrm{IC}}}{Z_{\mathrm{A}} + Z_{\mathrm{IC}}} \cdot \frac{V_{\mathrm{A}}}{Z_{\mathrm{A}} + Z_{\mathrm{IC}}} = \frac{V_{\mathrm{A}}^2 \cdot Z_{\mathrm{IC}}}{(Z_{\mathrm{A}} + Z_{\mathrm{IC}})^2} \tag{15.1}$$

From this formula it can be seen that the best power delivery will be when both impedances match:

$$Z_{\mathrm{A}} = Z_{\mathrm{IC}}^{*} \tag{15.2}$$

The impedance of the IC is fixed at the optimum frequency. To ensure an optimal fit between the impedances, the geometry of the tag antenna must be adjusted. The adjustment must be to account for both resistance as well as reactance. Adjusting the resistance is mostly done by the length of the antenna. On the other hand the energy transfer by the RF waves to the antenna is most efficient at the antenna's resonant frequency:

$$f = c/\lambda \text{ and } \lambda = n.\ell \tag{15.3}$$

with ℓ being the antenna length.

So there will be some discrepancy between both desired lengths. The inductive reactance can be manipulated by folding the antenna. To greatly drop the resistance of the antenna, a shorting stub can be introduced with the IC positioned inside this stub, as is shown in Fig. 15.12. It is time consuming to better fit the impedance in practice. Therefore, ComSol Multiphysics finite-element simulation is used. A lot of simulations are examined. Table 15.2 indexes the results of a selection of antennae designs. The type of design can be found in the name.

Table 15.2 Rated impedances, maximum reading ranges, and physical dimensions of simulated antennae

Antenna	Impedance (Ω)	Max. reading range (m)	Dimensions (m)
Dipole (stub)	120 + 102j	1.2	0.005 × 0.164
Rectangle (stub)	97 + 50j	<1.0	0.005 × 0.164
Telescope	117 + 98j	<1.1	0.01 × 0.164
Telescope (stub)	20 + 319j	<0.27	0.018 × 0.164
Bowtie	114 + 84j	<1.1	0.01 × 0.164
Bowtie (stub)	20 + 136j	<1.75	0.02 × 0.164
Square (stub)	10 + 144j	<1.05	0.04 × 0.04
Square (ind.loop)	17 + 356j	<0.25	0.04 × 0.04

The best reading distance according to the simulation should be the bowtie with a stub. But for this design, a lot of conductive inks will have to be printed. Several other designs offer the same simulated maximum reading range. But again, depending on the dimensions, less or more ink is needed for a correct design. The rectangle with stub design is chosen as the geometry for realization. The principle layout of this rectangle RFID tag, the real printed RFID tag, and the simulations are shown in Figs. 15.13 and 15.14.

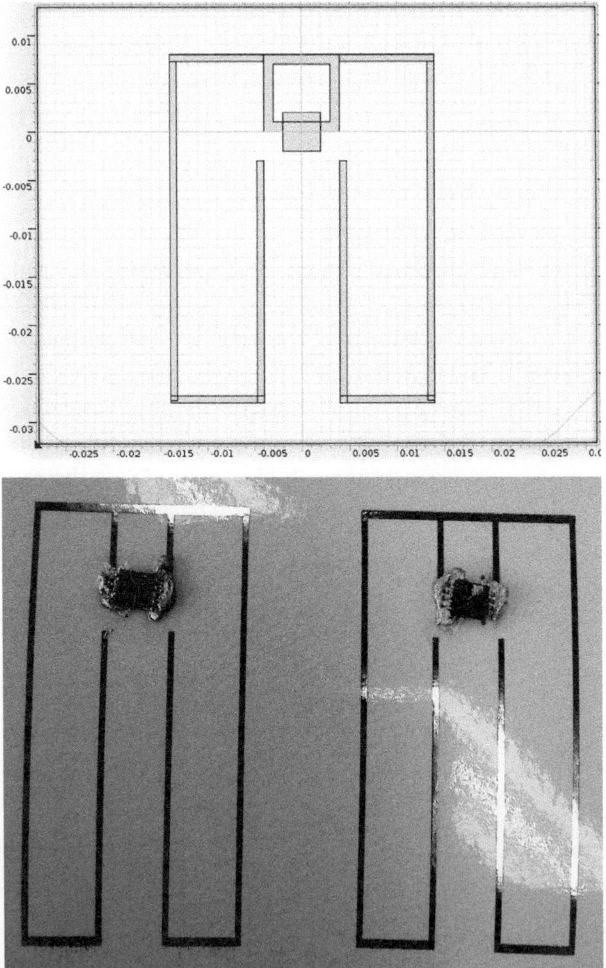

Figure 15.13 Geometry rectangle antenna with an added shorting stub. Outer dimensions 2.1 × 4.1 cm. Top: COMSOL; bottom: printed.

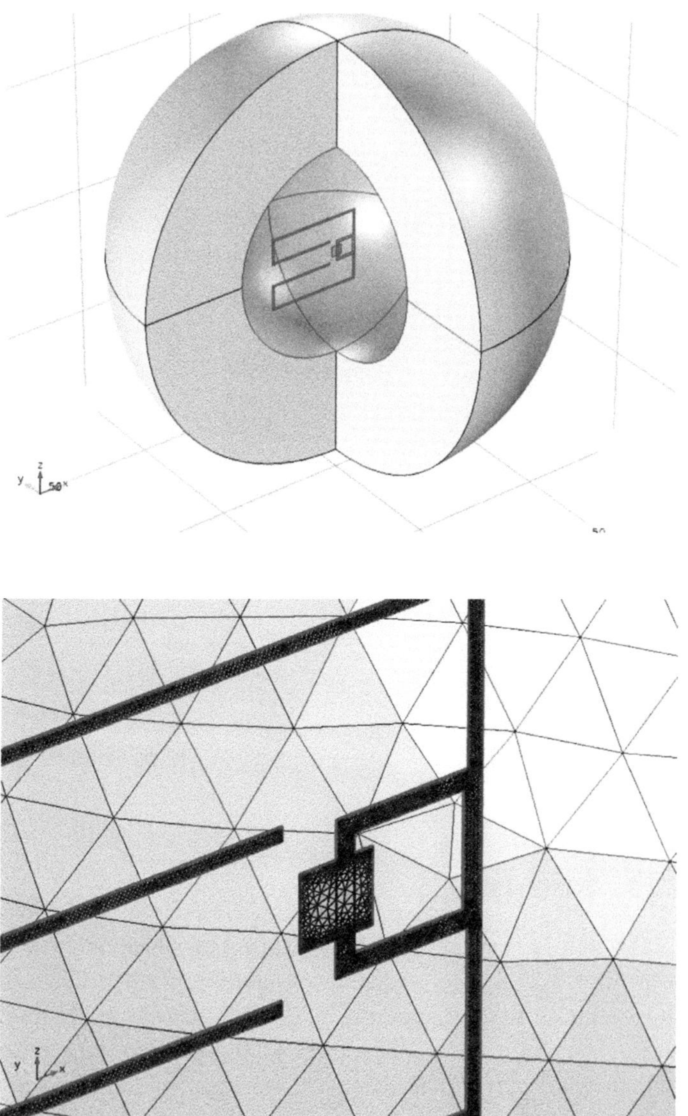

Figure 15.14 Geometry and mesh for COMSOL simulation. The outmost layer of the sphere is modeled as a perfectly matched layer.

The simulated distance over which an RFID works is shown in Fig. 15.15. The polar plot of a rectangular dipole antenna with an added stub is shown.

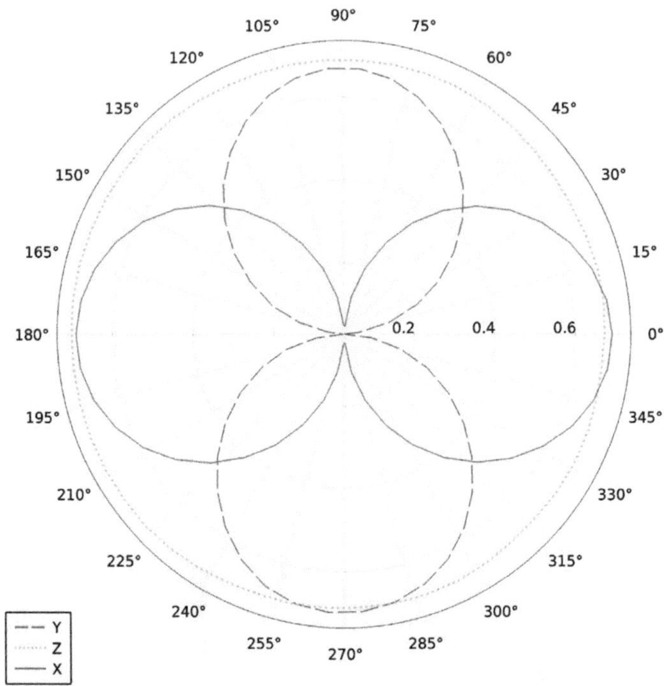

Figure 15.15 Radiation pattern of a rectangular dipole antenna with an added stub.

15.2.1 Materials

The antenna is realized by inkjet printing of silver nanoparticles. The printer used is the Pixdro LP50 made by Roth & Rau using a Konica Minolta KM512S printhead. The ink used is a nanoparticle solution from Suntronics in an ethanol base with a concentration of silver nanoparticles of 20%. The particle size is smaller than 150 nm. With the LP50 the ink was inkjetted with 600 × 600 dpi in a single layer. The substrate used was Epson premium glossy photopaper. After cleaning the paper with an UV-ozone plasma the antenna was printed. The NXP IC was placed in the stub and with RiteLok SL65 Epoxy glued to the antenna. Sintering was performed in an oven at 150°C for 90 minutes. Finally, several tags were readable up to 1.0 m. The best reading distance was at 865 MHz.

It is also possible to use an Epson Photo P50 as the inkjet printer to print the antennae. Examples of printed lines are shown in Fig. 15.16. The disadvantage using this printer is that it quickly becomes clogged.

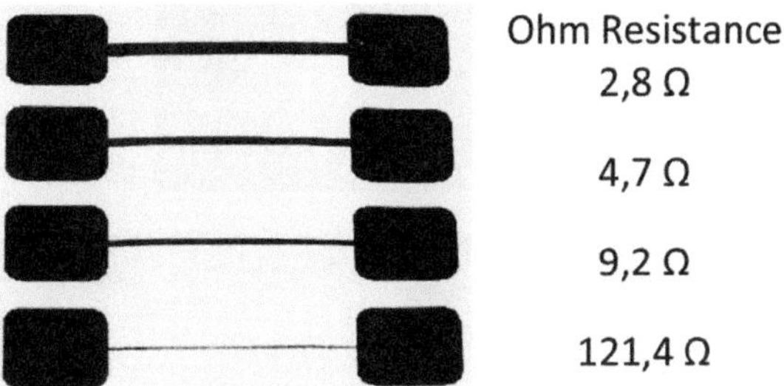

Figure 15.16 Simple conducting lines printed with an Epson Photo P50.

15.3 Applications of μPlasma Printing

On a macroscale, well-known plasmas are lightning bolts. On a microscale plasma is created by applying a high AC voltage between two metal electrodes. The two electrodes are in this case the needles in a so-called μPlasma printhead, see Fig. 15.17, and a substrate holder table. A dielectric insulating layer on the table acts as a dielectrical barrier. Typical layers are made out of glass or plastic. An electric discharge etches the surface or terminates free radicals of this dielectricum. A precursor makes it possible to functionalize the barrier into a desired direction. Integration of an Innophysics μPlasma printhead in a Roth & Rau Pixdro LP50 inkjet printer makes it suitable to combine atmospheric plasma treatment with an on-demand print technology. In Fig. 15.17, photos are shown of a μPlasma discharge, a bottom view of a 24-needle Innophysics μPlasma printhead, and a Printer LP 50 set up with a plasma printhead (see also van Dongen, 2014).

With this equipment it is possible to use dielectric barrier gap discharge plasmas for patterned plasma treatment without the use of masks. Patterns as small as 300 μm can be created. Several applications for experiments are:

a. Modifying the local wetting behavior of the dielectricum barrier

b. Deteriorating the local wetting behavior of the dielectricum barrier
c. Functionalizing a surface of the dielectricum barrier

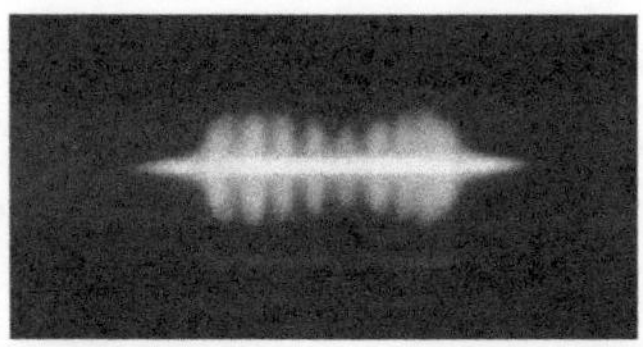

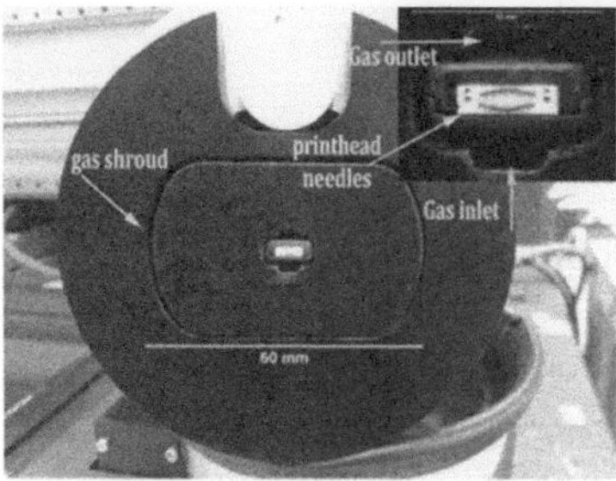

Figure 15.17 (Top) μPlasma discharge, (bottom left) Innophysics μPlasma printhead, with an inset of 24 needles, and (bottom right) LP 50 Roth & Rau Printer with μPlasma printhead and substrate table.

15.3.1 Changing Wetting Behavior and Functionalizing a Surface

The experiment described here is changing the local wetting behavior by creating hydrophilic channels in a hydrophobic environment. Air as plasma gas makes a glass surface hydrophilic. This is caused by adding oxygen groups to the surface. If a plasma precursor gas like hexamethyldisiloxane (HMDSO) is applied, a hydrophobic layer is deposited on the glass surface. In this experiment the hydrophobic layer was achieved with a chemical method.

First the glass slides were cleaned in successive steps:

a. Washing with soap and rinsing thoroughly with deionized water
b. UV ozone plasma cleaning
c. Placing the glass slides in a Piranha solution of 3:1 concentrated H_2SO_4 and a 35 wt% H_2O_2 solution

In Step c potassium hydoxide solution in an ethanol/water mix can be used as well. This step is performed to bring hydroxyl groups onto the clean substrate.

After this cleaning procedure the glass slides can be made hydrophobic in different ways. The best results were achieved by placing the clean glass slides in a solution of 0.001 M dodecyl-trichlorosilane (DTS) in toluene for 60 min at a temperature of 3°C. The water contact angle (WCA) on the silanized glass was measured at 110° as opposed to <5° for untreated clean glass. Other, simple, possibilities to achieve slight hydrophobic surfaces is by placing the cleaned glass slides in a car window cleaning solution. This cleaning solution showed to give a WCA of 95°. Another simple method to achieve strong hydrophobic surfaces is spraying the clean glasses with Ultra Ever Dry. Now the WCA was over 130°. The previously mentioned physical method with the μPlasma printhead and HMDSO takes at least 12 treatments on glass to ensure a WCA of at least 90°. In Fig. 15.18 the WCA on glass after several μPlasma treatments is displayed. This method is too time consuming and is therefore not described any further. In all cases the hydrophobic quality was demonstrated by measuring the WCA with a Dataphysics OCA-30 device.

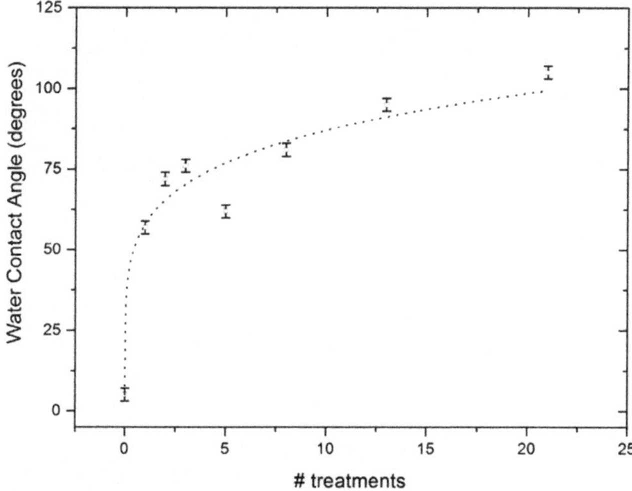

Figure 15.18 WCA on glass after several treatments with the Innophysics μPlasma printhead.

On the DTS-treated slides, μPlasma treatment removes the exposed DTS. The original glass surface reappears and the hydrophilic nature rises again. Multiple structures were printed this way. Some examples are shown in Fig. 15.19. These structures were printed at 20 mm.s^{-1} with an applied AC voltage of 4.5 kV max. and with a 50 μm charge gap. To visualize the hydrophilic structure (colored) water or a water-based ink is spread over the area. For the Fontys logo water was dripped on the surface, water droplets are formed on the hydrophobic environment and roll off, while on the plasma-printed pattern the dripped water wets the pattern. It behaves like a hydrophilic channel with hydrophobic quays. For Fig. 15.19b two glass slides with DTS are used and two Y-shaped patterns, which are mirror images, are applied by μPlasma printing on each of them. The two glasses are stacked with the two patterns facing each other. Now a solution at the beginning of the two patterns is sucked by capillary forces into the pattern. With the use of different water colors it is shown that the two solutions mix. The third example (Fig. 15.19c) is a hidden QR code made visible with ink.

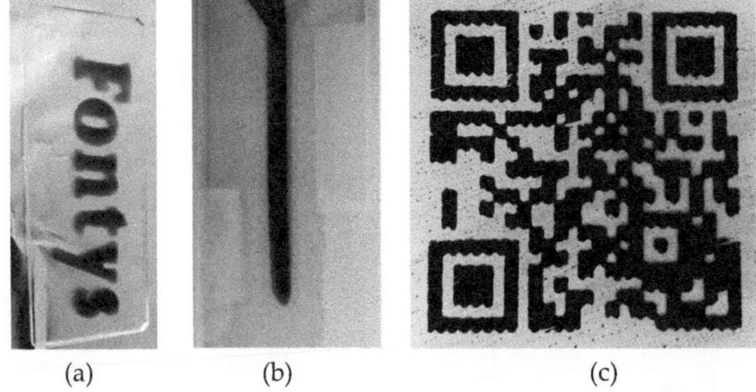

Figure 15.19 Examples of a hydrophilic structure in a hydrophobic environment: (a) Fontys logo, (b) capillary mixing of two liquids, and (c) QR code.

The Y-shaped structures on the two glass slides of Fig. 15.19b are suitable to be used together as a simple microcapillary reactor. More advanced patterns can also be used. The working principle is simple. This microreactor consists of two identical treated glass slides in which a pathway is burned by the plasma. The width of the pathway is 3.0 mm. The two treated glass slides are placed on top of each other and gapped by Scotch adhesive tape 0.10 mm thick. Both slides are tightly bound to each other again by Scotch adhesive tape. In this way a simple microreactor 3.0 × 0.10 mm and 6.0 cm long is constructed. The speed of the solutions after the node depends, among other things, on the dimensions of the reactor. In this case the hydraulic diameter equals 194 μm. The speed of the frontline of the two mixing solutions was followed by a simple camera with 25 fps. The results were compared with the Lucas Washburn equation:

$$x^2 = \frac{\gamma . D_{\mathrm{H}} \cos(\theta)}{4.\eta} \cdot t = a \cdot t \tag{15.4}$$

with γ being the surface tension, η dynamical viscosity, D_{H} hydraulic diameter, and $\theta < 5°$. For water, at 20°C, the coefficient in this equation equals $a = 35\ \mathrm{cm^2.s^{-1}}$.

The experimental results are shown in Fig. 15.20 (Vercoulen, 2013a). From the experiments it followed that the coefficient $a = 36.2\ \mathrm{cm^2.s^{-1}}$.

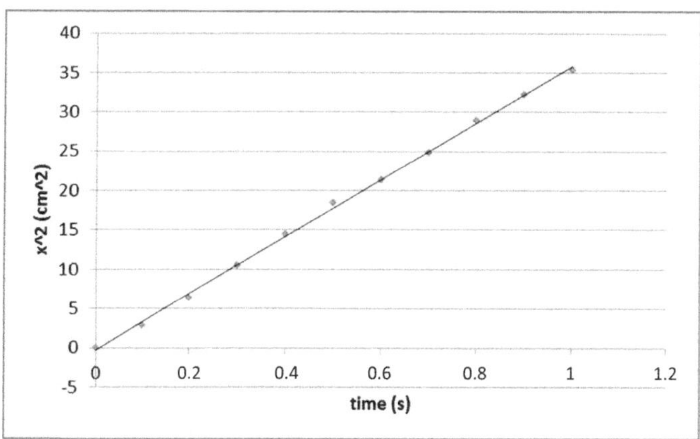

Figure 15.20 Capillary flow measurements compared with the Washburn equation.

15.4 Manufacturing a Piezo Single-Nozzle Drop-on-Demand Device

Inkjet printing is a promising thin-film technology. The best way to learn more about inkjet printing is to learn bottom-up. This means from manufacturing an inkjet head up to making use of an inkjet printer. The first insights are available in patents. The United States Patent of Krause "Ink Jet Printer with Ink System Filter Means" tells us, according to the abstract, "An ink jet printing apparatus in which a pressure generated ink stream is produced from an opening through a piezoelectric transducer which produces axial vibration when energized to produce a stream of uniform drops." This patent shows that the main aspects are:

- Piezoelectric transducers
- Pressure generation and prolongation of acoustic waves inside the ink capillary
- The layout of a drop-on-demand (DOD) glass inkjet printer

A piezo-driven DOD device can be obtained using two different approaches: either by interrupting a continuous liquid stream on command or by generating single drops by drop on demand.

The continuous inkjet method can be achieved by a low-cost manufacturing method. The construction consists of a glass orifice plate with a small hole, fused at the end of a glass tube. Around the tube a piezoelectric cylinder is glued stiff onto the glass. A mere start is to weld or fuse an orifice with a diameter in the range of 10–15 μm on a Pyrex Pasteur pipette. In a commercial Pasteur pipette the capillary has not the form of a straight-lined cylinder. For that reason rectangular-sized piezo actuators are used. The result is illustrated in Fig. 15.21.

Figure 15.21 Continuous jet, fabricated with a thin circular orifice of diameter ca. 12 μm, epoxy-glued on a Pyrex Pasteur pipette and provided with two thin rectangular piezo elements.

The benefit of this method is the rate of producing drops, at a megahertz rate, with a minimum of fluid volumes. The jet is broken up into drops by means of an acoustic excitation. However, despite the simple construction and the commercially available geometry, the practical disadvantage is the need for high pressure on the fluid. Depending mostly on the size of the aperture, the hydraulic pressure can go up to 80 bar. By chemical means the aperture could be enlarged, resulting in a lower pressure and a bigger drop volume. The chemicals to be used are not environmental friendly and the procedure can only be carried out by a specialist. The illustrated construction did not work because of the shape and size of the actuators and because of a nonstiff assemblage.

More confidence was given to variations on the piezoelectrically driven single DOD device.

15.4.1 Drop-on-Demand Device

There are single-nozzle systems in different assemblies but all working with a piezo activator. In principle, such a system consists of a cylindrical cavity surrounded by a cylindrical piezo actuator, an ink reservoir with a flexible tube, which gives a small underpressure, and one or two obstructions.

The construction of the capillary consists of two obstructions, in which the orifice has a length-to-diameter ratio of 2:1. The capillary size could vary in length and vary in diameter and the second obstruction should have a bigger diameter than the orifice (Fig. 15.22 and Fig. 15.23).

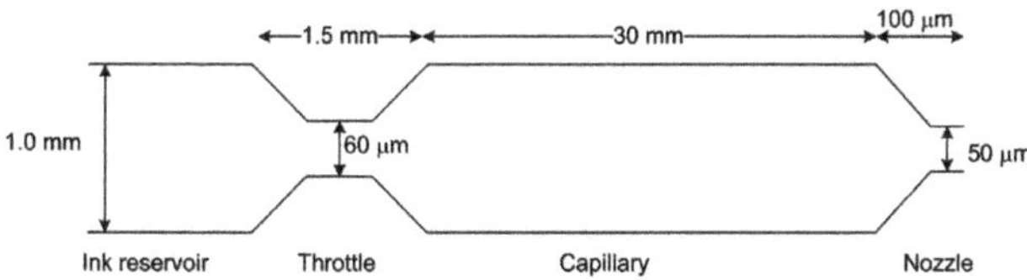

Figure 15.22 Principle design of the glass capillary in the single-nozzle DOD.

The diameters of the capillaries and orifices are measured with a microscope.

Different types of capillaries can be made: with one or with two obstructions, different lengths, or with or without a glass ink reservoir.

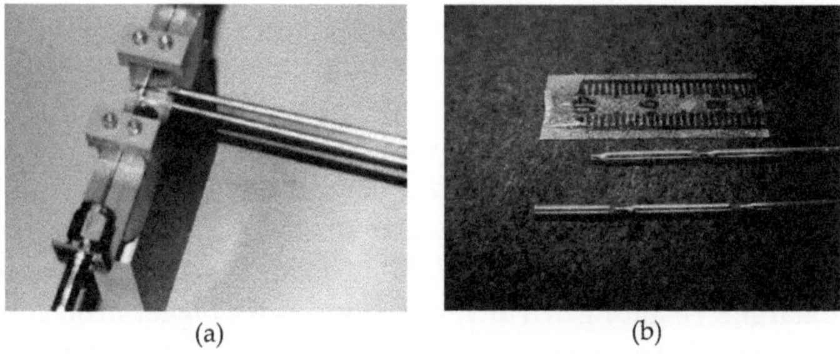

(a) (b)

Figure 15.23 (a) Local heating of the glass pipe; (b) two different results.

Finally the nozzles were polished. The results are shown in Fig. 15.24.

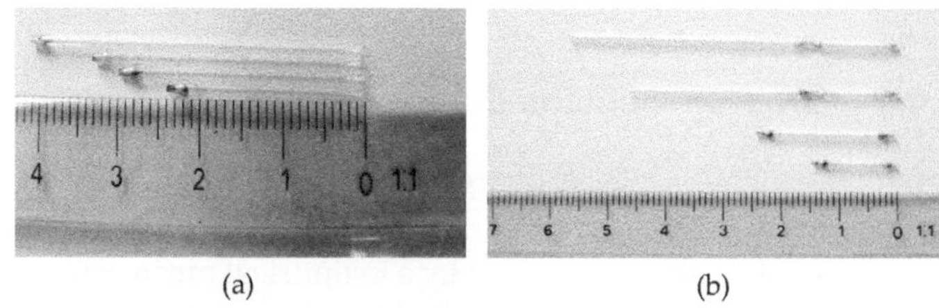

(a) (b)

Figure 15.24 Capillaries of different lengths with (a) one obstruction and (b) two obstructions with or without a glass ink reservoir. The scale bar is in centimeters.

Different types of piezo elements can be applied. for example, cylindrical actuators and thin rectangular piezo actuators. The best results were obtained by using a cylindrical piezo actuator. Such an activator has two solder connections at the outside of the actuator. The rectangular activators have a narrow connected surface to the capillary. Examples are shown in Fig. 15.25.

The two different types of piezo activators made it necessary to have different power supplies available.

With the home-made arrangement one can vary the pulse height and pulse duration to the desired level, in our case between 0 and 200 V and between 20 μs and 120 μs, respectively. The frequency was limited to 70 Hz. The rise time and fall time of the pulse shape differ depending on whether a rectangular piezo element was used or a cylindrical one. The rise times were 2.5 μs and 1.5 μs, respectively. The fall times were 0.50 μs and 1.0 μs, respectively. Results are shown in Fig. 15.26.

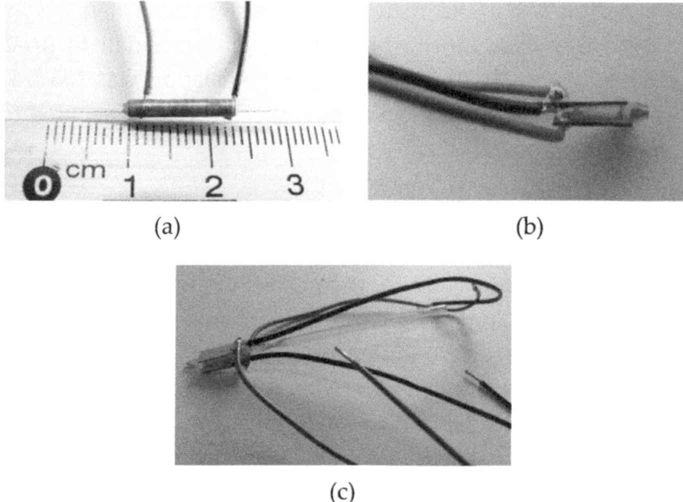

Figure 15.25 Piezoelectric drop-on-demand ink jetters, all with two obstructions: (a) cylindrical piezo element with the solder connections outside, (b) two rectangular piezo elements without an ink reservoir, and (c) two rectangular piezo elements with an ink reservoir.

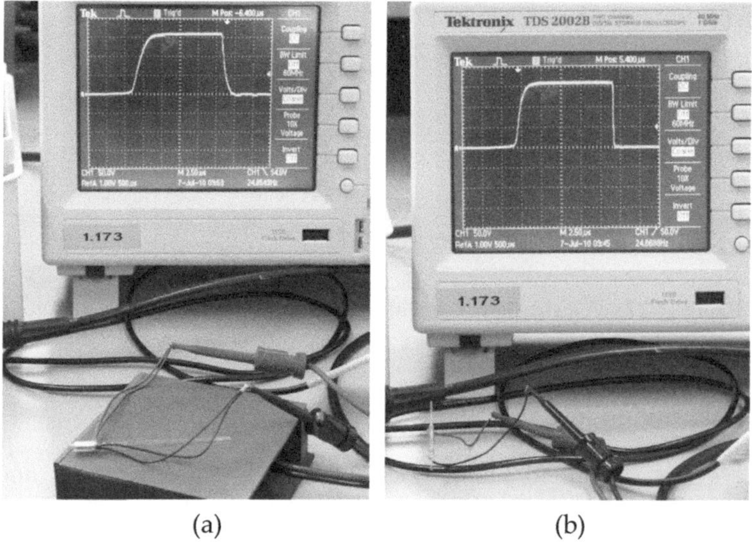

Figure 15.26 Pulse shapes for the piezo elements: (a) rectangular piezo element on a capillary with ink reservoir and (b) cylindrical piezo element around a capillary without ink reservoir.

This shows that the rectangular constructions had a contact area with the glass tube, which was much too small to produce good and repeatedly drops. Also the construction was too fragile. The construction with the cylindrical piezo activator seemed to be assistant and student proof.

15.4.2 Simulation

Simulations can be very useful to improve the understanding of the fluid flow and to predict or check the optimal design of an inkjet head. Therefore we were grateful that Comsol supplied us with a demonstrator called "dynamic model of a drop shot from an inkjet printer." This is a model to be solved with Comsol Multiphysics. For example, the velocity field was calculated at various times (results are shown in Fig. 15.27 for $t = 40$ μs).

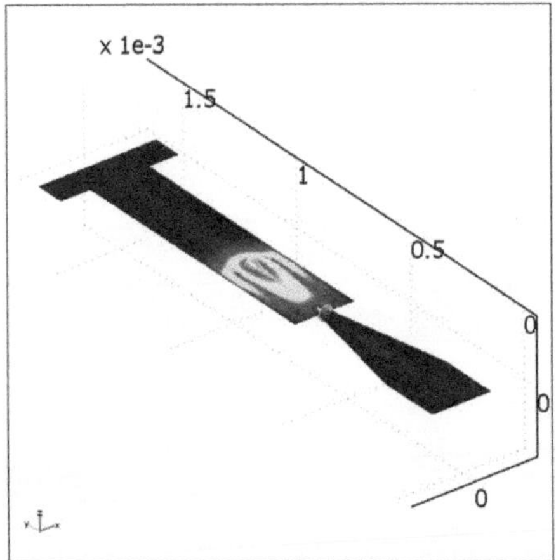

Figure 15.27 Velocity field at $t = 40$ μs (Source: Comsol Multiphysics).

Acknowledgments

We are grateful to Innophysics B.V., Eindhoven; employees and students of Thin Films and Functional Materials, Fontys University of Applied Sciences; COMSOL Inc.; Frans Kuypers, Eindhoven University of Technology; Pim Groen, Morgan Advanced Materials PLC; Pim Pernot, PSI; and W. Meyer, Microdrop Technologies GmbH.

Suggested Readings

Gao, F. G., and Bard, A. J. (2000). Solid-state organic light-emitting diodes based on tris(2,2′-bipyridine)ruthenium(II) complexes. *J. Am. Chem. Soc*, **122**(30), pp. 7426–7427, http://education.mrsec.wisc.edu/nanolab/index.html.

Müller, S., Rudmann, H., Rubner, M. F., and Sevian, H. (2004). Using organic light-emitting electrochemical thin-film devices to teach materials science. *J. Chem. Educ.*, **81**(11), pp. 1620–1623.

Simons, C., et al. (2013). *Labviev Programme for V,L Charaterisation*. Internal report, Fontys University of Applied Sciences.

van Dongen, M., (2014). *μPlasma Patterning and Inkjet Printing to Enhance Localized Wetting and Mixing Behaviour*. Thesis, University of Twente.

Vercoulen, T., (2013a). *The Design of Microchannels through Coating Hydrofilic and Hydrofobic Layers on Glass*. Internal report, Fontys University of Applied Sciences.

Vercoulen, T., (2013b). *Process Equipment*. Internal report, Fontys University of Applied Sciences.

Abbreviations

μCP	microcontact printing
AC	alternating current
ACA	anisotropic conductive adhesive
ACF	anisotropic conductive film
ADF	amplitude distribution function
AE	auxiliary electrode
AFM	atomic force microscope
AHR	aryl hydrocarbon receptor
ALD	atomic layer deposition
ALE	atomic layer epitaxy
AM1.5	air mass 1.5; refers to ASTM standards for photovoltaic efficiency measurements
AMLCD	active matrix liquid crystal display
AMOLED	active matrix organic light-emitting diode
ANOVA	analysis of variance
a-Si:H	hydrogenated amorphous silicon
ASIC	application-specific integrated circuit
BET	Brunauer–Emmett–Teller (analysis)
BIPV	building integrated photovoltaic
BOS	balance of system
CAB	cellulose acetate butyrate
Cat	catalase

CdSe	cadmium selenide
CdTe	cadmium telluride
CED	cumulative energy demand
CELIV	charge extraction in a linearly increasing voltage
CGL	charge-generation layer
CIE	International Commission on Illumination
CIGS	copper indium gallium selenide
CML	Centrum voor Milieukunde (CML), University of Leiden
CMOS	complementary metal–oxide–semiconductor
CNTs	carbon nanotubes
COLAE	Commercialisation of Organic and Large Area Electronics (EU project name)
CRI	color-rendering index
CRT	cathode ray tube
CSL	current source load
CTE	coefficient of thermal expansion
CV	cyclic voltammogram
CVD	chemical vapor deposition
DARPA	Defense Advanced Research Projects Agency (USA)
DBSA	dodecylbenzenesulfonic acid
DC	direct current
DL	diode load
DMD	dielectric/metal/dielectric
DMSO	dimethyl sulfoxide
DMTA	dynamic mechanical thermal analysis
DNA	deoxyribonucleic acid
DOD	drop-on-demand
DOS	density of states
dpi	dots per inch
DTS	dodecyltrichlorosilane

EA	electron affinity
EC	electrochemical capacitors
ECD	electrochromic displays
ECR	electron cyclotron resonance
ECS	electrochemical sensors
EDA	electronic design automation
EDL	electrochemical double layer
EDLC	electrochemical double-layer capacitor
EGOFET	electrolyte-gated OFET
EIL	electron injection layer
EML	emissive layer
EPBT	energy pay-back time
EPDM	ethylene propylene diene rubber
EQE	external quantum efficiency
ESR	equivalent series resistance
ETFE	ethyletetetrafluorethylene
ETL	electron transport layer
EVA	ethylene vinyl acetate
FET	field-effect transistor
FF	fill factor
FIB	focused ion beam
FP	Framework Programme (EU)
FPC	flexible printed circuit (board)
FTIR	Fourier transform infrared spectroscopy
GaAs	gallium arsenide
GDM	gaussian disorder model
GND	ground
HIJ	hole-injection layer
HMDSN	hexamethyldisilazane
HMDSO	hexamethyldisiloxane
HOMO	highest occupied molecular orbital

HPE	hybrid printed electronics
hpf	hour(s) post fertilization
HTL	hole-transport layer
HWF	high work function
IC	integrated circuit
ICA	isotropic conductive adhesive
ICP	inductively coupled plasma
IDE	interdigitated electrodes
IEC	international electrotechnical commission
IEEE	Institute of Electrical and Electronics Engineering
III-V	semiconductors made of elements from groups III and V of the periodic table
InSb	indium antimonide
IP	ionization potential
IPCC	Intergovernmental Panel on Climate Change
IQE	internal quantum efficiency
IR	infrared
IS	impedance spectroscopy
ISC	intersystem crossing
ISE	ion-selective electrode
ITN	initial training network
ITO	indium tin oxide
JAPERA	Japan Advanced Printed Electronics Technology Research Association
J_{sc}	short-circuit current density
KoPEA	Korean Printed Electronics Association
kWp	kilowatt peak
LA-ICP-MS	laser ablation inductively coupled plasma mass spectrometry
LCA	life cycle assessment
LCD	liquid crystal display
LCI	life cycle inventory

LCIA	life cycle impact assessment
LD_{50}	lethal dose, 50% of the tested population dies
LED	light-emitting diode
LIFT	laser-induced forward transfer
LLE	Landau–Levich equation
lm/W	lumen/watt
LUMO	lowest unoccupied molecular orbital
LVLS	liquid–vapor–liquid–solid
LWF	low work function
MDMO-PPV	poly[2-methoxy-5-(3′,7′-dimethyloctyloxy)-1,4-phenylene vinylene]
MEH-PPV	poly[2-methoxy,5-(2′-ethyl-hexyloxy)-*p*-phenylene vinylene)
MEMS	micro electro-mechanical systems
MIS	metal insulator semiconductor
MISFET	metal insulator semiconductor FET
MLA	microlens array structure
MOSFET	metal oxide semiconductor FET
mRNA	messenger ribonucleic acid
MS	mass spectrometer
MT	metallothioneins
NA	numerical aperture
NC	non-contact (mode)
NCA	non-conductive adhesive
NEDO	new energy and industrial technology development organization (Japan)
NFC	near-field communication
NIL	nanoimprint lithography
NIP	non-impact printing
NOEC	no observed effect concentrations
NOM	natural organic matter
NSE	Navier–Stokes equation

NTC	negative temperature coefficient
NV-RAM	non-volatile RAM
OE-A	organic (and printed) electronic association
OFET	organic field-effect transistors
OLED	organic light-emitting diode
OPE	organic and printed electronics
OPVs	organic photovoltaics
OSC	organic semiconductor
OTFT	organic thin film transistor
OTR	oxygen transmission rate
OWRK	Owens–Wendt–Rabel–Kaelble
P3HT	poly-(3-hexylthiophene)
P4	Pragmatic Printing Pilot Program
PANI	polyaniline
PBS	plasma beam source
PCB	printable circuit board
PCBM	phenyl-C_{61}-butyric acid methyl ester
PCE	power conversion efficiency
PCPDTBT	poly[2,6-(4,4-bis-(2-ethylhexyl)-4H-cyclopenta[2,1-b;3,4-b]-dithiophene)-alt-4,7-(2,1,3-benzothiadiazole)]
PCR	polymerase chain reaction
PDA	personal digital assistant
PDMS	polydimethylsiloxane
PE	printed electronics
PEA	Printed Electronics Arena
PEC	predicted environmental concentrations
PECVD	plasma enhanced chemical vapor deposition
PEDOT:PSS	poly (3,4-ethylenedioxythiophene):poly(styrene sulfonate)
PEEK	polyether-etherketone
PEN	polyethylene naphthalate

PET	polyethylene terephthalate
PFTP	pentafluorothiophenol
PI	polyimide
PIB	polyisobutylene
PLED	polymer light-emitting diode
PML	polymer multilayer
PMMA	polymethyl methacrylate
PPV	poly(*p*-phenylene vinylene)
PS	polystyrene
PSA	pressure-sensitive adhesive
PSI	phase-shift interferometry
PTFE	polytetrafluoroethylene
PVs	photovoltaics
PVA	polyvinyl alcohol
PVD	physical vapor deposition
PVDF	polyvinylidene fluoride
qRT-PCR	quantitative real-time PCR
R&D	research and development
RH	relative humidity
R2R	roll-to-roll /reel-to-reel
RAM	random access memory
RE	reference electrode
REACH	registration, evaluation, authorization and restriction of chemical substances
ReCiPe	life cycle impact assessment methodology
REE	rare earth elements
RF	radio frequency
RFID	radio frequency identification
RISC	reverse intersystem crossing
rms	root mean square
RNA	ribonucleic acid
RO	ring oscillator

RTD	resistance temperature detector
S2S	sheet-to-sheet
SAM	self-assembled monolayer
SCLC	space charge limited currents
SEM	scanning electron microscopy
SF	screen frequency
SME	small- and medium-size enterprise
SOC	spin–orbit coupling
SOD	superoxide dismutase
SOHO	small offices and home offices
STM	scanning tunneling microscope
SWNT	single-wall carbon nanotubes
TADF	thermally activated delayed fluorescence
TCO	transparent conducting oxide
TCR	temperature coefficient of resistance
TEM	transmission electron microscopy
TEOS	tetraethyl orthosilicate
TFE	thin-film encapsulation
TFE	Thin Film Electronics (corporation name)
TFT	thin-film transistor
TFV	through foil vias
T_g	glass transition temperature
TGA	thermogravimetric analysis
TMA	trimethylaluminium
TME	trimethyloethane
TMS	tetramethylsilane
TNT	trinitrotoluene
TRSP-ICP MS	time-resolved single-particle inductively coupled plasma mass spectrometry
TSC	thermal stimulated currents
U_{cc}	close-circuit voltage
UHF	ultra-high frequency

UMEM	unified model and extraction methodology
U_{oc}	open-circuit voltage
USEtox	scientific consensus model for chemical toxicity assessment
UV	ultraviolet
UV-VIS	ultraviolet-visible
VDP	vapor phase deposition polymerization
V_{oc}	open-circuit voltage
VSI	vertical scanning interferometry
VTC	voltage-transfer curve
WCA	water contact angle
WE	working electrode
WHO	World Health Organization
WMO	World Meteorological Organisation
WOLED	white OLED
WORM	write once read many
WVTR	water vapor transmission rate
XPS	X-ray photoelectron spectroscopy
ZTO	zinc tin oxide

Index

T - #0034 - 160425 - C42 - 229/152/32 [34] - CB - 9789814669740 - Gloss Lamination